Statistics for Economics, Accounting and Business Studies

Contents

Guided tour of the book

Setting the scene

Chapter introductions set the scene for learning and link the chapters together.

Chapter contents guide you through the chapter, highlighting key topics and showing you where to find them.

Learning outcomes summarise what you should have learned by the end of the chapter.

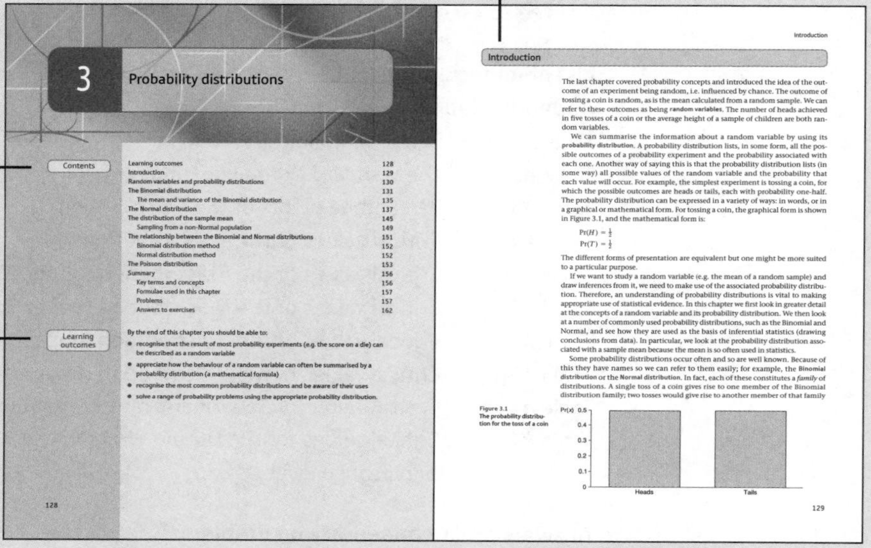

3 Probability distributions

Contents

Learning outcomes	128
Introduction	129
Random variables and probability distributions	130
The Binomial distribution	131
The mean and variance of the Binomial distribution	135
The Normal distribution	137
The distribution of the sample mean	145
Sampling from a non-Normal population	149
The relationship between the Binomial and Normal distributions	151
Binomial distribution method	152
Normal distribution method	152
The Poisson distribution	153
Summary	156
Key terms and concepts	156
Formulae used in this chapter	157
Problems	157
Answers to exercises	162

Learning outcomes

By the end of this chapter you should be able to:

- recognise that the result of most probability experiments (e.g. the score on a die) can be described as a random variable
- appreciate how the behaviour of a random variable can often be summarised by a probability distribution (a mathematical formula)
- recognise the most common probability distributions and be aware of their uses
- solve a range of probability problems using the appropriate probability distribution.

128

Introduction

Introduction

The last chapter covered probability concepts and introduced the idea of the outcome of an experiment being random, i.e. influenced by chance. The outcome of tossing a coin is random, as is the mean calculated from a random sample. We can refer to these outcomes as being **random variables**. The number of heads achieved in five tosses of a coin or the average height of a sample of children are both random variables.

We can summarise the information about a random variable by using its **probability distribution**. A probability distribution lists, in some form, all the possible outcomes of a probability experiment and the probability associated with each one. Another way of saying this is that the probability distribution lists (in some way) all possible values of the random variable and the probability that each value will occur. For example, the simplest experiment is tossing a coin, for which the possible outcomes are heads or tails, each with probability one-half. The probability distribution can be expressed in a variety of ways: in words, or in a graphical or mathematical form. For tossing a coin, the graphical form is shown in Figure 3.1, and the mathematical form is:

$$\Pr(H) = \tfrac{1}{2}$$
$$\Pr(T) = \tfrac{1}{2}$$

The different forms of presentation are equivalent but one might be more suited to a particular purpose.

If we want to study a random variable (e.g. the mean of a random sample) and draw inferences from it, we need to make use of the associated probability distribution. Therefore, an understanding of probability distributions is vital to making appropriate use of statistical evidence. In this chapter we first look in greater detail at the concepts of a random variable and its probability distribution. We then look at a number of commonly used probability distributions, such as the Binomial and Normal, and see how they are used as the basis of inferential statistics (drawing conclusions from data). In particular, we look at the probability distribution associated with a sample mean because the mean is so often used in statistics.

Some probability distributions occur often and so are well known. Because of this they have names so we can refer to them easily; for example, the **Binomial distribution** or the **Normal distribution**. In fact, each of these constitutes a *family* of distributions. A single toss of a coin gives rise to one member of the Binomial distribution family; two tosses would give rise to another member of that family

Figure 3.1 The probability distribution for the toss of a coin

129

Practising and testing your understanding

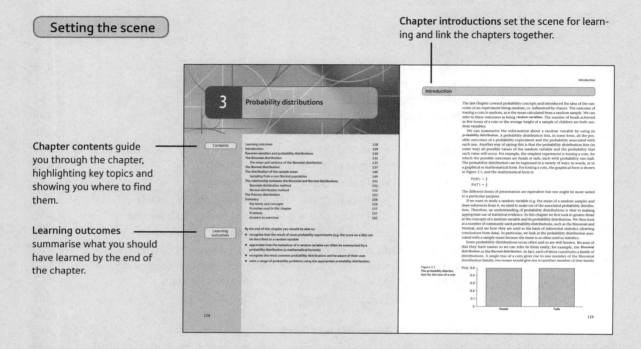

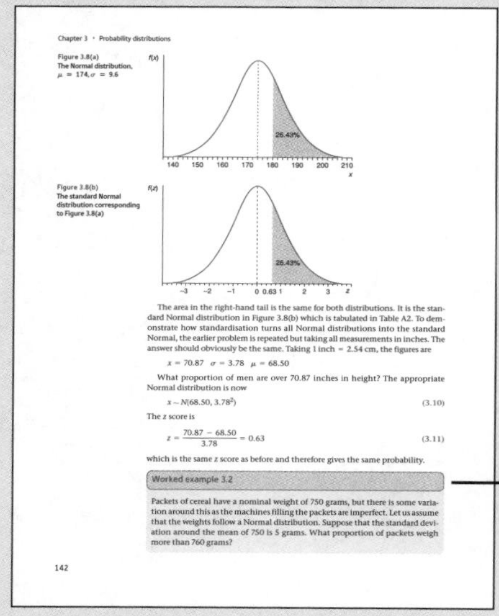

Chapter 3 • Probability distributions

Figure 3.8(a) The Normal distribution, $\mu = 174, \sigma = 9.6$

Figure 3.8(b) The standard Normal distribution corresponding to Figure 3.8(a)

The area in the right-hand tail is the same for both distributions. It is the standard Normal distribution in Figure 3.8(b) which is tabulated in Table A2. To demonstrate how standardisation turns all Normal distributions into the standard Normal, the earlier problem is repeated but taking all measurements in inches. The answer should obviously be the same. Taking 1 inch = 2.54 cm, the figures are

$$x = 70.87 \quad \sigma = 3.78 \quad \mu = 68.50$$

What proportion of men are over 70.87 inches in height? The appropriate Normal distribution is now

$$x \sim N(68.50, 3.78^2) \tag{3.10}$$

The z score is

$$z = \frac{70.87 - 68.50}{3.78} = 0.63 \tag{3.11}$$

which is the same z score as before and therefore gives the same probability.

Worked example 3.2

Packets of cereal have a nominal weight of 750 grams, but there is some variation around this as the machines filling the packets are imperfect. Let us assume that the weights follow a Normal distribution. Suppose that the standard deviation around the mean of 750 is 5 grams. What proportion of packets weigh more than 760 grams?

142

Worked examples break down statistical techniques step-by-step and illustrate how to apply an understanding of statistical techniques to real life.

For students:

- The associated website contains numerous exercises (with answers) for the topics covered in this text. Many of these contained randomised values so that you can try out the tests several times and keep track of you progress and understanding.

Mathematics requirements and suggested texts

No more than elementary algebra is assumed in this text, any extensions being covered as they are needed in the text. It is helpful to be comfortable with manipulating equations, so if some revision is needed, I recommend one of the following books:

Jacques, I., *Mathematics for Economics and Business*, 8th edn, Pearson, 2015
Renshaw, G., *Maths for Economists*, 4th edn, Oxford University Press, 2016.

Acknowledgements

I would like to thank the reviewers who made suggestions for this new edition and to the many colleagues and students who have passed on comments or pointed out errors or omissions in previous editions. I would like to thank the editors at Pearson, especially Caitlin Lisle and Carole Drummond, who have encouraged me, responded to my various queries and gently reminded me of impending deadlines. I would also like to thank my family for giving me encouragement and time to complete this edition.

Custom publishing

Custom publishing allows academics to pick and choose content from one or more textbooks for their course and combine it into a definitive course text.

Here are some common examples of custom solutions which have helped over 800 courses across Europe:

- different chapters from across our publishing imprints combined into one book;
- lecturer's own material combined together with textbook chapters or published in a separate booklet;
- third-party cases and articles that you are keen for your students to read as part of the course;
- any combination of the above.

The Pearson custom text published for your course is professionally produced and bound – just as you would expect from a normal Pearson text. Since many of our titles have online resources accompanying them we can even build a Custom website that matches your course text.

If you are teaching an introductory statistics course for economics and business students, do you also teach an introductory mathematics course for economics and business students? If you do, you might find chapters from *Mathematics for Economics and Business, Sixth Edition* by Ian Jacques useful for your course. If you are teaching a year-long course, you may wish to recommend both texts. Some adopters have found, however, that they require just one or two extra chapters from one text or would like to select a range of chapters from both texts.

Custom publishing has allowed these adopters to provide access to additional chapters for their students, both online and in print. You can also customise the online resources.

If, once you have had time to review this title, you feel Custom publishing might benefit you and your course, please do get in contact. However minor, or major the change – we can help you out.

For more details on how to make your chapter selection for your course please go to: **www.pearsoned.co.uk/barrow**

You can contact us at: **www.pearsoncustom.co.uk** or via your local representative at: **www.pearsoned.co.uk/replocator**

written down). For a (statistically) unsophisticated audience the explanation is quite useful and might then be supplemented by a few examples.

Statistics can also be written well or badly. Two examples follow, concerning a confidence interval, which is explained in Chapter 4. Do not worry if you do not understand the statistics now.

Good explanation

The 95% confidence interval is given by

$$\bar{x} \pm 1.96 \times \sqrt{s^2/n}$$

Inserting the sample values $\bar{x} = 400$, $s^2 = 1600$ and $n = 30$ into the formula we obtain

$$400 \pm 1.96 \times \sqrt{1600/30}$$

yielding the interval

$$[385.7, 414.3]$$

Bad explanation

$95\%\text{ interval} = \bar{x} - 1.96\sqrt{s^2/n} =$

$\bar{x} + 1.96\sqrt{s^2/n} = 0.95$

$= 400 - 1.96\sqrt{1600/30}$ and

$= 400 + 1.96\sqrt{1600/30}$

so we have [385.7, 414.3]

In good statistical writing there is a logical flow to the argument, like a written sentence. It is also concise and precise, without too much extraneous material. The good explanation exhibits these characteristics whereas the bad explanation is simply wrong and incomprehensible, even though the final answer is correct. You should therefore try to note the way the statistical arguments are laid out in this text, as well as take in their content. Chapter 1 contains a short section on how to write good statistical reports.

When you do the exercises at the end of each chapter, try to get another student to read through your work. If they cannot understand the flow or logic of your work, then you have not succeeded in presenting your work sufficiently accurately.

How to use this book

For students:

You will not learn statistics simply by reading through this text. It is more a case of 'learning by doing' and you need to be actively involved by such things as doing the exercises and problems and checking your understanding. There is also material on the website, including further exercises, which you can make use of.

Here is a suggested plan for using the book.

- Take it section by section within each chapter. Do not try to do too much at one sitting.
- First, read the introductory section of the chapter to get an overview of what you are going to learn. Then read through the first section of the chapter trying to follow all the explanation and calculations. Do not be afraid to check the working of the calculations. You can type the data into Excel (it does not take long) to help with calculation.
- Check through the worked example which usually follows. This uses small amounts of data and focuses on the techniques, without repeating all the descriptive explanation. You should be able to follow this fairly easily. If not, work out where you got stuck, then go back and re-read the relevant text. (This is all obvious, in a way, but it's worth saying once.)

- Now have a go at the exercise, to test your understanding. Try to complete the exercise *before* looking at the answer. It is tempting to peek at the answer and convince yourself that you did understand and could have done it correctly. This is not the same as actually doing the exercise – really it is not.
- Next, have a go at the relevant problems at the end of the chapter. Answers to odd-numbered problems are at the back of the book. Your tutor will have answers to the even-numbered problems. Again, if you cannot do a problem, figure out what you are missing and check over it again in the text.
- If you want more practice you can go online and try some of the additional exercises.
- Then, refer back to the learning outcomes to see what you have learnt and what is still left to do.
- Finally – finally – take a deserved break.

Remember – you will probably learn most when you attempt and solve (or fail to) the exercises and problems. That is the critical test. It is also helpful to work with other students rather than only on your own. It is best to attempt the exercises and problems on your own first, but then discuss them with colleagues. If you cannot solve it, someone else probably did. Note also that you can learn a lot from your (and others') mistakes – seeing why a particular answer is wrong is often as informative as getting the right answer.

For lecturers and tutors:

You will obviously choose which chapters to use in your own course, it is not essential to use all of the material. Descriptive statistics material is covered in Chapters 1, 10 and 11; inferential statistics is covered in Chapters 4 to 8, building upon the material on probability in Chapters 2 and 3. Chapter 9 covers sampling methods and might be of interest to some but probably not all.

You can obtain PowerPoint slides to form the basis of you lectures if you wish, and you are free to customize them. The slides contain the main diagrams and charts, plus bullet points of the main features of each chapter.

Students can practise by doing the odd-numbered questions. The even-numbered questions can be set as assignments – the answers are available on request to adopters of the book.

Answer to the 'best' schools problem

A high proportion of small schools appear in the list simply because they are lucky. Consider one school of 20 pupils, another with 1000, where the average ability is similar in both. The large school is highly unlikely to obtain a 100% pass rate, simply because there are so many pupils and (at least) one of them will probably perform badly. With 20 pupils, you have a much better chance of getting them all through. This is just a reflection of the fact that there tends to be greater variability in smaller samples. The schools themselves, and the pupils, are of similar quality.

Figure 1.5
Educational qualifications of those in work

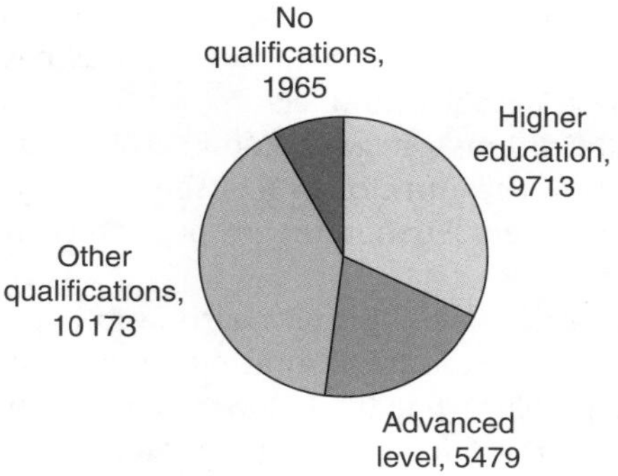

Note: If you have to draw a pie chart by hand, the angle of each slice can be calculated as follows:

$$angle = \frac{frequency}{total\,frequency} \times 360.$$

The angle of the first slice, for example, is

$$\frac{9713}{27330} \times 360 = 127.9°.$$

The pie chart

Another common way of presenting information graphically is the **pie chart**, which is a good way to describe how a variable is distributed between different categories. For example, from Table 1.1 we have the distribution of educational qualifications for those in work (the first row of the table). This can alternatively be shown as a pie chart, as in Figure 1.5.

The area (and angle) of each slice is proportional to the respective frequency, and the pie chart is an alternative means of presentation to the bar chart shown in Figure 1.1. The numbers falling into each education category have been added around the chart, but this is not essential. For presentational purposes, it is best not to have too many slices in the chart: beyond about six the chart tends to look crowded. It might be worth amalgamating less important categories to make such a chart look clearer.

The chart reveals, as did the original bar chart, that 'higher education' and 'other qualifications' are the two biggest categories. However, it is more difficult to compare them accurately; it is more difficult to compare angles than it is to compare heights. The results may be contrasted with Figure 1.6 which shows a similar

Figure 1.6
Educational qualifications of the unemployed

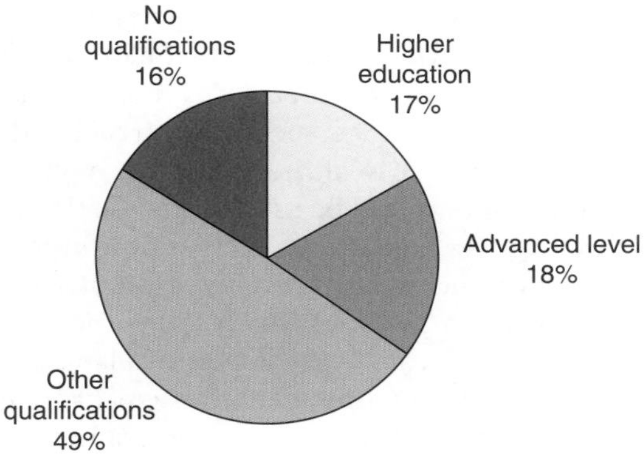

pie chart for the unemployed (the second row of Table 1.1). This time, we have put the proportion in each category in the labels (Excel has an option which allows this), rather than the absolute number.

The 'other qualifications' category is now substantially larger and the 'no qualifications' group now accounts for 16% of the unemployed, a bigger proportion than for those employed. Further, the proportion with a degree approximately halves from 35% to 17%.

Notice that we would need three pie charts (another for the 'inactive' group) to convey the same information as the multiple bar chart in Figure 1.2. It is harder to look at the three pie charts than it is to look at one bar chart, so in this case the bar chart is the better method of presenting the data.

Exercise 1.1

?

The following table shows the total numbers (in millions) of tourists visiting each country and the numbers of English tourists visiting each country:

	France	Germany	Italy	Spain
All tourists	12.4	3.2	7.5	9.8
English tourists	2.7	0.2	1.0	3.6

Adapted from data from the Office for National Statistics licensed under the Open Government Licence v.3.0.
Source: Office for National Statistics.

(a) Draw a bar chart showing the total numbers visiting each country.

(b) Draw a stacked bar chart which shows English and non-English tourists making up the total visitors to each country.

(c) Draw a pie chart showing the distribution of all tourists between the four destination countries. Do the same for English tourists and compare results.

Experiment with the presentation of each graph to see which works best. Try a horizontal (rather than vertical) bar chart, try different colours, make all text horizontal (including the title of the vertical axis and the labels on the horizontal axis), place the legend in different places, etc.

Looking at cross-section data: wealth in the United Kingdom in 2005

Frequency tables and charts

We now move on to examine data in a different form. The data on employment and education consisted simply of frequencies, where a characteristic (such as higher education) was either present or absent for a particular individual. We now look at the distribution of wealth, a variable which can be measured on a **ratio scale** so that a different value is associated with each individual. For example, one person might have £1000 of wealth, and another might have £1 million. Different presentational techniques will be used to analyse this type of data. We use these techniques to investigate questions such as how much wealth does the average person have and whether wealth is evenly distributed or not.

The data are given in Table 1.3 which shows the distribution of wealth in the United Kingdom for the year 2005 (the latest available at the time of writing), available at http://webarchive.nationalarchives.gov.uk/+/http://www.hmrc.gov.uk/stats/personal_wealth/archive.htm. This is an example of a **frequency table**. Wealth

Figure 1.10
The relative frequency distribution of wealth in the United Kingdom, 2005

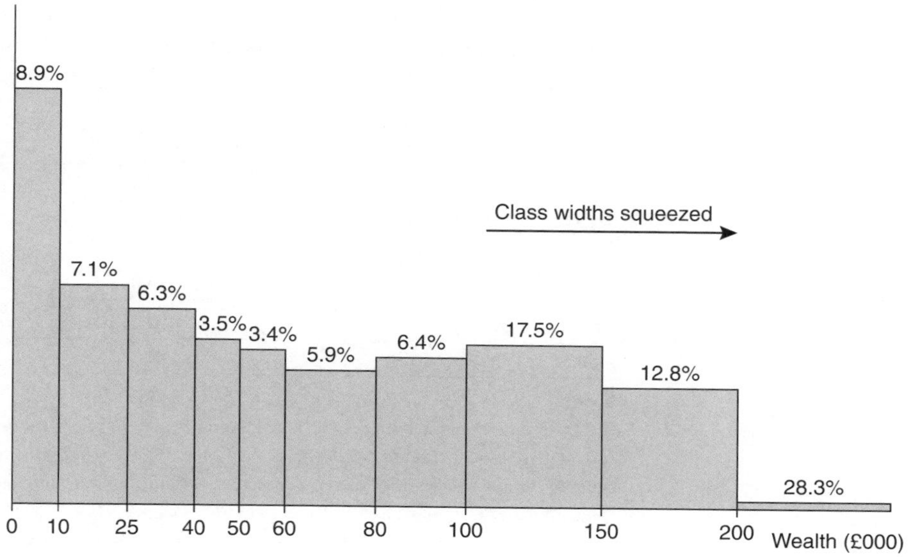

Figure 1.11
The cumulative frequency distribution of wealth in the United Kingdom, 2005

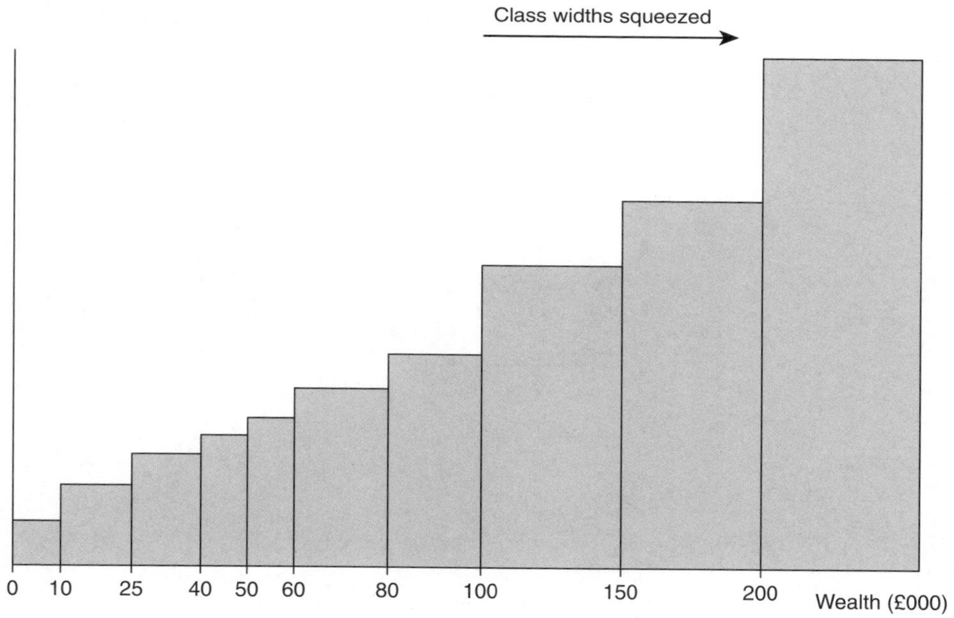

Note: The y-axis coordinates are obtained by cumulating the frequency densities in Table 1.4. For example, the first two y coordinates are 0.1668, 0.2547.

Worked example 1.1

There is a mass of detail in the sections above, so this worked example is intended to focus on the essential calculations required to produce the summary graphs. Simple artificial data are deliberately used to avoid the distraction of a lengthy interpretation of the results and their meaning. The data on →

the variable X and its frequencies f are shown in the following table, with the calculations required:

X	Frequency, f	Relative frequency	Cumulative frequency, F
10	6	0.17	6
11	8	0.23	14
12	15	0.43	29
13	5	0.14	34
14	1	0.03	35
Total	35	1.00	

Notes:

The X values are unique but could be considered the mid-point of a range, as earlier.

The relative frequencies are calculated as $0.17 = 6/35$, $0.23 = 8/35$, etc. Note that these are expressed as decimals rather than percentages; either form is acceptable.

The cumulative frequencies are calculated as $14 = 6 + 8$, $29 = 6 + 8 + 15$, etc.

The symbol F usually denotes the cumulative frequency in statistical work.

The resulting bar chart and cumulative frequency distribution are:

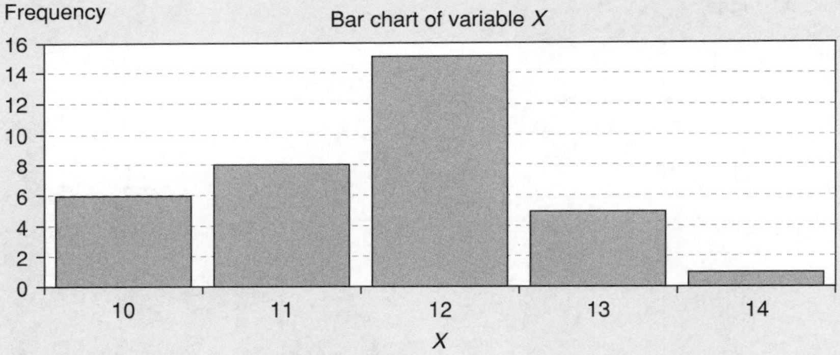

and

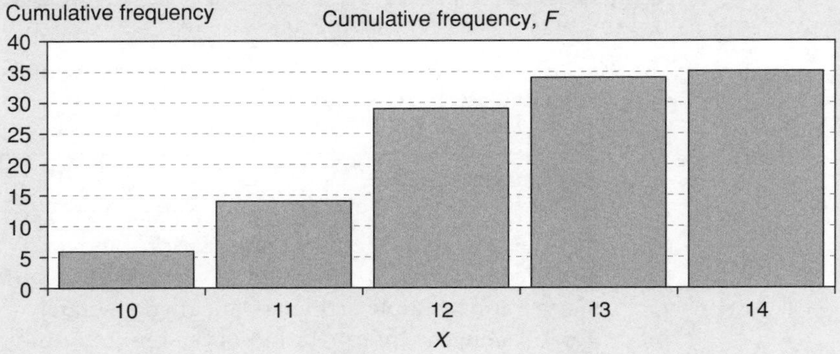

Table 1.6 The calculation of average wealth

Range	x	f	fx
0–	5.0	1 668	8 340
10 000–	17.5	1 318	23 065
25 000–	32.5	1 174	38 155
40 000–	45.0	662	29 790
50 000–	55.0	627	34 485
60 000–	70.0	1 095	76 650
80 000–	90.0	1 195	107 550
100 000–	125.0	3 267	408 375
150 000–	175.0	2 392	418 600
200 000–	250.0	2 885	721 250
300 000–	400.0	1 480	592 000
500 000–	750.0	628	471 000
1 000 000–	1 500.0	198	297 000
2 000 000–	3 000.0	88	264 000
Total		18 677	3 490 260

Note: The fx column gives the product of the values in the f and x columns (so, for example, $5.0 \times 1668 = 8340$, which is the total wealth held by those in the first class interval). The sum of the fx values gives total wealth.

histogram clearly shows most people have wealth below this point (approximately two-thirds of individuals are below the mean, in fact). The mean does not seem to be typical of the wealth that most people have. The reason the mean has such a high value is that there are some individuals whose wealth is way above the figure of £186 875 – up into the £millions, in fact. The mean is the 'balancing point' of the distribution – if the histogram were a physical model, it would balance on a fulcrum placed at 186 875. The few very high wealth levels exert a lot of leverage and counterbalance the more numerous individuals below the mean.

Worked example 1.3

Suppose we have 10 families with a single television in their homes, 12 families with two televisions each and three families with three. You can probably work out in your head that there are 43 televisions in total $(10 + 24 + 9)$ owned by the 25 families $(10 + 12 + 3)$. The average number of televisions per family is therefore $43/25 = 1.72$.

Setting this out formally, we have (as for the wealth distribution, but simpler):

x	f	fx
1	10	10
2	12	24
3	3	9
Totals	25	43

This gives our resulting mean as 1.72. The data are discrete values in this case and we have the actual values, not a broad class interval. Note that no single family could actually have 1.72 television sets; it is the average over all families.

The mean as the expected value

We also refer to the mean as the **expected value** of x and write:

$$E(x) = \mu = 186875 \tag{1.7}$$

$E(x)$ is read 'E of x' or 'the expected value of x'. The mean is the expected value in the sense that if we selected a household at random from the population, we would 'expect' its wealth to be £186 875. It is important to note that this is a *statistical* expectation, rather than the everyday use of the term. Most of the random individuals we encounter have wealth substantially below this value. Most people might therefore 'expect' a lower value because that is their everyday experience; but statisticians are different; they refer to the mean as the expected value.

The expected value notation is particularly useful in keeping track of the effects upon the mean of certain data transformations (e.g. dividing wealth by 1000 also divides the mean by 1000); Appendix 1B provides a detailed explanation. Use is also made of the E operator in inferential statistics, to describe the properties of estimators (see Chapter 4).

The sample mean and the population mean

Very often we have only a sample of data (as in worked example 1.3), and it is important to distinguish this case from the one where we have all the possible observations. For this reason, the sample mean is given by:

$$\bar{x} = \frac{\sum x}{n} \quad \text{or} \quad \bar{x} = \frac{\sum fx}{\sum f} \quad \text{for grouped data} \tag{1.8}$$

Note the distinctions between μ (the population mean) and $\bar{x}$ (the sample mean), and between N (the size of the population) and n (the sample size). Otherwise, the calculations are identical. It is a convention to use Greek letters, such as μ, to refer to the population and Roman letters, such as $\bar{x}$, to refer to a sample.

The weighted average

Sometimes observations have to be given different weightings in calculating the average, as in the following example. Consider the problem of calculating the average spending per pupil by an education authority. Some figures for spending on primary (ages 5–11), secondary (11–16) and post-16 pupils are given in Table 1.7.

Clearly, significantly more is spent on secondary and post-16 pupils (a general pattern throughout England and most other countries) and the overall average should lie somewhere between 1750 and 3820. However, taking a simple average of these three values would give the wrong answer, because there are different numbers of children in the three age ranges. The numbers and proportions of children in each age group are given in Table 1.8.

Table 1.7 Cost per pupil in different types of school (£ p.a.)

	Primary	Secondary	Post-16
Unit cost	1750	3100	3820

Table 1.8 Numbers and proportions of pupils in each age range

	Primary	Secondary	Post-16	Total
Numbers	8000	7000	3000	18 000
Proportion	44.4%	38.9%	16.7%	

Since there are relatively more primary schoolchildren than secondary, and relatively fewer post-16 pupils, the primary unit cost should be given greatest weight in the averaging process and the post-16 unit cost the least. The **weighted average** is obtained by multiplying each unit cost figure by the proportion of children in each category and summing. The weighted average is therefore

$$0.444 \times 1750 + 0.389 \times 3100 + 0.167 \times 3820 = 2620.8 \tag{1.9}$$

The weighted average gives an answer closer to the primary unit cost than does the simple average of the three figures (2890 in this case), which would be misleading. The formula for the weighted average is

$$\bar{x}_w = \sum_i w_i x_i \tag{1.10}$$

where w represents the weights, *which must sum to one*, i.e.

$$\sum_i w_i = 1 \tag{1.11}$$

and x represents the unit cost figures.

Notice that what we have done is equivalent to multiplying each unit cost by its frequency (8000, etc.) and then dividing the sum by the grand total of 18 000. This is the same as the procedure we used for the wealth calculation. The difference with weights is that we first divide 8000 by 18 000 (and 7000 by 18 000, etc.) to get the weights, which must then sum to one, and use these weights in formula (1.10).

STATISTICS IN PRACTICE

Calculating your degree result

If you are a university student your final degree result will probably be calculated as a weighted average of your marks on the individual courses. The weights may be based on the credits associated with each course or on some other factors. For example, in my university the average mark for a year is a weighted average of the marks on each course, the weights being the credit values of each course.

The grand mean G, on which classification is based, is then a weighted average of the averages for the different years, as follows:

$$G = \frac{0 \times \textit{Year}\,1 + 40 \times \textit{Year}\,2 + 60 \times \textit{Year}\,3}{100}$$

i.e. the year 3 mark has a weight of 60%, year 2 is weighted 40% and the first year is not counted at all.

For students taking a year abroad the formula is slightly different:

$$G = \frac{0 \times Year1 + 40 \times Year2 + 25 \times Yabroad + 60 \times Year3}{125}$$

Note that, to accommodate the year abroad mark, the weights on years 2 and 3 are effectively reduced (to $40/125 = 32\%$ and $60/125 = 48\%$, respectively).

The median

Returning to the study of wealth, the unrepresentative result for the mean suggests that we may prefer a measure of location which is not so strongly affected by outliers (extreme observations) and skewness.

The **median** is a measure of location which is more robust to such extreme values; it may be defined by the following procedure. Imagine everyone in a line from poorest to wealthiest. Go to the individual located halfway along the line. Ask her what her wealth is. Her answer is the median. The median is clearly unaffected by extreme values, unlike the mean: if the wealth of the richest person were doubled (with no reduction in anyone else's wealth), there would be no effect upon the median. The calculation of the median is not so straightforward as for the mean, especially for grouped data. The following worked example first shows how to calculate the median for ungrouped data.

Worked example 1.4 The median

Calculate the median of the following values: 45, 12, 33, 80, 77.

First we put them into ascending order: 12, 33, 45, 77, 80.

It is then easy to see that the middle value is 45. This is the median. Note that if the value of the largest observation changes to, say, 150, the value of the median is unchanged. This is not the case for the mean, which would change from 49.4 to 63.4.

If there is an even number of observations, then there is no middle observation. The solution is to take the average of the two middle observations. For example:

Find the median of 12, 33, 45, 63, 77, 80.

Note the new observation, 63, making six observations. The median value is halfway between the third and fourth observations, i.e. $(45 + 63)/2 = 54$.

For grouped data there are two stages to the calculation: first we must identify the class interval which contains the median person, and then we must calculate where in the interval that person lies.

(1) To find the appropriate class interval: since there are 18 677 000 observations, we need the wealth of the person who is 9 338 500 in rank order. The table of cumulative frequencies (see Table 1.5) is the most suitable for this. There are

7 739 000 individuals with wealth of less than £100 000 and 11 006 000 with wealth of less than £150 000. The middle person therefore falls into the £100 000–150 000 class. Furthermore, given that 9 338 500 falls roughly halfway between 7 739 000 and 11 006 000, it follows that the median should be close to the middle of the class interval. We now go on to make this statement more precise.

(2) To find the position in the class interval, we can now use formula (1.12):

$$median = x_L + (x_U - x_L)\frac{\left\{\frac{N+1}{2} - F\right\}}{f} \tag{1.12}$$

where:

x_L = the lower limit of the class interval containing the median
x_U = the upper limit of this class interval
N = the number of observations (using $N + 1$ rather than N in the formula is only important when N is relatively small)
F = the cumulative frequency of the class intervals up to (but not including) the one containing the median
f = the frequency for the class interval containing the median.

For the wealth distribution we have:

$$median = 100\,000 + (150\,000 - 100\,000)\left\{\frac{\frac{18\,677\,000}{2} - 7\,739\,000}{3\,267\,000}\right\}$$
$$= £124\,480$$

This alternative measure of location gives a very different impression: it is around two-thirds of the mean. Nevertheless, it is an equally valid statistic, despite having a different meaning. It demonstrates that the person 'in the middle' has wealth of £124 480 and in this sense is typical of the UK population. Before going on to compare these measures further we examine a third, the mode.

STATISTICS IN PRACTICE

Generalising the median – quantiles

The idea of the median as the middle of the distribution can be extended: **quartiles** divide the distribution into 4 equal parts, **quintiles** into 4, **deciles** into 10, and finally **percentiles** divide the distribution into 100 equal parts. Generically they are known as **quantiles**. We shall illustrate the idea by examining deciles (quartiles are covered below).

The first decile occurs one-tenth of the way along the line of people ranked from poorest to wealthiest. This means we require the wealth of the person ranked 1 867 700 (= $N/10$) in the distribution. From the table of cumulative frequencies, this person lies in the second class interval. Adapting formula (1.12), we obtain:

$$first\ decile = 10\,000 + (25\,000 - 10\,000) \times \left\{\frac{1\,867\,700 - 1\,668\,000}{1\,318\,000}\right\} = £12\,273$$

Thus we estimate that any household with less than £12 273 of wealth falls into the bottom 10% of the wealth distribution. In a similar fashion, the ninth decile can be found by calculating the wealth of the household ranked 16 809 300 (= $N \times 9/10$) in the distribution.

The mode

The **mode** is defined as that level of wealth which occurs with the greatest frequency, in other words the value that occurs most often. It is most useful and easiest to calculate when one has all the data and there are relatively few distinct observations. This is the case in the simple example below.

Suppose we have the following data on sales of dresses by a shop, according to size:

Size	Sales
8	7
10	25
12	36
14	11
16	3
18	1

The modal size is 12. There are more women buying dresses of this size than any other. This may be the most useful form of average as far as the shop is concerned. Although it needs to stock a range of sizes, it knows it needs to order more dresses in size 12 than any other size. The mean would not be so helpful in this case (it is $\bar{x} = 11.7$), as it is not an actual dress size.

In the case of grouped data, matters are more complicated. The modal class interval is required, once the intervals have been corrected for width (otherwise a wider class interval is unfairly compared with a narrower one). For this, we can again make use of the frequency densities. From Table 1.4 it can be seen that it is the first interval, from £0 to £10 000, which has the highest frequency density. It is 'typical' of the distribution because it is the one which occurs most often (using the frequency densities, *not* frequencies). The wealth distribution is most concentrated at this level, and more people are like this in terms of wealth than anything else. Once again, it is notable how different the mode is from both the median and the mean.

The three measures of location give different messages because of the skewness of the distribution: if it were symmetric, then they would all give approximately the same answer. Here we have a rather extreme case of skewness, but it serves to illustrate how the different measures of location compare. When the distribution is skewed to the right, as here, they will be in the order mode, median, mean; if skewed to the left, the order is reversed. If the distribution has more than one peak, then this rule for orderings may not apply.

Which of the measures is 'correct' or most useful? In this particular case the mean is not very useful: it is heavily influenced by extreme values. The median is therefore often used when discussing wealth (and income) distributions. Where inequality is even more pronounced, as in some less developed countries, the mean is even less informative. The mode is also quite useful in telling us about a large section of the population, although it can be sensitive to how the class intervals are arranged. If the data were arranged such that there was a class interval of £5000 to £15 000, then this might well be the modal class, conveying a slightly different impression.

Figure 1.12
The histogram with the mean, median and mode marked

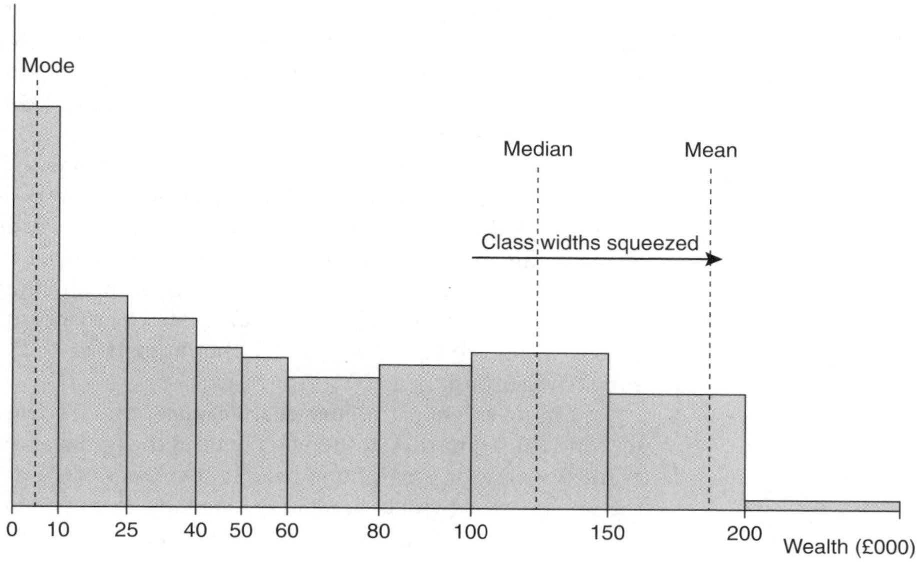

The three different measures of location are marked on the histogram in Figure 1.12. This brings out the substantial difference between the measures for a skewed distribution, such as for wealth.

Exercise 1.3

(a) For the data in Exercise 1.2, calculate the mean, median and mode of the data.

(b) Mark these values on the histogram you drew for Exercise 1.2.

Measures of dispersion

Two different distributions (e.g. wealth in two different countries) might have the same mean yet look very different, as shown in Figure 1.13 (the distributions have

Figure 1.13
Two distributions with different degrees of dispersion

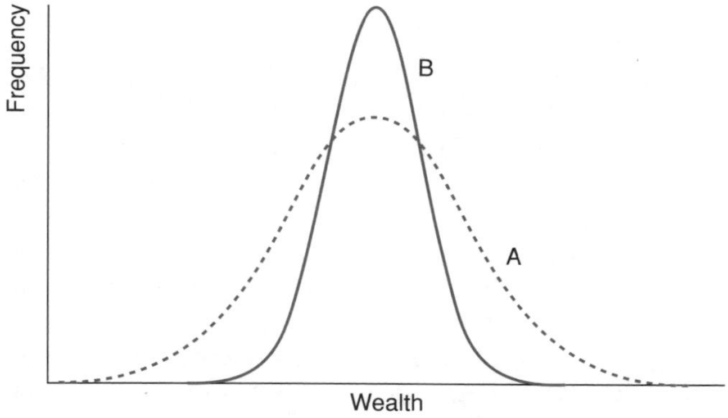

Note: Distribution A has a greater degree of dispersion than B, where everyone has similar levels of wealth.

been drawn using smooth curves rather than bars to improve clarity). In one country, everyone might have a similar level of wealth (curve B). In another, although the average is the same, there might be extremes of great wealth and poverty (curve A). A measure of dispersion is a number which allows us to distinguish between these two situations.

The simplest measure of dispersion is the **range**, which is the difference between the smallest and largest observations. It is impossible to calculate accurately from the table of wealth holdings since the largest observation is not available. In any case, it is not a very useful figure since it relies on two extreme values and ignores the rest of the distribution. In simpler cases, it might be more informative. For example, in an exam the marks may range from a low of 28% to a high of 74%. In this case the range is $74 - 28 = 46$ and this tells us something useful.

An improvement is the **inter-quartile range**, which is the difference between the first and third quartiles. It therefore defines the limits of wealth of the middle half of the distribution and ignores the very extremes of the distribution. To calculate the first quartile (which we label Q_1) we have to go one-quarter of the way along the line of wealth holders (ranked from poorest to wealthiest) and ask the person in that position what their wealth is. Their answer is the first quartile. The calculation is as follows:

- one-quarter of 18 677 observations is 4669.25;
- the person ranked 4669.25 is in the £40000–50000 class;
- adapting formula (1.12):

$$Q_1 = 40\,000 + (50\,000 - 40\,000)\left\{\frac{4669.25 - 4160}{662}\right\} = 47\,692.6 \qquad (1.13)$$

The third quartile is calculated in similar fashion:

- three-quarters of 18 677 is 14 007.75;
- the person ranked 14 007.75 is in the £200 000 to 300 000 class;
- again using (1.12):

$$Q_3 = 200\,000 + (300\,000 - 200\,000)\left\{\frac{14\,007.75 - 13\,398}{2885}\right\} = 221\,135.1$$

and therefore the inter-quartile range is $Q_3 - Q_1 = 221\,135 - 47\,693 = 173\,442$. This might be reasonably rounded to £175 000 given the approximations in our calculation, and is a much more memorable figure. Thus, the 50% of people in the middle of the distribution have wealth between £48 000 and £221 000 of wealth, approximately.

This gives one summary measure of the dispersion of the distribution: the higher the value the more spread out is the distribution. Therefore, two different wealth distributions might be compared according to their inter-quartile ranges, with the country having the larger figure exhibiting greater inequality. Note that the figures would have to be expressed in a common unit of currency for this comparison to be valid.

Figure 1.14
Descriptive statistics calculated using Excel

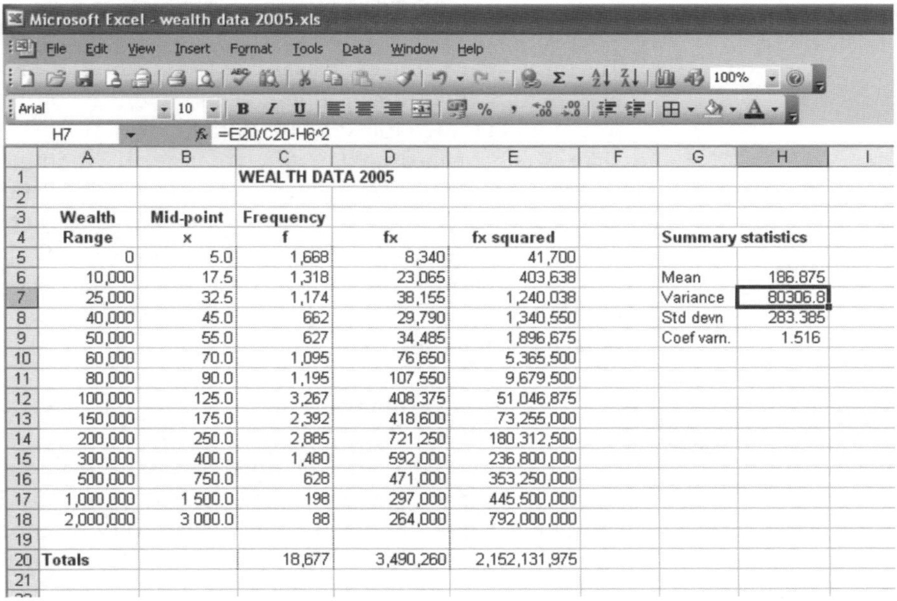

Microsoft Excel - wealth data 2005.xls

H7 =E20/C20-H6^2

	A	B	C	D	E	F	G	H
1			**WEALTH DATA 2005**					
2								
3	**Wealth**	**Mid-point**	**Frequency**					
4	**Range**	x	f	fx	**fx squared**		**Summary statistics**	
5	0	5.0	1,668	8,340	41,700			
6	10,000	17.5	1,318	23,065	403,638		Mean	186.875
7	25,000	32.5	1,174	38,155	1,240,038		Variance	80306.8
8	40,000	45.0	662	29,790	1,340,550		Std devn	283.385
9	50,000	55.0	627	34,485	1,896,675		Coef varn.	1.516
10	60,000	70.0	1,095	76,650	5,365,500			
11	80,000	90.0	1,195	107,550	9,679,500			
12	100,000	125.0	3,267	408,375	51,046,875			
13	150,000	175.0	2,392	418,600	73,255,000			
14	200,000	250.0	2,885	721,250	180,312,500			
15	300,000	400.0	1,480	592,000	236,800,000			
16	500,000	750.0	628	471,000	353,250,000			
17	1,000,000	1 500.0	198	297,000	445,500,000			
18	2,000,000	3 000.0	88	264,000	792,000,000			
19								
20	**Totals**		18,677	3,490,260	2,152,131,975			
21								

or, for grouped data,

$$s^2 = \frac{\sum fx^2 - n\bar{x}^2}{n - 1} \qquad \textbf{(1.23)}$$

The standard deviation may of course be obtained as the square root of these formulae.

Using a calculator or computer for calculation

Electronic calculators and (particularly) computers have simplified the calculation of the mean, etc. Figure 1.14 shows how to set out the above calculations in Microsoft Excel, including some of the appropriate cell formulae.

The variance in this case is calculated using the formula $\sigma^2 = \frac{\sum fx^2}{\sum f} - \mu^2$ which is the formula given in equation (1.21). Note that it gives the same result as that calculated in the text.

The following formulae are contained in the cells:

D5:	= C5*B5	to calculate f times x
E5:	= D5*B5	to calculate f times x^2
C20:	= SUM(C5:C18)	to sum the frequencies, Σf
H6:	= D20/C20	calculates $\Sigma fx/\Sigma f$
H7:	= E20/C20 − H6^2	calculates $\Sigma fx^2/\Sigma f - \mu^2$
H8:	= SQRT(H7)	calculates Σ
H9:	= H8/H6	calculates σ/μ

The coefficient of variation

The measures of dispersion examined so far are all measures of **absolute dispersion** and, in particular, their values depend upon the units in which the variable is measured. It is therefore difficult to compare the degrees of dispersion of two variables which are measured in different units. For example, one could not compare wealth in the United Kingdom with that in Germany if the former uses £s and the latter euros for measurement. Nor could one compare the wealth distribution in one country between two points in time because inflation alters the value of the currency over time. The solution is to use a measure of **relative dispersion**, which is independent of the units of measurement. One such measure is the **coefficient of variation**, defined as:

$$\textit{Coefficient of variation} = \frac{\sigma}{\mu} \tag{1.24}$$

i.e. the standard deviation divided by the mean. Whenever the units of measurement are changed, the effect upon the mean and the standard deviation is the same; hence the coefficient of variation is unchanged. For the wealth distribution its value is $283.385/186.875 = 1.516$, i.e. the standard deviation is 152% of the mean. This may be compared directly with the coefficient of variation of a different wealth distribution to see which exhibits a greater relative degree of dispersion.

Independence of units of measurement

It is worth devoting a little attention to this idea, that some summary measures are independent of the units of measurement and some are not, as it occurs quite often in statistics and is not often appreciated at first. A statistic which is independent of the units of measurement is one which is unchanged, even when the units of measurement are changed. It is therefore more useful in general than a statistic which is not independent, since one can use it to make comparisons, or judgements, without worrying too much about how it was measured.

The mean is not independent of the units of measurement. If we are told the average income in the United Kingdom is 30 000, for example, we need to know whether it is measured in pounds sterling, euros or even dollars. The underlying level of income is the same, of course, but it is measured differently. By contrast, the rate of growth (described in detail shortly) is independent of the units of measurement. If we are told it is 3% p.a., it would be the same whether the calculation was based on pound, euro or dollar figures. If told that the rate of growth in the United States is 2% p.a., we can immediately conclude that the United Kingdom is growing faster, and no further information is needed.

Most measures we have encountered so far, such as the mean and variance, do depend on units of measurement. The coefficient of variation is one that does not. We now go on to describe another means of measuring dispersion that avoids the units of measurement problem.

The standard deviation of the logarithm

Another solution to the problem of different units of measurement is to use the logarithm[4] of wealth rather than the actual value. The reason why this works can best be illustrated by an example. Suppose that between 1997 and 2005 each individual's wealth doubled, so that

$$X_i^{2005} = 2X_i^{1997}$$

where X_i^t indicates the wealth of individual i in year t. It follows that the standard deviation of wealth in 2005 is exactly twice that of 1997 (and hence the coefficient of variation is unchanged). Taking logs, we have $\ln X_i^{2005} = \ln 2 + \ln X_i^{1997}$, so it follows that the distribution of $\ln X^{2005}$ is the same as that of $\ln X^{1997}$ except that it is shifted to the right by ln 2 units. The variances (and hence standard deviations) of the two logarithmic distributions must therefore be the same, indicating no change in the *relative* dispersion of the two wealth distributions.

The use of logarithms in data analysis is very common, so it is worth making sure you understand the principles and mechanics of using them.

The standard deviation of the logarithm of wealth is calculated from the data in Table 1.10. The variance turns out to be:

$$\sigma^2 = \frac{\sum fx^2}{\sum f} - \mu^2 = \frac{417\,772.5}{18\,677} - \left(\frac{84\,840.9}{18\,677}\right)^2 = 1.734$$

and the standard deviation $\sigma = 1.317$. The larger this figure is, the greater the dispersion. On its own the number is difficult to interpret; it is only really useful when compared to another such figure.

Table 1.10 The calculation of the standard deviation of the logarithm of wealth

Range	Mid-point, *x*	ln (*x*)	Frequency, *f*	*fx*	*fx* squared
0–	5.0	1.609	1 668	2 684.5	4 320.6
10 000–	17.5	2.862	1 318	3 772.4	10 797.3
25 000–	32.5	3.481	1 174	4 087.0	14 227.7
40 000–	45.0	3.807	662	2 520.0	9 592.8
50 000–	55.0	4.007	627	2 512.6	10 068.8
60 000–	70.0	4.248	1 095	4 652.1	19 764.4
80 000–	90.0	4.500	1 195	5 377.3	24 196.7
100 000–	125.0	4.828	3 267	15 774.1	76 162.3
150 000–	175.0	5.165	2 392	12 354.2	63 806.6
200 000–	250.0	5.521	2 885	15 929.4	87 953.6
300 000–	400.0	5.991	1 480	8 867.4	53 128.5
500 000–	750.0	6.620	628	4 157.4	27 522.3
1 000 000–	1 500.0	7.313	198	1 448.0	10 589.7
2 000 000–	3 000.0	8.006	88	704.6	5 641.0
Totals			18 677	84 840.9	417 772.5

Notes: Use the 'ln' key on your calculator or the = LN() function in a spreadsheet to obtain natural logarithms of the data. You should obtain ln 5 = 1.609, ln 17.5 = 2.862, etc.
The column headed '*fx*' is the product of the *f* and ln(*x*) columns.

[4]See Appendix 1C if you are unfamiliar with logarithms. Note that we use the natural logarithm here, but the effect would be the same using logs to base 10.

For comparison, the standard deviation of log wealth in 1979 (discussed in more detail later on) is 1.310, so there appears to have been little change in relative dispersion over this time period. Thus we have found two different ways of measuring relative dispersion. In a later chapter we will meet a third, the Gini coefficient.

Measuring deviations from the mean: *z* scores

Imagine the following problem. A man and a woman are arguing over their career records. The man says he earns more than she does, so he is more successful. The woman replies that women are discriminated against and that, relative to other women, she is doing better than the man is, relative to other men. Can the argument be resolved?

Suppose the data are as follows: the average male salary is £19 500 and the average female salary £16 800. The standard deviation of male salaries is £4750 and for women it is £3800. The man's salary is £31 375, while the woman's is £26 800. The man is therefore £11 875 above the mean, and the woman is £10 000 above. However, women's salaries are less dispersed than men's, so the woman has done well to get to £26 800.

One way to resolve the problem is to calculate the **z score**, which gives the salary in terms of the *number of standard deviations from the mean*. Thus for the man, the z score is

$$z = \frac{X - \mu}{\sigma} = \frac{31\,375 - 19\,500}{4750} = 2.50 \tag{1.25}$$

Thus the man is 2.5 standard deviations above the male mean salary, i.e. $31\,375 = 19\,500 + 2.5 \times 4750$. For the woman the calculation is

$$z = \frac{268\,00 - 16\,800}{3800} = 2.632 \tag{1.26}$$

The woman is 2.632 standard deviations above her mean and therefore wins the argument – she is nearer the top of her distribution than is the man and so is more of an outlier. Actually, this probably won't end the argument, but is the best the statistician can do. The z score is an important concept which will be used again (see Chapter 5) when we cover hypothesis testing.

Chebyshev's inequality

Use of the z score leads on naturally to **Chebyshev's inequality**, which tells us about the proportion of observations that fall into the tails of any distribution, regardless of its shape. The theorem is expressed as follows:

At least $(1 - 1/k^2)$ of the observation in any distribution lie within k standard deviations of the mean (1.27)

If we take the female wage distribution given above, we can ask what proportion of women lie beyond 2.632 standard deviations from the mean (in both tails of the distribution). Setting $k = 2.632$, then

$$\left(1 - \frac{1}{k^2}\right) = \left(1 - \frac{1}{2.632^2}\right) = 0.8556.$$

So at least 85% of women have salaries within $\pm$ 2.632 standard deviations of the mean, i.e. between £6800 and £26 800 (16 800 $\pm$ 2.632 $\times$ 3800). Fifteen percent of women therefore lie outside this range.

Chebyshev's inequality is a very conservative rule since it applies to *any* distribution; if we know more about the shape of a particular distribution (for example, men's heights follow a Normal distribution – see Chapter 3), then we can make a more precise statement. In the case of the Normal distribution, over 99% of men are within 2.632 standard deviations of the average height because there is a concentration of observations near the centre of the distribution.

We can also use Chebyshev's inequality to investigate the inter-quartile range. The formula (1.27) implies that 50% of observations lie within $\sqrt{2} = 1.41$ standard deviations of the mean, a more conservative value than our previous 1.3.

Exercise 1.4

?

(a) For the data in Exercise 1.2, calculate the inter-quartile range, the variance and the standard deviation.

(b) Calculate the coefficient of variation.

(c) Check if the relationship between the IQR and the standard deviation stated in the text (worked example 1.6) is approximately true for this distribution.

(d) Approximately how much of the distribution lies within one standard deviation either side of the mean? How does this compare with the prediction from Chebyshev's inequality?

Measuring skewness

The **skewness** of a distribution is the third characteristic that was mentioned earlier, in addition to location and dispersion. The wealth distribution is heavily skewed to the right, or **positively skewed**; it has its long tail in the right-hand end of the distribution. A measure of skewness gives a numerical indication of how asymmetric is the distribution.

One measure of skewness, known as the **coefficient of skewness**, is

$$\frac{\sum f(x-\mu)^3}{N\sigma^3} \tag{1.28}$$

and it is based upon *cubed* deviations from the mean. The result of applying formula (1.28) is positive for a right-skewed distribution (such as wealth), zero for a symmetric one, and negative for a left-skewed one. Table 1.11 shows the calculation for the wealth data (some rows are omitted for brevity).

Table 1.11 Calculation of the skewness of the wealth data

Range	Mid-point x	Frequency f	$x-\mu$	$(x-\mu)^3$	$f(x-\mu)^3$
0–	5.0	1 668	−181.9	−6 016 132	−10 034 907 815
10 000–	17.5	1 318	−169.4	−4 858 991	−6 404 150 553
:	:	:	:	:	:
1 000 000–	1 500.0	198	1 313.1	2 264 219 059	448 315 373 613
2 000 000–	3 000.0	88	2 813.1	22 262 154 853	1 959 069 627 104
Total		18 677	3 898.8	24 692 431 323	2 506 882 551 023

From this we obtain:

$$\frac{\sum f(x-\mu)^3}{N} = \frac{2\,506\,882\,551\,023}{18\,677} = 134\,222\,977.5$$

and dividing by Σ^3 gives $\dfrac{134\,222\,977.5}{22\,757\,714} = 5.898$ which is positive, as expected.

The measure of skewness is much less useful in practical work than measures of location and dispersion, and even knowing the value of the coefficient does not always give much idea of the shape of the distribution: two quite different distributions can share the same coefficient. In descriptive work, it is probably better to draw the histogram itself.

Comparison of the 2005 and 1979 distributions of wealth

Some useful lessons may be learned about these measures by comparing the 2005 distribution with its counterpart from 1979. This covers the period of Conservative government starting with Mrs Thatcher in 1979 and much of the following Labour administration. This shows how useful the various summary statistics are when it comes to comparing two different distributions. The wealth data for 1979 are given in Problem 1.5, where you are asked to confirm the following calculations.

Average wealth in 1979 was £16 399, about one-eleventh of its 2005 value. The average increased substantially therefore (at about 10% p.a., on average), but some of this was due to inflation rather than a real increase in the quantity of assets held. In fact, between 1979 and 2005 the retail price index rose from 52.0 to 217.9, i.e. it increased approximately four times. Thus the nominal[5] increase (i.e. in cash terms, before any adjustment for rising prices) in wealth is made up of two parts: (a) an inflationary part which more than quadrupled measured wealth and (b) a real part, consisting of a 2.75-fold increase (thus $4 \times 2.75 = 11$, approximately). Price indexes are covered in Chapter 10 where it is shown more formally how to divide a nominal increase into price and real (quantity) components. It is likely that the extent of the real increase in wealth is overstated here due to the use of the retail price index rather than an index of asset prices. A substantial part of the increase in asset values over the period is probably due to the very rapid rise in house prices (houses form a significant part of the wealth of many households).

The standard deviation is similarly affected by inflation. The 1979 value is 25 552 compared to 2005's 283 385, which is about 11 times larger (as was the mean). The spread of the distribution appears to be about the same therefore (even if we take account of the general price effect). Looking at the coefficient of variation reveals a similar finding: the value has changed from 1.56 to 1.52, which is a modest difference. The spread of the distribution *relative to its mean* has not changed by much. This is confirmed by calculating the standard deviation of the logarithm: for 1979 this gives a figure of 1.310, almost identical to the 2005 figure (1.317).

The measure of skewness for the 1979 data comes out as 5.723, only slightly smaller than the 2005 figure (5.898). This suggests that the 1979 distribution is similarly skewed to the 2005 one. Again, these two figures can be directly com-

[5]This is a different meaning of the term 'nominal' from that used earlier to denote data measured on a nominal scale, i.e. data grouped into categories without an obvious ordering. Unfortunately, both meanings of the word are in common (statistical) usage, though it should be obvious from the context which use is meant.

pared because they do not depend upon the units in which wealth is measured. However, the relatively small difference is difficult to interpret in terms of how the shape of the distribution has changed.

The box and whiskers diagram

Having calculated these various summary statistics, we can now return to a useful graphical method of presentation. This is the **box and whiskers diagram** (sometimes called a **box plot**) which shows the median, quartiles and other aspects of a distribution on a single diagram. Figure 1.15 shows the box plot for the wealth data.

Wealth is measured on the vertical axis. The rectangular box stretches (vertically) from the first to third quartile and therefore encompasses the middle half of the distribution. The horizontal line through it is at the median and lies slightly less than halfway up the box. This tells us that there is a degree of skewness even within the central half of the distribution, though it does not appear very severe. The two 'whiskers' extend above and below the box as far as the highest and lowest observations, *excluding outliers*. An outlier is defined to be any observation which is more than 1.5 times the inter-quartile range (which is the same as the height of the box) above or below the box. Earlier we found the IQR to be 173 443 and the upper quartile to be 221 135, so an (upper) outlier lies beyond 221 135 + 1.5 × 173 443 = 481 300. There are no outliers below the box as wealth cannot fall below zero. The top whisker is thus substantially longer than the bottom one, and indicates the extent of dispersion towards the tails of the distribution. The crosses indicate the outliers and in reality extend far beyond those shown in the diagram.

A simple diagram thus reveals a lot of information about the distribution.

Figure 1.15
Box plot of the wealth distribution

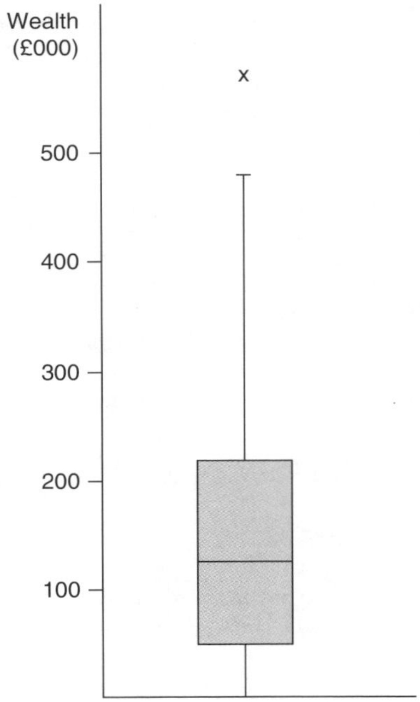

Other boxes and whiskers could be placed alongside in the same diagram (perhaps representing other countries), making comparisons straightforward. Some statistical software packages, such as SPSS and STATA, can generate box plots from the original data, without the need for the user to calculate the median, etc. However, spreadsheet packages do not yet have this useful facility.

Time-series data: investment expenditures 1977–2009

The data on the wealth distribution give a snapshot of the situation at particular points in time, and comparisons can be made between the 1979 and 2005 snapshots. Often, however, we wish to focus on the time-path of a variable and therefore we use **time-series data**. The techniques of presentation and summarising are slightly different than for cross-section data. As an example, we use data on investment in the United Kingdom for the period 1977–2009. These data are available from the Office of National Statistics (ONS) website. However, even after its recent redesign, it is almost impossible to find the data that you want. To save a lot of frustration, use the Econstats website (http://www.econstats.com/uk/index.htm), a U.S. site which aggregates economic data from around the world. The data series used is total gross fixed capital formation (series NPQX), which is measured in current prices (i.e. not adjusted for inflation). We will refer to this series simply as 'investment'.

Investment expenditure is important to the economy because it is one of the primary determinants of growth. Until recent years, the UK economy's growth record had been poor by international standards and lack of investment may have been a cause. The variable studied here is total gross (i.e. before depreciation is deducted) domestic fixed capital formation, measured in £m. The data are shown in Table 1.12.

It should be remembered that the data are in current prices so that the figures reflect price increases as well as changes in the volume of physical investment. The series in Table 1.12 thus shows the actual amount of cash that was spent each year on investment. The techniques used below for summarising the investment

Table 1.12 UK investment, 1977–2009 (£m)

Year	Investment	Year	Investment	Year	Investment
1977	28 351	1988	97 956	1999	161 722
1978	32 387	1989	113 478	2000	167 172
1979	38 548	1990	117 027	2001	171 782
1980	43 612	1991	107 838	2002	180 551
1981	43 746	1992	103 913	2003	186 700
1982	47 935	1993	103 997	2004	200 415
1983	52 099	1994	111 623	2005	209 758
1984	59 278	1995	121 364	2006	227 234
1985	65 181	1996	130 346	2007	249 517
1986	69 581	1997	138 307	2008	240 361
1987	80 344	1998	155 997	2009	204 270

Note: Time-series data consist of observations on one or more variables over several time periods. The observations can be daily, weekly, monthly, quarterly or, as here, annually.

Source: Data adapted from the Office for National Statistics licensed under the Open Government Licence(OGL) v.3.0. http://www.nationalarchives.gov.uk/doc/open-government-licence/open-government

Overlapping the ranges of the data series

The graph below provides a nice example of how to compare different time periods on a single chart. The aim is to compare the recessions starting in 1973Q2, 1979Q2, 1990Q2 and 2008Q1 and the subsequent recoveries of real gross domestic product (GDP). Instead of plotting time on the horizontal axis, the number of quarters since the start of each recession is used, so that the series overlap. To aid comparison, all the series have been set to a value of 100 in the initial quarter. From this, one can see that the most recent recession, starting in 2008, was deeper and longer lasting than the earlier ones.

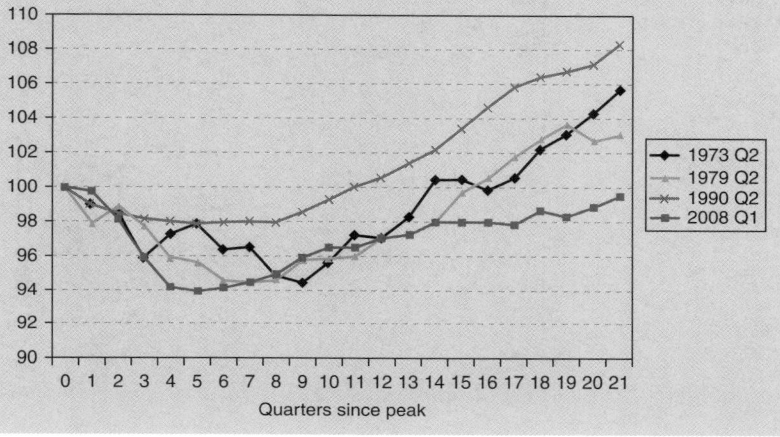

The investment categories may also be illustrated by means of an **area graph**, which plots the four series stacked one on top of the other, as illustrated in Figure 1.22.

This shows, for example, the 'dwellings' and 'machinery' categories each take up about one quarter of total investment. This is easier to see from the area graph than from the multiple series graph in Figure 1.20.

Figure 1.22
Area graph of investment categories, 1977–2009

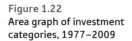

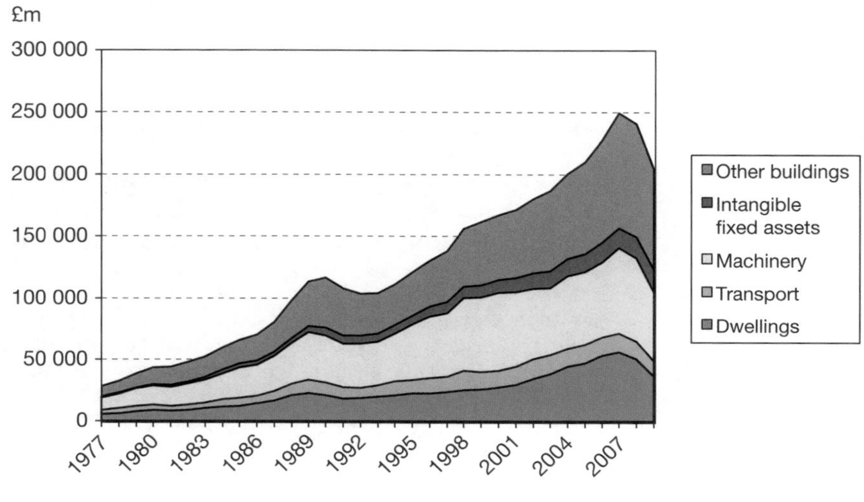

'Chart junk'

With modern computer software it is easy to get carried away and produce a chart that actually hides more than it reveals. There is a great temptation to add some 3-D effects, liven it up with a bit of colour, rotate and tilt the viewpoint, etc. This sort of stuff is generally known as 'chart junk'. As an example, look at Figure 1.23 which is an alternative to the area graph in Figure 1.22. It was fun to create, but it doesn't get the message across at all. Taste is, of course, personal, but moderation is usually an essential part of it.

Figure 1.23
Over-the-top graph of investment

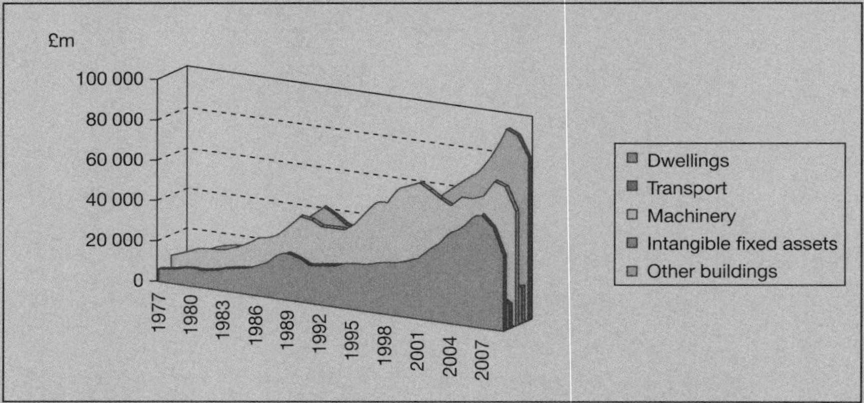

Exercise 1.5

Given the following data:

	1990	1991	1992	1993	1994	1995	1996	1997	1998	1999
Profit	50	60	25	−10	10	45	60	50	20	40
Sales	300	290	280	255	260	285	300	310	300	330

(a) Draw a multiple time-series graph of the two variables. Label both axes appropriately and provide a title for the graph.

(b) Adjust the graph by using the right-hand axis to measure profits, the left hand axis sales. What difference does this make?

Improving the presentation of graphs – example 2: time series

Earlier we showed how a bar chart might be improved. Here we look at a slightly curious presentation of time-series data, taken from the Office for National Statistics[7] and relating to the importance of the EU to UK trade. This chart shows the GDP of various

[7]See http://www.ons.gov.uk/ons/rel/international-transactions/outward-foreign-affiliates-statistics/how-important-is-the-european-union-to-uk-trade-and-investment-/sty-eu.html. This is the author's rendition, so the colours do not quite match the more vivid original.

countries or groups of countries, separated into EU and non-EU territories, from 1993 to 2013.

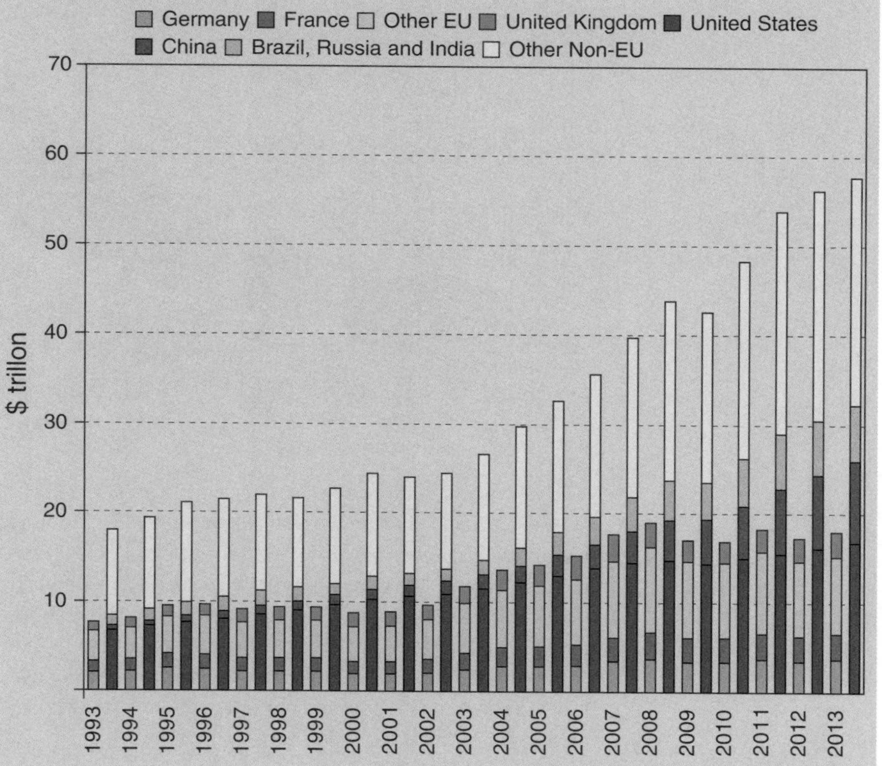

It actually takes a while to interpret this graph; it is not at all intuitive. Among the reasons are:

1. A bar chart is used to present time-series data.
2. There are actually two time series presented, summarised in the two sets of bars, one for EU countries, one for the rest.
3. The colours give information about individual countries, but it is difficult to go from the colour to the legend, and then back again to see the pattern. The author seems to have just used the Excel defaults for this type of graph.

How could this be improved? It presents a lot of information, so the answer is not immediately obvious. It helps to focus on what messages are being conveyed. From the text of the ONS document these are:

1. EU combined GDP is larger than that of any individual country, surpassing the USA in 2003, and
2. EU growth has been slower than non-EU growth.

→

We therefore do not need all of the information in the original graph; for example, nothing is said about individual EU economies. Hence, a better version of the chart would be as follows:

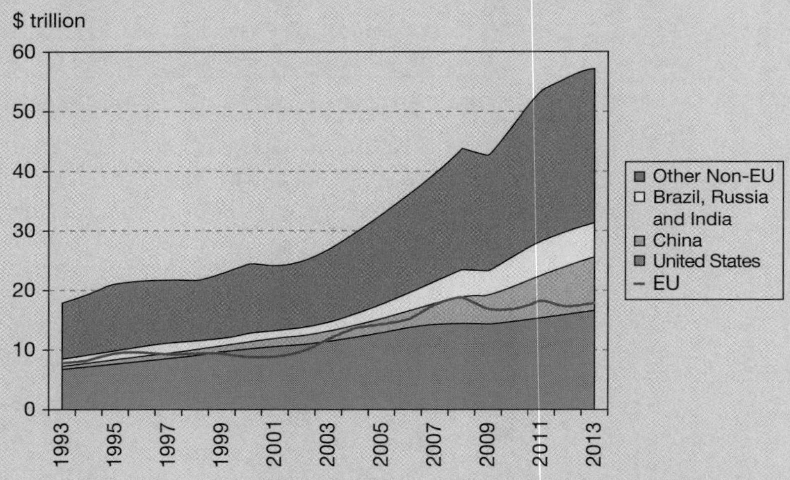

Now it is much easier to see the relevant features. The EU (blue line) overtakes the United States in 2003 (though it was also bigger pre-1999, hard to see from the original chart) and EU growth is slower than all non-EU territories, principally since 2008. One can also see the rapid growth of China, again difficult to see in the initial chart. The legend is put on the right-hand side, and it is very easy to match up the colours to the associated country or region.

This revised chart is what Excel calls a 'combo' chart, which shows two different types of chart in the same picture. The non-EU territories make up an area graph (hence filled with colour), while the EU is a line chart.

Numerical summary statistics

The graphs have revealed quite a lot about the data already, but we can also calculate numerical descriptive statistics as we did for the cross-section data. First, we consider the mean, and then the variance and standard deviation.

The mean of a time series

We could calculate the mean of investment itself, but this would not be very informative. Because the series is trended over time, it passes through the mean at some point between 1977 and 2005 but never returns to it. The mean of the series is actually £123.1bn, achieved in 1995–6, but this is not very informative since it tells nothing about its value today, for instance. The problem is that the variable is trended, so that the mean is not typical of the series. The annual increase in investment is also trended (though much less so), so it is subject to similar criticism (see Figure 1.17).

It is better in this case to calculate the **average growth rate**, since this is less likely to have a trend and hence more likely to be representative of the whole time period. The average growth rate of investment spending was calculated in equation (1.29) as 6.2% p.a. by measuring the slope of the graph of the log investment series, but this is only one way of measuring the growth rate. Furthermore, different methods give slightly different answers, although this is rarely important in practice.

The growth rate may be calculated based upon annual compounding or continuous compounding principles. The former is probably the more common and simpler method, so we explain this first.

Calculating the growth rate based upon annual compounding

This method likens growth to the way money grows in a savings account as interest is added annually. We can calculate the growth rate in the following way:

(1) Calculate the overall **growth factor** of the series, i.e. x_T/x_1 where x_T is the final observation and x_1 is the initial observation. This is:

$$\frac{x_T}{x_1} = \frac{204\,270}{28\,351} = 7.2050,$$

i.e. investment expenditure is 7.2 times larger in 2009 than in 1977.

(2) To get the annual figure, take the $T - 1$ root of the growth factor, where T is the number of observations. Since $T = 33$ we calculate $\sqrt[32]{7.205} = 1.0637$ (This can be performed on a scientific calculator by raising 7.205 to the power $7.205^{(1/32)} = 1.0637$.)

(3) Subtract 1 from the result in the previous step, giving the growth rate as a decimal. In this case we have $1.0637 - 1 = 0.0637$.

Thus the average growth rate of investment is 6.4% p.a., slightly different from the 6.2% calculated earlier (which, as we will see, is based on continuous compounding). The difference is small in practical terms, and neither is definitively the right answer. Both are estimates of the true growth rate. To emphasise this issue, note that since the calculated growth rate is based only upon the initial and final observations, it could be unreliable if either of these two values is an outlier (as in this case, the 2009 value). For example, if the growth rate is measured from 1977 to 2007, then the answer is 7.0%. With a sufficient span of time, however, such outliers are unlikely to be a serious problem.

STATISTICS IN PRACTICE

The power of compound growth

The *Economist* provided some amusing and interesting examples of how a \$1 investment can grow over time. They assumed that an investor (they named her Felicity Foresight, for reasons that become obvious) started with \$1 in 1900 and had the foresight or luck to invest, each year, in the best performing asset of the year. Sometimes she invested in equities, some years in gold and so on. By the end of the century she had amassed \$9.6 quintillion (9.6×10^{18}, more than world GDP, so impossible in practice). This is equivalent to an average annual growth rate of 55%. In contrast, Henry Hindsight did the same, but invested in the *previous year's* best asset. This might be thought more realistic. Unfortunately, his \$1 turned into only \$783, a still respectable annual growth rate of 6.9%. This, however, is beaten by the strategy of investing in the previous year's *worst* performing asset (what goes down must come up . . .). This turned \$1 into \$1730, a return of 7.7%. Food for thought!

Source: Based on *The Economist*, 12 February 2000, p. 111.

The geometric mean

In calculating the average growth rate of investment we have implicitly calculated the geometric mean of a series. If we have a series of n values, then their **geometric mean** is calculated as the nth root of the *product* of the values, i.e.

$$geometric\ mean = \sqrt[n]{\prod_{i=1}^{n} x_i} \tag{1.30}$$

The x values in this case are the growth factors in each year, as in Table 1.15 (the values in intermediate years are omitted). The 'Π' symbol is similar to the use of Σ, but means 'multiply together' rather than 'add up'.

The product of the 32 growth factors is 7.205 (the same as is obtained by dividing the final observation by the initial one – why?) and the 32nd root of this is 1.0637. This latter figure, 1.0637, is the geometric mean of the growth factors, and from it we can derive the growth rate of 6.37% p.a. by subtracting 1.

Whenever one is dealing with growth data (or any series that is based on a multiplicative process), one should ideally use the geometric mean rather than the arithmetic mean to get the answer. However, using the arithmetic mean in this case generally gives a similar answer, as long as the growth rate is reasonably small. If we take the arithmetic mean of the growth factors, we obtain:

$$\frac{1.142 + 1.190 + \cdots + 0.963 + 0.850}{32} = 1.0664$$

giving an estimate of the growth rate of $1.0664 - 1 = 0.0664 = 6.64\%$ p.a. – close to the correct value. Equivalently, one could take the average of the annual growth rates (0.142, 0.190, etc.), giving 0.0664, to get the same result. Use of the arithmetic mean is justified in this context if one needs only an approximation to the right answer and annual growth rates are reasonably small. It is usually quicker and easier to calculate the arithmetic rather than geometric mean, especially if one does not use a computer.

Table 1.15 Calculation of the geometric mean – annual growth factors

	Investment	Growth factors	
1977	28 351		
1978	32 387	1.142	(= 32387/28351)
1979	38 548	1.190	(= 38548/32387)
1980	43 612	1.131	etc.
2006	227 234	1.083	
2007	249 517	1.098	
2008	240 361	0.963	
2009	204 270	0.850	

Note: Each growth factor simply shows the ratio of that year's investment to the previous year's.

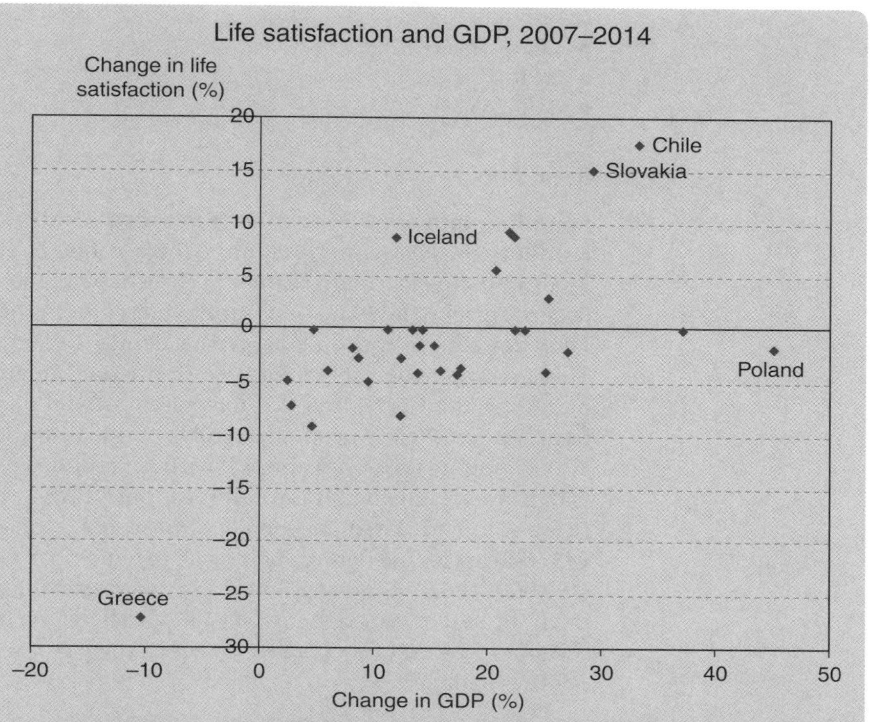

Now one can see a lot more, more clearly (and the scatter plot actually shows more countries). There seems to be a positive relation between the variables, and below about 20% growth (over seven years, so 2.6% p.a.) satisfaction generally falls. The exception to this is Iceland, recovering from its severe banking crisis. Chile and Slovakia show the greatest increase in life satisfaction and contrast with Poland which has even higher GDP growth, yet a fall in satisfaction.

One piece of information that is missing from the improved version is the country names, apart from a selected few. However, the aim is to show the general relationship so these are not all needed. One can add individual labels (as shown) if one wants to comment upon a particular country.

Data transformations

In analysing employment and investment data in the examples above, we have often changed the variables in some way in order to bring out the important characteristics. In statistics, one usually works with data that have been transformed in some way rather than using the original numbers. It is therefore worth summarising the main data transformations available, providing justifications for their use and exploring the implications of such adjustments to the original data. We briefly deal with the following transformations:

- rounding
- grouping
- dividing or multiplying by a constant

- differencing
- taking logarithms
- taking the reciprocal
- deflating.

Rounding

Rounding improves readability. Too much detail can confuse the message, so rounding the answer makes it more memorable. To give an example, the average wealth holding calculated earlier in this chapter is actually £186875.766 (to three decimal places). It would be absurd to present it in this form, however. We do not know for certain that this figure is accurate (in fact, it almost certainly is not). There is a spurious degree of precision which might mislead the reader. How much should this be rounded for presentational purposes? Remember that the figures have already been effectively rounded by allocation to classes of width 10000 or more (all observations have been rounded to the mid-point of the interval). However, much of this rounding is offsetting, i.e. numbers rounded up offset those rounded down, so the class mean is reasonably accurate. Rounding to £187000 makes the figure much easier to remember, and is only a change of 0.07% (187000/186874.766 = 1.00067), so is a reasonable compromise. In a report, it might be best to use the figure of £187000 therefore. In the text above, the answer was not rounded to such an extent since the purpose was to highlight the methods of calculation.

STATISTICS IN PRACTICE

Inflation in Zimbabwe

> *'Zimbabwe's rate of inflation surged to 3731.9%, driven by higher energy and food costs, and amplified by a drop in its currency, official figures show.'*
>
> BBC news online, 17 May 2007.

Whether official or not, it is impossible that the rate of inflation is known with such accuracy (to one decimal place), especially when prices are rising so fast. It would be more reasonable to report a figure of 3700% in this case. Sad to say, inflation rose even further in subsequent months.

Rounding is a 'trapdoor' function: you cannot obtain the original value from the transformed (rounded) value. Therefore, if you are going to need the original value in further calculations, you should not round your answer. Furthermore, small rounding errors can cumulate, leading to a large error in the final answer. Therefore, you should *never* round an intermediate answer, only the final one. Even if you only round the intermediate answer by a small amount, the final answer could be grossly inaccurate. Try the following: calculate $60.29 \times 30.37 - 1831$ both before and after rounding the first two numbers to integers. In the first case you get 0.0073, and in the second -31.

Grouping

When there is too much data to present easily, grouping solves the problem, although at the cost of hiding some of the information. The examples relating to education and unemployment and to wealth used grouped data. Using the raw data would have given us far too much information, so grouping is a first stage in data analysis. Grouping is another trapdoor transformation: once it's done you cannot recover the original information (unless you have access to the raw data, of course).

Dividing/multiplying by a constant

This transformation is carried out to make numbers more readable or to make calculation simpler by removing trailing zeros. The data on wealth were divided by 1000 to ease calculation; otherwise the fx^2 column would have contained extremely large values. Some summary statistics (e.g. the mean) will be affected by the transformation, but not all (e.g. the coefficient of variation). Try to remember which are affected. E and V operators (see Appendix 1B) can help. The transformation is easy to reverse.

Differencing

In time-series data there may be a trend, and it is better to describe the features of the data relative to the trend. The result may also be more economically meaningful, e.g. governments are often more concerned about the growth of output than about its level. Differencing is one way of eliminating the trend (see Chapter 11 for other methods of detrending data). Differencing was used for the investment data for both of these reasons. One of the implications of differencing is that information about the *level* of the variable is lost.

Taking logarithms

Taking logarithms is used to linearise a non-linear series, in particular one that is growing at a fairly constant rate. It is often easier to see the important features of such a series if the logarithm is graphed rather than the raw data. The logarithmic transformation is also useful in regression (see Chapter 9) because it yields estimates of **elasticities** (e.g. of demand). Taking the logarithm of the investment data linearised the series and tended to smooth it. The inverses of the logarithmic transformations are 10^x (for common logarithms) and e^x (for natural logarithms) so one can recover the original data.

Taking the reciprocal

The reciprocal of a variable might have a useful interpretation and provide a more intuitive explanation of a phenomenon. The reciprocal transformation will also turn a linear series into a non-linear one. The reciprocal of turnover in the labour market (i.e. the number leaving unemployment divided by the number unemployed) gives an idea of the duration of unemployment. If one-half of those unemployed find work each year (turnover $= 0.5$), then the average duration of unemployment is two years ($=1/0.5$). If a graph of turnover shows a linear decline over time, then the average duration of unemployment will be rising, at a faster and faster rate. Repeating the reciprocal transformation recovers the original data.

Deflating

Deflating turns a nominal series into a real one, i.e. one that reflects changes in quantities without the contamination of price changes. This is dealt with in more detail in Chapter 10. It is often more meaningful in economic terms to talk about a real variable than a nominal one. Consumers are more concerned about their real income than about their money income, for example.

Confusing real and nominal variables is dangerous. For example, someone's nominal (money) income may be rising yet their real income falling (if prices are rising faster than money income). It is important to know which series you are dealing with (this is a common failing among students new to statistics). An income series that is growing at 2 to 3 p.a. is probably a real series; one that is growing at 10% p.a. or more is likely to be nominal.

The information and data explosion

Recent developments in technology, especially those relating to the web, have led to a huge increase in data availability. Data files can now have tags allowing other websites to query the data, provide 'mashups' and exhibit the data in new ways. This is leading to a democratisation of data – anyone can now obtain them, draw graphs, interpret them and so on. Most of the data manipulation is done behind the scenes; the user does not need to have any understanding of the formulae nor of the calculations involved. This has some implications for those producing and using statistics, which we consider here.

First, however, we take a look at some examples of such 'data visualisation'. One of the most striking is Gapminder (www.gapminder.org), a site which allows you to construct interactive graphs of many variables. It is innovative in that it provides an enormous amount of information in each graph. Figure 1.26 shows a graph of CO_2 emissions against real GDP (an *XY* chart) for countries around the world in 1950.

As well as the two main variables, note the additional features:

- The data points are coloured bubbles. The colour represents the region and the size of the bubble represents the total emissions of the country (the variable on the *y*-axis is emissions *per capita*). Hence, we actually have four variables represented on the chart.
- Much of the graph can be customised. The axes can be linear or log scale (note that Gapminder has chosen a linear scale for emissions graphed against the log of income). The size of the bubble can be changed to represent alternative variables, e.g. an index of urbanisation.
- By hovering the cursor over a bubble, the country name is revealed (here, the United States) and the values for that observation are shown on the axes.
- By 'playing' the graph or using the slider, one can go forward through time to see the changes in a most vivid way. Unfortunately, this cannot be replicated in this text, but Figure 1.27 is the same graph for 2008.

The new graph shows how total emissions have grown (note China in particular) over the time period. A good feature of Gapminder is that the underlying data can be downloaded so that you can carry out your own further analysis if you wish to.

Figure 1.26
CO_2 emissions versus real GDP in 1950

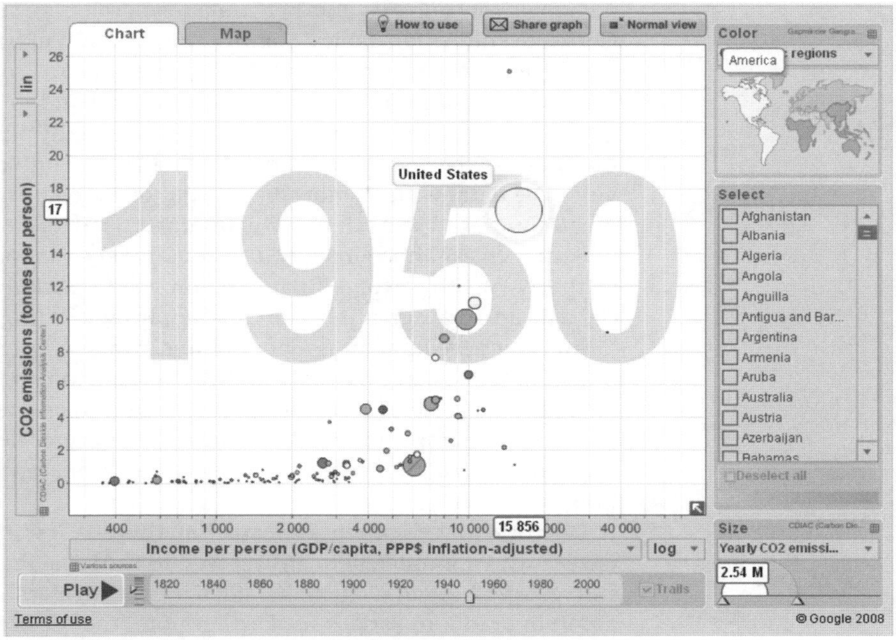

Source: From www.gapminder.org, Visualization formGapminder World, powered by Trendalyzer from www.gapminder.org.

Figure 1.27
CO_2 emissions versus real GDP in 2008

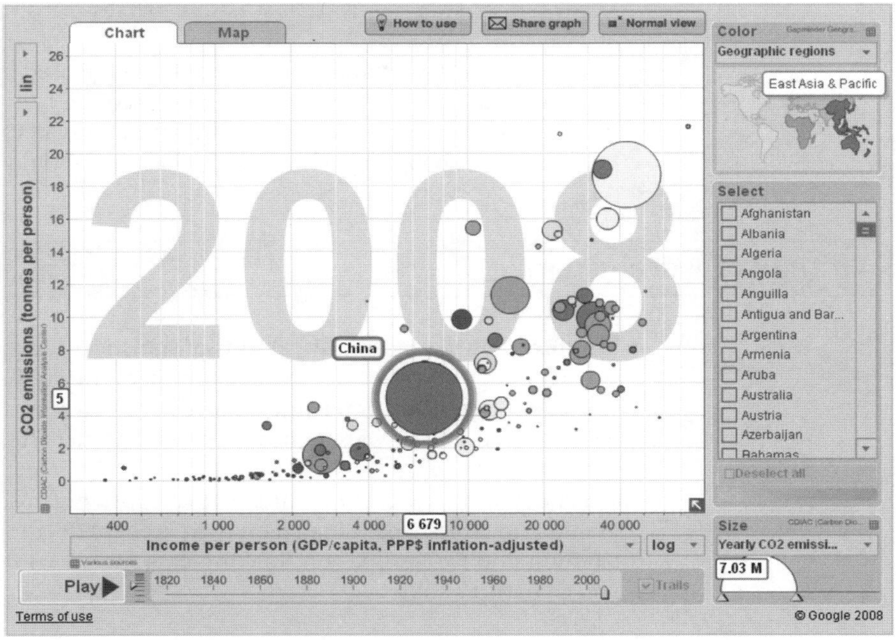

Source: From www.gapminder.org, Visualization from Gapminder World, powered by Trendalyzer from www.gapminder.org.

Google (of course) is another company developing such tools, such as Trends, Correlate and Fusion Tables. Some of these are still in beta, so will no doubt have developed by the time you read this. To locate them, use Google.

Another interesting example is the Mayor of London's Datastore (data.london.gov.uk/) which aims to open up London's data to the public, for free. It contains a multitude of data, some of it from other sites, and unlike Gapminder it only provides the data, with the idea being that other companies will create websites, mobile apps, and so forth that will interpret the data for you.

Sites such as Yahoo Finance (http://uk.finance.yahoo.com/) or Google Finance (http://www.google.co.uk/finance) allow you to examine a large amount of financial data and draw graphs of the data simply by selecting from a few menus.

Another development is 'data mining', which uses artificial intelligence techniques to 'mine' large databases of information, looking for trends and other features. For example, it is used by supermarkets to analyse their sales data with a view to spotting spending patterns and exploit them. Not only does one not need to perform calculations on the data, one does not even need to know what questions to ask of it.

Do these developments mean that there is less need to study statistics? Obviously, I am unlikely to answer 'yes' to this! The interpretation of the results still requires human judgement, aided by statistical tests to ensure one is not just observing random variations. Furthermore, the use of pictures (which many of such sites rely on) can be highly informative but it is difficult to convey that to another person without the picture. One can look at a graph of CO_2 emissions, but to convey the trend it is easier to pass on the average growth rate of those emissions. Looking at the Gapminder graphs above, how would you convey to another person (without showing them the graphs) how fast US emissions have risen? (The answer is only 0.2% p.a., somewhat surprisingly. China's have risen by 6.4% p.a.)

Easy access to such data also means it will be used indiscriminately by those unaware of its shortcomings, unsure how to correctly interpret them and eager to use them for support rather than illumination. It is all the more important, therefore, that more people are trained to understand and not be misled by statistics.

Writing statistical reports

This text presents most of the results of statistical analyses in a fairly formal way, since the aim is exposition and explanation of the methods. However, more often you might be writing or reading a short statistical report which requires a punchier and more concise presentation. This will attract attention, but it is important to maintain the accuracy of what is said.

Here is some advice on writing such a report, organised in sections on writing, graphs and tables. I have drawn on two excellent documents of the UN Economic Commission for Europe[9] on 'making data meaningful', and I recommend that you go to those sources.

[9]UNECE, *Making Data Meaningful*, parts 1 and 2, available at http://www.unece.org/stats/documents/writing/.

Writing

In a report, you should put the most important facts first, less important and supporting material afterwards. This is known as the 'inverted pyramid'. The opening paragraph should be concise and tell the story, making little use of numbers if possible.

Each paragraph should focus on one or two main points, explained using short sentences. Aim to avoid jargon, so-called elevator statistics ("... this went up, that went down ..."), acronyms and making the reader refer to tables in order to understand the point. The text should provide interpretation and context rather than repeating values which are in tables.

Break up the text with sub-headings, which should include a verb (to encourage you to make a point with the heading). For example, "More Britons finding work" is better than "Employment trends in Britain". Once you have completed the story, write the headline with the aim of catching the reader's attention.

As an example, consider the discussion earlier on the comparison of 1979 and 2005 wealth distributions in the United Kingdom. That was written in textbook style for the purpose of learning, but in a report might be better presented as follows.

> *Britons becoming wealthier but inequality persists*
>
> In 2005 the average Briton had wealth of around £187,000, about 11 times greater than in 1979. Adjusting for inflation, wealth has grown by nearly three times, or an average of 4% p.a. Despite this, the gap between rich and poor has remained much the same, with someone in the top 10% of the distribution owning 25 times more wealth than someone in the bottom 10%.

Tables

Tables in a report should be simple, sometimes called **presentation tables**, in contrast to **reference tables** which are best kept to an appendix. Many of the tables in this text are reference tables, such as the table calculating average wealth, Table 1.6. These are generally too complex to put into the text of a report and would have to be simplified in some way, according to the point that is being made.

Tables should be comprehensible in themselves, without the reader needing to look for further information. One way to ensure this is to ask if the table could be cut and pasted into another document and still make sense. Hence the table needs at least an informative title.

Another useful tip is to order the categories in a table according to frequency where this is appropriate. Consider the table used in Exercise 1.1(b) earlier, on tourist destinations. There is no natural ordering of the four countries, so why not order the countries by the number of tourists, as follows?

Tourists (millions) visiting European countries, 2013

	France	Spain	Italy	Germany
All tourists	12.4	9.8	7.5	3.2
English tourists	2.7	3.6	1.0	0.2
Non-English tourists	9.7	6.2	6.5	3.0

Note the difference in presentation from the exercise and how this makes it easier to read the table. In presentation tables it is important to keep decimal places as small as possible (also aligning the numbers properly) to aid readability. In this text, we often use more decimal places so that you can follow calculation of various statistics. This is not necessary for presentation tables.

Graphs

Much advice has already been given about drawing graphs already, but there are some additional points relevant to a report. Make sure that each graph tells a simple story with not more than one or two elements. An informative title helps with this, as shown below illustrating the tourism data.

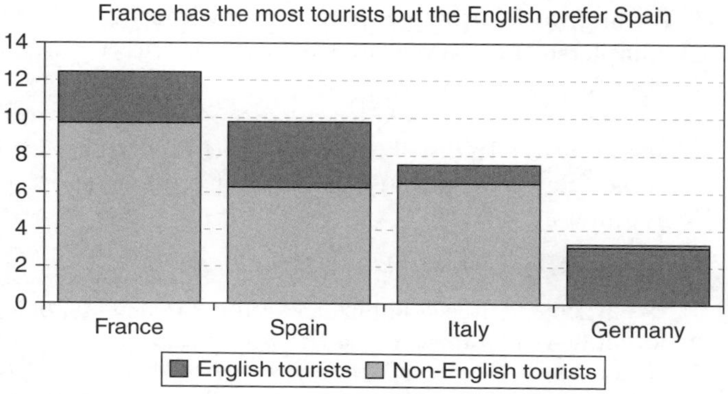

Compare this chart with the answer to Exercise 1.1, where the countries are not ordered by frequency. This one is easier to read, and the title spells out the essential messages.

Guidance to the student: how to measure your progress

Now you have reached the end of the chapter your work is not yet over. It is very unlikely that you have fully understood everything after one reading. What you should do now is:

- Check back over the learning outcomes at the start of the chapter. Do you feel you have achieved them? For example, can you list the various different data types you should be able to recognise (the first learning outcome)?
- Read the chapter summary below to help put things in context. You should recognise each topic and be aware of the main issues, techniques, etc., within them. There should be no surprises or gaps.
- Read the list of key terms. You should be able to give a brief and precise definition or description of each one. Do not worry if you cannot remember all the formulae (though you should try to memorise simple ones, such as that for the mean).
- Try out the problems (most important!). Answers to odd-numbered problems are at the back of the text, so you can check your answers. There is more detail for some of the answers on the text's website.

From all of this, you should be able to work out whether you have really mastered the chapter. Do not be surprised if you have not – it will take more than one reading. Go back over those parts where you feel unsure of your knowledge. Use these same learning techniques for each chapter.

Summary

- Descriptive statistics are useful for summarising large amounts of information, highlighting the main features but omitting the detail.
- Different techniques are suited to different types of data, e.g. bar charts for cross-section data and rates of growth for time series.
- Graphical methods, such as the bar chart, provide a picture of the data. These give an informal summary, but they are unsuitable as a basis for further analysis.
- Important graphical techniques include the bar chart, frequency distribution, relative and cumulative frequency distributions, histogram and pie chart. For time-series data, a time-series chart of the data is informative.
- Numerical techniques are more precise as summaries. Measures of location (such as the mean), of dispersion (the variance) and of skewness form the basis of these techniques.
- Important numerical summary statistics include the mean, median and mode; variance, standard deviation and coefficient of variation; coefficient of skewness.
- For bivariate data, the scatter diagram (or *XY* graph) is a useful way of illustrating the data.
- Data are often transformed in some way before analysis, e.g. by taking logs. Transformations often make it easier to see key features of the data in graphs and sometimes make summary statistics easier to interpret. For example, with time-series data the average rate of growth may be more appropriate than the mean of the series.

Key terms and concepts

absolute dispersion
area graph
arithmetic mean
average growth rate
bar chart
bivariate method
box and whiskers plot
Chebyshev's inequality
class interval
class width
coefficient of skewness
coefficient of variation
compound growth
compound interest
cross-section data
cross-tabulation
data transformation
decile
depreciation rate
elasticity
expected value
frequency
frequency density
frequency table
geometric mean
growth factor

→

heteroscedasticity	quantile solidus quantiles
histogram	quartile
homoscedasticity	quintile
inter-quartile range	range
logarithm	ratio scale
mean	reference tables
measure of dispersion	relative and cumulative frequencies
measure of location	relative dispersion
measure of skewness	scatter diagram (*XY* chart)
median	serial correlation
mid-point	skewness
mode	stacked bar chart
multiple bar chart	standard deviation
multiple time-series graph	standard width
multivariate method	time-series data
nominal scale	time-series graph
non-linear trend	transformed data
ordinal scale	trend
outliers	unbiased
percentile	univariate method
pie chart	variance
positively skewed	weighted average
presentation tables	*z* score

Reference

Atkinson, A. B., *The Economics of Inequality*, 2nd edn, Oxford University Press, 1983.

Formulae used in this chapter

Formula	Description	Notes
$\mu = \dfrac{\sum x}{N}$	Mean of a population	Use when all individual observations are available. *N* is the population size.
$\mu = \dfrac{\sum fx}{\sum f}$	Mean of a population	Use with grouped data. *f* represents the class or group frequencies, *x* represents the mid-point of the class interval
$\bar{x} = \dfrac{\sum x}{n}$	Mean of a sample	*n* is the number of observations in the sample
$\bar{x} = \dfrac{\sum fx}{\sum f}$	Mean of a sample	Use with grouped data
$m = x_L + (x_U - x_L)\left\{\dfrac{\frac{N+1}{2} - F}{f}\right\}$	Median (where data are grouped)	x_L and x_U represent the lower and upper limits of the interval containing the median. *F* represents the cumulative frequency up to (but excluding) the interval
$\sigma^2 = \dfrac{\sum (x-\mu)^2}{N}$	Variance of a population	*N* is the population size.
$\sigma^2 = \dfrac{\sum f(x-\mu)^2}{\sum f}$	Population variance (grouped data)	
$s^2 = \dfrac{\sum (x-\bar{x})^2}{n-1}$	Sample variance	
$s^2 = \dfrac{\sum f(x-\bar{x})^2}{n-1}$	Sample variance (grouped data)	
$c.v = \dfrac{\sigma}{\mu}$	Coefficient of variation	The ratio of the standard deviation to the mean. A measure of dispersion.
$z = \dfrac{x-\mu}{\sigma}$	*z* score	Measures the distance from observation *x* to the mean μ measured in standard deviations
$\dfrac{\sum f(x-\mu)^3}{N\sigma^3}$	Coefficient of skewness	A positive value means the distribution is skewed to the right (long tail to the right).
$g = \sqrt[t-1]{\dfrac{x_T}{x_1}} - 1$	Rate of growth	Measures the average annual rate of growth between years 1 and *T*
$\sqrt[n]{\Pi x}$	Geometric mean (of *n* observations on *x*)	
$1 - \dfrac{1}{k^2}$	Chebyshev's inequality	Minimum proportion of observations lying within *k* standard deviations of the mean of any distribution

Problems

Some of the more challenging problems are indicated by highlighting the problem number in colour.

1.1 The following data show the education and employment status of women aged 20–29:

	Higher education	A levels	Other qualification	No qualification	Total
In work	209	182	577	92	1060
Unemployed	12	9	68	32	121
Inactive	17	34	235	136	422
Sample	238	225	880	260	1603

(a) Draw a bar chart of the numbers in work in each education category (the first line of the table). Can this be easily compared with the similar diagram in the text, for both males and females (Figure 1.1)?

(b) Draw a stacked bar chart using all the employment states, similar to Figure 1.3. Comment upon any similarities and differences from the diagram in the text.

(c) Convert the table into (column) percentages and produce a stacked bar chart similar to Figure 1.4. Comment upon any similarities and differences.

(d) Draw a pie chart showing the distribution of educational qualifications of those in work and compare it to Figure 1.5 in the text.

1.2 The data below show the average hourly earnings (in £s) of those in full-time employment, by category of education (NVQ levels. NVQ 4 corresponds to a university degree).

	NVQ 4	NVQ 3	NVQ 2	Below NVQ 2	No qualification
Males	17.69	12.23	11.47	10.41	8.75
Females	14.83	9.57	9.40	9.24	7.43

(a) In what fundamental way do the data in this table differ from those in Problem 1.1?

(b) Construct a bar chart showing male and female earnings by education category. What does it show?

(c) Why would it be inappropriate to construct a stacked bar chart of the data? How should one graphically present the combined data for males and females? What extra information is necessary for you to do this?

1.3 Using the data from Problem 1.1:

(a) Which education category has the highest proportion of women in work? What is the proportion?

(b) Which category of employment status has the highest proportion of women with a degree? What is the proportion?

1.4 Using the data from Problem 1.2:

(a) What is the premium, in terms of average earnings, of a degree over A levels (NVQ 3)? Does this differ between men and women?

(b) Would you expect *median* earnings to show a similar picture? What differences, if any, might you expect?

1.5 The distribution of marketable wealth in 1979 in the United Kingdom is shown in the table below (adapted from *Inland Revenue Statistics, 1981*, contains public sector information licensed under the Open Government Licence (OGL) v3.0, http://www.nationalarchives.gov.uk/doc/open-government-licence/open-government:

Range	Number 000s	Amount £m
0–	1 606	148
1 000–	2 927	5 985
3 000–	2 562	10 090
5 000–	3 483	25 464
10 000–	2 876	35 656
15 000–	1 916	33 134
20 000–	3 425	104 829
50 000–	621	46 483
100 000–	170	25 763
200 000–	59	30 581

Draw a bar chart and histogram of the data (assume the final class interval has a width of 200 000). Comment on the differences between the two types of chart. Comment on any differences between this histogram and the latest one for 2005 given in the text of this chapter.

1.6 The data below show the number of enterprises in the United Kingdom in 2010, arranged according to employment:

Number of employees	Number of firms
1–	1 740 685
5–	388 990
10–	215 370
20–	141 920
50–	49 505
100–	25 945
250–	7 700
500–	2 795
1 000–	1 320

Draw a bar chart and histogram of the data (assume the mid-point of the last class interval is 2000). What are the major features apparent in each and what are the differences?

1.7 Using the data from Problem 1.5:

(a) Calculate the mean, median and mode of the distribution. Why do they differ?

(b) Calculate the inter-quartile range, variance, standard deviation and coefficient of variation of the data.

(c) Calculate the skewness of the distribution.

(d) From what you have calculated, and the data in the chapter, can you draw any conclusions about the degree of inequality in wealth holdings, and how this has changed?

(e) What would be the effect upon the mean of assuming the final class width to be £10m? What would be the effects upon the median and mode?

1.8 Using the data from Problem 1.6:

(a) Calculate the mean, median and mode of the distribution. Why do they differ?

(b) Calculate the inter-quartile range, variance, standard deviation and coefficient of variation of the data.

(c) Calculate the coefficient of skewness of the distribution.

1.9 A motorist keeps a record of petrol purchases on a long journey, as follows:

Petrol station	1	2	3
Litres purchased	33	40	25
Price per litre (pence)	134	139	137

Calculate the average petrol price for the journey.

1.10 Demonstrate that the weighted average calculation given in equation (1.9) is equivalent to finding the total expenditure on education divided by the total number of pupils.

1.11 On a test taken by 100 students, the average mark is 65, with variance 144. Student A scores 83; student B scores 47.

(a) Calculate the z scores for these two students.

(b) What is the maximum number of students with a score either better than A's or worse than B's?

(c) What is the maximum number of students with a score better than A's?

1.12 The average income of a group of people is £8000, and 80% of the group have incomes within the range £6000–10 000. What is the minimum value of the standard deviation of the distribution?

1.13 The following data show car registrations in the United Kingdom for 1987–2010:

Year	Registrations	Year	Registrations	Year	Registrations
1987	2212.6	1995	2024.0	2003	2820.7
1988	2437.0	1996	2093.3	2004	2784.7
1989	2535.2	1997	2244.3	2005	2603.6
1990	2179.9	1998	2367.0	2006	2499.1
1991	1708.5	1999	2342.0	2007	2539.3
1992	1694.4	2000	2430.0	2008	2188.3
1993	1853.4	2001	2710.0	2009	1959.1
1994	1991.7	2002	2816.0	2010	1994.6

(a) Draw a time-series graph of car registrations. Comment upon the main features of the series. (It looks daunting, but it will take you less than 10 minutes to type in these data.)

(b) Draw time-series graphs of the change in registrations, the (natural) log of registrations and the change in the ln. Comment upon the results.

1.14 The table below shows the different categories of investment in the United Kingdom over a series of years:

Year	Dwellings	Transport	Machinery	Intangible fixed assets	Other buildings
1977	5699	3248	9950	797	8657
1978	6325	4112	11709	760	9481
1979	7649	4758	13832	964	11289
1980	8674	4707	15301	1216	13680
1981	8138	4011	15454	1513	14603
1982	8920	4489	16734	2040	15730
1983	10447	4756	18377	2337	16157
1984	11932	5963	20782	2918	17708
1985	12219	6676	24349	3239	18648
1986	14140	6527	25218	3219	20477
1987	16548	7871	28226	3430	24269
1988	21097	9228	32615	4305	30713
1989	22771	10625	38419	4977	36689
1990	21048	10572	37776	6298	41334
1991	18339	9051	35094	6722	38632
1992	18826	8420	35426	6584	34657
1993	19886	9315	35316	6492	32988
1994	21155	11395	38426	6702	33945
1995	22448	11036	45012	7272	35596
1996	22516	12519	50102	7889	37320
1997	23928	12580	51465	8936	41398
1998	25222	16113	58915	9461	46286
1999	25700	14683	60670	10023	50646
2000	27394	13577	63535	10670	51996
2001	29806	14656	60929	11326	55065
2002	34499	16314	57152	12614	59972
2003	38462	15592	54441	13850	64355
2004	44298	14339	59632	14164	67982
2005	47489	14763	59486	14386	73634
2006	53331	14855	61497	15531	82020
2007	55767	15482	69411	16049	92808
2008	50292	14570	67837	16726	90936
2009	37044	12127	56411	17710	80978

Use appropriate graphical techniques to analyse the properties of any one of the investment series. Comment upon the results. (Although this seems a lot of data, it shouldn't take long to type in, even less time if two people collaborate and share their results.)

1.15 Using the data from Problem 1.13:

(a) Calculate the average rate of growth of the series.

(b) Calculate the standard deviation around the average growth rate.

(c) Does the series appear to be more or less volatile than the investment figures used in the chapter? Suggest reasons.

1.16 Using the data from Problem 1.14:

(a) Calculate the average rate of growth of the series for dwellings.

(b) Calculate the standard deviation around the average growth rate.

(c) Does the series appear to be more or less volatile than the investment figures used in the chapter? Suggest reasons.

1.17 How would you expect the following time-series variables to look when graphed? (e.g. Trended? Linear trend? Trended up or down? Stationary? Homoscedastic? Autocorrelated? Cyclical? Anything else?)

(a) Nominal national income.

(b) Real national income.

(c) The nominal interest rate.

1.18 How would you expect the following time-series variables to look when graphed?

(a) The price level.

(b) The inflation rate.

(c) The £/$ exchange rate.

1.19 (a) A government bond is issued, promising to pay the bearer £1000 in five years' time. The prevailing market rate of interest is 7%. What price would you expect to pay now for the bond? What would its price be after two years? If, after two years, the market interest rate jumped to 10%, what would the price of the bond be?

(b) A bond is issued which promises to pay £200 p.a. over the next five years. If the prevailing market interest rate is 7%, how much would you be prepared to pay for the bond? Why does the answer differ from the previous question? (Assume interest is paid at the end of each year.)

1.20 A firm purchases for £30 000 a machine which is expected to last for 10 years, after which it will be sold for its scrap value of £3000. Calculate the average rate of depreciation p.a., and calculate the written-down value of the machine after one, two and five years.

1.21 Depreciation of BMW and Mercedes cars is given in the following table of new and used car prices:

Age	BMW 525i	Mercedes 200E
Current	22 275	21 900
1 year	18 600	19 700
2 years	15 200	16 625
3 years	12 600	13 950
4 years	9 750	11 600
5 years	8 300	10 300

(a) Calculate the average rate of depreciation of each type of car.

(b) Use the calculated depreciation rates to estimate the value of the car after 1, 2, etc., years of age. How does this match the actual values?

(c) Graph the values and estimated values for each car.

1.22 A bond is issued which promises to pay £400 p.a. in perpetuity. How much is the bond worth now, if the interest rate is 5%? (Hint: the sum of an infinite series of the form

$$\frac{1}{1+r} + \frac{1}{(1+r)^2} + \frac{1}{(1+r)^3} + \cdots$$

is $1/r$, as long as $r > 0$.)

Note that this is different from:

$$\left(\sum x\right)^2 = (x_1 + x_2 + \cdots + x_5)^2 = 676$$

Part of the formula for the variance calls for the following calculation:

$$\sum fx^2 = f_1x_1^2 + f_2x_2^2 + \cdots + f_5x_5^2 = 2 \times 3^2 + 2 \times 5^2 + \cdots + 1 \times 8^2 = 324$$

Using Σ notation we can see the effect of transforming x by dividing by 1000, as was done in calculating the average level of wealth. Instead of working with x we used kx, where $k = 1/1000$. In finding the mean we calculated

$$\frac{\sum kx}{N} = \frac{kx_1 + kx_2 + \cdots}{N} = \frac{k(x_1 + x_2 + \cdots)}{N} = k\frac{\sum x}{N} \tag{1.34}$$

So to find the mean of the original variable x we had to divide by k again, i.e. multiply by 1000. In general, whenever each observation in a sum is multiplied by a constant, the constant can be taken outside the summation operator, as in (1.34) above.

Problems on Σ notation

1A.1 Given the following data on x_i : {4, 6, 3, 2, 5} evaluate:

$$\sum x_i, \sum x_i^2, \left(\sum x_i\right)^2, \sum (x_i - 3), \sum x_i - 3, \sum_{i=2}^{4} x_i$$

1A.2 Given the following data on x_i : {8, 12, 6, 4, 10} evaluate:

$$\sum x_i, \sum x_i^2, \left(\sum x_i\right)^2, \sum (x_i - 3), \sum x_i - 3, \sum_{i=2}^{4} x_i$$

1A.3 Given the following frequencies, f_i associated with the x values in Problem 1A.1: {5, 3, 3, 8, 5}, evaluate:

$$\sum fx, \sum fx^2, \sum f(x - 3), \sum fx - 3$$

1A.4 Given the following frequencies, f_i associated with the x values in Problem 1A.2: {10, 6, 6, 16, 10}, evaluate:

$$\sum fx, \sum fx^2, \sum f(x - 3), \sum fx - 3$$

1A.5 Given the pairs of observations on x and y,

x	4	3	6	8	12
y	3	9	1	4	3

evaluate $\sum xy, \sum x(y - 3), \sum (x + 2)(y - 1)$.

1A.6 Given the pairs of observations on x and y,

x	3	7	4	1	9
y	1	2	5	1	2

evaluate $\sum xy, \sum x(y - 2), \sum (x - 2)(y + 1)$.

1A.7 Demonstrate that

$$\frac{\Sigma f(x - k)}{\Sigma f} = \frac{\Sigma fx}{\Sigma f} - k \text{ where } k \text{ is a constant.}$$

1A.8 Demonstrate that

$$\frac{\Sigma f(x - \mu)^2}{\Sigma f} = \frac{\Sigma fx^2}{\Sigma f} - \mu^2$$

Appendix 1B E and V operators

These operators are an extremely useful form of notation that we shall make use of later in the text. It is quite easy to keep track of the effects of data transformations using them. There are a few simple rules for manipulating them that allow some problems to be solved quickly and elegantly.

$E(x)$ is the mean of a distribution and $V(x)$ is its variance. We showed above in (1.34) that multiplying each observation by a constant k multiplies the mean by k. Thus we have:

$$E(kx) = kE(x) \tag{1.35}$$

Similarly, if a constant is added to every observation, the effect is to add that constant to the mean (see Problem 1.23):

$$E(x + a) = E(x) + a \tag{1.36}$$

(Graphically, the whole distribution is shifted a units to the right and hence so is the mean.) Combining (1.35) and (1.36):

$$E(kx + a) = kE(x) + a \tag{1.37}$$

Similarly, for the variance operator it can be shown that:

$$V(x + k) = V(x) \tag{1.38}$$

Proof:

$$V(x + k) = \frac{\sum((x - k) - (\mu + k))^2}{N} = \frac{\sum((x - \mu) + (k - k))^2}{N} = \frac{\sum(x - \mu)^2}{N} = V(x)$$

(A shift of the whole distribution leaves the variance unchanged.) Also:

$$V(kx) = k^2V(x) \tag{1.39}$$

(See Problem 1.24.) This is why, when the wealth figures were divided by 1000, the variance became divided by 1000^2. Applying (1.38) and (1.39):

$$V(kx + a) = k^2V(x) \tag{1.40}$$

Finally, we should note that V itself can be expressed in terms of E:

$$V(x) = E(x - E(x))^2 \text{ or } E(x^2) - E(x)^2 \tag{1.41}$$

Appendix 1C Using logarithms

Logarithms are less often used now that cheap electronic calculators are available. Formerly, logarithms were an indispensable aid to calculation. However, the logarithmic transformation is useful in other contexts in statistics and economics, so its use is briefly set out here.

The logarithm (to the base 10) of a number x is defined as the power to which 10 must be raised to give x. For example, $10^2 = 100$, so the log of 100 is 2 and we write $\log_{10} 100 = 2$ or simply $\log 100 = 2$.

Similarly, the log of 1000 is 3 ($100 = 10^3$), of 10 000 it is 4, etc. We are not restricted to integer (whole number) powers of 10, so for example $10^{2.5} =$ 316.227 766 (try this if you have a scientific calculator). So the log of 316.227 766 is 2.5. Every number x can therefore be represented by its logarithm.

Multiplication of two numbers

We can use logarithms to multiply two numbers x and y, based on the property[10]

$$\log xy = \log x + \log y$$

For example, to multiply 316.227 766 by 10:

$$\begin{aligned}\log(316.227766 \times 10) &= \log 316.227\,766 + \log 10 \\ &= 2.5 + 1 \\ &= 3.5\end{aligned}$$

The *anti-log* of 3.5 is given by $10^{3.5} = 3162.27\,766$ which is the answer.

Taking the anti-log (i.e. 10 raised to a power) is the inverse of the log transformation. Schematically we have:

$$x \rightarrow \text{take logarithms} \rightarrow a (= \log x) \rightarrow \text{raise 10 to the power } a \rightarrow x$$

Division

To divide one number by another we subtract the logs. For example, to divide 316.227 766 by 100:

$$\begin{aligned}\log(316.2277766/100) &= \log 316.227766 - \log 100 \\ &= 2.5 - 2 \\ &= 0.5\end{aligned}$$

and $10^{0.5} = 3.16\,227\,766$.

Powers and roots

Logarithms simplify the process of raising a number to a power. To find the square of a number, multiply the logarithm by 2, e.g. to find $316.22\,7766^2$:

$$\log(316.27766^2) = 2\log(316.227\,766) = 5$$

and $10^5 = 100\,000$.

[10] This is equivalent to saying $10^x \times 10^y = 10^{x+y}$.

To find the square root of a number (equivalent to raising it to the power $\frac{1}{2}$), divide the log by 2. To find the nth root, divide the log by n. For example, in the text we have to find the 32nd root of 13.518:

$$\frac{\log(13.518)}{32} = \frac{1.1309}{32} = 0.0353$$

and $10^{0.0353} = 1.085$.

Common and natural logarithms

Logarithms to the base 10 are known as common logarithms but one can use any number as the base. *Natural* logarithms are based on the number $e(= 2.71\,828\ldots)$, and we write $\ln x$ instead of $\log x$ to distinguish them from common logarithms. So, for example,

$$\ln 316.227\,766 = 5.756\,462\,732$$
$$\text{since } e^{5.756\,462732} = 316.227766.$$

Natural logarithms can be used in the same way as common logarithms and have similar properties. Use the 'ln' key on your calculator just as you would the 'log' key, but remember that the inverse transformation is e^x rather than 10^x.

Problems on logarithms

1C.1 Find the common logarithms of: 0.15, 1.5, 15, 150, 1500, 83.7225, 9.15, −12.

1C.2 Find the log of the following values: 0.8, 8, 80, 4, 16, −37.

1C.3 Find the natural logarithms of: 0.15, 1.5, 15, 225, −4.

1C.4 Find the ln of the following values: 0.3, *e*, 3, 33, −1.

1C.5 Find the anti-log of the following values: −0.823 909, 1.1, 2.1, 3.1, 12.

1C.6 Find the anti-log of the following values: −0.09 691, 2.3, 3.3, 6.3.

1C.7 Find the anti-ln of the following values: 2.708 05, 3.708 05, 1, 10.

1C.8 Find the anti-ln of the following values: 3.496 508, 14, 15, −1.

1C.9 Evaluate: $\sqrt[2]{10}$, $\sqrt[4]{3.7}$, $4^{1/4}$, 12^{-3}, $25^{-3/2}$.

1C.10 Evaluate: $\sqrt[3]{30}$, $\sqrt[6]{17}$, $8^{1/4}$, 15^0, 12^0, $3^{-1/3}$.

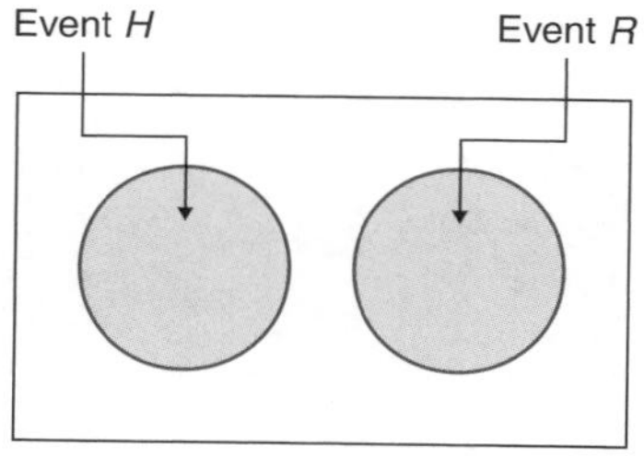

Figure 2.5
Mercedes winning the Grand Prix

Figure 2.6
The sample space for rolling a die

1	2	3	4	5	6
•	•	•	•	•	•

To reinforce the idea, consider getting a five or a six on a roll of a die[1]. This is

$$\Pr(5 \text{ or } 6) = \Pr(5) + \Pr(6) = 1/6 + 1/6 = 1/3 \tag{2.6}$$

This answer can be verified from the sample space, as shown in Figure 2.6. Each dot represents a simple event (one to six). The compound event is made up of two of the six points, shaded in Figure 2.6, so the probability is 2/6 or 1/3.

However, (2.4) is not a general solution to this type of problem, i.e. it does not *always* give the right answer, as can be seen from the following example. What is the probability of a queen or a spade in a single draw from a pack of cards? $\Pr(Q) = 4/52$ (four queens in the pack) and $\Pr(A) = 13/52$ (13 spades), so applying (2.4) gives

$$\Pr(Q \text{ or } S) = \Pr(Q) + \Pr(S) = 4/52 + 13/52 = 17/52 \tag{2.7}$$

However, if the sample space is examined the correct answer is found to be 16/52, as in Figure 2.7.

The problem is that one point in the sample space (the one representing the queen of spades) is double-counted in equation (2.7), once as a queen and again as a spade. The event 'drawing a queen *and* a spade' is possible, and gets double-counted. This issue can again be illustrated using a Venn diagram (see Figure 2.8). This is similar to Figure 2.5 except that the two circles overlap. The overlap area is called the **intersection** of the two sets Q and S and represents any card that is both a queen *and* a spade (i.e. just the queen of spades). The intersection of the sets is written $Q \cap S$, the symbol '$\cap$' meaning 'intersection'.

Therefore, we wish to consider all of the outcomes within the circles, counted *once only*. Formally this is known as the **union** of the two sets, written $Q \cup S$. But if

Figure 2.7
The sample space for drawing a queen or a spade

	A	K	Q	J	10	9	8	7	6	5	4	3	2
♠	•	•	•	•	•	•	•	•	•	•	•	•	•
♥	•	•	•	•	•	•	•	•	•	•	•	•	•
♦	•	•	•	•	•	•	•	•	•	•	•	•	•
♣	•	•	•	•	•	•	•	•	•	•	•	•	•

[1]We assume each outcome is equally likely so the probability of each the six numbers occurring is 1/6.

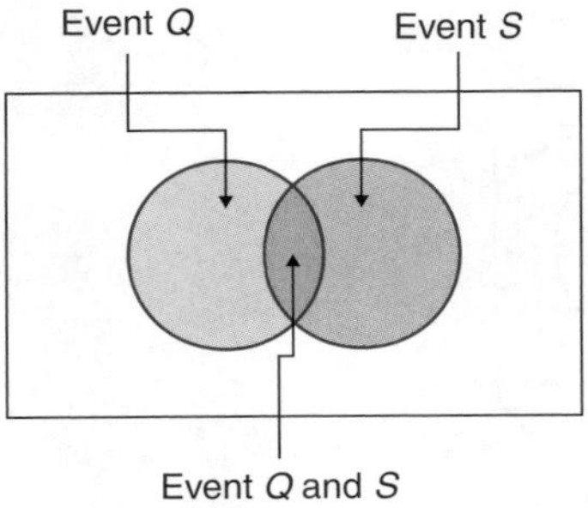

Figure 2.8
Drawing the queen of spades

we count all the outcomes in Q and then all those in S, we will double count those in the intersection. Hence, we need to subtract the intersection. In the language of sets we have:

$$Q \cup S = Q + S - (Q \cap S)$$

This carries over to probabilities. Equation (2.4) has to be modified by subtracting the probability of getting a queen *and* a spade, to eliminate this double counting. The correct answer is obtained from

$$\begin{aligned} \Pr(Q \text{ or } S) &= \Pr(Q) + \Pr(S) - \Pr(Q \text{ and } S) \\ &= 4/52 + 13/52 - 1/52 \\ &= 16/52 \end{aligned} \tag{2.8}$$

The general rule is therefore

$$\Pr(A \text{ or } B) = \Pr(A) + \Pr(B) - \Pr(A \text{ and } B) \tag{2.9}$$

Rule (2.4) worked for the earlier example because $\Pr(H \text{ and } R) = 0$ since it is impossible for both Hamilton and Rosberg to win the same race. The two sets, H and R, did not overlap. The possibility of double counting could not occur in the calculation of that probability.

In general, therefore, one should use (2.9), but when two events are mutually exclusive the rule simplifies to (2.4).

The multiplication rule

The **multiplication rule** is associated with use of the word 'and' to combine events. Consider the example of a mother with two children. What is the probability that they are both boys? This is really a compound event: that the first child is a boy *and* the second is also a boy. It corresponds to the intersection of the two sets in a Venn diagram, similar to that shown in Figure 2.8. Assume that in any single birth a boy or girl is equally likely, so Pr(boy) = Pr(girl) = 0.5. Denote by $\Pr(B1)$ the probability of a boy on the first birth and by $\Pr(B2)$ the probability of a boy on the second. Thus the question asks for $\Pr(B1 \text{ and } B2)$ and this is given by:

$$\Pr(B1 \text{ and } B2) = \Pr(B1) \times \Pr(B2) = 0.5 \times 0.5 = 0.25 \tag{2.10}$$

Intuitively, the multiplication rule can be understood as follows. One-half of mothers have a boy on their first birth and of these, one-half will again have a boy on the second. Therefore, a quarter (a half of one-half) of mothers have two boys.

Tree diagrams

This is a useful point at which to introduce another helpful visual aid – the **tree diagram.** Many people (including experienced researchers) have difficulty mastering probabilities and are easily deceived, but it has been found that if a problem is expressed in the form of a tree diagram, then many find it easier to follow. Furthermore, many find it easier to understand an issue when it is expressed in terms of frequencies rather than probabilities (a tree diagram can use either approach).

Consider the example just above of the mother with two children. We can think about this problem in the following way. Start with 1000 mothers (a convenient round figure) and consider the first child. We would expect 500 of them to be boys and 500 girls since a boy and a girl are equally likely. We can illustrate this as in Figure 2.9(a).

If we add the second birth, we get the extension of the diagram as in Figure 2.9(b). Of the 500 mothers with a boy, 250 have another boy while 250 have a girl, etc. Hence, 250 (out of 1000) mothers have two boys, or 0.25, as found using probabilities. Figure 2.9(c) shows how we could use the same tree diagram but label it with probabilities instead of frequencies. Either form of the diagram will do but the version using frequencies might be more intuitive for an audience.

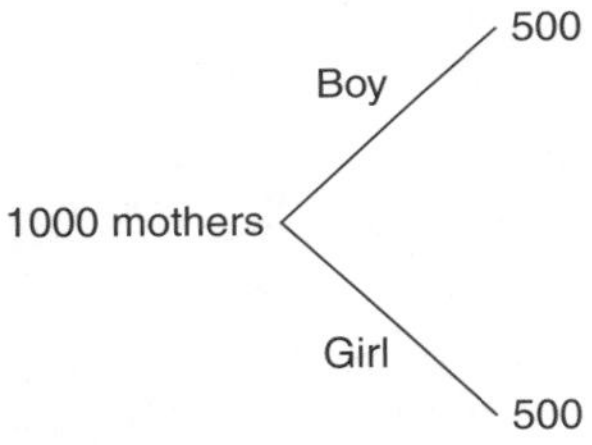

Figure 2.9(a)
Tree diagram for first child

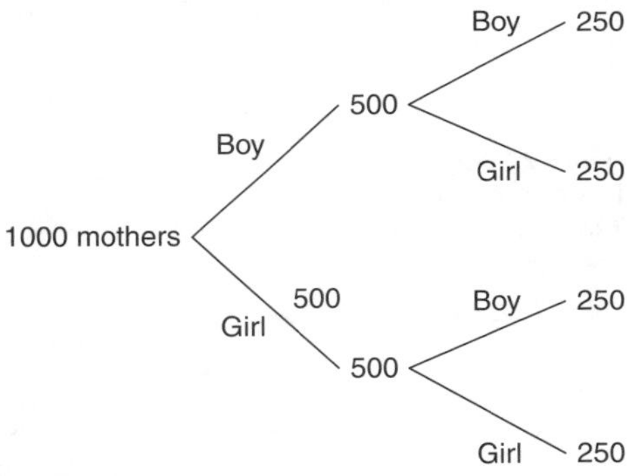

Figure 2.9(b)
Tree diagram for second child

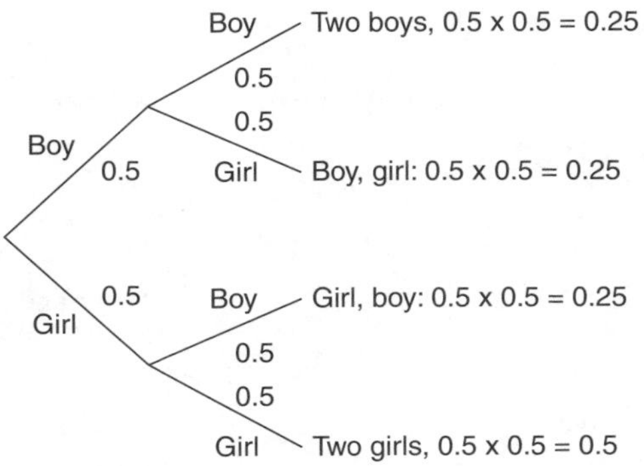

Figure 2.9(c)
Tree diagram using probabilities

The multiplication rule and independence

Like the addition rule, the multiplication rule requires slight modification before it can be applied generally and give the right answer in all circumstances. The example above assumes first and second births to be **independent events**, i.e. that having a boy on the first birth does not affect the probability of a boy on the second. This assumption is not always valid, so we now consider this.

Write $\Pr(B2 \mid B1)$ to indicate the probability of the event $B2$ *given* that the event $B1$ has occurred. This is known as the **conditional probability**, more precisely the probability of $B2$ conditional upon $B1$. In words, it means the probability of having a second boy after the first is a boy. Let us drop the independence assumption and suppose the following:

$$\Pr(B1) = \Pr(G1) = 0.5 \tag{2.11}$$

i.e. boys and girls are equally likely on the first birth (as previously assumed), but

$$\Pr(B2 \mid B1) = \Pr(G2 \mid G1) = 0.6 \tag{2.12}$$

i.e. a boy is more likely to be followed by another boy, and a girl by another girl. (It is easy to work out $\Pr(B2 \mid G1)$ and $\Pr(G2 \mid B1)$. What are they? Use a version of equation (2.2) to think about this.)

This new situation can again be usefully illustrated with a tree diagram, either using frequencies (Figure 2.10(a)) or probabilities (Figure 2.10(b)).

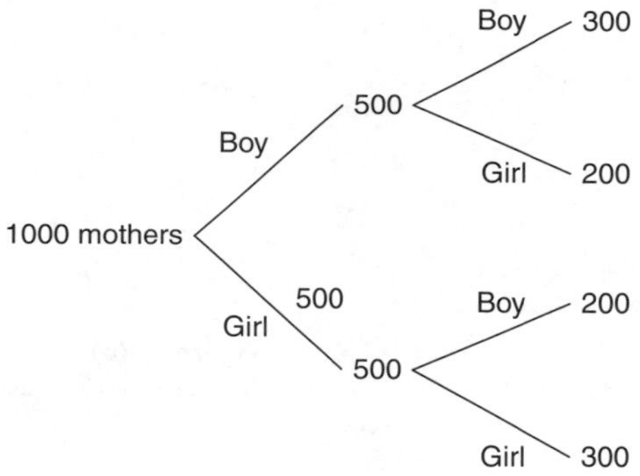

Figure 2.10(a)
Tree diagram, non-independence case (frequencies)

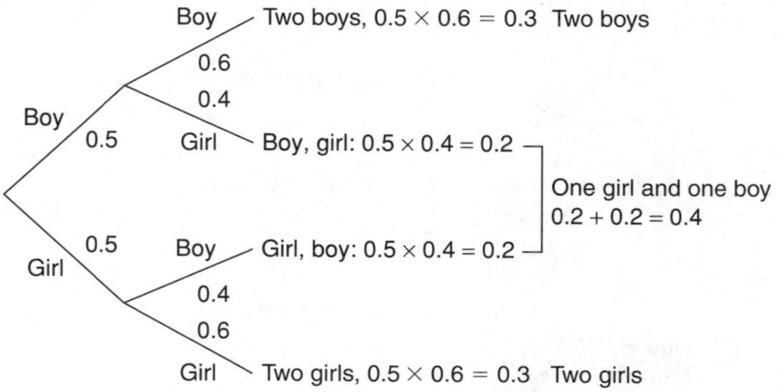

Figure 2.10(b)
Tree diagram, non-independence case (probabilities)

If there were four girls out of five children, then the number of orderings or combinations would be

$$5C4 = \frac{5!}{4! \times 1!} = \frac{5 \times 4 \times 3 \times 2 \times 1}{\{4 \times 3 \times 2 \times 1\} \times 1} = 5 \quad (2.20)$$

This gives five possible orderings, i.e. the single boy could be the first, second, third, fourth or fifth born.

Why does this formula work?

(If you are happy just to accept the combinatorial formula above, you can skip this section and go straight to the exercises below.) Consider five empty places to fill, corresponding to the five births in chronological order. Take the case of three girls (call them Amanda, Bridget and Caroline for convenience) who have to fill three of the five places. For Amanda there is a choice of five empty places. Having 'chosen' one, there remain four for Bridget, so there are $5 \times 4 = 20$ possibilities (i.e. ways in which these two could choose their places). Three remain for Caroline, so there are $60(=5 \times 4 \times 3)$ possible orderings in all (the two boys take the two remaining places). Sixty is the number of **permutations** of three *named* girls in five births. This is written as $5P3$ or, in general, nPr. Hence

$$5P3 = 5 \times 4 \times 3$$

or, in general,

$$nPr = n \times (n - 1) \times \cdots \times (n - r + 1) \quad (2.21)$$

A simpler formula is obtained by multiplying and dividing by $(n - r)!$

$$nPr = \frac{n \times (n - 1) \times \cdots \times (n - r + 1) \times (n - r)!}{(n - r)!} = \frac{n!}{(n - r)!} \quad (2.22)$$

What is the difference between nPr and nCr? The latter does not distinguish between the girls; the two cases Amanda, Bridget, Caroline, boy, boy and Bridget, Amanda, Caroline, boy, boy are effectively the same (three girls followed by two boys). So nPr is larger by a factor representing the number of ways of ordering the three girls. This factor is given by $r! = 3 \times 2 \times 1 = 6$ (any of the three girls could be first, either of the other two second, and then the final one). Thus to obtain nCr one must divide nPr by $r!$, giving (2.18).

Exercise 2.6

?

For this exercise we extend the analysis of Exercise 2.5 to a third shot by the archer.

(a) Extend the tree diagram (assuming independence, so $Pr(H) = 0.3$, $Pr(M) = 0.7$) to a third arrow. Use this to mark out the paths with two successful shots out of three. Calculate the probability of two hits out of three shots.

(b) Repeat part (a) for the case of non-independence. For this you may assume that a hit raises the problem of success with the next arrow to 50%. A miss lowers it to 20%.

Exercise 2.7

?

(a) Show how the answer to Exercise 2.6(a) may be arrived at using algebra, including the use of the combinatorial formula.

(b) Repeat part (a) for the non-independence case.

Bayes' theorem

Bayes' theorem is a factual statement about probabilities which in itself is uncontroversial. However, the use and interpretation of the result is at the heart of the difference between **classical** and **Bayesian statistics**. The theorem itself is easily derived from first principles. Equation (2.23) is similar to equation equation (2.14) covered earlier when discussing the multiplication rule:

$$\Pr(A \text{ and } B) = \Pr(A \mid B) \times \Pr(B) \tag{2.23}$$

hence,

$$\Pr(A \mid B) = \frac{\Pr(A \text{ and } B)}{\Pr(B)} \tag{2.24}$$

Expanding both top and bottom of the right-hand side,

$$\Pr(A \mid B) = \frac{\Pr(B \mid A) \times \Pr(A)}{\Pr(B \mid A) \times \Pr(A) + \Pr(B \mid \text{not } A) \times \Pr(\text{not } A)} \tag{2.25}$$

Equation (2.25) is known as **Bayes' theorem** and is a statement about the probability of the event A, conditional upon B having occurred. The following example demonstrates its use.

Two bags contain red and yellow balls. Bag A contains six red and four yellow balls, and bag B has three red and seven yellow balls. A ball is drawn at random from one bag and turns out to be red. What is the probability that it came from bag A? Since bag A has relatively more red balls to yellow balls than does bag B, it seems bag A ought to be favoured. The probability should be more than 0.5. We can check if this is correct.

Denoting:

$\Pr(A) = 0.5$ (the probability of choosing bag A at random) $= \Pr(B)$
$\Pr(R \mid A) = 0.6$ (the probability of selecting a red ball from bag A), etc.

we have

$$\Pr(A \mid R) = \frac{\Pr(R \mid A) \times \Pr(A)}{\Pr(R \mid A) \times \Pr(A) + \Pr(R \mid B) \times \Pr(B)} \tag{2.26}$$

using Bayes' theorem. Evaluating this gives

$$\Pr(A \mid R) = \frac{0.6 \times 0.5}{0.6 \times 0.5 + 0.3 \times 0.5} = {}^{2}/_{3} \tag{2.27}$$

As expected, this result is greater than 0.5. (You can check that $\Pr(B \mid R) = 1/3$ so that the sum of the probabilities is 1.)

It may help us understand this if we draw another tree diagram, as Figure 2.14. Once again, this shows frequencies, 1000 trials of taking a ball from a bag. In 500 of the trials we would expect to select bag A, in the other 500 trials we select bag $B(\Pr(A) = \Pr(B) = 0.5)$. This is the first stage of the diagram. Of the 500 occasions we select bag A, 300 times we get a red ball ($\Pr(R \mid A) = 0.6$) and 200 times we get yellow. From bag B we get 150 draws of a red ball ($\Pr(R \mid B) = 0.3$) and 350 yellows.

Figure 2.14
Bayes' theorem

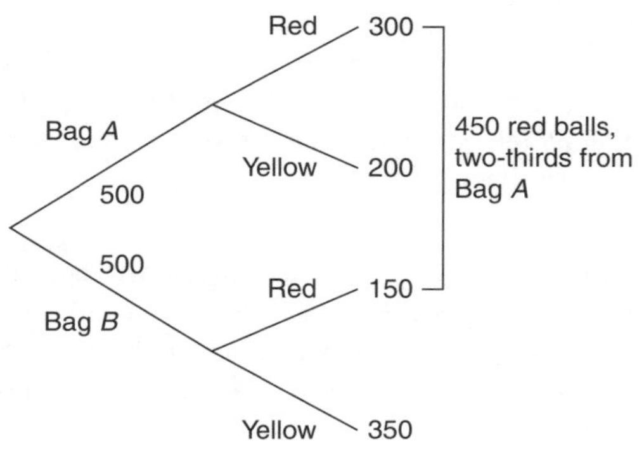

Hence, on 450 occasions we get a red ball. Two-thirds of those (300/450) came from bag A, which is $\Pr(A \mid R)$.

Bayes' theorem can be extended to cover more than two bags: if there are five bags, for example, labelled A to E, then

$$\Pr(A \mid R) = \frac{\Pr(R \mid A) \times \Pr(A)}{\Pr(R \mid A) \times \Pr(A) + \Pr(R \mid B) \times \Pr(B) + \cdots + \Pr(R \mid E) \times \Pr(E)} \tag{2.28}$$

In Bayesian language, $\Pr(A)$, $\Pr(B)$, etc., are known as the **prior** (to the drawing of the ball) probabilities, $\Pr(R \mid A)$, $\Pr(R \mid B)$, etc., are the **likelihoods** and $\Pr(A \mid R)$, $\Pr(B \mid R)$, etc., are the **posterior probabilities**. Bayes' theorem can alternatively be expressed as

$$\text{posterior probability} = \frac{\text{likelihood} \times \text{prior probability}}{\Sigma(\text{likelihood} \times \text{prior probability})} \tag{2.29}$$

This is illustrated below, by reworking the above example in a different format.

	Prior probabilities	Likelihoods	Prior × likelihood	Posterior probabilities
A	0.5	0.6	0.30	0.30/0.45 = 2/3
B	0.5	0.3	0.15	0.15/0.45 = 1/3
Total			0.45	

The general version of Bayes' theorem may be stated as follows. If there are n events labelled $E_1, \ldots, E_n$, then the probability of the event E_i occurring, given the sample evidence S, is

$$\Pr(E_i \mid S) = \frac{\Pr(S \mid E_i) \times \Pr(E_i)}{\Sigma(\Pr(S \mid E_i) \times \Pr(E_i))} \tag{2.30}$$

As stated earlier, debate arises over the interpretation of Bayes' theorem. In the above example, there is no difficulty because the probability statements can be interpreted as relative frequencies. If the experiment of selecting a bag at random and choosing a ball from it were repeated many times, then in two-thirds of those occasions when a red ball is selected, bag A will have been chosen. However, consider an alternative interpretation of the symbols:

A: a coin is fair
B: a coin is unfair
R: the result of a toss is a head

Then, given a toss (or series of tosses) of a coin, this evidence can be used to calculate the probability of the coin being fair. But this makes no sense according to the frequentist school: either the coin is fair or not; it is not a question of probability. The calculated value must be interpreted as a degree of belief and be given a subjective interpretation.

Exercise 2.8

?

(a) Repeat the 'balls in the bag' exercise from the text, but with bag A containing five red and three yellow balls, bag B containing one red and two yellow balls. The single ball drawn is red. Before doing the calculation, predict which bag is more likely to be the source of the drawn ball. Explain why. Then compare your prediction with the calculated answer.

(b) Bag A now contains 10 red and 6 yellow balls (i.e. twice as many as before, but in the same proportion). Does this alter the answer you obtained in part (a)?

(c) Set out your answer to part (b) in the form of prior probabilities and likelihoods, in order to obtain the posterior probability.

Decision analysis

The study of probability naturally leads to the analysis of decision-making where risk is involved. This is the realistic situation facing most firms, and the use of probability can help to illuminate the problem. To illustrate the topic, we use the example of a firm facing a choice of three different investment projects. The uncertainty which the firm faces concerns the interest rate at which to discount the future flows of income. If the interest/discount rate is high, then projects which have income far in the future become less attractive relative to projects with more immediate returns. A low rate reverses this conclusion. The question is: which project should the firm select? As we shall see, there is no unique, right answer to the question but, using probability theory, we can see why the answer might vary.

Table 2.1 provides the data required for the problem. The three projects are imaginatively labelled *A*, *B* and *C*. There are four possible **states of the world**, i.e. future scenarios, each with a different interest rate, as shown across the top of the table. This is the only source of uncertainty; otherwise the states of the world are identical. The figures in the body of the table show the present value of each income stream at the given discount rate.

Table 2.1 **Data for decision analysis: present values of three investment projects at different interest rates (£000)**

Project	Future interest rate			
	4%	5%	6%	7%
A	1475	1363	1200	1115
B	1500	1380	1148	1048
C	1650	1440	1200	810
Probability	0.1	0.4	0.4	0.1

- Bayes' theorem provides a formula for calculating a conditional probability, e.g. the probability of someone being a smoker, given they have been diagnosed with cancer. It forms the basis of Bayesian statistics, allowing us to calculate the probability of a hypothesis being true, based on the sample evidence and prior beliefs. Classical statistics disputes this approach.
- Probabilities can also be used as the basis for decision-making in conditions of uncertainty, using as decision criteria expected value maximisation, maximin, maximax or minimax regret.

Key terms and concepts

addition rule	maximax
axiomatic approach	maximin
Bayes' theorem	minimax
Bayesian statistics	minimax regret
classical statistics	multiplication rule
combinations	mutually exclusive
combinatorial formula	operator
complement	outcome or event
compound event	perfect information
conditional probability	permutations
degree of belief	posterior probabilities
event	prior belief
exhaustive	prior probabilities
expected value	probability
expected value of perfect information	proportion
experiment	sample space
frequentist view	states of the world
independent events	statistical inference
intersection	subjective view
joint probabilities	tree diagram
likelihoods	trial
marginal probabilities	union

Formulae used in this chapter

Formula	Description	Note
$nCr = \frac{n!}{r!(n-r)!}$	Combinatorial formula	$n! = n \times (n-1) \times \cdots \times 1$

Problems

Some of the more challenging problems are indicated by highlighting the problem number in colour.

2.1 Given a standard pack of cards, calculate the following probabilities:

(a) drawing an ace;

(b) drawing a court card (i.e. jack, queen or king);

(c) drawing a red card;

(d) drawing three aces without replacement;

(e) drawing three aces with replacement.

2.2 The following data give duration of unemployment by age.

Age	Duration of unemployment (weeks)				Total	Economically active
	≤ 8	8–26	26–52	>52	(000s)	(000s)
	(Percentage figures, rows sum to 100)					
16–19	27.2	29.8	24.0	19.0	273.4	1270
20–24	24.2	20.7	18.3	36.8	442.5	2000
25–34	14.8	18.8	17.2	49.2	531.4	3600
35–49	12.2	16.6	15.1	56.2	521.2	4900
50–59	8.9	14.4	15.6	61.2	388.1	2560
≥ 60	18.5	29.7	30.7	21.4	74.8	1110

The 'economically active' column gives the total of employed (not shown) plus unemployed in each age category.

(a) In what sense may these figures be regarded as probabilities? What does the figure 27.2 (top-left cell) mean following this interpretation?

(b) Assuming the validity of the probability interpretation, which of the following statements are true?

(i) The probability of an economically active adult aged 25–34, drawn at random, being unemployed is 531.4/3600.

(ii) If someone who has been unemployed for over one year is drawn at random, the probability that they are aged 16–19 is 19%.

(iii) For those aged 35–49 who became unemployed at least one year ago, the probability of their still being unemployed is 56.2%.

(iv) If someone aged 50–59 is drawn at random from the economically active population, the probability of their being unemployed for eight weeks or less is 8.9%.

(v) The probability of someone aged 35–49 drawn at random from the economically active population being unemployed for between 8 and 26 weeks is $0.166 \times 521.2/4900$.

(c) A person is drawn at random from the population and found to have been unemployed for over one year. What is the probability that they are aged between 16 and 19?

2.3 'Odds' in horserace betting are defined as follows: 3/1 (three-to-one against) means a horse is expected to win once for every three times it loses; 3/2 means two wins out of five races; 4/5 (five to four *on*) means five wins for every four defeats, etc.

(a) Translate the above odds into 'probabilities' of victory.

(b) In a three-horse race, the odds quoted are 2/1, 6/4 and 1/1. What makes the odds different from probabilities? Why are they different?

(c) Discuss how much the bookmaker would expect to win in the long run at such odds (in part (b)), assuming each horse is backed equally.

2.4 (a) Translate the following odds to 'probabilities': 13/8, 2/1 *on*, 100/30.

(b) In the 2.45 race at Plumpton the odds for the five runners were:

Philips Woody	1/1
Gallant Effort	5/2
Satin Noir	11/2
Victory Anthem	9/1
Common Rambler	16/1

Calculate the 'probabilities' and their sum.

(c) Should the bookmaker base his odds on the true probabilities of each horse winning, or adjust them depending upon the amount bet on each horse?

2.5 How might you estimate the probability of Peru defaulting on its debt repayments next year? What type of probability estimate is this?

2.6 How might you estimate the probability of a corporation reneging on its bond payments?

2.7 Judy is 33, unmarried and assertive. She is a graduate in political science, and involved in union activities and anti-discrimination movements. Which of the following statements do you think is more probable?

(a) Judy is a bank clerk.

(b) Judy is a bank clerk, active in the feminist movement.

2.8 A news item revealed that a London 'gender' clinic (which reportedly enables you to choose the sex of your child) had just set up in business. Of its first six births, two were of the 'wrong' sex. Assess this from a probability point of view.

2.9 A newspaper advertisement reads 'The sex of your child predicted, or your money back!' Discuss this advertisement from the point of view of (a) the advertiser and (b) the client.

2.10 'Roll six sixes to win a Mercedes!' is the announcement at a fair. You have to roll six dice. If you get six sixes you win the car, valued at £40 000. The entry ticket costs £1. What is your expected gain or loss on this game? If there are 400 people who try the game, what is the probability of the car being won? The organisers of the fair have to take out insurance against the car being won. This costs £400 for the day. Does this seem a fair premium? If not, why not?

2.11 At another stall, you have to toss a coin numerous times. If a head does not appear in 20 tosses you win £1 billion. The entry fee for the game is £100.

(a) What are your expected winnings?

(b) Would you play?

2.12 A four-engine plane can fly as long as at least two of its engines work. A two-engine plane flies as long as at least one engine works. The probability of an individual engine failure is 1 in 1000.

(a) Would you feel safer in a four- or two-engine plane, and why? Calculate the probabilities of an accident for each type.

(b) How much safer is one type than the other?

(c) What crucial assumption are you making in your calculation? Do you think it is valid?

2.13 Which of the following events are independent?

(a) Two flips of a fair coin.

(b) Two flips of a biased coin.

(c) Rainfall on two successive days.

(d) Rainfall on St Swithin's Day and rain one month later.

2.14 Which of the following events are independent?

(a) A student getting the first two questions correct in a multiple-choice exam.

(b) A driver having an accident in successive years.

(c) IBM and Dell earning positive profits next year.

(d) Arsenal Football Club winning on successive weekends.

How is the answer to (b) reflected in car insurance premiums?

2.15 Manchester United beat Liverpool 4–2 at soccer, but you do not know the order in which the goals were scored. Draw a tree diagram to display all the possibilities and use it to find (a) the probability that the goals were scored in the order L, MU, MU, MU, L, MU and (b) the probability that the score was 2–2 at some stage.

2.16 An important numerical calculation on a spacecraft is carried out independently by three computers. If all arrive at the same answer, it is deemed correct. If one disagrees, it is overruled. If there is no agreement, then a fourth computer does the calculation and, if its answer agrees with any of the others, it is deemed correct. The probability of an individual computer getting the answer right is 99%. Use a tree diagram to find:

(a) the probability that the first three computers get the right answer;

(b) the probability of getting the right answer;

(c) the probability of getting no answer;

(d) the probability of getting the wrong answer.

2.17 The French national lottery works as follows. Six numbers from the range 0 to 49 are chosen at random. If you have correctly guessed all six, you win the first prize. What are your chances of winning if you are allowed to choose only six numbers? A single entry like this costs one euro. For 210 euros you can choose 10 numbers, and you win if the 6 selected numbers are among them. Is this better value than the single entry?

2.18 The UK national lottery originally worked as follows. You choose six (different) numbers in the range 1 to 49. If all six come up in the draw (in any order), you win the first prize, generally valued at around £2m (which could be shared if someone else chooses the six winning numbers).

(a) What is your chance of winning with a single ticket?

(b) You win a second prize if you get five out of six right *and* your final chosen number matches the 'bonus' number in the draw (also in the range 1–49). What is the probability of winning a second prize?

(c) Calculate the probabilities of winning a third, fourth or fifth prize, where a third prize is won by matching five out of the six numbers, a fourth prize by matching four out of six and a fifth prize by matching three out of six.

(d) What is the probability of winning a prize?

(e) The prizes are as follows:

Prize	Value	
First	£2 million	(expected, possibly shared)
Second	£100 000	(expected, for each winner)
Third	£1500	(expected, for each winner)
Fourth	£65	(expected, for each winner)
Fifth	£10	(guaranteed, for each winner)

Comment upon the distribution of the fund between first, second, etc., prizes.

(f) Why is the fifth prize guaranteed whereas the others are not?

(g) In the first week of the lottery, 49 million tickets were sold. There were 1 150 000 winners, of which 7 won (a share of) the jackpot, 39 won a second prize, 2139 won a third prize and 76 731 a fourth prize. Are you surprised by these results or are they as you would expect?

2.19 A coin is either fair or has two heads. You initially assign probabilities of 0.5 to each possibility. The coin is then tossed twice, with two heads appearing. Use Bayes' theorem to work out the posterior probabilities of each possible outcome.

2.20 A test for AIDS is 99% successful, i.e. if you are HIV+, it will be detected in 99% of all tests, and if you are not, it will again be right 99% of the time. Assume that about 1% of the population are HIV+. You take part in a random testing procedure, which gives a positive result. What is the probability that you are HIV+? What implications does your result have for AIDS testing?

2.21 (a) Your initial belief is that a defendant in a court case is guilty with probability 0.5. A witness comes forward claiming he saw the defendant commit the crime. You know the witness is not totally reliable and tells the truth with probability p. Use Bayes' theorem to calculate the posterior probability that the defendant is guilty, based on the witness's evidence.

(b) A second witness, equally unreliable, comes forward and claims she/he saw the defendant commit the crime. Assuming the witnesses are not colluding, what is your posterior probability of guilt?

(c) If $p < 0.5$, compare the answers to (a) and (b). How do you account for this curious result?

2.22 A man is mugged and claims that the mugger had red hair. In police investigations of such cases, the victim was able correctly to identify the assailant's hair colour 80% of the time. Assuming that 10% of the population have red hair, what is the probability that the assailant in this case did, in fact, have red hair? Guess the answer first, and then find the right answer using Bayes' theorem. What are the implications of your results for juries' interpretation of evidence in court, particularly in relation to racial minorities?

2.23 A firm has a choice of three projects, with profits as indicated below, dependent upon the state of demand.

Project	Demand		
	Low	Middle	High
A	100	140	180
B	130	145	170
C	110	130	200
Probability	0.25	0.45	0.3

(a) Which project should be chosen on the expected value criterion?

(b) Which project should be chosen on the maximin and maximax criteria?

(c) Which project should be chosen on the minimax regret criterion?

(d) What is the expected value of perfect information to the firm?

2.24 A firm can build a small, medium or large factory, with anticipated profits from each dependent upon the state of demand, as in the table below.

Factory	Demand		
	Low	Middle	High
Small	300	320	330
Medium	270	400	420
Large	50	250	600
Probability	0.3	0.5	0.2

(a) Which project should be chosen on the expected value criterion?

(b) Which project should be chosen on the maximin and maximax criteria?

(c) Which project should be chosen on the minimax regret criterion?

(d) What is the expected value of perfect information to the firm?

2.25 There are 25 people at a party. What is the probability that there are at least two with a birthday in common? They do not need to have been born in the same year, just the same day and month of the year. Also, ignore leap year dates. (Hint: the *complement* is (much) easier to calculate.)

2.26 This problem is tricky, but amusing. Three gunmen, A, B and C, are shooting at each other. The probabilities that each will hit what they aim at are 1, 0.75 and 0.5, respectively. They take it in turns to shoot (in alphabetical order) and continue until only one is left alive. Calculate the probabilities of each winning the contest. (Assume they draw lots for the right to shoot first.)

Hint 1: Start with one-on-one gunfights, e.g. the probability of A beating B, or of B beating C. You need to solve this first, and *then* figure out the optimal strategies in the first stage when all three are alive.

Hint 2: You'll need the formula for the sum of an infinite series, given in Chapter 1.

Hint 3: To solve this, you need to realize that it might be in a gunman's best interest *not* to aim at one of his opponents . . .

Exercise 2.8

(a) Bag A has proportionately more red balls than bag B; hence it should be the favoured bag from which the single red ball was drawn. Performing the calculation:

$$\Pr(A|R) = \frac{\Pr(R|A) \times \Pr(A)}{\Pr(R|A) \times \Pr(A) + \Pr(R|B) \times \Pr(B)} = \frac{0.625 \times 0.5}{0.625 \times 0.5 + 1/3 \times 0.5} = 0.652$$

(b) The result is the same as $\Pr(R|A) = 0.625$ as before. The number of balls does not enter the calculation.

	Prior probabilities	Likelihoods	Prior × likelihood	Posterior probabilities
A	0.5	0.625	0.3125	0.3125/0.5625 = 0.556
B	0.5	0.5	0.25	0.25/0.5625 = 0.444
Total			0.5625	

Exercise 2.9

(a) $1200/(1 + r) - 1200/1.1 = 1090.91$.

(b) $1200/1.15 = 1043.48$. The *PV* has only changed by 4.3%. This is calculated as $1.1/1.15 - 1 = -0.043$.

(c) $1200/1.1^2 = 991.74$; $1200/1.1^5 = 745.11$.

(d) $PV = 500/1.1 + 500/1.1^2 + 500/1.1^3 = 1243.43$.

(e) At 10%: project A yields a *PV* of $300/1.1 + 600/1.1^2 = 768.6$. Project B yields $400/1.1 + 488/1.1^2 = 766.9$. At 20% the *PVs* are 666.7 and 672.2, reversing the rankings. A's large benefits in year 2 are penalised by the higher discount rate.

Exercise 2.10

(a)

Project	Expected value	Minimum	Maximum
A	0.3 × 100 + 0.4 × 80 + 0.3 × 70 = 83	70	100
B	0.3 × 90 + 0.4 × 85 + 0.3 × 75 = 83.5	75	90
C	0.3 × 120 + 0.4 × 60 + 0.3 × 40 = 72	40	120

The maximin is 75, associated with project *B* and the maximax is 120, associated with project *C*. The regret values are given by

	4%	6%	8%	Max
A	20	5	5	20
B	30	0	0	30
C	0	25	35	35
			Min	20

The minimax regret is 20, associated with project *A*.

(b) With perfect information the firm could earn $0.3 \times 120 + 0.4 \times 85 + 0.3 \times 75 = 92.5$. The highest expected value is 83.5, so the value of perfect information is $92.5 - 83.5 = 9$.

3 Probability distributions

Contents

Learning outcomes

By the end of this chapter you should be able to:

- recognise that the result of most probability experiments (e.g. the score on a die) can be described as a random variable
- appreciate how the behaviour of a random variable can often be summarised by a probability distribution (a mathematical formula)
- recognise the most common probability distributions and be aware of their uses
- solve a range of probability problems using the appropriate probability distribution.

Introduction

The last chapter covered probability concepts and introduced the idea of the outcome of an experiment being random, i.e. influenced by chance. The outcome of tossing a coin is random, as is the mean calculated from a random sample. We can refer to these outcomes as being **random variables**. The number of heads achieved in five tosses of a coin or the average height of a sample of children are both random variables.

We can summarise the information about a random variable by using its **probability distribution**. A probability distribution lists, in some form, all the possible outcomes of a probability experiment and the probability associated with each one. Another way of saying this is that the probability distribution lists (in some way) all possible values of the random variable and the probability that each value will occur. For example, the simplest experiment is tossing a coin, for which the possible outcomes are heads or tails, each with probability one-half. The probability distribution can be expressed in a variety of ways: in words, or in a graphical or mathematical form. For tossing a coin, the graphical form is shown in Figure 3.1, and the mathematical form is:

$$\Pr(H) = \tfrac{1}{2}$$
$$\Pr(T) = \tfrac{1}{2}$$

The different forms of presentation are equivalent but one might be more suited to a particular purpose.

If we want to study a random variable (e.g. the mean of a random sample) and draw inferences from it, we need to make use of the associated probability distribution. Therefore, an understanding of probability distributions is vital to making appropriate use of statistical evidence. In this chapter we first look in greater detail at the concepts of a random variable and its probability distribution. We then look at a number of commonly used probability distributions, such as the Binomial and Normal, and see how they are used as the basis of inferential statistics (drawing conclusions from data). In particular, we look at the probability distribution associated with a sample mean because the mean is so often used in statistics.

Some probability distributions occur often and so are well known. Because of this they have names so we can refer to them easily; for example, the **Binomial distribution** or the **Normal distribution**. In fact, each of these constitutes a *family* of distributions. A single toss of a coin gives rise to one member of the Binomial distribution family; two tosses would give rise to another member of that family

Figure 3.1
The probability distribution for the toss of a coin

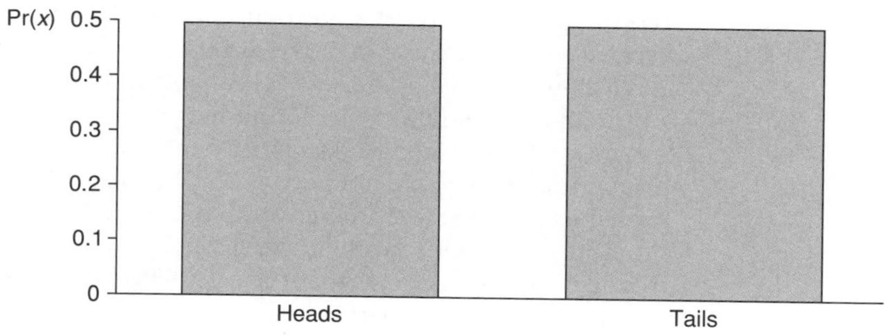

(where the possible outcomes are zero, one or two heads in two tosses). These two distributions differ in the number of tosses but are members of the same family. If a biased coin were tossed, this would lead to yet another Binomial distribution, but it would differ from the previous two because of the different probability of heads.

Members of the Binomial family of distributions are distinguished by the number of trials and by the probability of the outcome occurring. These are the two **parameters of the distribution** and tell us all we need to know about the distribution. Other distributions might have different numbers of parameters, with different meanings. Some distributions, for example, have only one parameter. We will come across examples of different types of distribution throughout the rest of this text.

In order to understand fully the idea of a probability distribution, we first introduce a new concept, that of a **random variable**. As will be seen later in the chapter, an important random variable is the sample mean, and to understand how to draw inferences from the sample mean, we must recognise it as a random variable.

Random variables and probability distributions

Examples of random variables have already been encountered in the previous chapter, for example, the result of the toss of a coin, or the number of boys in a family of five children. A random variable is one whose outcome or value is the result of chance and is therefore unpredictable, although the range of possible outcomes and the probability of each outcome may be known. It is impossible to know in advance the outcome of a toss of a coin, for example, but it must be either heads or tails, each with probability one-half. The number of heads in 250 tosses is another random variable, which can take any value between zero and 250, although values near 125 are the most likely. You are very unlikely to get 250 heads from tossing a fair coin.

Intuitively, most people would 'expect' to get 125 heads from 250 tosses of the coin, since heads comes up half the time on average. This suggests we could use the expected value notation introduced in Chapter 1 and write $E(X) = 125$, where X represents the number of heads obtained from 250 tosses. This usage is indeed valid and we will explore this further below. It is a very convenient shorthand notation.

The time of departure of a train is another example of a random variable. It may be timetabled to depart at 11.15, but it probably (almost certainly) won't leave at exactly that time. If a sample of 10 basketball players were taken, and their average height calculated, this would be a random variable. In this latter case, it is the process of taking a sample that introduces the variability which makes the resulting average a random variable. If the experiment were repeated, a different sample and a different value of the random variable would be obtained.

The above examples can be contrasted with some things which are *not* random variables. If one were to take *all* basketball players and calculate their average height, the result would not be a random variable. This time there is no sampling procedure to introduce variability into the result. If the experiment were repeated, the same result would be obtained, since the same people would be measured the second time (this assumes that the population does not change, of course). Just because the value of something is unknown does not mean it qualifies as a random variable. This is an important distinction to bear in mind, since it is legitimate to make probability statements about random variables ('the probability that the average height of a sample of basketball players is over 195 cm is 60%') but not about parameters ('the probability that the Pope is over 180 cm tall is 60%'). Here again, there is a difference of opinion between frequentist and subjective schools of thought. The latter group would argue that it is possible to make probability statements about the Pope's height. It is a way of expressing lack of knowledge about the true value. The frequentists would say the Pope's height is a fact that we do not happen to know; that does not make it a random variable.

The Binomial distribution

One of the simplest distributions which a random variable can have is the Binomial. The Binomial distribution arises whenever the underlying probability experiment has just two possible outcomes, e.g. heads or tails from the toss of a coin. Even if the coin is tossed many times (so one could end up with one, two, three, etc., heads in total), the *underlying* experiment (sometimes called a Bernoulli trial) has only two outcomes, so the Binomial distribution should be used. A counter-example would be the rolling of a die, which has six possible outcomes (in this case the Multinomial distribution, not covered in this text, would be used). Note, however, that if we were interested only in rolling a six or not, we *could* use the Binomial by defining the two possible outcomes as 'six' and 'not-six'. It is often the case in statistics that by suitable transformation of the data, we can use different distributions to tackle the same problem. We will see more of this later in the chapter.

The Binomial distribution can therefore be applied to the type of problem encountered in the previous chapter, concerning the sex of children. It provides a convenient formula for calculating the probability of r boys in n births or, in more general terms, the probability of r 'successes' in n trials[1]. We shall use it to calculate the probabilities of 0, 1, . . . , 5 boys in five births.

For the Binomial distribution to apply, we first need to assume independence of successive events and we shall assume that, for any birth:

$$\Pr(\text{boy}) = P = \tfrac{1}{2}$$

It follows that

$$\Pr(\text{girl}) = 1 - \Pr(\text{boy}) = 1 - P = \tfrac{1}{2}$$

Although we have $P = \frac{1}{2}$ in this example, the Binomial distribution can be applied for any value of P between 0 and 1.

First we consider the case of $r = 5, n = 5$, i.e. five boys in five births. This probability is found using the multiplication rule:

$$\Pr(r = 5) = P \times P \times P \times P \times P = P^5 = (\tfrac{1}{2})^5 = 1/32$$

The probability of four boys (and then implicitly one girl) is

$$\Pr(r = 4) = P \times P \times P \times P \times (1 - P) = 1/32$$

[1]The identification of a boy with 'success' is a purely formal one and is not meant to be pejorative.

But this gives only one possible ordering of the four boys and one girl. Our original statement of the problem did not specify a particular ordering of the children. There are five possible orderings (the single girl could be in any of five positions in rank order). Recall that we can use the combinatorial formula nCr to calculate the number of orderings, giving $5C4 = 5$. Hence the probability of four boys and one girl in any order is 5/32. Summarising, the formula for four boys and one girl is

$$\Pr(r = 4) = 5C4 \times P^4 \times (1 - P)$$

For three boys (and two girls) we obtain

$$\Pr(r = 3) = 5C3 \times P^3 \times (1 - P)^2 = 10 \times 1/8 \times 1/4 = 10/32$$

In a similar manner

$$\Pr(r = 2) = 5C2 \times P^2 \times (1 - P)^3 = 10/32$$
$$\Pr(r = 1) = 5C1 \times P^1 \times (1 - P)^4 = 5/32$$
$$\Pr(r = 0) = 5C0 \times P^0 \times (1 - P)^5 = 1/32$$

As a check on our calculations, we may note that the sum of the probabilities equals 1, as they should do, since we have enumerated all possibilities.

A fairly clear pattern emerges. The probability of r boys in n births is given by

$$\Pr(r) = nCr \times P^r \times (1 - P)^{n-r}$$

and this is known as the Binomial formula or distribution. The Binomial distribution is appropriate for analysing problems with the following characteristics:

- There is a number (n) of trials.
- Each trial has only two possible outcomes, 'success' (with probability P) and 'failure' (probability $1 - P$) and the outcomes are independent between trials.
- The probability P does not change between trials.

The probabilities calculated by the Binomial formula may be illustrated in a diagram, as shown in Figure 3.2. This is very similar to the relative frequency distribution introduced in Chapter 1. That distribution was based on empirical data (to do with wealth) while the Binomial probability distribution is a theoretical construction, built up from the basic principles of probability theory.

As stated earlier, the Binomial is, in fact, a family of distributions, each member of which is distinguished by two **parameters**, n and P. The Binomial is thus a distribution with two parameters, and once their values are known the distribution is completely determined (i.e. $\Pr(r)$ can be calculated for all values of r). To illustrate

Figure 3.2
Probability distribution of the number of boys in five children

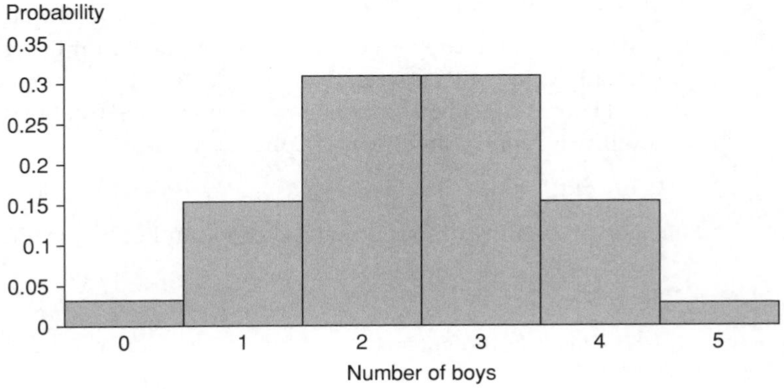

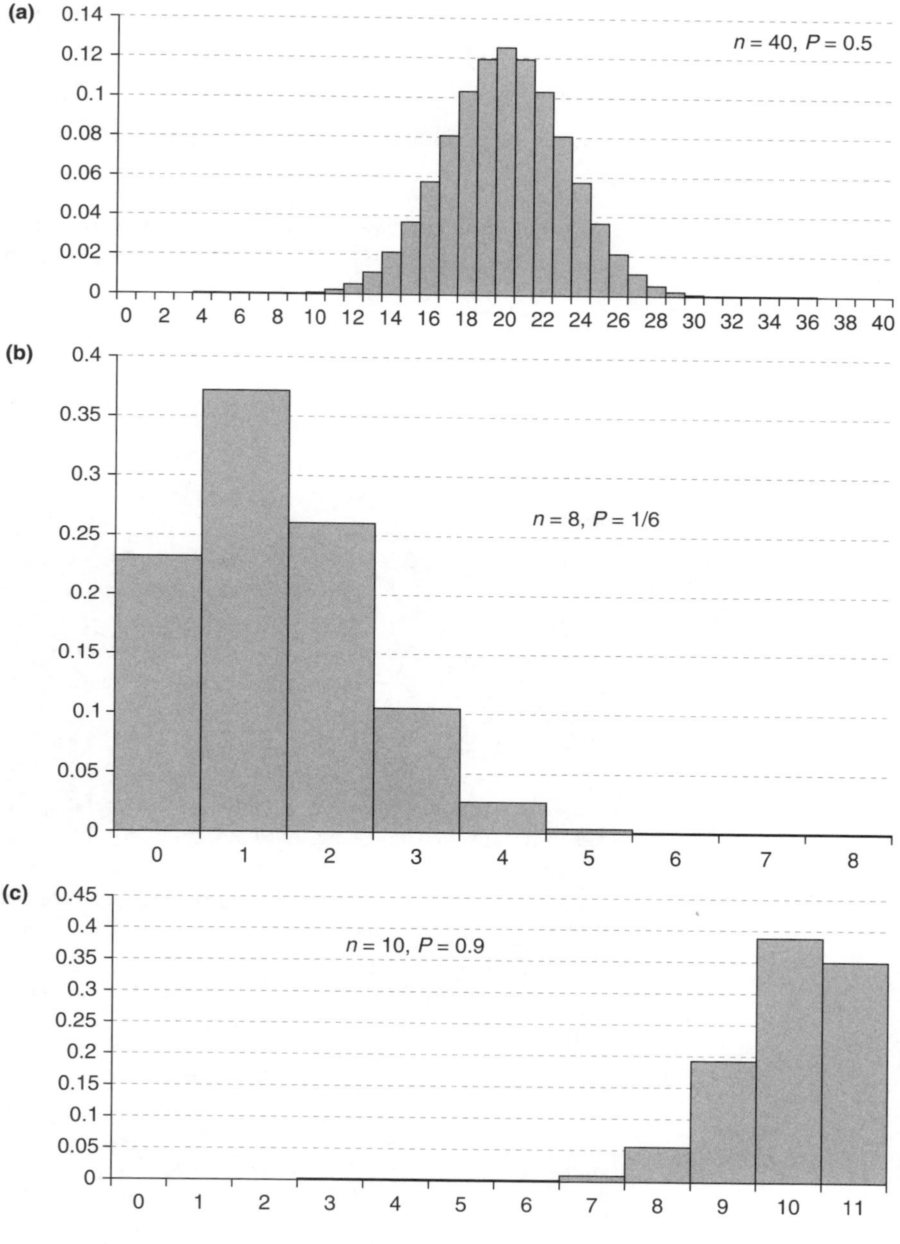

Figure 3.3
Binomial distributions with different parameter values

the difference between members of the family of the Binomial distribution, Figure 3.3 presents three other Binomial distributions, for different values of P and n. It can be seen that for the value of $P = \frac{1}{2}$ the distribution is symmetric, while for all other values it is skewed to either the left or the right. Part (b) of the figure illustrates the distribution relating to the worked example of rolling a die, described below.

Since the Binomial distribution depends only upon the two values n and P, we can use a shorthand notation rather than the formula itself. A random variable r, which has a Binomial distribution with the parameters n and P, can be written in general terms as

$$r \sim B(n, P) \tag{3.1}$$

Thus for the previous example of children, where r represents the number of boys,

$$r \sim B(5, \tfrac{1}{2})$$

This is simply a brief and convenient way of writing down the information available; it involves no new problems of a conceptual nature. Writing

$$r \sim B(n, P)$$

is just a shorthand for

$$\Pr(r) = nCr \times P^r \times (1 - P)^{n-r}$$

STATISTICS IN PRACTICE

Teenage weapons

A story entitled 'One in five teens carry weapon' (link on main BBC news website 23/7/2007) provides a nice example of how knowledge of the binomial distribution can help our interpretation of events in the news. The headline is somewhat alarming but reading the story reveals that this is what young teenagers report of their friends. It then reveals that some are only 'fairly sure' and that it applies to boys, not girls. By now our suspicions should be aroused. What is the truth?

Notice, incidentally, how the story subtly changes. The headline suggests 20% of teenagers carry a weapon. The text then says this is what *young* teenagers report of their *friends*. It then reveals that some are only 'fairly sure' and that it applies to boys, not girls. By now our suspicions should be aroused. What is the truth?

Note that you are more likely to know someone who carries a weapon than to carry one yourself. Let p be the proportion who truly carry a weapon. Assume also that each person has 10 friends. What is the probability that a person, selected at random, has no friends who carry a weapon? Assuming independence, this is given by $[1 - p]^{10}$. Hence the probability of at least one friend with a weapon is $1 - [1 - p]^{10}$. This is the proportion of people who will report having at least one friend with a weapon. How does this vary with p? This is set out in the table:

p	$p(\geq 1$ friend with weapon) $1 - [1 - p]^{10}$
0.0%	0%
0.5%	5%
1.0%	10%
1.5%	14%
2.0%	18%
2.5%	22%
3.0%	26%
3.5%	30%
4.0%	34%

Thus, a true proportion of just over 2% carrying weapons will generate a report suggesting 20% know someone carrying a weapon! This is much less alarming (and newsworthy) than in the original story.

You might like to test the assumptions. What happens if there are more than 10 friends assumed? What happens if events are not independent, i.e. having one friend with a weapon increases the probability of another friend with a weapon?

Update: The 2009/10 British Crime Survey estimated that 1% of 13–15-year-olds carried a knife for their own protection. It also reported 13% reporting they knew someone who carried a knife for protection. These numbers seem entirely consistent with our calculation above, carried out well before the 2009/10 survey.

The mean and variance of the Binomial distribution

In Chapter 1 we calculated the mean and variance of a set of data, of the distribution of wealth. The picture of that distribution (Figure 1.9) looks not too dissimilar to one of the Binomial distributions shown in Figure 3.3. This suggests that we can calculate the mean and variance of a Binomial distribution, just as we did for the empirical distribution of wealth. Calculating the mean would provide the answer to a question such as 'If we have a family with five children, how many do we expect to be boys?' Intuitively the answer seems clear, 2.5 (even though such a family could not exist). The Binomial formula allows us to confirm this intuition.

The mean and variance are most easily calculated by drawing up a relative frequency table based on the Binomial frequencies. This is shown in Table 3.1 for the values $n = 5$ and $P = \frac{1}{2}$. Note that r is equivalent to x in our usual notation and $\Pr(r)$, the relative frequency, is equivalent to $f(x)/\Sigma f(x)$. The mean of this distribution is given by

$$E(r) = \frac{\sum r \times \Pr(r)}{\sum \Pr(r)} = \frac{80/32}{32/32} = 2.5 \tag{3.2}$$

and the variance is given by

$$V(r) = \frac{\sum r^2 \times \Pr(r)}{\sum \Pr(r)} - \mu^2 = \frac{240/32}{32/32} - 2.5^2 = 1.25 \tag{3.3}$$

The mean value tells us that in a family of five children we would expect, on average, two and a half boys. Obviously no single family can be like this; it is the average over all such families. The variance is more difficult to interpret intuitively, but it tells us something about how the number of boys in different families will be spread around the average of 2.5.

Table 3.1 Calculating the mean and variance of the Binomial distribution

r	$\Pr(r)$	$r \times \Pr(r)$	$r^2 \times \Pr(r)$
0	1/32	0	0
1	5/32	5/32	5/32
2	10/32	20/32	40/32
3	10/32	30/32	90/32
4	5/32	20/32	80/32
5	1/32	5/32	25/32
Totals	32/32	80/32	240/32

There is a quicker way to calculate the mean and variance of the Binomial distribution. It can be shown that the mean can be calculated as nP, i.e. the number of trials times the probability of success. For example, in a family with five children and an equal probability that each child is a boy or a girl, we expect $nP = 5 \times {}^1/_2 = 2.5$ to be boys.

The variance can be calculated as $nP(1 - P)$. This gives $5 \times {}^1/_2 \times {}^1/_2 = 1.25$, as found above by extensive calculation.

Worked example 3.1 Rolling a die

If a die is thrown four times, what is the probability of getting two or more sixes? This is a problem involving repeated experiments (rolling the die) with but two types of outcome for each roll: success (a six) or failure (anything but a six). Note that we combine several possibilities (scores of 1, 2, 3, 4 or 5) together and represent them all as failure. The probability of success (one-sixth) does not vary from one experiment to another, and so use of the Binomial distribution is appropriate. The values of the parameters are $n = 4$ and $P = 1/6$. Denoting by r the random variable 'the number of sixes in four rolls of the die', then

$$r \sim B(4, \tfrac{1}{6}) \tag{3.4}$$

Hence

$$\Pr(r) = nCr \times P^r(1 - P)^{(n-r)}$$

where $P = \frac{1}{6}$ and $n = 4$. The probabilities of two, three and four sixes are then given by

$$\Pr(r = 2) = 4C2(\tfrac{1}{6})^2(\tfrac{5}{6})^2 = 0.116$$
$$\Pr(r = 3) = 4C3(\tfrac{1}{6})^3(\tfrac{5}{6})^1 = 0.115$$
$$\Pr(r = 4) = 4C4(\tfrac{1}{6})^4(\tfrac{5}{6})^0 = 0.00077$$

Since these events are mutually exclusive, the probabilities can simply be added together to get the desired result, which is 0.132, or 13.2%. This is the probability of two or more sixes in four rolls of a die.

This result can be illustrated diagrammatically as part of the area under the appropriate Binomial distribution, shown in Figure 3.4.

Figure 3.4
Probability of two or more sixes in four rolls of a die

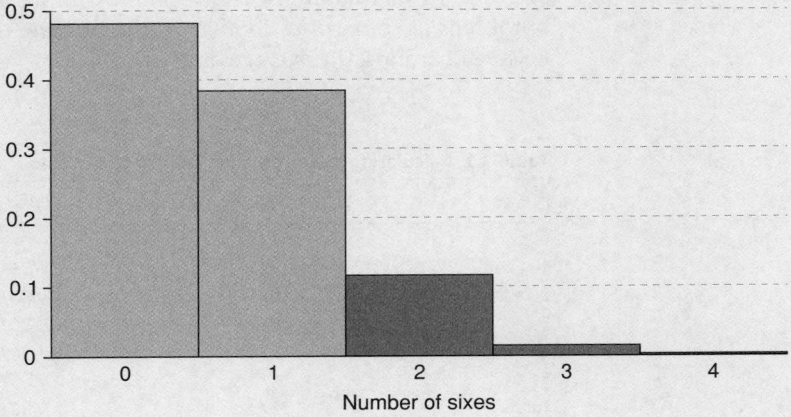

The darker-shaded areas represent the probabilities of two or more sixes and together their area represents 13.2% of the whole distribution. This illustrates an important principle: that probabilities can be represented by areas under an appropriate probability distribution. We shall see more of this later.

Exercise 3.1

(a) If the probability of a randomly drawn individual having blue eyes is 0.6, what is the probability that four people drawn at random all have blue eyes?

(b) What is the probability that two of the sample of four have blue eyes?

(c) For this particular example, write down the Binomial formula for the probability of r blue-eyed individuals, $r = 0 \ldots 4$. Confirm that the calculated probabilities sum to one.

Exercise 3.2

(a) Calculate the mean and variance of the number of blue-eyed individuals in the previous exercise.

(b) Draw a graph of this Binomial distribution and on it mark the mean value and the mean value $+/-$ one standard deviation.

Having introduced the concept of probability distributions using the Binomial, we now move on to the most important of all probability distributions, the Normal.

The Normal distribution

The Binomial distribution applies when there are two possible outcomes to an experiment, but not all problems fall into this category. For instance, the (random) arrival time of a train is a continuous variable and cannot be analysed using the Binomial. There are many probability distributions in statistics, developed to analyse different types of problem. Several of them are covered in this text, and the most important of them is the Normal distribution, to which we now turn. It was discovered by the German mathematician Gauss in the nineteenth century (hence it is also known as the Gaussian distribution), in the course of his work on regression (see Chapter 7).

Many random variables turn out to be Normally distributed. Men's (or women's) heights are Normally distributed. IQ (the measure of intelligence) is also Normally distributed. Another example is of a machine producing (say) bolts with a nominal length of 5 cm which will actually produce bolts of slightly varying length (these differences would probably be extremely small) due to factors such as wear in the machinery, slight variations in the pressure of the lubricant, etc. These would result in bolts whose length varies, in accordance with the Normal distribution. This sort of process is extremely common, with the result that the Normal distribution often occurs in everyday situations.

The Normal distribution tends to arise when a random variable is the result of many independent, random influences added together, none of which dominates the others. A man's height is the result of many genetic influences, plus environmental factors such as diet, etc. As a result, height is Normally distributed. If one takes the height of men and women together, the result is not a Normal distribution, however. This is because one influence dominates the others: gender. Men are, on average, taller than women. Many variables familiar in economics are not

Normal, however – incomes, for example (although the logarithm of income is approximately Normal). We shall learn techniques to deal with such circumstances in due course.

Now that we have introduced the idea of the Normal distribution, what does it look like? It is presented below in graphical and then mathematical forms. Unlike the Binomial, the Normal distribution applies to continuous random variables such as height, and a typical Normal distribution is illustrated in Figure 3.5. Since the Normal distribution is a continuous one, it can be evaluated for any values of x, not just for integers as was the case for the Binomial distribution. The figure illustrates the main features of the distribution:

- It is unimodal, having a single, central peak. If this were men's heights, it would illustrate the fact that most men are clustered around the average height, with a few very tall and a few very short people.
- It is symmetric, the left and right halves being mirror images of each other.
- It is bell-shaped.
- It extends continuously over all the values of x from minus infinity to plus infinity, although the value of $f(x)$ becomes extremely small as these values are approached (the presentation of this figure being of only finite width, this last characteristic is not faithfully reproduced). This also demonstrates that most empirical distributions (such as men's heights) can only be an approximation to the theoretical ideal, although the approximation is close and good enough for practical purposes.

Note that we have labelled the y-axis '$f(x)$' rather than 'Pr(x)' as we did for the Binomial distribution. This is because it is *areas under the curve* that represent probabilities, not the heights. With the Binomial, which is a discrete distribution, one can legitimately represent probabilities by the heights of the bars. For the Normal, although $f(x)$ does not give the probability per se, it does give an indication: you are more likely to encounter values from the middle of the distribution (where $f(x)$ is greater) than from the extremes.

In mathematical terms, the formula for the Normal distribution is (x is the random variable)

$$f(x) = \frac{1}{\sigma\sqrt{2\pi}} e^{-\frac{1}{2}\left(\frac{x-\mu}{\sigma}\right)^2} \tag{3.5}$$

The mathematical formulation is not so formidable as it appears. μ and σ are the parameters of the distribution, like n and P for the Binomial (even though they have different meanings); π is 3.1416 and e is 2.7183. If the formula is

Figure 3.5
The Normal distribution

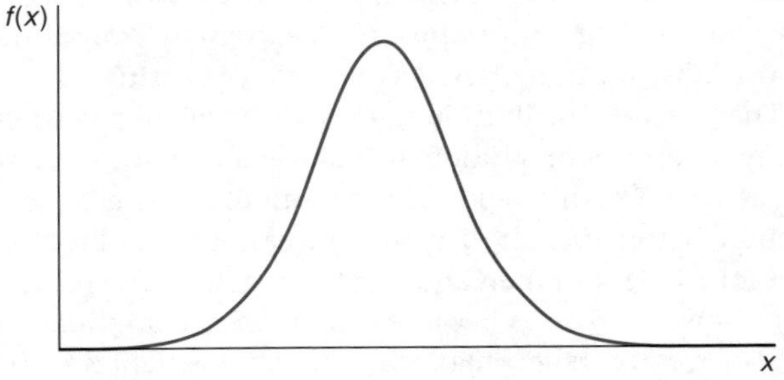

evaluated using different values of x, the values of $f(x)$ obtained will map out a Normal distribution. Fortunately, as we shall see, we do not need to use the mathematical formula in most practical problems.

Like the Binomial, the Normal is a family of distributions differing from one another only in the values of the parameters μ and σ. Several Normal distributions are drawn in Figure 3.6 for different values of the parameters.

Whatever value of μ is chosen turns out to be the centre of the distribution. Since the distribution is symmetric, μ is its mean. The effect of varying σ is to narrow (small σ) or widen (large σ) the distribution. σ turns out to be the standard deviation of the distribution. The Normal is another two-parameter family of distributions like the Binomial, and once the mean μ and the standard deviation σ

Figure 3.6(a)
The Normal distribution, $\mu = 20, \sigma = 5$

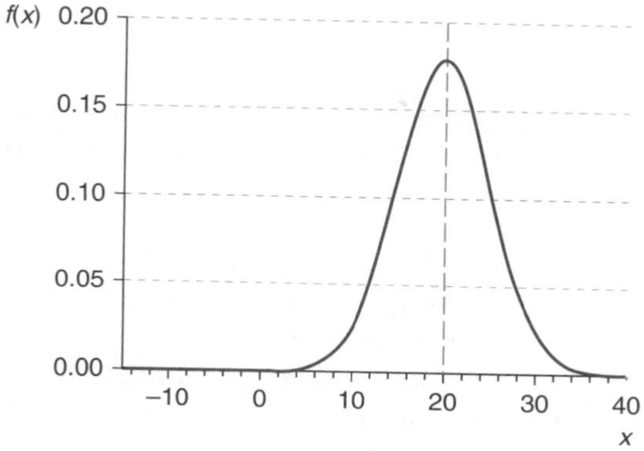

Figure 3.6(b)
The Normal distribution, $\mu = 15, \sigma = 2$

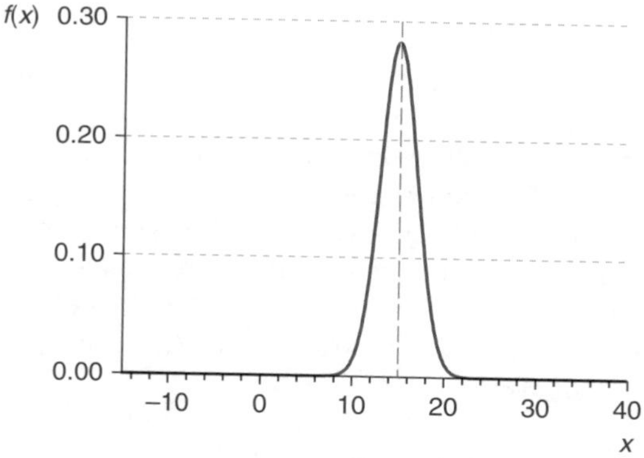

Figure 3.6(c)
The Normal distribution, $\mu = 0, \sigma = 4$

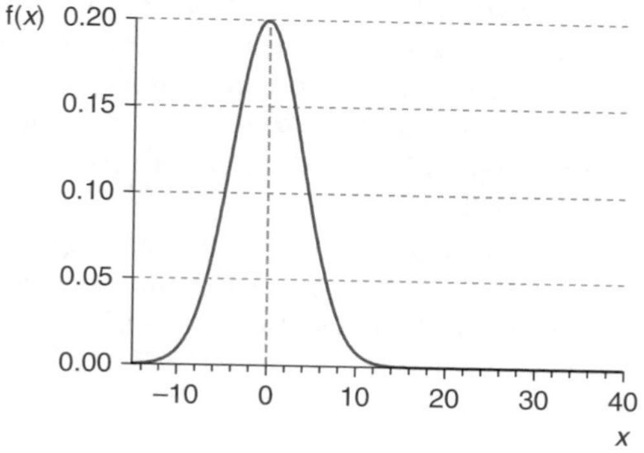

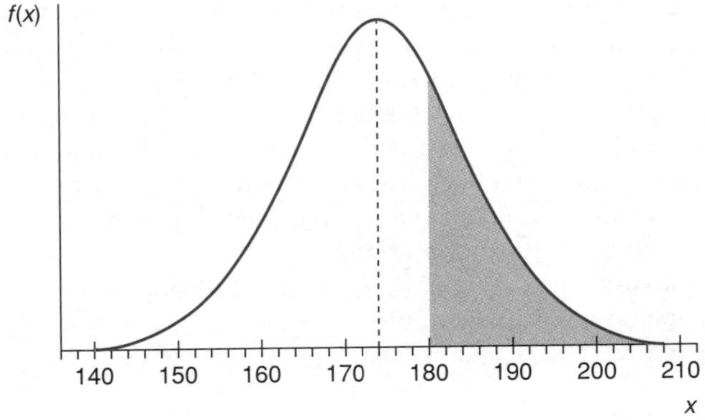

Figure 3.7
The height distribution of men

(or equivalently the variance, σ^2) are known, the whole of the distribution can be drawn. The shorthand notation for a Normal distribution is

$$x \sim N(\mu, \sigma^2) \tag{3.6}$$

meaning 'the variable x is Normally distributed with mean μ and variance σ^2'. This is similar in form to the expression for the Binomial distribution, although the meanings of the parameters are different.

Use of the Normal distribution can be illustrated using a simple example. The height of adult males is Normally distributed with mean height $\mu = 174$ cm and standard deviation $\sigma = 9.6$ cm. Let x represent the height of adult males; then

$$x \sim N(174, 92.16) \tag{3.7}$$

and this is illustrated in Figure 3.7. Note that (3.7) contains the variance rather than the standard deviation.

What is the probability that a randomly selected man is taller than 180 cm? If all men are equally likely to be selected, this is equivalent to asking what proportion of men are over 180 cm in height. This is given by the area under the Normal distribution, to the right of $x = 180$, i.e. the shaded area in Figure 3.7. The further from the mean of 174, the smaller the area in the tail of the distribution. One way to find this area would be to use equation (3.5), but this requires the use of sophisticated mathematics.

Since this is a frequently encountered problem, the answers have been set out in the tables of the **standard Normal distribution**. We can simply look up the solution. However, since there is an infinite number of Normal distributions (one for every combination μ and σ^2), it would be impossible to tabulate them all. The standard Normal distribution, which has a mean of zero and variance of one, is therefore used to represent all Normal distributions. Before the table can be consulted, therefore, the data have to be transformed so that they accord with the standard Normal distribution.

The required transformation is the z score, which was introduced in Chapter 1. This measures the distance between the value of interest (180) and the mean, measured in terms of standard deviations. Therefore, we calculate

$$z = \frac{x - \mu}{\sigma} \tag{3.8}$$

and z is a Normally distributed random variable with mean 0 and variance 1, i.e. $z \sim N(0, 1)$.

This transformation shifts the original distribution μ units to the left and then adjusts the dispersion by dividing through by σ, resulting in a mean of 0 and variance 1. z is Normally distributed because x is Normally distributed. The transformation in (3.8) retains the Normal distribution shape, despite the changes to mean and variance. If x followed some other distribution, then z would not be Normal either.

It is easy to verify the mean and variance of z using the rules for E and V operators encountered in Chapter 1:

$$\mathrm{E}(z) = \mathrm{E}\left(\frac{x - \mu}{\sigma}\right) = \frac{1}{\sigma}(\mathrm{E}(x) - \mu) = 0 \quad (\text{since } \mathrm{E}(x) = \mu)$$

$$\mathrm{V}(z) = \mathrm{V}\left(\frac{x - \mu}{\sigma}\right) = \frac{1}{\sigma^2}\mathrm{V}(x)\frac{\sigma^2}{\sigma^2} = 1$$

Evaluating the z score from our data, we obtain

$$z = \frac{180 - 174}{9.6} = 0.63 \tag{3.9}$$

This shows that 180 is 0.63 standard deviations above the mean, 174, of the distribution. This is a measure of how far 180 is from 174 and allows us to look up the answer in tables. The task now is to find the area under the standard Normal distribution to the right of 0.63 standard deviations above the mean. This answer can be read off directly from the table of the standard Normal distribution, included as Table A2 in the Appendix. An excerpt from Table A2 is presented in Table 3.2.

The left-hand column gives the z score to one place of decimals. The appropriate row of the table to consult is the one for $z = 0.6$, which is shaded. For the second place of decimals (0.03) we consult the appropriate column, also shaded. At their intersection we find the value 0.2643, which is the desired area and therefore probability. In other words, 26.43% of the distribution lies to the right of 0.63 standard deviations above the mean. Therefore 26.43% of men are over 180 cm in height.

Use of the standard Normal table is possible because, although there is an infinite number of Normal distributions, they are all fundamentally the same, so that the area to the right of 0.63 standard deviations above the mean is the same for all of them. As long as we measure the distance in terms of standard deviations, then we can use the standard Normal table. The process of standardisation turns all Normal distributions into a standard Normal distribution with a mean of zero and a variance of one. This process is illustrated in Figure 3.8.

Table 3.2 Areas of the standard Normal distribution (excerpt from Table A2)

z	0.00	0.01	0.02	0.03	...	0.09
0.0	0.5000	0.4960	0.4920	0.4880	...	0.4641
0.1	0.4602	0.4562	0.4522	0.4483	...	0.4247
⋮	⋮	⋮	⋮	⋮	...	⋮
0.5	0.3085	0.3050	0.3015	0.2981	...	0.2776
0.6	0.2743	0.2709	0.2676	0.2643	...	0.2451
0.7	0.2420	0.2389	0.2358	0.2327	...	0.2148

Figure 3.8(a)
The Normal distribution, $\mu = 174, \sigma = 9.6$

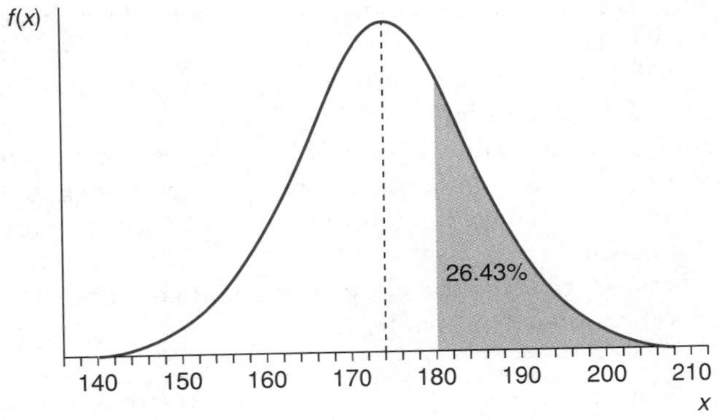

Figure 3.8(b)
The standard Normal distribution corresponding to Figure 3.8(a)

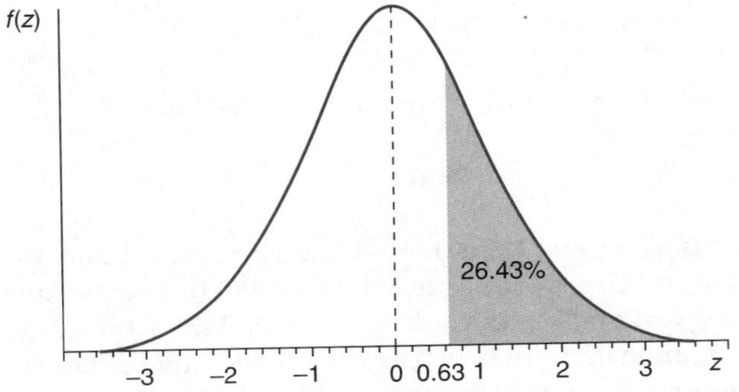

The area in the right-hand tail is the same for both distributions. It is the standard Normal distribution in Figure 3.8(b) which is tabulated in Table A2. To demonstrate how standardisation turns all Normal distributions into the standard Normal, the earlier problem is repeated but taking all measurements in inches. The answer should obviously be the same. Taking 1 inch = 2.54 cm, the figures are

$$x = 70.87 \quad \sigma = 3.78 \quad \mu = 68.50$$

What proportion of men are over 70.87 inches in height? The appropriate Normal distribution is now

$$x \sim N(68.50, 3.78^2) \tag{3.10}$$

The z score is

$$z = \frac{70.87 - 68.50}{3.78} = 0.63 \tag{3.11}$$

which is the same z score as before and therefore gives the same probability.

Worked example 3.2

Packets of cereal have a nominal weight of 750 grams, but there is some variation around this as the machines filling the packets are imperfect. Let us assume that the weights follow a Normal distribution. Suppose that the standard deviation around the mean of 750 is 5 grams. What proportion of packets weigh more than 760 grams?

Summarising our information, we have $x \sim N(750, 25)$ where x represents the weight. We wish to find $\Pr(x > 760)$. To be able to look up the answer, we need to measure the distance between 760 and 750 in terms of standard deviations. This is

$$z = \frac{760 - 750}{5} = 2.0$$

Looking up $z = 2.0$ in Table A2 reveals an area of 0.0228 in the tail of the distribution. Thus 2.28% of packets weigh more than 760 grams.

Since a great deal of use is made of the standard Normal tables, it is worth working through a couple more examples to reinforce the method. We have so far calculated that $\Pr(z > 0.63) = 0.2643$. Since the total area under the graph equals one (i.e. the sum of probabilities must be one), the area to the left of $z = 0.63$ must equal 0.7357, i.e. 73.57% of men are under 180 cm. It is fairly easy to manipulate areas under the graph to arrive at any required area. For example, what proportion of men are between 174 and 180 cm in height? It is helpful to refer to Figure 3.9 at this point.

The size of area A is required. Area B has already been calculated as 0.2643. Since the distribution is symmetric, the area A + B must equal 0.5, since 174 is at the centre (mean) of the distribution. Area A is therefore $0.5 - 0.2643 = 0.2357$. Therefore, 23.57% is the desired result.

STATISTICS IN PRACTICE

Using software to find areas under the standard Normal distribution

Using a spreadsheet program, you can look up the *z*-distribution directly and hence dispense with tables. In Excel, for example, the function '= *NORM.S.DIST*(0.63, TRUE)' gives the answer 0.7357, i.e. the area to the *left* of the *z* score. The area in the right-hand tail is then obtained by subtracting this value from 1, i.e. 1 − 0.7357 = 0.2643. Entering the formula '= 1 − *NORM.S.DIST*(0.63, TRUE)' in a cell will give the area in the right-hand tail directly.

As a final exercise consider the question of what proportion of men are between 166 and 178 cm tall. As shown in Figure 3.10, area C + D is wanted. The only way

Figure 3.9
The proportion of men between 174 cm and 180 cm in height

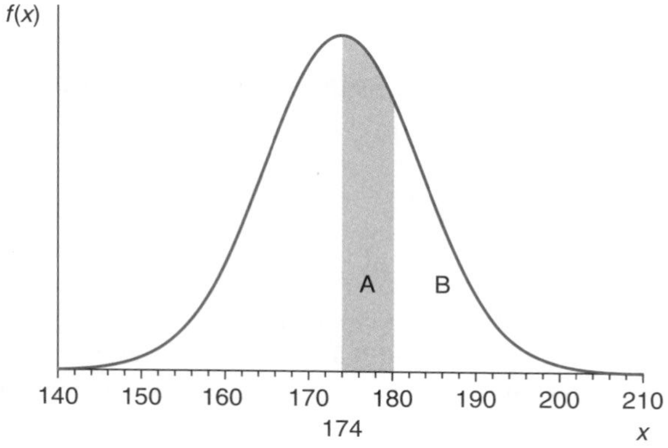

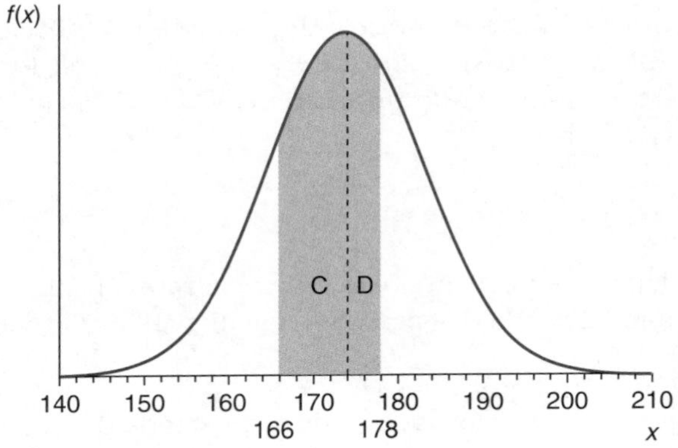

Figure 3.10
The proportion of men between 166 and 178 cm in height

to find this is to calculate the two areas separately and then add them together. For area D the z score associated with 178 is:

$$z_D = \frac{178 - 174}{9.6} = 0.42 \tag{3.12}$$

Table A2 indicates that the area in the right-hand tail, beyond $z = 0.42$, is 0.3372, so area D = 0.5 − 0.3372 = 0.1628. For C, the z score is

$$z_C = \frac{166 - 174}{9.6} = -0.83 \tag{3.13}$$

The minus sign indicates that it is the left-hand tail of the distribution, below the mean, which is being considered. Since the distribution is symmetric, it is the same as if it were the right-hand tail, so the minus sign may be ignored when consulting the table. Looking up $z = 0.83$ in Table A2 gives an area of 0.2033 in the tail, so area C is therefore 0.5 − 0.2033 = 0.2967. Adding areas C and D gives 0.2967 + 0.1628 = 0.4595. So nearly half of all men are between 166 and 178 cm in height.

An alternative interpretation of the results obtained above is that if a man is drawn at random from the adult population, the probability that he is over 180 cm tall is 26.43%. This is in line with the frequentist school of thought. Since 26.43% of the population is over 180 cm in height, that is the probability of a man over 180 cm being drawn at random.

Exercise 3.3

?

(a) The random variable x is distributed Normally, $x \sim N(40, 36)$. Find the probability that $x > 50$.

(b) Find $\Pr(x < 45)$.

(c) Find $\Pr(36 < x < 45)$.

Exercise 3.4

?

The mean +/−0.67 standard deviations cuts off 25% in each tail of the Normal distribution. Hence, the middle 50% of the distribution lies within +/−0.67 standard deviations of the mean. Use this fact to calculate the inter-quartile range for the distribution $x \sim N(200, 256)$.

Exercise 3.5

?

As suggested in the text, the logarithm of income is approximately Normally distributed. Suppose the log (to the base 10) of income has the distribution $x \sim N(4.18, 256)$. Calculate the inter-quartile range for x and then take anti-logs to find the inter-quartile range of income.

The distribution of the sample mean

One of the most important concepts in statistical inference is the probability distribution of the mean of a random sample, since we often use the sample mean to tell us something about an associated population. Suppose that, from the population of adult males, a random sample of size $n = 36$ is taken, their heights measured and the mean height of the sample calculated. What can we infer from this about the true average height of the population? To do this, we need to know about the statistical properties of the sample mean. The sample mean is a random variable because of the chance element of random sampling (different samples would yield different values of the sample mean). Since the sample mean is a random variable, it must have associated with it a probability distribution. We also refer to this as the **sampling distribution** of the sample mean, since the randomness is due to sampling.

We therefore need to know, first, what is the appropriate distribution and, second, what are its parameters. From the definition of the sample mean we have

$$\bar{x} = \frac{1}{n}(x_1 + x_2 + \cdots + x_n) \tag{3.14}$$

where each observation, x_i, is itself a Normally distributed random variable, with $x_i \sim N(\mu, \sigma^2)$, because each comes from the parent distribution with such characteristics. (We stated earlier that men's heights are Normally distributed.) We now make use of the following theorem to demonstrate first that $\bar{x}$ is Normally distributed:

Theorem

Any linear combination of independent, Normally distributed random variables is itself Normally distributed.

A linear combination of two variables x_1 and x_2 is of the form $w_1x_1 + w_2x_2$ where w_1 and w_2 are constants. This can be generalised to any number of x values. It is clear that the sample mean satisfies these conditions and is a linear combination of the individual x values (with the weight on each observation equal to $1/n$). As long as the observations are independently drawn, therefore, the sample mean is Normally distributed.

We now need the parameters (mean and variance) of the distribution. For this we use the E and V operators once again:

$$\begin{aligned} E(\bar{x}) &= \frac{1}{n}(E(x_1) + E(x_2) + \cdots + E(x_n)) \\ &= \frac{1}{n}(\mu + \mu + \cdots + \mu) \\ &= \frac{1}{n}n\mu \\ &= \mu \end{aligned} \tag{3.15}$$

$$
\begin{aligned}
V(\bar{x}) &= V\left(\frac{1}{n}[x_1 + x_2 + \cdots + x_n]\right) \\
&= \frac{1}{n^2}(V(x_1) + V(x_2) + \cdots + V(x_n)) \\
&= \frac{1}{n^2}(\sigma^2 + \sigma^2 + \cdots + \sigma^2) \\
&= \frac{1}{n^2}n\sigma^2 = \frac{\sigma^2}{n} \qquad (3.16)
\end{aligned}
$$

Putting all this together, we have[2]

$$\bar{x} \sim N\left(\mu, \frac{\sigma^2}{n}\right) \qquad (3.17)$$

This we may summarise in the following theorem:

Theorem

The mean, $\bar{x}$ of a random sample drawn from a population which has a Normal distribution with mean μ and variance σ^2, has a sampling distribution which is Normal, with mean μ and variance σ^2/n, where n is the sample size.

The meaning of this theorem is as follows. First, it is assumed that the population from which the samples are to be drawn is itself Normally distributed (this assumption will be relaxed in a moment), with mean μ and variance σ^2. From this population many samples are drawn, each of sample size n, and the mean of each sample is calculated. The samples are independent, meaning that the observations selected for one sample do not influence the selection of observations in the other samples. This gives many sample means, $\bar{x}_1$, $\bar{x}_2$, etc. If these sample means are treated as a new set of observations, then the probability distribution of these observations can be derived. The theorem states that this distribution is Normal, with the sample means centred around μ, the population mean, and with variance σ^2/n. The argument is set out diagrammatically in Figure 3.11.

Intuitively this theorem can be understood as follows. If the height of adult males is a Normally distributed random variable with mean $\mu = 174$ cm and variance $\sigma^2 = 92.16$, then it would be expected that a random sample of (say) nine males would yield a sample mean height of around 174 cm, perhaps a little more, perhaps a little less. In other words, the sample mean is centred around 174 cm, or the mean of the distribution of sample means is 174 cm.

The larger is the size of the individual samples (i.e. the larger n), the closer the sample mean would tend to be to 174 cm. For example, if the sample size is only two, a sample of two very tall people is quite possible, with a high sample mean as a result, well over 174 cm, e.g. 182 cm. But if the sample size were 20, it is very unlikely that 20 very tall males would be selected and the sample mean is likely to be much closer to 174. This is why the sample size n appears in the formula for the variance of the distribution of the sample mean, σ^2/n.

Note that, once again, we have transformed one (or more) random variables, the x_i's, with a particular probability distribution into another random variable, $\bar{x}$, with a (slightly) different distribution. This is common practice in statistics:

[2]Don't worry if you didn't follow the derivation of this formula, just accept that it is correct.

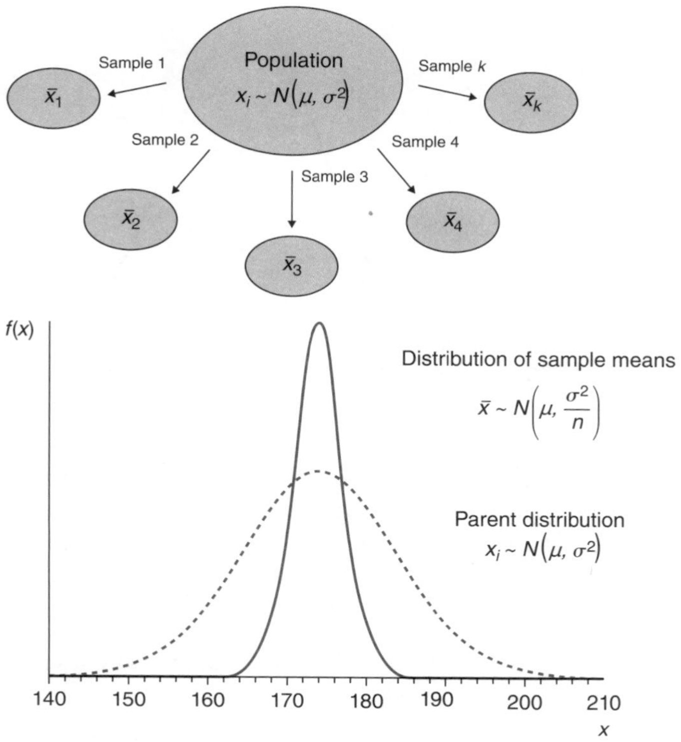

Figure 3.11
The parent distribution and the distribution of sample means

transforming a variable will often put it into a more useful form, e.g. one whose probability distribution is well known.

The above theorem can be used to solve a range of statistical problems. For example, what is the probability that a random sample of nine men will have a mean height greater than 180 cm? The height of all men is known to be Normally distributed with mean $\mu = 174$ cm and variance $\sigma^2 = 92.16$. The theorem can be used to derive the probability distribution of the sample mean. For the population we have:

$$x \sim N(\mu, \sigma^2), \text{ i.e } x \sim N(174, 92.16)$$

Hence for the sample mean:

$$\bar{x} \sim N(\mu, \sigma^2/n), \text{ i.e } \bar{x} \sim N(174, 92.16/9)$$

This is shown diagrammatically in Figure 3.12.

To answer the question posed, the area to the right of 180, shaded in Figure 3.12, has to be found. This should by now be a familiar procedure. First the z score is calculated:

$$z = \frac{\bar{x} - \mu}{\sqrt{\sigma^2/n}} = \frac{180 - 174}{\sqrt{92.16/9}} = 1.88 \qquad (3.18)$$

Note that the z score formula is subtly different because we are dealing with the sample mean $\bar{x}$ rather than x itself. In the numerator we use $\bar{x}$ rather than x and in the denominator we use σ^2/n, not σ^2. This is because $\bar{x}$ has a variance σ^2/n, not σ^2, which is the population variance. $\sqrt{\sigma^2/n}$ is known as the **standard error**, to

Figure 3.12
The proportion of sample means greater than $\bar{x} = 180$

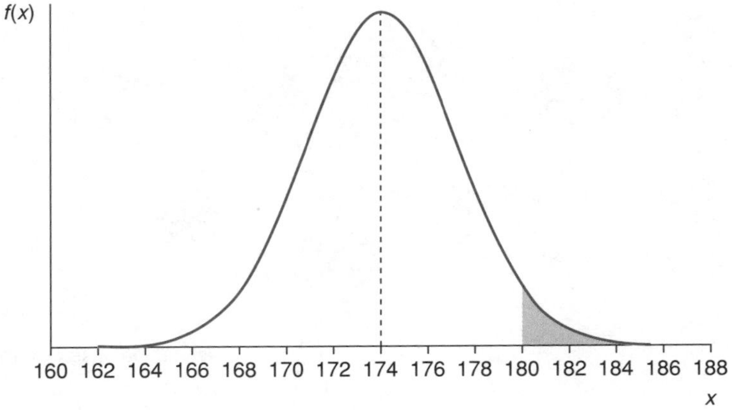

distinguish it from σ, the standard deviation of the population. The principle behind the z score is the same, however: it measures how far a sample mean of 180 is from the population mean of 174, measured in terms of standard deviations.

Looking up the value of $z = 1.88$ in Table A2 gives an area of 0.0311 in the right-hand tail of the Normal distribution. Thus, 3.11% of sample means will be greater than or equal to 180 cm when the sample size is nine. The desired probability is therefore 3.11%.

Since this probability is quite small, we might consider the reasons for this. There are two possibilities:

(a) through bad luck, the sample collected is not very representative of the population as a whole, or
(b) the sample *is* representative of the population, but the population mean is not 174 cm after all.

Only one of these two possibilities can be correct. How to decide between them will be taken up in Chapter 5 on hypothesis testing.

It is interesting to examine the difference between the answer for a sample size of nine (3.11%) and the one obtained earlier for a single individual (26.43%). The latter may be considered as a sample of size one from the population. The examples illustrate the fact that the larger the sample size, the closer the sample mean is likely to be to the population mean. Thus larger samples tend to give better estimates of the population mean.

STATISTICS IN PRACTICE

Oil reserves

An interesting application of probability distributions is to the estimation of oil reserves. The quantity of oil in an oil field is not known for certain, but is subject to uncertainty. The *proven* reserve of an oil field is the amount recoverable with probability of 90% (known as P90 in the oil industry). One can then add up the proven oil reserves around the world to get a total of proven reserves.

However, using probability theory, we can see this might be misleading. Suppose we have 50 fields, where the recoverable quantity of oil is distributed as $x \sim N[100, 81]$ in each. From tables we note that $\bar{x} - 1.28s$ cuts off the bottom 10% of the Normal distribution, 88.48 in this case. This is the proven reserve for a field. Summing across the 50 fields gives 4424 as total proven reserves. But is there a 90% probability of recovering at least this amount?

The total quantity of oil y is distributed Normally, with mean $E(y) = E(x_1) + \cdots + E(x_{50}) = 5000$ and variance $V(y) = V(x_1) + \cdots + V(x_{50}) = 4050$, assuming independence of the oil fields. Hence we have. $y \sim N(5000, 4050)$. Again, the bottom 10% is cut off by $\bar{y} - 1.28s$, which is 4919. This is 11% larger than the 4424 calculated above. Adding up the proven reserves of each field individually underestimates the true total proven reserves. In fact, the probability of total proven reserves being greater than 4424 is almost 100%. The reason the first calculation fails is that it effectively assumes the worst case in every oil field. But this is very unlikely – it is much more likely that some fields will yield plenty of oil, some will yield little.

Note that the numbers given here are for illustration purposes and do not reflect the actual state of affairs. The principle of the calculation is correct, however.

Sampling from a non-Normal population

The previous theorem and examples relied upon the fact that the population followed a Normal distribution. But what happens if it is not Normal? After all, it is not known for certain that the heights of all adult males are exactly Normally distributed, and there are many populations which are not Normal (e.g. wealth, as shown in Chapter 1). What can be done in these circumstances? The answer is to use another theorem about the distribution of sample means (presented without proof). This is known as the **Central Limit Theorem**:

Theorem

The mean, $\bar{x}$, of a random sample drawn from a population with mean μ and variance σ^2, has a sampling distribution which approaches a Normal distribution with mean μ and variance σ^2/n, as the sample size n approaches infinity.

This is very useful, since it drops the assumption that the population is Normally distributed. Note that, according to this theorem, the distribution of sample means is only Normal as long as the sample size is infinite; for any finite sample size the distribution is only approximately Normal. However, the approximation is close enough for practical purposes if the sample size is larger than 25 or so observations. If the population distribution is itself nearly Normal, then a smaller sample size would suffice. If the population distribution is particularly skewed, then more than 25 observations might be desirable. Twenty-five observations constitute a rule of thumb that is adequate in most circumstances. This is another illustration of statistics as an inexact science. It does not provide absolutely clear-cut answers to questions but, used carefully, helps us to arrive at sensible conclusions.

As an example of the use of the Central Limit Theorem, we return to the wealth data of Chapter 1. Recall that the mean level of wealth was 186.875 (measured in £000) and the variance 80 306. Suppose that a sample of $n = 50$ people were drawn from this population. What is the probability that the sample mean is greater than 200 (i.e. £200 000)?

On this occasion we know that the parent distribution is highly skewed, so it is fortunate that we have 50 observations. This should be ample for us to justify applying the Central Limit Theorem. The distribution of $\bar{x}$ is therefore

$$\bar{x} \sim N(\mu, \sigma^2/n) \tag{3.19}$$

and, inserting the parameter values, this gives[3]

$$\bar{x} \sim N(186.875, 80\,306/50) \tag{3.20}$$

To find the area beyond a sample mean of 200, the z score is first calculated:

$$z = \frac{200 - 186.875}{\sqrt{80\,306/50}} = 0.33 \tag{3.21}$$

Referring to the standard Normal tables, the area in the tail is then found to be 37.07%. This is the desired probability. So there is a probability of 37.07% of finding a mean of £200 000 or greater with a sample of size 50. This demonstrates that there is quite a high probability of getting a sample mean which is relatively far from £186 875. This is a consequence of the high degree of dispersion in the distribution of wealth.

Extending this example, we can ask: what is the probability of the sample mean lying within, say, £78 000 either side of the true mean of £186 875 (i.e. between £108 875 and £264 875)? Figure 3.13 illustrates the situation, with the desired area shaded. By symmetry, areas A and B must be equal, so we only need find one of them. For B, we calculate the z score:

$$z = \frac{264.875 - 186.875}{\sqrt{80\,306/50}} = 1.946 \tag{3.22}$$

From the standard Normal table, this cuts off approximately 2.5% in the upper tail,[4] so area B = 0.475. Areas A and B together make up approximately 95% of the distribution, therefore. There is thus a 95% probability of the sample mean falling within the range [108 875, 264 875], and we call this the 95% **probability interval** for the sample mean. We write this:

$$\Pr(108\,875 \leq \bar{x} \leq 264\,875) = 0.95 \tag{3.23}$$

or, in terms of the formulae we have used:[5]

$$\Pr(\mu - 1.96\sqrt{\sigma^2/n} \leq \bar{x} \leq \mu + 1.96\sqrt{\sigma^2/n}) = 0.95 \tag{3.24}$$

Figure 3.13
The probability of $\bar{x}$ lying within £78 000 either side of £186 875

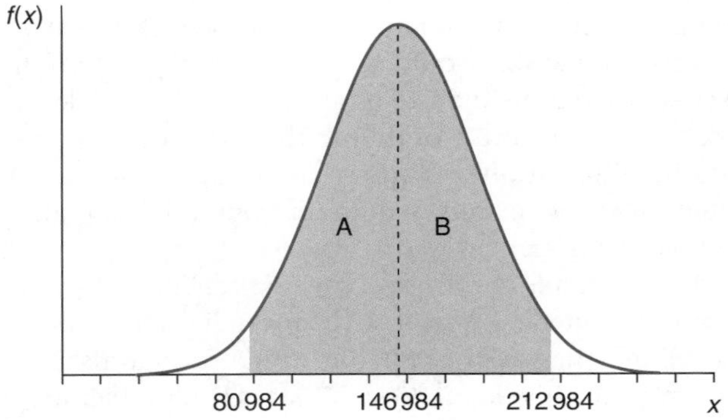

[3]Note that if we used 186 875 for the mean, we would have 80 306 000 000 as the variance. Using £000 keeps the numbers more manageable. The z score is the same in both cases.
[4]The precise figure is 2.58% but it is convenient to round this down to 2.5% in the discussion.
[5]1.96 is the precise value cutting off 2.5% in each tail.

The 95% probability interval and the related concept of the 95% confidence interval (introduced in Chapter 4) play important roles in statistical inference. We deliberately designed the example above to arrive at an answer of 95% for this reason.

Exercise 3.6

?

(a) If x is distributed as $x \sim N(50, 64)$ and samples of size $n = 25$ are drawn, what is the distribution of the sample mean $\bar{x}$?

(b) If the sample size doubles to 50, how is the standard error of $\bar{x}$ altered?

(c) Using the sample size of 25, (i) what is the probability that $\bar{x} > 51$? (ii) What is $\Pr(\bar{x} < 48)$? (iii) What is $\Pr(49 < \bar{x} < 50.5)$?

The relationship between the Binomial and Normal distributions

Many statistical distributions are related to one another in some way. This means that many problems can be solved by a variety of different methods (using different distributions), although usually one is more convenient or more accurate than the others. This point may be illustrated by looking at the relationship between the Binomial and Normal distributions.

Recall the experiment of tossing a coin repeatedly and noting the number of heads. We said earlier that this can be analysed via the Binomial distribution. But note that the number of heads, a random variable, is influenced by many independent random events (the individual tosses) added together. Furthermore, each toss counts equally, none dominates. These are just the conditions under which a Normal distribution arises, so it looks like there is a connection between the two distributions.

This idea is correct. Recall that if a random variable r follows a Binomial distribution, then

$$r \sim B(n, P) \tag{3.1}$$

and the mean of the distribution is nP and the variance $nP(1 - P)$. It turns out that as n gets larger, the Binomial distribution takes on approximately the same shape as a Normal distribution with mean nP and variance $nP(1 - P)$. This approximation is sufficiently accurate as long as $nP > 5$ and $n(1 - P) > 5$, so the approximation may not be very good (even for large values of n) if P is very close to zero or one. For the coin tossing experiment, where $P = 0.5$, 10 tosses should be sufficient. Note that this approximation is good enough with only 10 observations even though the underlying probability distribution is nothing like a Normal distribution.

To demonstrate, the following problem is solved using both the Binomial and Normal distributions. Forty students take an exam in statistics which is simply graded pass/fail. If the probability, P, of any individual student passing is 60%, what is the probability of at least 30 students passing the exam?

The sample data are:

$$\begin{aligned} P &= 0.6 \\ 1 - P &= 0.4 \\ n &= 40 \end{aligned}$$

Binomial distribution method

To solve the problem using the Binomial distribution it is necessary to find the probability of exactly 30 students passing, plus the probability of 31 passing, plus the probability of 32 passing, etc., up to the probability of 40 passing (the fact that the events are mutually exclusive allows this). The probability of 30 passing is

$$\begin{aligned}\Pr(r = 30) &= nCr \times P^r(1 - P)^{n-r} \\ &= 40C^{30} \times 0.6^{30} \times 0.4^{10} \\ &= 0.020\end{aligned}$$

(*Note:* This calculation assumes that the probabilities are independent, i.e. no copying.) This by itself is quite a tedious calculation, but Pr(31), Pr(32), etc., still have to be calculated. Calculating these and summing them gives the result of 3.52% as the probability of at least 30 passing. (It would be a useful exercise for you to do, if only to appreciate how long it takes.)

Normal distribution method

As stated above, the Binomial distribution can be approximated by a Normal distribution with mean nP and variance. $nP(1 - P)$. nP in this case is 24(=40 × 0.6) and $n(1 - P)$ is 16, both greater than 5, so the approximation can be safely used. Thus

$$r \sim N(nP, nP(1 - P))$$

and inserting the parameter values gives

$$r \sim N(24, 9.6)$$

The usual methods are then used to find the appropriate area under the distribution. However, before doing so, there is one adjustment to be made (this only applies when approximating the Binomial distribution by the Normal). The Normal distribution is a continuous one while the Binomial is discrete. Thus, 30 in the Binomial distribution is represented by the area under the Normal distribution between 29.5 and 30.5. The value 31 is represented by 30.5 to 31.5, etc. Thus, it is the area under the Normal distribution to the right of 29.5, not 30, which must be calculated. This is known as the **continuity correction**. Calculating the z score gives

$$z = \frac{29.5 - 24}{\sqrt{9.6}} = 1.78 \tag{3.25}$$

This gives an area of 3.75%, not far off the correct answer as calculated by the Binomial distribution. The time saved and ease of calculation would seem to be worth the slight loss in accuracy.

Other examples can be constructed to test this method, using different values of P and n. Small values of n, or values of nP or $n(1 - P)$ less than 5, will give poor results, i.e. the Normal approximation to the Binomial will not be very good.

Exercise 3.7

?

(a) A coin is tossed 20 times. What is the probability of more than 14 heads? Perform the calculation using both the Binomial and Normal distributions, and compare results.

(b) A biased coin, for which $\Pr(H) = 0.7$, is tossed six times. What is the probability of more than four heads? Compare Binomial and Normal methods in this case. How accurate is the Normal approximation?

(c) Repeat part (b) but for the probability of more than five heads.

The Poisson distribution

The section above showed how the Binomial distribution could be approximated by a Normal distribution under certain circumstances. The approximation does not work particularly well for very small values of P, when nP is less than 5. In these circumstances the Binomial may be approximated instead by a different distribution, the Poisson, which is given by the formula

$$\Pr(x) = \frac{\mu^x e^{-\mu}}{x!} \qquad (3.26)$$

The only parameter of this distribution is μ, which is the mean of the distribution (like μ for the Normal distribution and nP for the Binomial). Like the Binomial, but unlike the Normal, the Poisson is a discrete probability distribution and equation (3.26) is defined only for integer values of x. Furthermore, it is applicable to a series of trials which are independent, as in the Binomial case.

The use of the **Poisson distribution** is appropriate when the probability P of 'success' is very small and the number of trials n is large. Because of this it is sometimes referred to as the 'rare event' distribution. Its use is illustrated by the following example. A manufacturer gives a two-year guarantee on the TV screens it makes. From past experience it knows that 0.5% of its screens will be faulty and fail within the guarantee period. What is the probability that, of a consignment of 500 screens, (a) none will be faulty, (b) more than three are faulty?

To use the Poisson, we just need its parameter, the mean. In this case it is $\mu = nP = 2.5$, like the Binomial. In words, we expect 0.5% of the sample to be faulty on average, and 0.5% of 500 is 2.5. Therefore

$$\Pr(x = 0) = \frac{2.5^0 e^{-2.5}}{0!} = 0.0821 \qquad (3.27)$$

gives a probability of 8.2% of no failures. The answer to this problem via the Binomial method is

$$\Pr(r = 0) = 0.995^{500} = 0.0816$$

Thus, the Poisson method gives a reasonably accurate answer. Using the Poisson approximation to the Binomial is satisfactory if nP is less than about 7.

The probability of more than three screens expiring is calculated as

$$\Pr(x > 3) = 1 - \Pr(x = 0) - \Pr(x = 1) - \Pr(x = 2) - \Pr(x = 3)$$

$$\Pr(x = 1) = \frac{2.5^1 e^{-2.5}}{1!} = 0.205$$

$$\Pr(x = 2) = \frac{2.5^2 e^{-2.5}}{2!} = 0.256$$

$$\Pr(x = 3) = \frac{2.5^3 e^{-2.5}}{3!} = 0.214$$

So

$$\Pr(x > 3) = 1 - 0.082 - 0.205 - 0.256 - 0.214$$
$$= 0.242$$

Thus, there is a probability of about 24% of more than three failures. The Binomial calculation is much more tedious, but gives an answer of 24.2% also.

The Poisson distribution is also used in problems where events occur over time, such as goals scored in a football match (see Problem 3.25) or queuing-type problems (e.g. arrivals at a bank cash machine). In these problems, there is no natural 'number' of trials but it is clear that, if we take a short interval of time, the probability of an event occurring is small. We can then consider the number of trials to be the number of time intervals. In such problems we cannot identify n and P separately, and hence the Binomial distribution cannot be used. However, we might know the product nP, in which case the Poisson can be used.

This is illustrated by the following example. A football team scores, on average, two goals every game (you can vary the example by using your own favourite team plus their scoring record). We therefore have the value of μ, although we do not know the separate values of n and P. Indeed, it is not clear to what n and P would refer in this context. How can we calculate the probability of the team scoring zero or one goal during a game?

The mean of the distribution is 2, so we have, using the Poisson distribution:

$$\Pr(x = 0) = \frac{2^0 e^{-2}}{0!} = 0.135$$

$$\Pr(x = 0) = \frac{2^1 e^{-2}}{1!} = 0.271$$

So $\Pr(x = 0 \text{ or } 1) = 0.406$. You should continue to calculate the probabilities of 2 or more goals and verify that the probabilities sum to 1. (In principle you have to calculate the probabilities of 2, 3, 4, . . . 100, . . . which is impossible, but you will find that the probabilities soon decline to very small values which can be ignored for practical purposes.)

STATISTICS IN PRACTICE

Oops!

The formula1.com website reported that 'The (Grand Prix motor) race in Singapore has seen a total of five safety car deployments in three races, and carries a 100% safety car probability based on historical data.'

That does not sound right, and indeed it is not. We can find the right answer via the Poisson distribution. In this case, $\mu = 5/3 = 1.67$ deployments per race, on average. Hence, the probability of no safety car deployment during a race is

$$\Pr(x = 0) = \frac{1.67^0 e^{-1.67}}{0!} = 0.189 \text{ or about } 19\%.$$

The correct answer is an 81% chance of a safety car period during the race.

Another type of problem amenable to analysis via the Poisson distribution is queuing. For example, if a shop receives, on average, 20 customers per hour, what is the probability of no customers within a five-minute period while the owner takes a coffee break?

The average number of customers per five-minute period is $20 \times 5/60 = 1.67$. The probability of a free five-minute spell is therefore

$$\Pr(x = 0) = \frac{1.67^0 e^{1.67}}{0!} = 0.189$$

a probability of about 19%. Note again that this problem cannot be solved by the Binomial method since n and P are not known separately, only their product.

Exercise 3.8

(a) The probability of winning a prize in a lottery is 1 in 50. If you buy 50 tickets, what is the probability that (i) 0 tickets win, (ii) 1 ticket wins, (iii) 2 tickets win. (iv) What is the probability of winning at least one prize?

(b) On average, a person buys a lottery ticket in a supermarket every 5 minutes. What is the probability that 10 minutes will pass with no buyers?

STATISTICS IN PRACTICE

Railway accidents

Andrew Evans of University College London used the Poisson distribution to examine the numbers of fatal railway accidents in Britain between 1967 and 1997. Since railway accidents are, fortunately, rare, the probability of an accident in any time period is very small, and so use of the Poisson distribution is appropriate. He found that the average number of accidents has been falling over time and by 1997 had reached 1.25 p.a. This figure is therefore used as the mean μ of the Poisson distribution, and we can calculate the probabilities of 0, 1, 2, etc., accidents each year. Using $\mu = 1.25$ and inserting this into equation (3.26), we obtain the following table:

Number of accidents	0	1	2	3	4	5	6
Probability	0.287	0.358	0.224	0.093	0.029	0.007	0.002

and this distribution can be graphed:

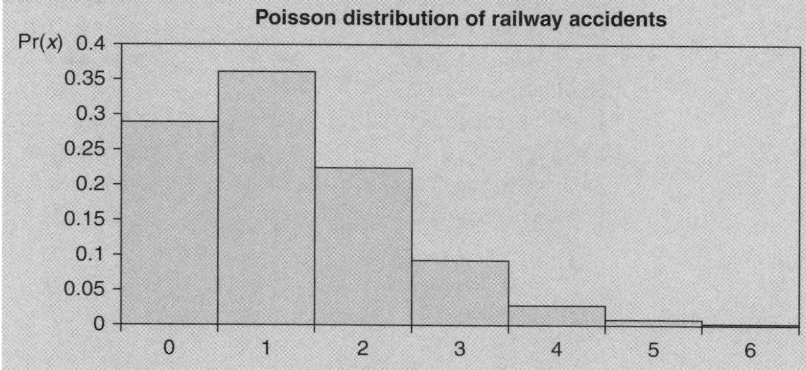

Thus the most likely outcome is one fatal accident per year, and anything over four is extremely unlikely. In fact, Evans found that the Poisson was not a perfect fit to the data: the actual variation was less than that predicted by the model.

Source: A. W. Evans, Fatal train accidents on Britain's mainline railways, *J. Royal Statistical Society*, Series A, vol. 163 (1), 2000.

Summary

- The behaviour of many random variables (e.g. the result of the toss of a coin) can be described by a probability distribution (in this case, the Binomial distribution).
- The Binomial distribution is appropriate for problems where there are only two possible outcomes of a chance event (e.g. heads/tails, success/failure) and the probability of success is the same each time the experiment is conducted.
- The Normal distribution is appropriate for problems where the random variable has the familiar bell-shaped distribution. This often occurs when the variable is influenced by many, independent factors, none of which dominates the others. An example is men's heights, which are Normally distributed.
- The Poisson distribution is used in circumstances where there is a very low probability of 'success' and a high number of trials.
- Each of these distributions is actually a family of distributions, differing in the parameters of the distribution. Both the Binomial and Normal distributions have two parameters: n and P in the former case, μ and σ^2 in the latter. The Poisson distribution has one parameter, its mean μ.
- The mean of a random sample follows a Normal distribution, because it is influenced by many independent factors (the sample observations), none of which dominates in the calculation of the mean. This statement is always true if the population from which the sample is drawn follows a Normal distribution.
- If the population is not Normally distributed, then the Central Limit Theorem states that the sample mean is Normally distributed in large samples. In this case 'large' means a sample of about 25 or more.

Key terms and concepts

Binomial distribution
Central Limit Theorem
continuity correction
Normal distribution
parameters
parameters of a distribution
Poisson distribution
probability distribution
probability interval
random variable
sampling distribution
standard error
standard Normal distribution

Formulae used in this chapter

Formula	Description	Notes
$nCr = \frac{n!}{r!(n-r)!}$	Combinatorial formula	$n! = n \times (n-1) \times \cdots \times 1$
$\Pr(r) = nCr \times P^r \times (1-P)^{n-r}$	Binomial distribution	In shorthand notation, $r \sim B(n, P)$
$\Pr(x) = \frac{1}{\sigma\sqrt{2\pi}} e^{-\frac{1}{2}\left\{\frac{x-\mu}{\sigma}\right\}^2}$	Normal distribution	In shorthand notation, $x \sim N(\mu, \sigma^2)$
$z = \frac{\bar{x} - \mu}{\sigma^2/n}$	z score for the sample mean	Used to test hypotheses about the sample mean
$\Pr(x) = \frac{\mu^x e^{-\mu}}{x!}$	Poisson distribution	Used when the probability of success is very small. The 'rare event' distribution

Problems

Some of the more challenging problems are indicated by highlighting the problem number in colour.

3.1 Two dice are thrown and the sum of the two scores is recorded. Draw a graph of the resulting probability distribution of the sum and calculate its mean and variance. What is the probability that the sum is 9 or greater?

3.2 Two dice are thrown and the absolute difference of the two scores is recorded. Graph the resulting probability distribution and calculate its mean and variance. What is the probability that the absolute difference is 4 or more?

3.3 Sketch the probability distribution for the likely time of departure of a train. Locate the timetabled departure time on your chart.

3.4 A train departs every half hour. You arrive at the station at a completely random moment. Sketch the probability distribution of your waiting time. What is your expected waiting time?

3.5 Sketch the probability distribution for the number of accidents on a stretch of road in one day.

3.6 Sketch the probability distribution for the number of accidents on the same stretch of road in one year. How and why does this differ from your previous answer?

3.7 Six dice are rolled and the number of sixes is noted. Calculate the probabilities of 0, 1, . . . , 6 sixes and graph the probability distribution.

3.8 If the probability of a boy in a single birth is $\frac{1}{2}$ and is independent of the sex of previous babies, then the number of boys in a family of 10 children follows a Binomial distribution with mean 5 and variance 2.5. In each of the following instances, describe how the distribution of the number of boys differs from the Binomial described above.

(a) The probability of a boy is 6/10.

(b) The probability of a boy is $\frac{1}{2}$ but births are not independent. The birth of a boy makes it more than an even chance that the next child is a boy.

(c) As (b) above, except that the birth of a boy makes it less than an even chance that the next child will be a boy.

(d) The probability of a boy is 6/10 on the first birth. The birth of a boy makes it a more than even chance that the next baby will be a boy.

3.9 A firm receives components from a supplier in large batches, for use in its production process. Production is uneconomic if a batch containing 10% or more defective components is used. The firm checks the quality of each incoming batch by taking a sample of 15 and rejecting the whole batch if more than one defective component is found.

(a) If a batch containing 10% defectives is delivered, what is the probability of its being accepted?

(b) How could the firm reduce this probability of erroneously accepting bad batches?

(c) If the supplier produces a batch with 3% defective, what is the probability of the firm sending back the batch?

(d) What role does the assumption of a 'large' batch play in the calculation?

3.10 The UK record for the number of children born to a mother is 39, 32 of them girls. Assuming the probability of a girl in a single birth is 0.5 and that this probability is independent of previous births:

(a) Find the probability of 32 girls in 39 births (you'll need a scientific calculator or a computer to help with this).

(b) Does this result cast doubt on the assumptions?

3.11 Using equation (3.5) describing the Normal distribution and setting $\mu = 0$ and $\sigma^2 = 1$, graph the distribution for the values $x = -2, -1.5, -1, -0.5, 0, 0.5, 1, 1.5, 2$.

3.12 Repeat the previous problem for the values $\mu = 2$ and $\sigma^2 = 3$, Use values of x from -2 in to $+6$ increments of 1.

3.13 For the standard Normal variable z, find

(a) $\Pr(z > 1.64)$

(b) $\Pr(z > 0.5)$

(c) $\Pr(z > -1.5)$

(d) $\Pr(-2 < z < 1.5)$

(e) $\Pr(z = -0.75)$.

For (a) and (d), shade in the relevant areas on the graph you drew for Problem 3.11.

3.14 Find the values of z which cut off

(a) the top 10%

(b) the bottom 15%

(c) the middle 50%

of the standard Normal distribution.

3.15 If $x \sim N(10, 9)$, find

(a) $\Pr(x > 12)$

(b) $\Pr(x < 7)$

(c) $\Pr(8 < x < 15)$

(d) $\Pr(x = 10)$

3.16 IQ (the intelligence quotient) is Normally distributed with mean 100 and standard deviation 16.

(a) What proportion of the population has an IQ above 120?

(b) What proportion of the population has an IQ between 90 and 110?

(c) In the past, about 10% of the population went to university. Now the proportion is about 30%. What was the IQ of the 'marginal' student in the past? What is it now?

3.17 Ten adults are selected at random from the population and their IQ measured. (Assume a population mean of 100 and standard deviation of 16 as in Problem 3.16.)

(a) What is the probability distribution of the sample average IQ?

(b) What is the probability that the average IQ of the sample is over 110?

(c) If many such samples were taken, in what proportion would you expect the average IQ to be over 110?

(d) What is the probability that the average IQ lies within the range 90 to 110? How does this answer compare to the answer to part (b) of Problem 3.16? Account for the difference.

(e) What is the probability that a random sample of 10 university students has an average IQ greater than 110?

(f) The first adult sampled has an IQ of 150. What do you expect the average IQ of the sample to be?

3.18 The average income of a country is known to be £10 000 with standard deviation £2500. A sample of 40 individuals is taken and their average income calculated.

(a) What is the probability distribution of this sample mean?

(b) What is the probability of the sample mean being over £10 500?

(c) What is the probability of the sample mean being below £8000?

(d) If the sample size were 10, why could you not use the same methods to find the answers to (a)–(c)?

3.19 A coin is tossed 10 times. Write down the distribution of the number of heads,

(a) exactly, using the Binomial distribution,

(b) approximately, using the Normal distribution.

(c) Find the probability of four or more heads, using both methods. How accurate is the Normal method, with and without the continuity correction?

3.20 A machine producing electronic circuits has an average failure rate of 15% (they are difficult to make). The cost of making a batch of 500 circuits is £8400 and the good ones sell for £20 each. What is the probability of the firm making a loss on any one batch?

3.21 An experienced invoice clerk makes an error once in every 100 invoices, on average.

(a) What is the probability of finding a batch of 100 invoices without error?

(b) What is the probability of finding such a batch with more than two errors?

Calculate the answers using both the Binomial and Poisson distributions. If you try to solve the problem using the Normal method, how accurate is your answer?

3.22 A firm employing 100 workers has an average absenteeism rate of 4%. On a given day, what is the probability of (a) no workers, (b) one worker, (c) more than six workers being absent?

3.23 **(Computer project)** This problem demonstrates the Central Limit Theorem at work. In your spreadsheet, use the = *RAND*( ) function to generate a random sample of 25 observations (I suggest entering this function in cells A4:A28, for example). Copy these cells across 100 columns, to generate 100 samples. In row 29, calculate the mean of each sample. Now examine the distribution of these sample means. (Hint: you will find the *RAND*() function recalculates automatically every time you perform an operation in the spreadsheet. This makes it difficult to complete the analysis. The solution is to copy and then use 'Edit, Paste Special, Values' to create a copy of the values of the sample means. These will remain stable.)

(a) What distribution would you expect them to have?

(b) What is the parent distribution from which the samples are drawn?

(c) What are the parameters of the parent distribution and of the sample means?

(d) Do your results accord with what you would expect?

(e) Draw up a frequency table of the sample means and graph it. Does it look as you expected?

(f) Experiment with different sample sizes and with different parent distributions to see the effect that these have.

3.24 **(Project)** An extremely numerate newsagent (with a spreadsheet program, as you will need) is trying to work out how many copies of a newspaper he should order. The cost to him per copy is 40 pence, which he then sells at £1.20. Sales are distributed Normally with an average daily sale of 250 and variance 625. Unsold copies cannot be returned for credit or refund; he has to throw them away.

(a) What do you think the seller's objective should be?

(b) How many copies should he order?

(c) What happens to the *variance* of profit as he orders more copies?

(d) Calculate the probability of selling *more than X* copies. (Create an extra column in the spreadsheet for this.) What is the value of this probability at the optimum number of copies ordered?

(e) What would the price–cost ratio have to be to justify the seller ordering *X* copies?

(f) The wholesaler offers a sale or return deal, but the cost per copy is 45p. Should the seller take up this new offer?

(g) Are there other considerations which might influence the seller's decision?

Hints:

Set up your spreadsheet as follows:

col. A:	(cells A10:A160) 175, 176, . . . up to 325 in unit increments (to represent sales levels).
col. B:	(cells B10:B160) the probability of sales falling between 175 and 176, between 176 and 177, etc., up to 325–326. (Excel has the '= NORMDIST()' function to do this – see the help facility.)
col. C:	(cells C10:C160) total cost (= 0.40 × number ordered. Put the latter in cell F3 so you can reference it and change its value).
col. D:	(cells D10:D160) total revenue ('= MIN(sales, number ordered) × 1.20').
col. E:	profit (revenue–cost).
col. F:	Profit × probability (i.e. col. E × col. B).
cell F161:	the sum of F10:F160 (this is the expected profit).

Now vary the number ordered (cell F3) to find the maximum value in F161. You can also calculate the variance of profit fairly simply, using an extra column.

3.25 **(Project)** Using a weekend's football results from the Premier (or other) League, see if the number of goals per game can be adequately modelled by a Poisson process. First calculate the average number of goals per game for the whole league, and then derive the distribution of goals per game using the Poisson distribution. Do the actual numbers of goals per game follow this distribution? You might want to take several weeks' results to obtain more reliable results.

3.26 **(Project)** A report in *The Guardian* newspaper (20 June 2010, http://www.guardian.co.uk/education/2010/jun/20/internet-plagiarism-rising-in-schools) reports 'Half of university students also prepared to submit essays bought off the internet, according to research.' According to the article, a survey (sample size not specified) found that 45% of students were certain that a peer had cheated during an essay, report, test or exam in the past year. Evaluate this evidence from a statistical point of view. Find a formula to calculate the true proportion, given the reported proportion (adapting the formula in the example in this chapter) and use this to evaluate the claim in the article. Using your formula, what is the true proportion based on the reported figure? If the reported proportion is 100%, what is the estimated true figure? Does this seem right? If not, why not?

The article also reported that they (the students) 'would be prepared to pay more than £300 for a first class essay'. Comment also upon this finding. What precisely does it mean?

Answers to exercises

Exercise 3.1

(a) $0.6^4 = 0.1296$ or 12.96%.

(b) $0.6^2 \times 0.4^2 \times 4C2 = 0.3456$.

(c) $\Pr(r) = 0.6^r \times 0.4^{4-r}\, 4Cr$. The probabilities of $r = 0 \ldots 4$ are, respectively, 0.0256, 0.1536, 0.3456, 0.3456, 0.1296, which sum to one.

Exercise 3.2

(a)

r	$P(r)$	$r \times P(r)$	$r^2 \times P(r)$
0	0.0256	0	0
1	0.1536	0.1536	0.1536
2	0.3456	0.6912	1.3824
3	0.3456	1.0368	3.1104
4	0.1296	0.5184	2.0736
Totals	1	2.4	6.72

The mean $= 2.4/1 = 2.4$ and the variance $= 6.72/1 - 2.4^2 = 0.96$. Note that these are equal to nP and $nP(1 - P)$.

(b)

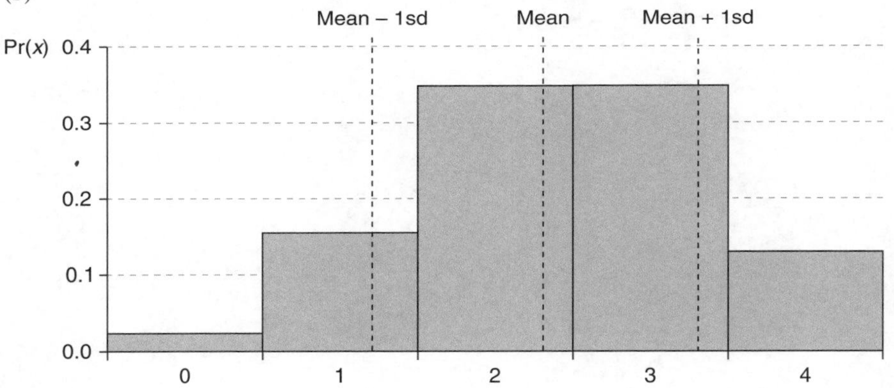

Exercise 3.3

(a) $z = (50 - 40)/\sqrt{36} = 1.67$ and from Table A2 the area beyond $z = 1.67$ is 4.75.

(b) $z = -0.83$ so area is 20.33%.

(c) This is symmetric around the mean, $z = \pm 0.67$, and the area within these two bounds is 49.72%.

Exercise 3.4

To obtain the IQR we need to go 0.67 standard deviations above and below the mean, giving $200 \pm 0.67 \times 16 = [189.28, 210.72]$.

Exercise 3.5

The IQR (in logs) is within $4.18 \pm 0.67 \times \sqrt{2.56} = [3.11, 5.25]$. Translated out of logs (using 10^x) yields [1288.2, 177 827.9]. Thus, we would expect 50% of the population to have incomes within this range of values.

Exercise 3.6

(a) $e \sim N(50, 64/25)$.

(b) The standard error gets smaller. It is $1/\sqrt{2}$ times its previous value.

(c) (i) $z = (51 - 50)/\sqrt{64/25} = 0.625$. Hence area in tail = 26.5. (ii) $z = -1.25$, hence area = 10.56. (iii) z values are −0.625 and +0.3125, giving tail areas of 26.5% and 37.8%, totalling 64.3%. The area between the limits is therefore 35.7%.

Exercise 3.7

(a) For this problem we have $p = 0.5$ and $n = 20$. Binomial method: $\Pr(r) = 0.5^r \times 0.5^{(20-r)} \times 20Cr$. This gives the probabilities of 15, 16, etc., heads as 0.0148, 0.0046, etc., which total 0.0207 or 2.1%. By the Normal approximation, $r \sim N(nP, Np(1 - P))$. $= N(10, 5)$ and $z = (14.5 - 10)/\sqrt{5} = 2.01$. The area in the tail is then 2.22%, not far off the correct value (a 10% error). Note that $np = 10 = n(1 - p)$.

(b) We have $P = 0.7, n = 6$. Binomial method: Pr(5 or 6 heads) $= 0.302 + 0.118 = 0.420$ or 42%. By the Normal, $r \sim N(4.2, 1.26), z = 0.267$ and the area is 39.36%, still reasonably close to the correct answer despite the fact that $n(1 - P) = 1.8$.

(c) By similar methods the answers are 11.8% (Binomial) and 12.3% (Normal).

Exercise 3.8

(a) (i) $\mu = 1$ in this case $(1/50 \times 50)$ so $\Pr(x = 0) = \mu^x e^{-\mu}/x! = 1^0 e^{-1}/0! = 0.368$. (ii) $\Pr(x = 1) = 1^1 e^{-1}/1! = 0.368$. (iii) $1^2 e^{-1}/2! = 0.184$. (iv) $1 - 0.368 = 0.632$.

(b) The average number of customers per 10 minutes is 2 (=10/5). Hence, $\Pr(x = 0) = 2^0 e^{-2}/0! = 0.135$.

4 Estimation and confidence intervals

Contents

Learning outcomes

By the end of this chapter you should be able to:

- recognise the importance of probability theory in drawing valid inferences (or deriving estimates) from a sample of data
- understand the criteria for constructing a good estimate
- construct estimates of parameters of interest from sample data, in a variety of different circumstances
- appreciate that there is uncertainty about the accuracy of any such estimate
- provide measures of the uncertainty associated with an estimate
- recognise the relationship between the size of a sample and the precision of an estimate derived from it.

three estimators suggested above. Taking the sample mean first, we have already learned (see equation (3.15)) that its expected value is μ, i.e.

$$E(\bar{x}) = \mu$$

which immediately shows that the sample mean is an unbiased estimator.

The second estimator (the smallest observation in the sample) can easily be shown to be biased, using the result derived above. Since the smallest sample observation must be less than the sample mean, its expected value must be less than μ. Denote the smallest observation by x_S, then

$$E(x_S) < \mu$$

so this estimator is biased downwards. It underestimates the population mean. The size of the bias is simply the difference between the expected value of the estimator and the value of the parameter, so the bias in this case is

$$\text{Bias} = E(x_S) - \mu \tag{4.1}$$

For the sample mean $\bar{x}$ the bias is obviously zero.

Turning to the third rule (the first sample observation), this can be shown to be another unbiased estimator. Choosing the first observation from the sample is equivalent to taking a random sample of size one from the population in the first place. Thus, the single observation may be considered as the sample mean from a random sample of size one. Since it is a sample mean, it is unbiased, as demonstrated earlier.

Precision

Two of the estimators above were found to be unbiased, and, in fact, there are many unbiased estimators (the sample median is another, for example). Some way of choosing between the set of all unbiased estimators is therefore required, which is where the criterion of precision helps. Unlike bias, precision is a relative concept, comparing one estimator to another. Given two estimators A and B, A is more precise than B if the estimates A yields (from all possible samples) are less spread out than those of estimator B. A precise estimator will tend to give similar estimates for all possible samples.

Consider the two unbiased estimators found above: how do they compare on the criteria of precision? It turns out that the sample mean is the more precise of the two, and it is not difficult to understand why. Taking just a single sample observation means that it is quite likely to be unrepresentative of the population as a whole, and thus leads to a poor estimate of the population mean. The single observation might be an extreme value from the population, purely by chance. The sample mean, on the other hand, is based on all the sample observations, and it is unlikely that all of them are unrepresentative of the population. The sample mean is therefore a good estimator of the population mean, being more precise than the single observation estimator.

Just as bias was related to the expected value of the estimator, so precision can be defined in terms of the variance. One estimator is more precise than another if it has a smaller variance. Recall that the probability distribution of the sample mean is

$$\bar{x} \sim N(\mu, \sigma^2/n) \tag{4.2}$$

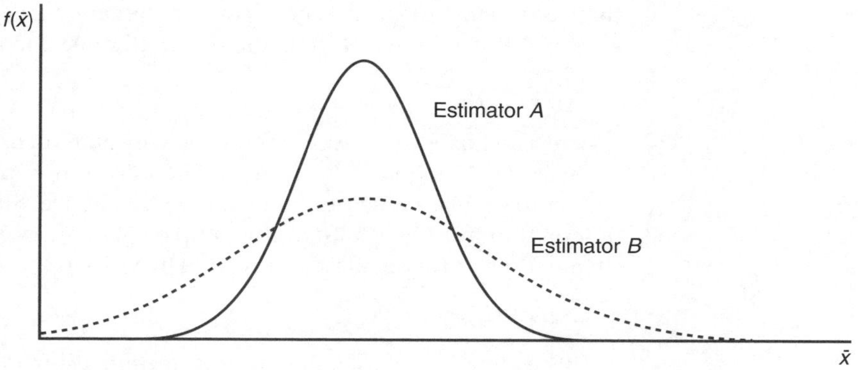

Figure 4.1
The sampling distributions of two estimators

in large samples, so the variance of the sample mean is

$$V(\bar{x}) = \sigma^2/n$$

As the sample size n gets larger the variance of the sample mean becomes smaller, so the estimator becomes more precise. For this reason, large samples give better estimates than small samples, and so the sample mean is a better estimator than taking just one observation from the sample. The two estimators can be compared in a diagram (see Figure 4.1) which draws the probability distributions of the two estimators.

It is easily seen that Estimator A (the larger sample) yields estimates which are *on average* closer to the population mean.

Mean squared error

The two measures, bias and precision, can be combined in the **mean squared error** (MSE) of the estimate. This is defined, for an estimator θ, as

$$\text{MSE} = \text{E}((\hat{\theta} - \theta)^2)$$

where $\hat{\theta}$ is an estimate of θ. The larger the value of the MSE, the further the estimate is likely to be from the true value and hence the poorer the estimator. It can be shown the MSE is equal to the variance of the estimator plus the square of the bias:

$$\text{MSE} = \text{variance} + \text{bias}^2$$

The MSE therefore captures both variance and bias and estimators can be compared using this new concept. The estimator with the smaller MSE is considered superior.

Most of the estimators covered in this text turn out to be unbiased, so the MSE is then simply equal to the variance. A related concept using the variance is that of **efficiency**. The efficiency of one unbiased estimator, relative to another, is given by the ratio of their sampling variances[3]. Thus, the efficiency of the first observation estimator, relative to the sample mean, is given by

$$\text{Efficiency} = \frac{var(\bar{x})}{var(x_1)} = \frac{\sigma^2/n}{\sigma^2} = \frac{1}{n} \tag{4.3}$$

[3]For biased estimators we can take the ratio of their MSEs.

Thus the efficiency is determined by the relative sample sizes in this case. Other things being equal, a more efficient estimator is to be preferred.

Similarly, the variance of the median can be shown to be (for a Normal distribution) $\pi/2 \times \sigma^2/n$ in large samples. The efficiency of the median is therefore $2/\pi \approx 64\%$ (compared to using the sample mean) and so on this basis the sample mean is a preferred estimator.

The trade-off between bias and precision: the Bill Gates effect

It should be noted that just because an estimator is biased does not necessarily mean that it is imprecise. Sometimes there is a trade-off between an unbiased, but imprecise, estimator and a biased, but precise, one. Figure 4.2 illustrates this.

Although estimator A is biased (it is not centred around μ), it will nearly always yield an estimate which is fairly close to the true value; even though the estimate is expected to be wrong, it is not likely to be far wrong. Estimator B, although unbiased, can give estimates which are far away from the true value, so that A might be the preferred estimator.

As an example of this, suppose we are trying to estimate the average wealth of the US population. Consider the following two estimators:

(1) Use the mean wealth of a random sample of Americans.
(2) Use the mean wealth of a random sample of Americans but, if Bill Gates is in the sample, omit him from the calculation.

Bill Gates, the former Chairman of Microsoft, is one of the world's richest men. He is a dollar billionaire (about \$50bn or more according to recent reports – it varies with the stock market). His presence in a sample of, say, 30 observations would swamp the sample and give a highly misleading result. Assuming Bill Gates has \$50bn and the others each have \$200 000 of wealth, the average wealth would be estimated at about \$1.6bn, which is surely wrong.

The first rule could therefore give us a wildly incorrect answer, although the rule is unbiased. The second rule is clearly biased but does rule out the possibility of such an unlucky sample. We can work out the approximate bias. It is the difference between the average wealth of all Americans and the average wealth of all

Figure 4.2
The trade-off between bias and precision

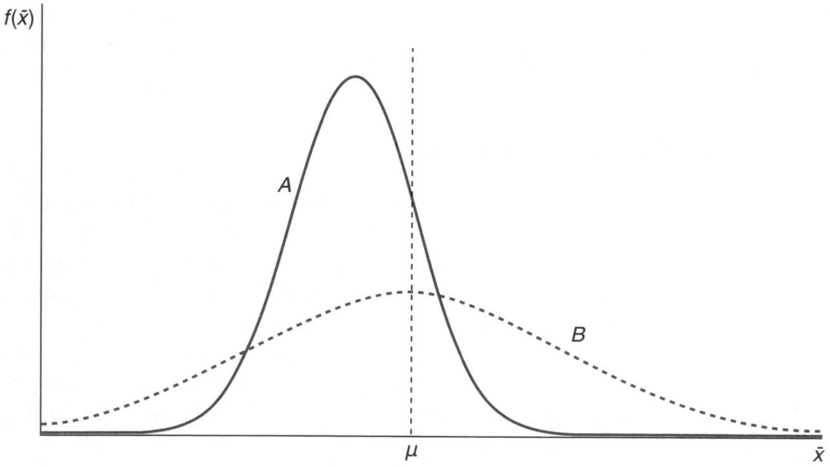

Americans except Bill Gates. If the true average of all 250 million Americans is \$200 000, then total wealth is \$50 000bn. Subtracting Bill's \$50bn leaves \$49 950bn shared amongst the rest, giving \$199 800 each, a difference of 0.1%. This is what we would expect the bias to be.

It might seem worthwhile, therefore, to accept this degree of bias in order to improve the precision of the estimate. Furthermore, if we did use the biased rule, we could adjust the sample mean upwards by 0.1% or so to compensate (if only approximately).

Of course, this point applies to any exceptionally rich person, not just Bill Gates. It points to the need to ensure that the rich are not over- (nor under-) represented in the sample. Chapter 9 on sampling methods investigates this point in more detail. In the rest of this text only unbiased estimators are considered, the most important being the sample mean.

Estimation with large samples

For the type of problem encountered in this chapter, the method of estimation differs according to the size of the sample. 'Large' samples, by which is meant sample sizes of 25 or more, are dealt with first, using the Normal distribution. Small samples are considered in a later section, where the *t* distribution is used instead of the Normal. The differences are relatively minor in practical terms and there is a close theoretical relationship between the *t* and Normal distributions.

With large samples there are three types of estimation problem we will consider.

(1) The estimation of a mean from a sample of data.
(2) The estimation of a proportion on the basis of sample evidence. This would consider a problem such as estimating the proportion of the population intending to buy an iPhone, based on a sample of individuals. Each person in the sample would simply indicate whether they have bought, or intend to buy, an iPhone. The principles of estimation are the same as in the first case but the formulae used for calculation are slightly different.
(3) The estimation of the difference of two means (or proportions), for example, a problem such as estimating the difference between men's and women's expenditure on clothes. Once again, the principles are the same, the formulae different.

Estimating a mean

To demonstrate the principles and practice of estimating the population mean, we shall take the example of estimating the average wealth of the UK population, the full data for which were given in Chapter 1. Suppose that we did not have this information but were required to estimate the average wealth from a sample of data. In particular, let us suppose that the sample size is $n = 100$, the sample mean is $\bar{x} = 180$ (in £000) and the sample variance is $s^2 = 75\,000$. Evidently, this sample has got fairly close to the true values (see Chapter 1), but we could not know that from the sample alone. What can we infer about the population mean μ from the sample data alone?

For the point estimate of μ the sample mean is a good candidate since it is unbiased, and it is generally more precise than other sample statistics such as the median. The point estimate of μ is simply £180 000, therefore.

The point estimate does not give an idea of the uncertainty associated with the estimate. We are not *absolutely* sure that the mean is £180 000 (in fact, it is not – it is £186 875). The interval estimate in contrast gives some idea of the uncertainty. It is centred on the sample mean, but gives a range of values to express the uncertainty.

To obtain the interval estimate we first require the probability distribution of $\bar{x}$, first established in Chapter 3 (equation (3.17)):

$$\bar{x} \sim N(\mu, \sigma^2/n) \tag{4.4}$$

From this, it was calculated that there is a 95% probability of the sample mean lying within 1.96 standard errors of μ[4], i.e.

$$\Pr(\mu - 1.96\sqrt{\sigma^2/n} \leq \bar{x} \leq \mu + 1.96\sqrt{\sigma^2/n}) = 0.95$$

We can manipulate each of the inequalities within the brackets to make μ the subject of the expression:

$$\mu - 1.96\sqrt{\sigma^2/n} \leq \bar{x} \quad \text{implies} \quad \mu \leq \bar{x} + 1.96\sqrt{\sigma^2/n}$$

Similarly

$$\bar{x} \leq \mu + 1.96\sqrt{\sigma^2/n} \quad \text{implies} \quad \bar{x} - 1.96\sqrt{\sigma^2/n} \leq \mu$$

Combining these two new expressions, we obtain

$$[\bar{x} - 1.96\sqrt{\sigma^2/n} \leq \mu \leq \bar{x} + 1.96\sqrt{\sigma^2/n}] \tag{4.5}$$

We have transformed the probability interval. Instead of saying $\bar{x}$ lies within 1.96 standard errors of μ, we now say μ lies within 1.96 standard errors of $\bar{x}$. Figure 4.3 illustrates this manipulation. Figure 4.3(a) shows μ at the centre of a probability interval for $\bar{x}$. Figure 4.3(b) shows a sample mean $\bar{x}$ at the centre of an interval relating to the possible positions of μ.

The interval shown in equation (4.5) is called the **95% confidence interval**, and this is the interval estimate for μ. In this formula the value of σ^2 is unknown, but in large ($n \geq 25$) samples it can be replaced by s^2 from the sample. s^2 is here used as an estimate of σ^2 which is unbiased and sufficiently precise in large ($n \geq 25$ or so) samples. The 95% confidence interval is therefore

$$\begin{aligned} &[\bar{x} - 1.96\sqrt{s^2/n} \leq \mu \leq \bar{x} + 1.96\sqrt{s^2/n}] \qquad (4.6) \\ &= [180 - 1.96\sqrt{75\,000/100}, 180 + 1.96\sqrt{75\,000/100}] \\ &= [126.3, 233.7] \end{aligned}$$

Thus the 95% confidence interval estimate for the true average level of wealth ranges between £126 300 and £233 700. Note that £180 000 lies exactly at the centre of the interval[5] (because of the symmetry of the Normal distribution).

[4]See equation (3.24) in Chapter 3 to remind yourself of this. Remember that ±1.96 is the *z* score which cuts off 2.5% in each tail of the Normal distribution.

[5]The two values are the lower and upper limits of the interval, separated by a comma. This is the standard way of writing a confidence interval.

Figure 4.3(a)
The 95% probability interval for $\bar{x}$ around the population mean μ

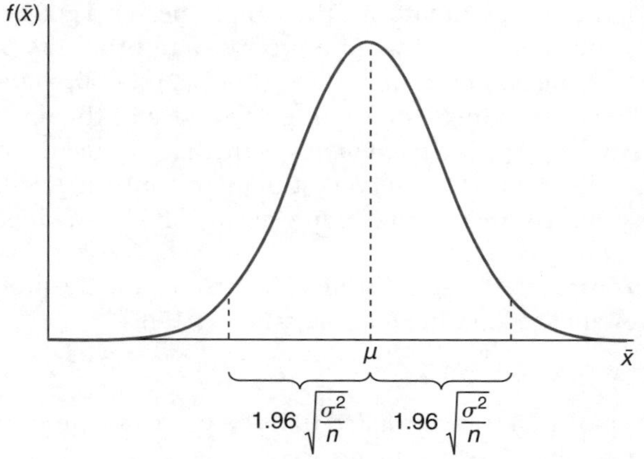

Figure 4.3(b)
The 95% confidence interval for μ around the sample mean $\bar{x}$

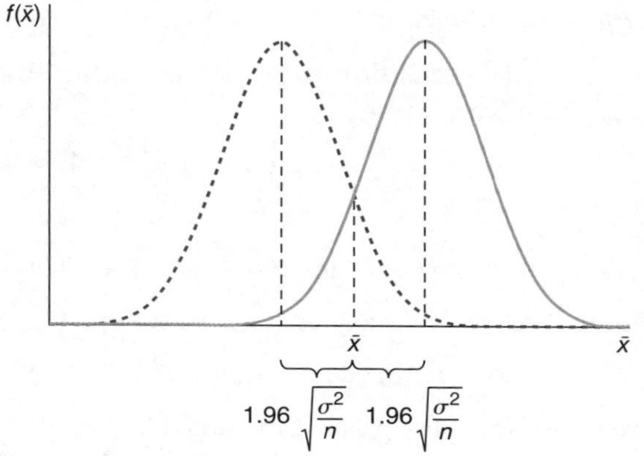

A more compact way of writing the confidence interval for μ, instead of equation (4.6), is

$$\bar{x} \pm 1.96\sqrt{s^2/n} \tag{4.6b}$$

which highlights the interval lying between the sample mean plus and minus 1.96 standard errors. This is easy to remember and can be used for different types of problem, as we show below.

By examining equation (4.6) or equation (4.6b), one can see that the confidence interval is wider

(1) the smaller the sample size,
(2) the greater the standard deviation of the sample.

The greater uncertainty which is associated with smaller sample sizes is manifested in a wider confidence interval estimate of the population mean. This occurs because a smaller sample has more chance of being unrepresentative (just because of an unlucky sample).

Greater variation in the sample data also leads to greater uncertainty about the population mean and a wider confidence interval. Greater sample variation suggests greater variation in the population so, again, a given sample could include observations which are a long way off the mean. Note that in this example there is great variation of wealth in the population and hence in the sample also. This means that a sample of 100 is not very informative (the confidence interval is quite wide). We would need a substantially larger sample to obtain a more precise estimate.

Note that the width of the confidence interval does *not* depend upon the population size – a sample of 100 observations reveals as much about a population of 10 000 as it does about a population of 10 000 000. In fact, this is not *quite* correct: if the sample were a large proportion of the population (of say 200 in this case), then the confidence interval should be narrower. However, in most cases this does not apply, and it is the sample size that really matters. This point will be discussed in more detail in Chapter 9 on sampling methods. This is a result that often surprises people, who generally believe that a larger sample is required if the population is larger.

Worked example 4.1

A sample of 50 school students found that they spent 45 minutes doing homework each evening, with a standard deviation of 15 minutes. Estimate the average time spent on homework by all students.

The sample data are $\bar{x} = 45$, $s = 15$ and $n = 50$. If we can assume the sample is representative, we may use $\bar{x}$ as an unbiased estimate of μ, the population mean. The point estimate is therefore 45 minutes.

The 95% confidence interval is given by equation (4.6b):

$$\begin{aligned} &\bar{x} \pm 1.96\sqrt{s^2/n} \\ &= 45 \pm 1.96\sqrt{15^2/50} \\ &= 45 \pm 4.2 \text{ or } [40.8, 49.2] \end{aligned}$$

The 95% confidence interval lies between 40.8 and 49.2 minutes. This might then be reasonably expressed as 'between 41 and 49 minutes'.

Exercise 4.1

?

(a) A sample of 100 is drawn from a population. The sample mean is 25 and the sample standard deviation is 50. Calculate the point and 95% confidence interval estimates for the population mean.

(b) If the sample size were 64, how would this alter the point and interval estimates?

Exercise 4.2

?

A sample of size 40 is drawn with sample mean 50 and standard deviation 30. Is it likely that the true population mean is 60?

Precisely what is a confidence interval?

There is often confusion over what a 95% confidence interval actually means. This is not really surprising since the obvious interpretation turns out to be wrong. It does *not* mean that there is a 95% chance that the true mean lies within the interval. We cannot make such a probability statement because of our definition of probability (based on the frequentist view of a probability). That view states that one can make a probability statement about a random variable (such as $\bar{x}$) but not about a parameter (such as μ). μ either lies within the interval or it does not – it cannot lie 95% within it. Unfortunately, we just do not know what the truth is.

It is for this reason that we use the term 'confidence interval' rather than 'probability interval'. Unfortunately, words are not as precise as numbers or algebra, and so most people fail to recognise the distinction. A precise explanation of the 95% confidence interval runs as follows. If we took many samples (all the same size) from a population with mean μ and calculated a confidence interval from each sample, we would find that μ lies within 95% of the calculated intervals. Of course, in practice we do not take many samples, usually just one. We do not know (and cannot know) if our one sample is one of the 95% or one of the 5% that miss the mean.

Figure 4.4 illustrates the point. It shows 95% confidence intervals calculated from 20 samples drawn from a population with a mean of 5. As expected, we see that 19 of these intervals contain the true mean, while the interval calculated from sample 9 does not contain the true value. This is the expected result, but is not guaranteed. You might obtain all 20 intervals containing the true mean, or fewer than 19. In the long run (with lots of estimates), we would expect 95% of the calculated intervals to contain the true mean.

A second question is, why use a probability (and hence a **confidence level**) of 95%? In fact, one can choose any confidence level, and thus confidence interval. The 90% confidence interval can be obtained by finding the z score which cuts off the outer 10% of the Normal distribution (5% in each tail). From Table A2 (see page 450) this is $z = 1.64$, so the 90% confidence interval is given by the sample mean plus and minus 1.64 standard errors:

$$\begin{aligned} & \bar{x} \pm 1.64\sqrt{s^2/n} \qquad (4.7) \\ &= 180 \pm 1.64\sqrt{75\,000/100} \\ &= 180 \pm 44.9 \text{ or } [135.1, 224.9] \end{aligned}$$

Figure 4.4
Confidence intervals calculated from 20 samples

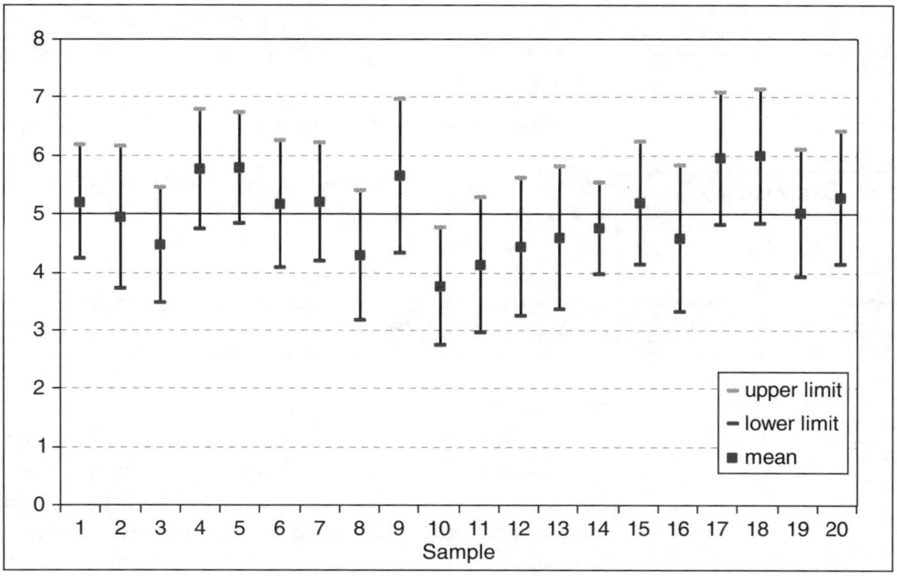

Notice that this is narrower than the 95% confidence level. The greater the degree of confidence desired, the wider the interval has to be. Any confidence level may be chosen, and by careful choice of this level the confidence interval can be made as wide or as narrow as wished. This would seem to undermine the purpose of calculating the confidence interval, which is to obtain some idea of the uncertainty attached to the estimate. This is not the case, however, because the reader of the results can interpret them appropriately, as long as the confidence level is made clear. To simplify matters, the 95% and 99% confidence levels are the most commonly used and serve as conventions. Beware of the researcher who calculates the 76% confidence interval – this may have been chosen in order to obtain the desired answer rather than in the spirit of scientific enquiry. The general formula for the $(100 - \alpha)\%$ confidence interval is

$$\bar{x} \pm z_{\alpha}\sqrt{s^2/n} \tag{4.8}$$

where z_{α} is the z score which cuts off the extreme $\alpha\%$ (in both tails, hence $\alpha/2$ in each tail) of the Normal distribution.

Exercise 4.3 will test your understanding of what a confidence interval really is.

Exercise 4.3

?

A study finds that the 95% confidence interval estimate for the mean of a population ranges from 0.1 to 0.4. Which of the following statements are true and which false?

(a) The probability that the true mean is greater than 0 is at least 95%.

(b) The probability that the true mean is 0 is less than 5%.

(c) The hypothesis that the true mean is 0 is unlikely to be correct.

(d) There is a 95% probability that the true mean lies between 0.1 and 0.4.

(e) We can be 95% confident that the true mean lies between 0.1 and 0.4.

(f) If we were to repeat the experiment many times, then 95% of the time the true mean lies between 0.1 and 0.4.

Estimating a proportion

It is often the case that we wish to estimate the **proportion** of the population that has a particular characteristic (e.g. is unemployed), rather than wanting an average. Given what we have already learned, this is fairly straightforward and is based on similar principles. Suppose that, following Chapter 1, we wish to estimate the proportion of educated people who are unemployed. We have a random sample of 200 individuals, of whom 15 are unemployed. What can we infer?

The sample data are:

$$n = 200, \text{ and}$$
$$p = 0.075\ (= 15/200)$$

where p is the (sample) proportion unemployed, 7.5% in this case. We denote the population proportion by the Greek letter π and it is this that we are trying to estimate using data from the sample.

The key to solving this problem is recognising p as a random variable just like the sample mean. This is because its value depends upon the sample drawn and will vary from sample to sample. Once the probability distribution of this random variable is established, the problem is quite easy to solve, using the same methods as were used for the mean. The sampling distribution of p is[6]

$$p \sim N\left(\pi, \frac{\pi(1-\pi)}{n}\right) \tag{4.9}$$

This tells us that the sample proportion is centred on the true value but will vary around it, varying from sample to sample. This variation is expressed by the variance of p, whose formula is $\pi(1-\pi)/n$. Having derived the probability distribution of p, we can use the same methods of estimation as for the sample mean. Since the expected value of p is π, the sample proportion is an unbiased estimate of the population parameter. The point estimate of π is simply p, therefore. Thus, it is estimated that 7.5% of all educated people are unemployed.

Given the sampling distribution for p in equation (4.9), the formula for the 95% confidence interval for π can immediately be written down as:

$$p \pm 1.96\sqrt{\frac{\pi(1-\pi)}{n}} \tag{4.10}$$

or alternatively

$$\left[p - 1.96\sqrt{\frac{\pi(1-\pi)}{n}}, p + 1.96\sqrt{\frac{\pi(1-\pi)}{n}}\right]$$

As usual, the 95% confidence interval limits are given by the point estimate plus and minus 1.96 standard errors.

Since the value of π is unknown, the confidence interval cannot yet be calculated, so the sample value of 0.075 has to be used instead of the unknown π. Like the substitution of s^2 for σ^2 in the case of the sample mean above, this is acceptable in large samples. Thus, the 95% confidence interval becomes

$$\begin{aligned} &0.075 \pm 1.96\sqrt{\frac{0.075(1-0.075)}{200}} \\ &= 0.075 \pm 0.037 \\ &= [0.038, 0.112] \end{aligned} \tag{4.11}$$

We say that the 95% confidence interval estimate for the true proportion of unemployed people lies between 3.8% and 11.2%.

It can be seen that these two cases apply a common method. The 95% confidence interval is given by the point estimate plus or minus 1.96 standard errors. For a different confidence level, 1.96 would be replaced by the appropriate value from the standard Normal distribution.

With this knowledge, two further cases can be swiftly dealt with.

Worked example 4.2 Music down the phone

Do you get angry when you try to phone an organisation and you get an automated reply followed by music while you hang on? Well, you are not alone. Mintel (a consumer survey company) asked 1946 adults what they thought of

[6] See the Appendix to this chapter (page 193) for the derivation of this formula.

music played to them while they were trying to get through on the phone; 36% reported feeling angered by the music played to them and more than one in four were annoyed by the automated voice response.

With these data we can calculate a confidence interval for the true proportion of people who dislike the music. First, we assume that the sample is a truly random one. This is probably not strictly true, so our calculated confidence interval will only be an approximate one. With $p = 0.36$ and $n = 1946$ we obtain the following 95% interval:

$$p \pm 1.96 \times \sqrt{\frac{p(1-p)}{n}} = 0.36 \pm 1.96 \times \sqrt{\frac{0.36(1-0.36)}{1946}}$$
$$= 0.36 \pm 0.021 = [0.339, 0.381]$$

Mintel further estimated that 2800 million calls were made by customers to call centres per year, so we can be (approximately) 95% confident that between 949 million and 1067 million of those calls have an unhappy customer on the line.

Source: *The Times*, 10 July 2000.

Estimating the difference between two means

We now move on to estimating differences. In this case we have two samples and want to know whether there is a difference between their respective populations. One sample might be of men, the other of women, or we could be comparing two different countries, etc. A point estimate of the difference is easy to obtain, but once again there is some uncertainty around this figure, because it is based on samples. Hence, we measure that uncertainty via a confidence interval. All we require are the appropriate formulae. Consider the following example.

Thirty-five pupils from school 1 scored an average mark of 70% in an exam, with a standard deviation of 12%; 60 pupils from school 2 scored an average of 62% with standard deviation 18%. Estimate the true difference between the two schools in the average mark obtained.

This is a more complicated problem than those previously treated since it involves two samples rather than one. An estimate has to be found for $\mu_1 - \mu_2$ (the true difference in the mean marks of the schools), in the form of both point and interval estimates. The pupils taking the exams may be thought of as samples of all pupils in the schools who could potentially take the exams.

Notice that this is a problem about sample means, not proportions, even though the question deals in percentages. The point is that each observation in the sample (i.e. each student's mark) can take a value between 0 and 100, and one can calculate the standard deviation of the marks. For this to be a problem of sample proportions, the mark for each pupil would each have to be of the pass/fail type, so that one could only calculate the proportion who passed.

One might think that the way to approach this problem is to derive one confidence interval for each sample (along the lines set out above), and then to somehow combine them; for example, the degree of overlap of the two confidence intervals could be assessed. This is not the best approach, however. It is sometimes a good strategy, when faced with an unfamiliar problem to solve, to translate it into a more familiar problem and then solve it using known methods. This procedure will be followed here. The essential point is to keep in mind the concept of a random variable and its probability distribution.

Problems involving a single random variable have already been dealt with above. The current problem deals with two samples and therefore there are two random variables to consider, i.e. the two sample means $\bar{x}_1$ and $\bar{x}_2$. Since the aim is to estimate $\mu_1 - \mu_2$, an obvious candidate for an estimator is the difference between the two sample means, $\bar{x}_1 - \bar{x}_2$. We can think of this as a single random variable (even though two means are involved) and use the methods we have already learned. We therefore need to establish the sampling distribution of $\bar{x}_1 - \bar{x}_2$. This is derived in the Appendix to this chapter (see page 193) and results in equation (4.12):

$$\bar{x}_1 - \bar{x}_2 \sim N\left(\mu_1 - \mu_2, \frac{\sigma_1^2}{n_1} + \frac{\sigma_2^2}{n_2}\right) \tag{4.12}$$

This equation states that the difference in sample means will be centred on the difference in the two population means, with some variation around this as measured by the variance. One assumption behind the derivation of (4.12) is that the two samples are independently drawn. This is likely in this example; it is difficult to see how the samples from the two schools could be connected. However, one must always bear this possibility in mind when comparing samples. For example, if one were comparing men's and women's heights, it would be dangerous to take samples of men and their wives as they are unlikely to be independent. People tend to marry partners of a similar height to themselves, so this might bias the results.

The distribution of $\bar{x}_1 - \bar{x}_2$ is illustrated in Figure 4.5. Equation (4.12) shows that $\bar{x}_1 - \bar{x}_2$ is an unbiased estimator of $\mu_1 - \mu_2$. The difference between the sample means will therefore be used as the point estimate of $\mu_1 - \mu_2$. Thus, the point estimate of the true difference between the schools is

$$\bar{x}_1 - \bar{x}_2 = 70 - 62 = 8\%$$

The 95% confidence interval estimate is derived in the same manner as before, making use of the standard error of the random variable. The formula is[7]

$$(\bar{x}_1 - \bar{x}_2) \pm 1.96\sqrt{\frac{s_1^2}{n_1} + \frac{s_2^2}{n_2}} \tag{4.13}$$

Since the values of σ^2 are unknown, they have been replaced in equation (4.13) by their sample values. As in the single sample case, this is acceptable in large samples. The 95% confidence interval for $\mu_1 - \mu_2$ is therefore

$$(70 - 62) \pm 1.96\sqrt{\frac{12^2}{35} + \frac{18^2}{60}}$$
$$= [1.95, 14.05]$$

The estimate is that school 1's average mark is between 1.95 and 14.05 percentage points above that of school 2. Notice that the confidence interval does not include the value zero, which would imply possible equality of the two schools' marks. Equality of the two schools can thus be ruled out with 95% confidence.

[7]The term under the square root sign is the standard error for $\bar{x}_1 - \bar{x}_2$.

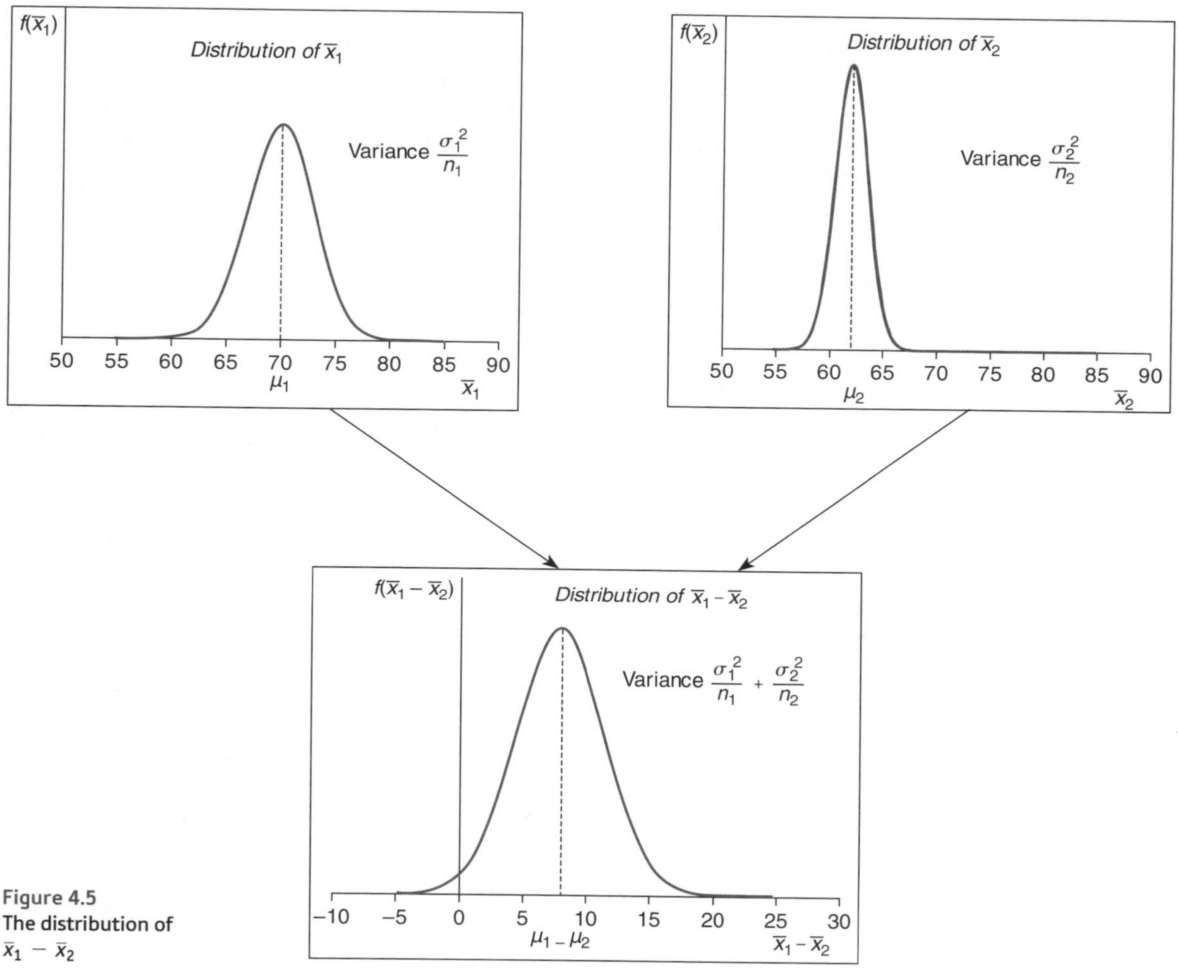

Figure 4.5
The distribution of $\bar{x}_1 - \bar{x}_2$

Worked example 4.3

A survey of holidaymakers found that on average women spent 3 hours per day sunbathing, men spent 2 hours. The sample sizes were 36 in each case and the standard deviations were 1.1 and 1.2 hours, respectively. Estimate the true difference between men and women in sunbathing habits. Use the 99% confidence level.

The point estimate is simply one hour, the difference of sample means. For the confidence interval we have:

$$(\bar{x}_1 - \bar{x}_2) \pm 2.57\sqrt{\frac{s_1^2}{n_1} + \frac{s_2^2}{n_2}}$$
$$= (3 - 2) \pm 2.57\sqrt{\frac{1.1^2}{36} + \frac{1.2^2}{36}}$$
$$= 1 \pm 0.70 = [0.30, 1.70]$$

→

This evidence suggests women do spend more time sunbathing than men (zero is not in the confidence interval). Note that we might worry the samples might not be independent here – it could represent 36 couples. If so, the evidence is likely to underestimate the true difference, if anything, as couples are likely to spend time sunbathing together.

Estimating the difference between two proportions

We move again from means to proportions. We use a simple example to illustrate the analysis of this type of problem. Suppose we wish to compare the market share of Apple Mac computers in the United States and the United Kingdom. A survey of 1000 American computer users shows that 160 use Macs while a similar survey of 500 Britons shows 65 using Macs. What is our estimate of the true difference between the two countries?

Here the aim is to estimate $\pi_1 - \pi_2$, the difference between the two population proportions, so the probability distribution of $p_1 - p_2$ is needed, the difference of the sample proportions. The derivation of this follows similar lines to those set out above for the difference of two sample means, so is not repeated. The probability distribution is

$$p_1 - p_2 \sim N\left(\pi_1 - \pi_2, \frac{\pi_1(1 - \pi_1)}{n_1} + \frac{\pi_2(1 - \pi_2)}{n_2}\right) \tag{4.14}$$

Again, the two samples must be independently drawn for this to be correct.

Since the difference between the sample proportions is an unbiased estimate of the true difference, this will be used for the point estimate. The point estimate is therefore

$$\begin{aligned} p_1 - p_2 &= 160/1000 - 65/500 \\ &= 0.16 - 0.13 = 0.03 \text{ or } 3\%. \end{aligned}$$

Note that this means a three percentage point difference in market share, not that the US market is 3% bigger. The 95% confidence interval is given by

$$p_1 - p_2 \pm 1.96\sqrt{\frac{\pi_1(1 - \pi_1)}{n_1} + \frac{\pi_2(1 - \pi_2)}{n_2}} \tag{4.15}$$

π_1 and π_2 are unknown so have to be replaced by p_1 and p_2 for purposes of calculation, so the interval becomes

$$0.16 - 0.13 \pm 1.96\sqrt{\frac{0.16 \times 0.84}{1000} + \frac{0.13 \times 0.87}{500}} \tag{4.16}$$

$$\begin{aligned} &= 0.03 \pm 0.0372 \\ &= [-0.0072, 0.0672] \end{aligned}$$

The 95% confidence interval indicates that the US market share is between −0.7 and 6.7 percentage points larger than in the United Kingdom. Note that this interval includes the value of zero, so we cannot be 95% confident the US share is bigger.

These data are for the purpose of illustrating the methods and are not real. However, they are closely based on figures from StatCounter (http://gs.statcounter.com/)

and are collected automatically based on visitor statistics to 'more than three million web sites'. The market shares for December 2011 are reported as 16.5% and 13.3%. StatCounter does not give sample sizes, so what are we to make of these numbers?

The 'three million' might suggest a huge sample size and hence a much smaller confidence interval. (If there were one million in each country, then the width of the confidence interval would be ± 0.0005.) However, there are likely to be many multiple visits by the same user, so the number of users (as opposed to visits) could be much smaller, we simply do not know. Furthermore, we should think whether there might be any kind of bias to the figures, for example if more US websites were dedicated to Apple customers.

Exercise 4.4

?

(a) Seven people out of a sample of 50 are left-handed. Estimate the true proportion of left-handed people in the population, finding both point and interval estimates.

(b) Repeat part (a) but find the 90% confidence interval. How does the 90% interval compare with the 95% interval?

(c) Calculate the 99% interval and compare to the others.

Exercise 4.5

?

Given the following data from two samples, estimate the true difference between the means. Use the 95% confidence level.

$$\bar{x}_1 = 25 \qquad \bar{x}_2 = 30$$
$$s_1 = 18 \qquad s_2 = 25$$
$$n_1 = 36 \qquad n_2 = 49$$

Exercise 4.6

?

A survey of 50 16-year-old girls revealed that 40% had a boyfriend. A survey of 100 16-year-old boys revealed 20% with a girlfriend. Estimate the true difference in proportions between the sexes.

Estimation with small samples: the t distribution

So far only large samples (defined as sample sizes in excess of 25) have been dealt with, which means that (by the Central Limit Theorem) the sampling distribution of $\bar{x}$ follows a Normal distribution, whatever the distribution of the parent population. Remember, from the two theorems of Chapter 3, that

- if the population follows a Normal distribution, $\bar{x}$ is also Normally distributed, and
- if the population is not Normally distributed, $\bar{x}$ is approximately Normally distributed in large samples ($n \geq 25$).

In both cases, confidence intervals can be constructed based on the fact that

$$\frac{\bar{x} - \mu}{\sqrt{\sigma^2/n}} \sim N(0, 1) \tag{4.17}$$

and so the standard Normal distribution is used to find the values which cut off the extreme 5% of the distribution ($z = \pm 1.96$). In practical examples, we had to replace σ by its estimate, s. Thus the confidence interval was based on the fact that

$$\frac{\bar{x} - \mu}{\sqrt{s^2/n}} \sim N(0, 1) \tag{4.18}$$

in large samples. For small sample sizes, equation (4.18) is no longer true. Instead, the relevant distribution is the t distribution, and we have[8]

$$\frac{\bar{x} - \mu}{\sqrt{s^2/n}} \sim t_{n-1} \tag{4.19}$$

The random variable defined in equation (4.19) has a t distribution with $n - 1$ degrees of freedom. As the sample size gets larger, the t distribution approaches the standard Normal, so the latter can be used for large samples.

The t distribution was derived by W.S. Gossett in 1908 while conducting tests on the average strength of Guinness beer (who says statistics has no impact on the real world?). He published his work under the pseudonym 'Student', since the company did not allow its employees to publish under their own names, so the distribution is sometimes also known as the Student distribution.

The t distribution is in many ways similar to the standard Normal, insofar as it is

- unimodal
- symmetric
- centred on zero
- bell-shaped
- extends from minus infinity to plus infinity.

The differences are that it is more spread out (has a larger variance) than the standard Normal distribution, and has only one parameter rather than two: the **degrees of freedom**, denoted by the Greek letter ν (pronounced 'nu'[9]). In problems involving the estimation of a sample mean, the degrees of freedom are given by the sample size minus one, i.e. $\nu = n - 1$.

The t distribution is drawn in Figure 4.6 for various values of the parameter ν. Note that the fewer the degrees of freedom (smaller sample size), the more dispersed is the distribution.

To summarise the argument so far, when

- the sample size is small, *and*
- the sample variance is used to estimate the population variance,

then the t distribution should be used for constructing confidence intervals, not the standard Normal. This results in a slightly wider interval than would be obtained using the standard Normal distribution, which reflects the slightly greater uncertainty involved when s^2 is used as an estimate of σ^2 when the sample size is small.

---

[8]We also require the assumption that the parent population is Normally distributed for (4.19) to be true.

[9]Once again, the Greeks pronounce this differently, as 'ni'. They also pronounce π 'pee' rather than 'pie' as in English. This makes statistics lectures in English hard for Greeks to understand.

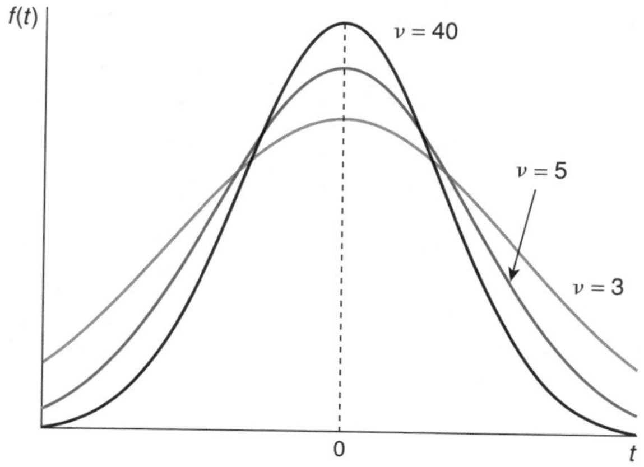

Figure 4.6
The t distribution drawn for different degrees of freedom

Apart from this, the methods are as before and are illustrated by the examples below. We look first at estimating a single mean, and then at estimating the difference of two means. The *t* distribution cannot be used for small sample proportions (explained below), so these cases are not considered.

Estimating a mean

The following example would seem to be appropriate. A sample of 15 bottles of beer showed an average specific gravity (a measure of alcohol content) of 1035.6, with standard deviation 2.7. Estimate the true specific gravity of the brew.

The sample information may be summarised as

$$\bar{x} = 1035.6$$
$$s = 2.7$$
$$n = 15$$

The sample mean is still an unbiased estimator of μ (this is true regardless of the distribution of the population) and serves as point estimate of μ. The point estimate of μ is therefore 1035.6.

Since σ is unknown, the sample size is small and it can be assumed that the specific gravity of all bottles of beer is Normally distributed (numerous small random factors affect the specific gravity), we should use the *t* distribution. Thus

$$\frac{\bar{x} - \mu}{\sqrt{s^2/n}} \sim t_{n-1} \quad (4.20)$$

The 95% confidence interval estimate is given by

$$\bar{x} \pm t_{n-1}\sqrt{s^2/n} \quad (4.21)$$

where t_{n-1} is the value of the *t* distribution which cuts off the extreme 5% (2.5% in each tail) of the *t* distribution with ν degrees of freedom. Table A3 (see page 451) gives percentage points of the *t* distribution, and part of it is reproduced in Table 4.1.

The structure of the *t* distribution table is different from that of the standard Normal table. The first column of the table gives the degrees of freedom. In this example we want the row corresponding to $\nu = n - 1 = 14$. The appropriate

Table 4.1 Percentage points of the *t* distribution (excerpt from Table A3)

	Area in each tail						
ν	0.4	0.25	0.10	0.05	0.025	0.01	0.005
1	0.325	1.000	3.078	6.314	12.706	31.821	63.656
2	0.289	0.816	1.886	2.920	4.303	6.965	9.925
⋮	⋮	⋮	⋮	⋮	⋮	⋮	⋮
13	0.259	0.694	1.350	1.771	2.160	2.650	3.012
14	0.258	0.692	1.345	1.761	2.145	2.624	2.977
15	0.258	0.691	1.341	1.753	2.131	2.602	2.947

Note: The appropriate *t* value for constructing the confidence interval is found at the intersection of the shaded row and column.

column of the table is the one headed '0.025' which indicates the area cut off in *each* tail. At the intersection of this row and column we find the appropriate value, $t_{14} = 2.145$. Therefore, the confidence interval is given by

$$\begin{aligned} &1035.6 \pm 2.145\sqrt{2.7^2/15} \\ &= 1035.6 \pm 1.5 \\ &= [1034.10, 1037.10] \end{aligned}$$

We are 95% confident that the true specific gravity lies within this range. If the Normal distribution had (incorrectly) been used for this problem, then the *t* value of 2.145 would have been replaced by a *z* score of 1.96, giving a confidence interval of

$$[1034.23, 1036.97]$$

This underestimates the true confidence interval and gives the impression of a more precise estimate than is actually the case. Use of the Normal distribution leads to a confidence interval which is about 9% too narrow in this case.

Estimating the difference between two means

As in the case of a single mean, the *t* distribution needs to be used in small samples when the population variances are unknown. Again, both parent populations must be Normally distributed, and in addition it must be assumed that the population variances are equal, i.e. $\sigma_1^2 = \sigma_2^2$ (this is required in the mathematical derivation of the *t* distribution). This latter assumption was not required in the large-sample case using the Normal distribution. Consider the following example as an illustration of the method.

A sample of 20 Labour-controlled local authorities shows that they spend an average of £175 per taxpayer on administration with a standard deviation of £25. A similar survey of 15 Conservative-controlled authorities finds an average figure of £158 with standard deviation of £30. Estimate the true difference in expenditure between Labour and Conservative authorities.

The sample information available is

$$\begin{aligned} \bar{x}_1 &= 175 \qquad & \bar{x}_2 &= 158 \\ s_1 &= 25 & s_2 &= 30 \\ n_1 &= 20 & n_2 &= 15 \end{aligned}$$

We wish to estimate $\mu_1 - \mu_2$. The point estimate of this is $\bar{x}_1 - \bar{x}_2$ which is an unbiased estimate. This gives $175 - 158 = 17$ as the expected difference between the two sets of authorities.

For the confidence interval, the *t* distribution has to be used since the sample sizes are small and the population variances unknown. It is assumed that the populations are Normally distributed and that the samples have been independently drawn. We also assume that the population variances are equal, which seems justified since s_1 and s_2 do not differ by much (this kind of assumption is tested in Chapter 6). The confidence interval is given by the formula:

$$(\bar{x}_1 - \bar{x}_2) \pm t_\nu \sqrt{\frac{S^2}{n_1} + \frac{S^2}{n_2}} \tag{4.22}$$

where

$$S^2 = \frac{(n_1 - 1)s_1^2 + (n_2 - 1)s_2^2}{n_1 + n_2 - 2} \tag{4.23}$$

is known as the **pooled variance** and

$$\nu = n_1 + n_2 - 2$$

gives the degrees of freedom associated with the *t* distribution.

S^2 is an estimate of (common value of) the population variances. It would be inappropriate to have the differing values s_1^2 and s_2^2 in the formula for this *t* distribution, for this would be contrary to the assumption that $\sigma_1^2 = \sigma_2^2$, which is essential for the use of the *t* distribution. The estimate of the common population variance is just the weighted average of the sample variances, using degrees of freedom as weights. Each sample has $n - 1$ degrees of freedom, and the total number of degrees of freedom for the problem is the sum of the degrees of freedom in each sample. The degrees of freedom are thus $20 + 15 - 2 = 33$ and hence the value $t = 2.042$ cuts off the extreme 5% of the distribution. The *t* table in the appendix does not give the value for $\nu = 33$ so instead we used $\nu = 30$ which will give a close approximation.

To evaluate the 95% confidence interval, we first calculate S^2:

$$S^2 = \frac{(20 - 1) \times 25^2 + (15 - 1) \times 30^2}{20 + 15 - 2} = 741.6$$

Inserting this into equation (4.22) gives

$$17 \pm 2.042\sqrt{\frac{741.6}{20} + \frac{741.6}{15}} = [-1.99, 35.99]$$

Thus the true difference is quite uncertain and the evidence is even consistent with Conservative authorities spending more than Labour authorities. The large degree of uncertainty arises because of the small sample sizes and the quite wide variation within each sample.

One should be careful about the conclusions drawn from this test. The greater expenditure on administration could be either because of inefficiency or because of a higher level of services provided. To find out which is the case would require further investigation. The statistical test carried out here examines the levels of expenditure, but not whether they are productive or not.

Estimating proportions

Estimating proportions when the sample size is small cannot be done with the t distribution. Recall that the distribution of the sample proportion p was derived from the distribution of r (the number of successes in n trials), which followed a Binomial distribution (see the Appendix to this chapter (page 193)). In large samples the distribution of r is approximately Normal, thus giving a Normally distributed sample proportion. In small samples it is inappropriate to approximate the Binomial distribution with the t distribution, and indeed is unnecessary since the Binomial itself can be used. Small-sample methods for the sample proportion should be based on the Binomial distribution, therefore, as set out in Chapter 3. These methods are not discussed further here, therefore.

Exercise 4.7

?

A sample of size $n = 16$ is drawn from a population which is known to be Normally distributed. The sample mean and variance are calculated as 74 and 121. Find the 99% confidence interval estimate for the true mean.

Exercise 4.8

?

Samples are drawn from two populations to see if they share a common mean. The sample data are:

$$\bar{x}_1 = 45 \qquad \bar{x}_2 = 55$$
$$s_1 = 18 \qquad s_2 = 21$$
$$n_1 = 15 \qquad n_2 = 20$$

Find the 95% confidence interval estimate of the difference between the two population means.

Summary

- Estimation is the process of using sample information to make good estimates of the value of population parameters, e.g. using the sample mean to estimate the mean of a population.
- There are several criteria for finding a good estimate. Two important ones are the (lack of) bias and precision of the estimator. Sometimes there is a trade-off between these two criteria – one estimator might have a smaller bias but be less precise than another.
- An estimator is unbiased if it gives a correct estimate of the true value on average. Its expected value is equal to the true value.
- The precision of an estimator can be measured by its sampling variance (e.g. s^2/n for the mean of a sample).
- Estimates can be in the form of a single value (point estimate) or a range of values (confidence interval estimate). A confidence interval estimate gives some idea of how reliable the estimate is likely to be.
- For unbiased estimators, the value of the sample statistic (e.g. $\bar{x}$) is used as the point estimate.

- In large samples the 95% confidence interval is given by the point estimate plus or minus 1.96 standard errors (e.g. $\bar{x} \pm 1.96\sqrt{s^2/n}$ for the mean).
- For small samples the *t* distribution should be used instead of the Normal (i.e. replace 1.96 by the critical value of the *t* distribution) to construct confidence intervals of the mean.

Key terms and concepts

95% confidence interval
bias
confidence interval
confidence level
degrees of freedom
efficiency
estimation
estimator
inference
interval estimate
mean squared error
point estimate
pooled variance
precision
proportion
testing hypothesis
unbiased

Formulae used in this chapter

Formula	Description	Notes
$\bar{x} \pm 1.96\sqrt{s^2/n}$	95% confidence interval for the mean	Large samples, using Normal distribution
$\bar{x} \pm t_\nu\sqrt{s^2/n}$	95% confidence interval for the mean	Small samples, using t distribution. t_ν is the critical value of the t distribution for $\nu = n - 1$ degrees of freedom
$p \pm 1.96\sqrt{\frac{p(1-p)}{n}}$	95% confidence interval for a proportion	Large samples only
$(\bar{x}_1 - \bar{x}_2) \pm 1.96\sqrt{\frac{s_1^2}{n_1} + \frac{s_2^2}{n_2}}$	95% confidence interval for the difference of two means	Large samples
$(\bar{x}_1 - \bar{x}_2) \pm t_\nu\sqrt{\frac{S^2}{n_1} + \frac{S^2}{n_2}}$	95% confidence interval for the difference of two means	Small samples. The pooled variance is given by $S^2 = \frac{((n_1 - 1)s_1^2 + (n_2 - 1)s_2^2)}{n_1 + n_2 - 2}$, $\nu = n_1 + n_2 - 2$.

Problems

Some of the more challenging problems are indicated by highlighting the problem number in colour.

4.1 (a) Why is an interval estimate better than a point estimate?

(b) What factors determine the width of a confidence interval?

4.2 Is the 95% confidence interval (a) twice as wide, (b) more than twice as wide and (c) less than twice as wide, as the 47.5% interval? Explain your reasoning.

4.3 Explain the difference between an estimate and an estimator. Is it true that a good estimator always leads to a good estimate?

4.4 Explain why an unbiased estimator is not always to be preferred to a biased one.

4.5 A random sample of two observations, x_1 and x_2, is drawn from a population. Prove that $w_1x_1 + w_2x_2$ gives an unbiased estimate of the population mean as long as $w_1 + w_2 = 1$. (Hint: Prove that $E(w_1x_1 + w_2x_2) = \mu$.)

4.6 Following the previous question, prove that the most precise unbiased estimate is obtained by setting $w_1 = w_2 = \frac{1}{2}$. (Hint: Minimise $V(w_1x_1 + w_2x_2)$ with respect to w_1 after substituting $w_2 = 1 - w_1$. You will need a knowledge of calculus to solve this.)

4.7 Given the sample data

$$\bar{x} = 40 \quad s = 10 \quad n = 36$$

calculate the 99% confidence interval estimate of the true mean. If the sample size were 20, how would the method of calculation and width of the interval be altered?

4.8 (a) A random sample of 100 record shops found that the average weekly sale of a particular CD was 260 copies, with standard deviation of 96. Find the 95% confidence interval to estimate the true average sale for all shops.

(b) To compile the CD chart it is necessary to know the correct average weekly sale to within 5% of its true value. How large a sample size is required?

4.9 Given the sample data $p = 0.4$, $n = 50$, calculate the 99% confidence interval estimate of the true proportion.

4.10 A political opinion poll questions 1000 people. Some 464 declare they will vote Conservative. Find the 95% confidence interval estimate for the Conservative share of the vote.

4.11 Given the sample data

$$\bar{x}_1 = 25 \qquad \bar{x}_2 = 22$$
$$s_1 = 12 \qquad s_2 = 18$$
$$n_1 = 80 \qquad n_2 = 100$$

estimate the true difference between the means with 95% confidence.

4.12 (a) A sample of 200 women from the labour force found an average wage of £26 000 p.a. with standard deviation £3500. A sample of 100 men found an average wage of £28 000 with standard deviation £2500. Estimate the true difference in wages between men and women.

(b) A different survey, of men and women doing similar jobs, obtained the following results:

$$\bar{x}_W = £27\,200 \qquad \bar{x}_M = £27\,600$$
$$s_W = £2225 \qquad s_M = £1750$$
$$n_W = 75 \qquad n_M = 50$$

Estimate the difference between male and female wages using these new data. What can be concluded from the results of the two surveys?

4.13 Sixty-seven percent out of 150 pupils from school A passed an exam; 62% of 120 pupils at school B passed. Estimate the 99% confidence interval for the true difference between the proportions passing the exam.

4.14 (a) A sample of 954 adults in early 1987 found that 23% of them held shares. Given a UK adult population of 41 million and assuming a proper random sample was taken, find the 95% confidence interval estimate for the number of shareholders in the United Kingdom.

(b) A 'similar' survey the previous year had found a total of 7 million shareholders. Assuming 'similar' means the same sample size, find the 95% confidence interval estimate of the increase in shareholders between the two years.

4.15 A sample of 16 observations from a Normally distributed population yields a sample mean of 30 with standard deviation 5. Find the 95% confidence interval estimate of the population mean.

4.16 A sample of 12 families in a town reveals an average income of £25 000 with standard deviation £6000. Why might you be hesitant about constructing a 95% confidence interval for the average income in the town?

4.17 Two samples were drawn, each from a Normally distributed population, with the following results:

$$\bar{x}_1 = 45 \quad s_1 = 8 \quad n_1 = 12$$
$$\bar{x}_2 = 52 \quad s_2 = 5 \quad n_2 = 18$$

Estimate the difference between the population means, using the 95% confidence level.

4.18 The heights of 10 men and 15 women were recorded, with the following results:

	Mean	Variance
Men	173.5	80
Women	162	65

Estimate the true difference between men's and women's heights. Use the 95% confidence level.

4.19 **(Project)** Estimate the average weekly expenditure upon alcohol by students. Ask a (reasonably) random sample of your fellow students for their weekly expenditure on alcohol. From this, calculate the 95% confidence interval estimate of such spending by all students.

Answers to exercises

Exercise 4.1

(a) The point estimate is 25 and the 95% confidence interval is $25 \pm 1.96 \times 50/\sqrt{100} = 25 \pm 9.8 = [15.2, 34.8]$.

(b) The CI becomes larger as the sample size reduces. In this case we would have $25 \pm 1.96 \times 50/\sqrt{64} = 25 \pm 12.25 = [12.75, 37.25]$. Note that the width of the CI is inversely proportional to the square root of the sample size.

Exercise 4.2

The 95% CI is $50 \pm 1.96 \times 30/\sqrt{40} = 50 \pm 9.30 = [40.70, 59.30]$. The value of 60 lies (just) outside this CI so is unlikely to be the true mean.

Exercise 4.3

All the statements are false. The first four all make probability statements about the population mean, so are invalid according to the classical view of probability. Statement (e) sounds more plausible, but it refers to the CI for this specific sample and we simply do not know if the true mean lies inside or outside the CI for any individual sample. However, it is defensible to use this form of wording as long as we recognise the principles of constructing a CI. In (f), repeated experiments would have different CIs, so we cannot say anything about the range 0.1 to 0.4 in particular. This example comes from Hoekstra *et al.* (2014, DOI 10.3758/s13423-013-0572-3) who surveyed researchers and students. The respondents, on average, believed (erroneously) about 3.5 out of the 6 statements to be true. Interestingly, there was little difference between experienced researchers and novice students in the results.

Exercise 4.4

(a) The point estimate is 14% (7/50). The 95% CI is given by

$$0.14 \pm 1.96 \times \sqrt{\frac{0.14 \times (1 - 0.14)}{50}} = 0.14 \pm 0.096.$$

(b) Use 1.64 instead of 1.96, giving 0.14 ± 0.080.

(c) 0.14 ± 0.126.

Exercise 4.5

$\bar{x}_1 - \bar{x}_2 = 25 - 30 = -5$ is the point estimate. The interval estimate is given by

$$(\bar{x}_1 - \bar{x}_2) \pm 1.96\sqrt{\frac{s_1^2}{n_1} + \frac{s_2^2}{n_2}} = -5 \pm 1.96\sqrt{\frac{18^2}{36} + \frac{25^2}{49}}$$
$$= -5 \pm 9.14 = [-14.14, 4.14]$$

Exercise 4.6

The point estimate is $40 - 20 = 20\%$ or 0.2. The interval estimate is

$$0.2 \pm 1.96 \times \sqrt{\frac{0.4 \times 0.6}{50} + \frac{0.2 \times 0.8}{100}} = 0.2 \pm 0.157 = [0.043, 0.357]$$

Exercise 4.7

The 99% CI is given by $74 \pm t^* \times \sqrt{121/16} = 74 \pm 2.947 \times 2.75 = 74 \pm 8.10 = [65.90, 82.10]$.

Exercise 4.8

The pooled variance is given by

$$S^2 = \frac{(15 - 1) \times 18^2 + (20 - 1) \times 21^2}{15 + 20 - 2} = 391.36$$

The 95% CI is therefore

$$(45 - 55) \pm 2.042 \times \sqrt{\frac{391.36}{15} + \frac{391.36}{20}} = -10 \pm 13.80 = [-3.8, 23.8]$$

Appendix Derivations of sampling distributions

Derivation of the sampling distribution of p

The sampling distribution of p is fairly straightforward to derive, given what we have already learned. The sampling distribution of p can be easily derived from the distribution of r, the number of successes in n trials of an experiment, since $p = r/n$. The distribution of r for large n is approximately Normal (from Chapter 3):

$$r \sim N(nP, nP(1 - P)) \tag{4.24}$$

Knowing the distribution of r, is it possible to find that of p? Since p is simply r multiplied by a constant, $1/n$, it is also Normally distributed. The mean and variance of the distribution can be derived using the E and V operators. The expected value of p is

$$\mathrm{E}(p) = \mathrm{E}(r/n) = \frac{1}{n}\mathrm{E}(r) = \frac{1}{n}nP = P = \pi \tag{4.25}$$

The expected value of the sample proportion is equal to the population proportion (note that the probability P and the population proportion π are the same thing and may be used interchangeably). The sample proportion therefore gives an unbiased estimate of the population proportion.

For the variance:

$$\mathrm{V}(p) = \mathrm{V}\left(\frac{r}{n}\right) = \frac{1}{n^2}\mathrm{V}(r) = \frac{1}{n^2}nP(1 - P) = \frac{\pi(1 - \pi)}{n} \tag{4.26}$$

Hence, the distribution of p is given by

$$p \sim N\left(\pi, \frac{\pi(1 - \pi)}{n}\right) \tag{4.27}$$

Derivation of the sampling distribution of $\bar{x}_1 - \bar{x}_2$

This is the difference between two random variables so is itself a random variable. Since any linear combination of Normally distributed, independent random variables is itself Normally distributed, the difference of sample means follows a Normal distribution. The mean and variance of the distribution can be found using the E and V operators. Letting

$$\mathrm{E}(\bar{x}_1) = \mu_1, \mathrm{V}(\bar{x}_1) = \sigma_1^2/n_1 \text{ and}$$
$$\mathrm{E}(\bar{x}_2) = \mu_2, \mathrm{V}(\bar{x}_2) = \sigma_2^2/n_2$$

then

$$\mathrm{E}(\bar{x}_1 - \bar{x}_2) = \mathrm{E}(\bar{x}_1) - \mathrm{E}(\bar{x}_2) = \mu_1 - \mu_2 \tag{4.28}$$

And

$$V(\bar{x}_1 - \bar{x}_2) = V(\bar{x}_1) + V(\bar{x}_2) = \frac{\sigma_1^2}{n_1} + \frac{\sigma_2^2}{n_2} \tag{4.29}$$

Equation (4.29) assumes $\bar{x}_1$ and $\bar{x}_2$ are independent random variables. The probability distribution of $\bar{x}_1 - \bar{x}_2$ can therefore be summarised as:

$$\bar{x}_1 - \bar{x}_2 \sim N\left(\mu_1 - \mu_2, \frac{\sigma_1^2}{n_1} + \frac{\sigma_2^2}{n_2}\right) \tag{4.30}$$

This is equation (4.12) in the text.

5 Hypothesis testing

Contents

Learning outcomes

By the end of this chapter you should be able to:

- understand the philosophy and scientific principles underlying hypothesis testing
- appreciate that hypothesis testing is about deciding whether a hypothesis is true or false on the basis of a sample of data
- recognise the type of evidence which leads to a decision that the hypothesis is false
- carry out hypothesis tests for a variety of statistical problems
- recognise the relationship between hypothesis testing and a confidence interval
- recognise the shortcomings of hypothesis testing.

Introduction

This chapter deals with issues very similar to those of the previous chapter on estimation, but examines them in a different way. The estimation of population parameters and the testing of hypotheses about those parameters are similar techniques (indeed they are formally equivalent in a number of respects), but there are important differences in the interpretation of the results arising from each method. The process of estimation is appropriate when measurement is involved, such as measuring the true average expenditure on food; hypothesis testing is relevant when decision-making is involved, such as whether to accept that a supplier's products are up to a specified standard. Hypothesis testing is also used to make decisions about the truth or otherwise of different theories, such as whether rising prices are caused by rising wages; and it is here that the issues become contentious. It is sometimes difficult to interpret correctly the results of hypothesis tests in these circumstances. This is discussed further later in this chapter.

The concepts of hypothesis testing

In many ways hypothesis testing is analogous to a criminal trial. In a trial there is a defendant who is *initially presumed innocent*. The *evidence* against the defendant is then presented and, if the jury finds this convincing *beyond all reasonable doubt*, he or she is found guilty; the presumption of innocence is overturned. Of course, mistakes are sometimes made: an innocent person is convicted or a guilty person set free. Both of these errors involve costs (not only in the monetary sense), either to the defendant or to society in general, and the errors should be avoided if at all possible. The laws under which the trial is held may help avoid such errors. The rule that the jury must be convinced 'beyond all reasonable doubt' helps to avoid convicting the innocent, for instance.

The situation in hypothesis testing is similar. First, there is a **maintained** or **null hypothesis** which is initially *presumed* to be true. The empirical evidence, usually data from a random sample, is then gathered and assessed. If the evidence seems inconsistent with the null hypothesis, i.e. it has a low probability of occurring *if* the hypothesis were true, then the null hypothesis is *rejected* in favour of an alternative. Once again, there are two types of error one can make, either rejecting the null hypothesis when it is really true, or not rejecting it when in fact it is false. Ideally one would like to avoid both types of error.

An example helps to clarify the issues and the analogy. Suppose that you are thinking of taking over a small business franchise. The current owner claims the weekly turnover of each existing franchise averages £5000 and at this level you are willing to take on a franchise. You would be more cautious if the turnover is less than this figure. You examine the books of 26 franchises chosen at random and find that the average turnover was £4900 with standard deviation £280. What do you do?

The null hypothesis in this case is that average weekly turnover is £5000 (or more; that would be even more to your advantage). The **alternative hypothesis** is that turnover is strictly less than £5000 per week. We may write these more succinctly as follows:

$$H_0: \mu = 5000$$
$$H_1: \mu < 5000$$

H_0 is conventionally used to denote the null hypothesis, H_1 the alternative. Initially, H_0 is presumed to be true and this presumption will be tested using the sample evidence. Note that the sample evidence is *not* used in forming the null or alternative hypotheses.

You have to decide whether the owner's claim is correct (H_0) or not (H_1). The two types of error you could make are as follows:

- **Type I error** – reject H_0 when it is in fact true. This would mean missing a good business opportunity.
- **Type II error** – not rejecting H_0 when it is in fact false. You would go ahead and buy the business and then find out that it is not as attractive as claimed. You would have overpaid for the business.

The situation is set out in Figure 5.1.

Obviously a good decision rule would give a good chance of making a correct decision and rule out errors as far as possible. Unfortunately, it is impossible to completely eliminate the possibility of errors. As the decision rule is changed to reduce the probability of a Type I error, the probability of making a Type II error inevitably increases. The skill comes in balancing these two types of error.

Again a diagram is useful in illustrating this. Assuming that the null hypothesis is true, then the sample observations are drawn from a population with mean 5000 and some variance, which we shall assume is accurately measured by the sample variance. The distribution of $\bar{x}$ is then given by

$$\bar{x} \sim N(\mu, \sigma^2/n) \text{ or}$$
$$\bar{x} \sim N(5000, 280^2/26) \tag{5.1}$$

Under the alternative hypothesis the distribution of $\bar{x}$ would be the same except that it would be centred on a value less than 5000. These two situations are illustrated in Figure 5.2. The distribution of $\bar{x}$ under H_1 is shown by a dashed curve to signify that its exact position is unknown, only that it lies to the left of the distribution under H_0.

A **decision rule** amounts to choosing a point or dividing line on the horizontal axis in Figure 5.2. If the sample mean lies to the left of this point, then H_0 is rejected (the sample mean is too far away from H_0 for it to be credible) in favour of H_1 and you do not buy the franchise. If $\bar{x}$ lies above this decision point, then H_0 is not rejected and you go ahead with the purchase. Such a decision point is shown

Figure 5.1
The two different types of error

		True situation	
		H_0 true	H_0 false
Decision	Accept H_0	Correct decision	Type II error
	Reject H_0	Type I error	Correct decision

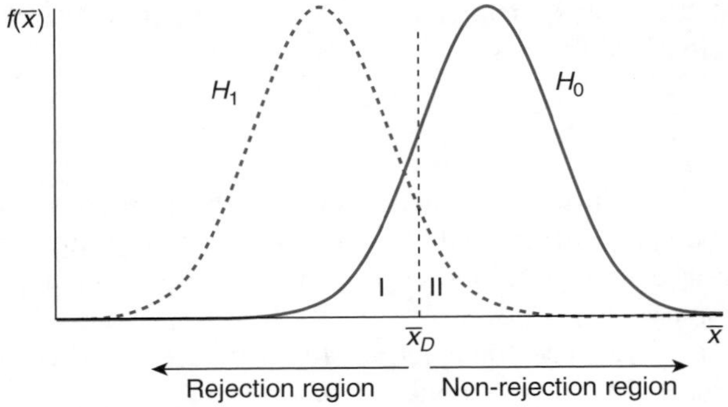

Figure 5.2
The sampling distributions of $\bar{x}$ under H_0 and H_1

in Figure 5.2, denoted by $\bar{x}_D$. To the left of $\bar{x}_D$ lies the **rejection** (of H_0) **region**; to the right lies the **non-rejection region**.

Based on this point, we can see the probabilities of Type I and Type II errors. The area under the H_0 distribution to the left of $\bar{x}_D$, labelled I, shows the probability of rejecting H_0 given that it is in fact true: a Type I error. The area under the H_1 distribution to the right of $\bar{x}_D$, labelled II, shows the probability of a Type II error: not rejecting H_0 when it is in fact false (and H_1 is true).

Shifting the decision line to the right or left alters the balance of these probabilities. Moving the line to the right increases the probability of a Type I error but reduces the probability of a Type II error. Moving the line to the left has the opposite effect.

The Type I error probability can be calculated for any value of $\bar{x}_D$. Suppose we set $\bar{x}_D$ to a value of 4950. Using the distribution of $\bar{x}$ given in equation (5.1), the area under the distribution to the left of 4950 is obtained using the z score:

$$z = \frac{\bar{x}_D - \mu}{\sqrt{s^2/n}} = \frac{4950 - 5000}{\sqrt{280^2/26}} = -0.91 \tag{5.2}$$

From the tables of the standard Normal distribution we find that the probability of a Type I error is 18.1%. Unfortunately, the Type II error probability cannot be established because the exact position of the distribution under H_1 is unknown. Therefore, we cannot decide on the appropriate position of $\bar{x}_D$ by some balance of the two error probabilities.

The convention therefore is to set the position of $\bar{x}_D$ by using a Type I error probability of 5%, known as the **significance level**[1] of the test. In other words, we are prepared to accept a 5% probability of rejecting H_0 when it is, in fact, true. This allows us to establish the position of $\bar{x}_D$. From Table A2 (see page 450) we find that $z = -1.64$ cuts off the bottom 5% of the distribution, so the decision line should be 1.64 standard errors below 5000. The value -1.64 is known as the **critical value** of the test. We therefore obtain

$$\bar{x}_D = 5000 - 1.64\sqrt{280^2/26} = 4910 \tag{5.3}$$

Since the sample mean of 4900 lies below 4910, we reject H_0 *at the 5% significance level* or equivalently we reject *with 95% confidence*. The significance level is

[1]The term **size** of the test is also used, not to be confused with the sample size. We use the term 'significance level' in this text.

generally denoted by the symbol α and the complement of this, given by $1 - \alpha$, is known as the confidence level (as used in the confidence interval).

An equivalent procedure would be to calculate the z score associated with the sample mean, known as the **test statistic**, and then compare this to the critical value of the test. This allows the hypothesis testing procedure to be broken down into five neat steps:

(1) Write down the null and alternative hypotheses:

$$H_0\colon \mu = 5000$$
$$H_1\colon \mu < 5000$$

(2) Choose the significance level of the test, conventionally $\alpha = 0.05$ or 5%.
(3) Look up the critical value of the test from statistical tables, based on the chosen significance level. $z^* = 1.64$ is the critical value in this case.
(4) Calculate the test statistic:

$$z = \frac{\bar{x} - \mu}{\sqrt{s^2/n}} = \frac{-100}{\sqrt{280^2/26}} = -1.82 \tag{5.4}$$

(5) Decision rule. Compare the test statistic with the critical value: if $z < -z^*$ reject H_0 in favour of H_1. Since $-1.82 < -1.64$, H_0 is rejected with 95% confidence. Note that we use $-z^*$ here (rather than $+z^*$) because we are dealing with the left-hand tail of the distribution.

Worked example 5.1

A sample of 100 workers found the average overtime hours worked in the previous week was 7.8, with standard deviation 4.1 hours. Test the hypothesis that the average for all workers is 5 hours or less.

We can set out the five steps of the answer as follows:

(1) $H_0\colon \mu = 5$
$H_1\colon \mu > 5$
(2) Significance level, $\alpha = 5\%$.
(3) Critical value $z^* = 1.64$.
(4) Test statistic:

$$z = \frac{\bar{x} - \mu}{\sqrt{s^2 n}} = \frac{7.8 - 5}{\sqrt{4.1^2/100}} = 6.8$$

(5) Decision rule: $6.8 > 1.64$ so we reject H_0 in favour of H_1. Note that in this case we are dealing with the right-hand tail of the distribution (positive values of z and z^*). Only high values of $\bar{x}$ reject H_0.

One-tail and two-tail tests

In the above example the rejection region for the test consisted of one tail of the distribution of $\bar{x}$, since the buyer was only concerned about turnover being less than claimed. For this reason, it is known as a **one-tail test**. Suppose now that an accountant is engaged to sell the franchise and wants to check the claim about

Figure 5.3
A two-tail hypothesis test

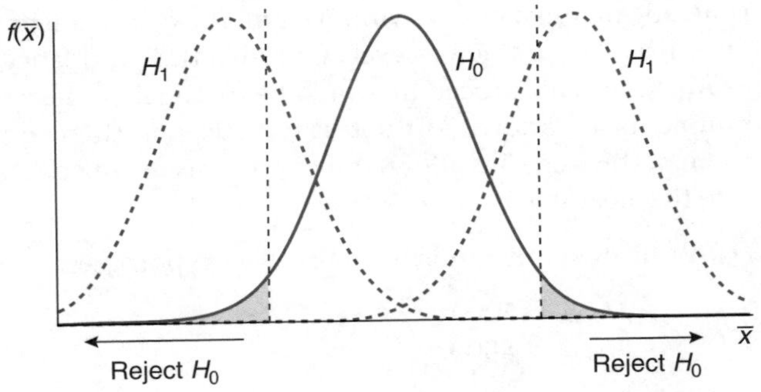

turnover before advertising the business for sale. In this case he or she would be concerned about turnover being either below *or* above 5000.

This would now become a **two-tail test** with the null and alternative hypotheses being

$$H_0: \mu = 5000$$
$$H_1: \mu \neq 5000$$

Now there are two rejection regions for the test. Either a very low sample mean *or* a very high one will serve to reject the null hypothesis. The situation is presented graphically in Figure 5.3.

The distribution of $\bar{x}$ under H_0 is the same as before, but under the alternative hypothesis the distribution could be *either* to the left *or* to the right, as depicted. If the significance level is still chosen to be 5%, then the complete rejection region consists of the *two* extremes of the distribution under H_0, containing 2.5% in each tail (hence 5% in total). This gives a Type I error probability of 5% as before. In other words, we would make a Type I error if the sample mean falls too far above or below the hypothesised value.

The critical value of the test therefore becomes $z^* = 1.96$, the values which cut off 2.5% in each tail of the standard Normal distribution. Only if the test statistic falls into one of the rejection regions beyond 1.96 standard errors from the mean is H_0 rejected.

Using data from the previous example, the test statistic remains $z = -1.82$ so that the null hypothesis cannot be rejected in this case, as -1.82 does not fall beyond -1.96. To recap, the five steps of the test are:

(1) $H_0: \mu = 5000$
$H_1: \mu \neq 5000$
(2) Choose the significance level: $\alpha = 0.05$.
(3) Look up the critical value: $z^* = 1.96$.
(4) Evaluate the test statistic:

$$z = \frac{-100}{\sqrt{280^2/26}} = -1.82$$

(5) Compare test statistic and critical values: if $z < -z^*$ or $z > z^*$ reject H_0 in favour of H_1. In this case $-1.82 > -1.96$, so H_0 cannot be rejected with 95% confidence.

One- and two-tail tests therefore differ only at steps 1 and 3. Note that we have come to different conclusions according to whether a one- or two-tail test was used, with the same sample evidence. There is nothing wrong with this, however, for there are different interpretations of the two results. If the investor always uses his or her rule, he or she will miss out on 5% of good investment opportunities, when sales are (by chance) low. He or she will never miss out on a good opportunity because the investment appears too good (i.e. sales by chance are very high). For the accountant, 5% of the firms with sales averaging £5000 will not be advertised as such, *either* because sales appear too low *or* because they appear too high.

Another way of interpreting the difference between one- and two-tail tests is to say that the former includes some **prior information**, i.e. that the true value cannot lie above the hypothesised value (or that we are not interested in that region). Hence, although the sample evidence is the same, the overall evidence is not quite the same due to our prior knowledge. This additional knowledge allows us to sometimes reject a null via a one-tail test but not via a two-sided test.

It is tempting on occasion to use a one-tail test because of the sample evidence. For example, the accountant might look at the sample evidence above and decide that the franchise operation can only have true sales less than or equal to 5000. Therefore, she/he uses a one-tail test. This is a dangerous practice, since the sample evidence is being used to help formulate the hypothesis, which is then tested on that same evidence. This is going round in circles; the hypothesis should be chosen *independently* of the evidence which is then used to test it[2]. Presumably the accountant would also use a one-tail test (with $H_1\colon \mu > 5000$ as the alternative hypothesis) if she/he noticed that the sample mean was *above* the hypothesised value. Taking these possibilities together, she/he would in effect be using the 10% significance level, not the 5% level, since there would be 5% in each tail of the distribution. She/he would make a Type I error on 10% of all occasions rather than 5%.

It is acceptable to use a one-tail test when you have *independent* information about what the alternative hypothesis should be, or when you are not concerned about one side of the distribution (like the investor) and can effectively add that in to the null hypothesis. Otherwise, it is safer to use a two-tail test.

Exercise 5.1

(a) Two political parties are debating crime figures. One party says that crime has increased compared to the previous year. The other party says it has not. Write down the null and alternative hypotheses.

(b) Explain the two types of error that could be made in this example and the possible costs of each type of error.

Exercise 5.2

?

(a) We test the hypothesis $H_0\colon \mu = 100$ against $H_1\colon \mu > 100$ by rejecting H_0 if our sample mean is greater than 108. If in fact $\bar{x} \sim N(100, 900/25)$, what is the probability of making a Type I error?

(b) If we wanted a 5% Type I error probability, what decision rule (value of $\bar{x}$) should we adopt?

(c) If we knew that μ could only take on the values 100 (under H_0) or 112 (under H_1) what would be the Type II error probability using the decision rule in part (a)?

Exercise 5.3

?

Test the hypothesis $H_0\colon \mu = 500$ versus $H_1\colon \mu \neq 500$ using the evidence $\bar{x} = 530, s = 90$ from a sample of size $n = 30$.

[2]Alternatively, we could say this is assuming the sample evidence provides the additional prior information that might justify a one-tail test. However, it is *not* additional evidence, and it would be wrong to use the sample evidence for two purposes in this way.

The choice of significance level

The explanation above put hypothesis testing into a framework of decision-making between hypotheses, balancing Type I and Type II errors. This was first developed by the statisticians Neyman and Pearson in the 1930s. It is fine for a situation where both null and alternative hypotheses are well defined (as, for example, in a court of law) but, as noted, is not helpful when the alternative hypothesis is only vaguely specified ('H_1: $\mu \neq 5000$', for instance) and we cannot calculate the Type II error probability.

In an ideal world we would have precisely specified null *and* alternative hypotheses (e.g. we would test H_0: $\mu = 5000$ against H_1: $\mu = 4500$, these being the only possibilities). Then we could calculate the probabilities of both Type I *and* Type II errors, for any given decision rule. We could then choose the optimal decision rule, which gives the best compromise between the two types of error. This is reflected in a court of law. In criminal cases, the jury must be convinced of the prosecution's case beyond reasonable doubt, because of the cost of committing a Type I error. In a civil case (libel, for example) the jury need only be convinced *on the balance of probabilities*. In a civil case, the costs of Type I and Type II error are more evenly balanced and so the burden of proof is lessened.

However, in statistics we usually do not have the luxury of two well-specified hypotheses. As in the earlier worked example, the null hypothesis is precisely specified (it has to be or the test could not be carried out) but the alternative hypothesis is imprecise (sometimes called a **composite hypothesis** because it encompasses a range of values). Statistical inference is often used not so much as an aid to decision-making but to provide evidence for or against a particular theory, to alter one's degree of belief in the truth of the theory. For example, a researcher might believe large firms are more profitable than small ones and wishes to test this. The null and alternative hypotheses would be:

H_0: large and small firms are equally profitable
H_1: large firms are more profitable

(Note that the null has to be 'equally profitable', since this is a precise statement. 'More profitable' is too vague to be the null hypothesis.). Data could be gathered to test this hypothesis, but it is not possible to calculate the Type II error probability (because 'more profitable' is too vague). Hence we cannot find the optimal balance of Type I and Type II errors in order to make our decision to accept or reject H_0. Another statistician, R.A. Fisher, proposed the use of the 5% significance level in this type of circumstance, arguing that a researcher could justifiably ignore any results that fail to reach this standard. Thus our procedures today are actually an uncomfortable mixture of two different approaches: Fisher did not agree with the decision-making framework; Neyman and Pearson did not propose a 5% convention for the Type I error probability.

STATISTICS IN PRACTICE

The five sigma level of certainty

In particle physics the accepted level of certainty for the results of an experiment to be considered a valid discovery is five sigma (i.e. five standard deviations). A distance of five standard deviations cuts off approximately 0.00003% in one tail of the Normal distribution and represents an extremely low significance level. The reason for this is two-fold: (i) scientists are very reluctant to accept new hypotheses which may later turn out to be false, and (ii) there is a lot of random noise in the results of their experiments, so it would be easy to confuse noise with a valid finding unless a low significance level is chosen.

At the time of writing (February 2011), physicists are getting close to uncovering the existence of the Higgs boson but so far their results are only significant at about the three sigma level.

Update: The existence of the Higgs boson has been confirmed. More data (i.e. larger sample) allowed this conclusion to be reached.

The 5% significance level really does depend upon convention; therefore, it cannot be justified by reference to the relative costs of Type I and Type II errors. The 5% convention does impose some sort of discipline upon research; it sets some kind of standard which all theories (hypotheses) should be measured against. Beware the researcher who reports that a particular hypothesis is rejected at the 8% significance level; it is likely that the significance level was chosen so that the hypothesis could be rejected, which is what the researcher was hoping for in the first place. As we shall see later, however, this 'discipline upon research' is not a strong one.

The Prob-value approach

Fisher later changed his mind about the 5% significance level rule. He argued that this was too rigid to apply mechanically in all situations. Instead, one should present the **Prob-value** (also known as the **P-value**), which is the actual significance level of the test statistic. In this way one presents information (the likelihood of a false positive) rather than imposing a decision. The reader could then make their own judgement.

The Prob-value is calculated as follows. The test statistic calculated earlier for the investor problem was $z = -1.82$ and the associated Prob-value is obtained from Table A2 as 3.44%, i.e. -1.82 cuts off 3.44% in one tail of the standard Normal distribution. This means that the null hypothesis could be rejected at the 3.44% significance level or, alternatively expressed, with 96.56% confidence.

Notice that Table A2 gives the Prob-value for a one-tail test; for a two-tail test the Prob-value should be doubled. Thus for the accountant, using the two-tail test, the significance level is 6.88%, and this is the level at which the null hypothesis can be rejected. Alternatively, we could say we reject the null with 93.12% confidence. This does not meet the standard 5% criterion (for the significance level) which is most often used, so would result in non-rejection of the null. (Notice that, despite using P-values, we have slipped back into decision-making mode. It is difficult to avoid doing this.)

An advantage of using the Prob-value approach is that many statistical software programs routinely provide the Prob-value of a calculated test statistic[3]. If one understands the use of Prob-values, then one does not have to look up tables (this applies to any distribution, not just the Normal), which can save time.

[3]It is sometimes referred to as the 'P-value' in the statistical results. Excel uses this notation.

To summarise, one rejects the null hypothesis if either:

- (Method 1) the test statistic *is greater than* the critical value, i.e. $z > z^*$, or
- (Method 2) the Prob-value associated with the test statistic *is less than* the significance level, i.e. $P < 0.05$ (if the 5% significance level is used).

I have found that many students initially find this confusing, because of the opposing inequality in the two versions (greater than and less than). For example, a program might calculate a hypothesis test and report the result as '$z = 1.4$ (P-value = 0.162)'. The first point to note is that most software programs report the Prob-value for a two-tail test by default. Hence, assuming a 5% significance level, in this case we cannot reject H_0 because $z = 1.4 < 1.96$ or equivalently because $0.162 > 0.05$, against a two-tailed alternative (i.e. H_1 contains $\neq$).

If you wish to conduct a one-tailed test, you have to halve the reported Prob-value, becoming 0.081 in this example. This is again greater than 5%, so the hypothesis is still accepted, even against a one-sided alternative (H_1 contains $>$ or $<$). Equivalently, one could compare 1.4 with the one-tail critical value, 1.64, showing non-rejection of the null, but one has to look up the standard Normal table with this method. Computers cannot guess whether a one- or two-sided test is wanted, so take the conservative option and report the two-sided value. The correction for a one-sided test has to be done manually.

Exercise 5.4

This is a useful exercise to test your understanding of hypothesis tests. You carry out a one-tail hypothesis test and obtain the result $z = 2.3$, P-value = 0.01. Which of the following statements are true?

(a) You have disproved the null hypothesis.

(b) You have found the probability that the null hypothesis is true.

(c) You have proved the alternative hypothesis to be true.

(d) You can calculate the probability of the alternative hypothesis being true.

(e) If you reject the null, you know the probability of having made the wrong decision.

(f) If this experiment were repeated many times, a statistically significant result would be obtained in 99% of trials.

These questions are adapted from an original questionnaire in Oakes (1986).

Significance, effect size and power

Researchers usually look for 'significant' results; it is the way to get attention and to get published. Academic papers report that 'the results are significant' or that 'the coefficient is significantly different from zero at the 5% significance level'. It is vital to realise that the word 'significant' is used here in the *statistical* sense and not in its everyday sense of being *important*. Something can be statistically significant yet still unimportant.

Suppose that we have some more data about the business examined earlier. Data for 100 franchises have been uncovered, revealing an average weekly turnover of £4975 with standard deviation £143. Can we reject the hypothesis that the average weekly turnover is £5000? The test statistic is

$$z = \frac{4975 - 5000}{\sqrt{143^2/100}} = -1.75$$

Since this is below $-z^* = -1.64$, the null is rejected with 95% confidence. True average weekly turnover is less than £5000. However, the difference is only £25 per week, which is 0.5% of £5000. Common sense would suggest that the difference may be unimportant, even if it is significant in the statistical sense. One should not interpret statistical results in terms of significance alone, therefore; one should also look at the size of the difference (sometimes known as the **effect size**) and ask whether it is important or not. Even experienced researchers make this mistake; a review of articles in the prestigious *American Economic Review* reported that 82% of them confused statistical significance for economic significance in some way (McCloskey and Ziliak, 2004).

This problem with hypothesis testing paradoxically gets worse as the sample size increases. For example, if 250 observations reveal average sales of 4985 with standard deviation 143, the null would (just) be rejected at 5% significance. In fact, given a large enough sample size we can virtually guarantee to reject the null hypothesis even before we have gathered the data. This can be seen from equation (5.4) for the z score test statistic: as n gets larger, the test statistic also inevitably gets larger.

A good way to remember this point is to appreciate that it is the *evidence* which is significant, not the size of the effect or the results of your research. Strictly, it is better to say 'there is significant evidence of difference between . . .' than 'there is a significant difference between . . .'.

A related way of considering the effect of increasing sample size is via the concept of the **power** of a test. This is defined as

$$\textbf{Power} \text{ of a test} = 1 - \Pr(\text{Type II error}) = 1 - \beta \qquad (5.5)$$

where β is the symbol conventionally used to indicate the probability of a Type II error. Since a Type II error is defined as not rejecting H_0 when false (equivalent to rejecting H_1 when true), power is the probability of rejecting H_0 when false (if H_0 is false, it must be *either* accepted *or* rejected; hence these probabilities sum to one). This is one of the correct decisions identified earlier, associated with the lower right-hand box in Figure 5.1, that of correctly rejecting a false null hypothesis. The power of a test is therefore given by the area under the H_1 distribution, to the left of the decision line, as illustrated (shaded) in Figure 5.4 (for a one-tail test).

Figure 5.4
The power of a test

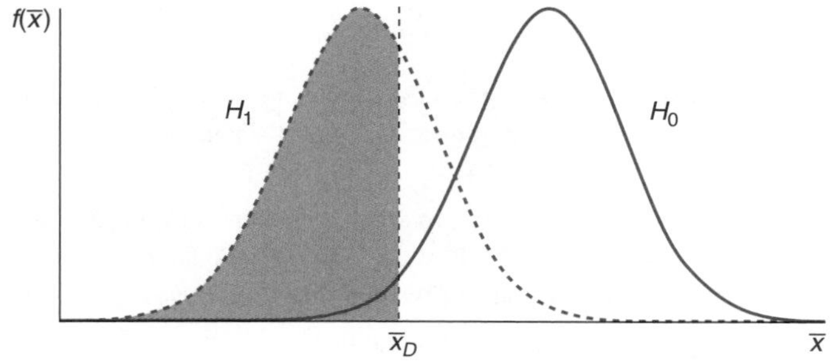

It is generally desirable to maximise the power of a test, as long as the probability of a Type I error is not raised in the process. There are essentially three ways of doing this:

- Avoid situations where the null and alternative hypotheses are very similar, i.e. the hypothesised means are not far apart (a small effect size).
- Use a large sample size. This reduces the sampling variance of $\bar{x}$ (under both H_0 and H_1) so the two distributions become more distinct.
- Use good sampling methods which have small sampling variances. This has a similar effect to increasing the sample size.

Worked example 5.2

This example shows how a larger sample size increases the power of a test. Suppose we wish to test the hypothesis H_0: $\mu = 500$ versus H_0: $\mu \leq 500$. We have a choice of two methods to test the hypothesis: (a) use a sample size of 100, or (b) use a sample size of 49. Which will give us a better test? Let us use the 5% significance level and assume. $\sigma = 200$ For test (a) we would then reject H_0 if $\bar{x} < 467.2$, and for test (b) we would reject if $\bar{x} < 453.1$. (You can check that these would lead to a z score of -1.64 in both cases, which cuts of the bottom 5% of the Normal distribution.)

Now what is the probability in each case of rejecting the null if it were false, i.e. what is the power of each test? Suppose the true mean were 470 (so H_0 above is false and should be rejected). For test (a) the probability of rejecting H_0 (i.e. $\bar{x} < 467.2$) given that the true mean is 470 is obtained by calculating the z score:

$$z = \frac{467.2 - 470}{200/\sqrt{100}} = -0.14$$

From Table A2, this cuts off 44% in the lower tail, so this is the power of the test. For test (b) the z score is -0.59, which cuts off only 28%. Hence, test (a) is preferred, having the greater power. Note that the same significance level is required for both tests.

Unfortunately, in economics and business the data (including sample size) are very often given in advance and there is little or no control possible over the sampling procedures. This leads to a neglect of consideration of power, unlike in psychology or biology, for example, where the experiment can often be designed by the researcher. The gathering of sample data will be covered in detail in Chapter 9.

Exercise 5.5

If a researcher believes the cost of making a Type I error is much greater than the cost of a Type II error, should they prefer a 5% or 1% significance level? Explain why.

Exercise 5.6

?

(a) A researcher uses Excel to analyse data and test a hypothesis. The program reports a test statistic of $z = 1.77$ (P-value $= 0.077$). Would you reject the null hypothesis if carrying out (i) a one-tailed test (ii) a two-tailed test? Use the 5% significance level.

(b) Repeat part (a) using a 1% significance level.

Exercise 5.7

A researcher wishes to test the hypothesis H_0: $\mu = 160$ versus H_0: $\mu > 160$. If the sample size is to be 400 and $\sigma = 50$ is assumed:

(a) What value of $\bar{x}$ should be used as the cutoff for rejecting H_0 at the 5% significance level?

(b) What is the power of the test if the true mean is (i) 163, (ii) 166?

Further hypothesis tests

We now consider a number of different types of hypothesis test, all involving the same principles but differing in details of their implementation. This is similar to the exposition in the last chapter covering, in turn, tests of a proportion, tests of the difference of two means and proportions, and finally problems involving small sample sizes.

Testing a proportion

A car manufacturer claims that no more than 10% of its cars should need repairs in the first three years of their life, the warranty period. A random sample of 50 three-year-old cars found that 8 had required attention. Does this contradict the maker's claim?

This problem can be handled very similarly to the methods used for a mean. The key, once again, is to recognise the sample proportion as a random variable with an associated probability distribution. From Chapter 4 equation (4.9)), the sampling distribution of the sample proportion in large samples is given by

$$p \sim N\left(\pi, \frac{\pi(1-\pi)}{n}\right) \tag{5.6}$$

In this case $\pi = 0.10$ (under the null hypothesis, the maker's claim). The sample data are

$$p = 8/50 = 0.16$$
$$n = 50$$

Thus 16% of the sample required attention within the warranty period. This is substantially higher than the claimed 10%, but is this just because of a non-representative sample or does it reflect the reality that the cars are badly built? The hypothesis test is set out along the same lines as for a sample mean:

(1) Set out the null and alternative hypotheses:
H_0: $\pi = 0.10$
H_1: $\pi > 0.10$

(The only concern is the manufacturer not matching its claim; hence a one-tail test is appropriate.)

(2) Significance level: $\alpha = 0.05$.

(3) The critical value of the one-tail test at the 5% significance level is $z^* = 1.64$, obtained from the standard Normal table.

(4) The test statistic is

$$z = \frac{p - \pi}{\sqrt{\frac{\pi(1-\pi)}{n}}} = \frac{0.16 - 0.10}{\sqrt{\frac{0.1 \times 0.9}{50}}} = 1.41$$

(5) Since the test statistic is less than the critical value, it falls into the non-rejection region. The null hypothesis is not rejected by the data. The manufacturer's claim is not unreasonable.

Note that for this problem, the rejection region lies in the *upper* tail of the distribution because of the 'greater than' inequality in the alternative hypothesis. The null hypothesis is therefore rejected in this case if $z > z^*$.

STATISTICS IN PRACTICE

Do children prefer branded goods only because of the name?

Researchers at Johns Hopkins Bloomberg School of Public Health in Maryland found young children were influenced by the packaging of foods. Sixty-three children were offered two identical meals, save that one was still in its original packaging (from McDonald's). Seventy-six per cent of the children preferred the branded French fries.

Is this evidence significant? The null hypothesis is H_0: $\pi = 0.5$ versus H_1: $\pi > 0.5$. The test statistic for this hypothesis test is

$$z = \frac{p - \pi}{\sqrt{\frac{\pi(1 - \pi)}{n}}} = \frac{0.76 - 0.50}{\sqrt{\frac{0.5 \times 0.5}{63}}} = 4.12$$

which is greater than the critical value of $z^* = 1.64$. Hence we conclude children are influenced by the packaging or brand name.

Source: New Scientist, 11 August 2007.

Testing the difference of two means

Suppose a car company wishes to compare the performance of its two factories producing an identical model of car. The factories are equipped with the same machinery but their outputs might differ due to managerial ability, labour relations, etc. Senior management wishes to know if there is any difference between the two factories. Output is monitored for 30 days, chosen at random, with the following results:

	Factory 1	Factory 2
Average daily output	420	408
Standard deviation of daily output	25	20

Does this produce sufficient evidence of a real difference between the factories, or does the difference between the samples simply reflect random differences such as minor breakdowns of machinery? The information at our disposal may be summarised as

$$\begin{aligned} \bar{x}_1 &= 420 & \bar{x}_2 &= 408 \\ s_1 &= 25 & s_2 &= 20 \\ n_1 &= 30 & n_2 &= 30 \end{aligned}$$

The hypothesis test to be conducted concerns the difference between the factories' outputs, so the appropriate random variable to examine is $\bar{x}_1 - \bar{x}_2$. From Chapter 4 (equation (4.12)), this has the following distribution, in large samples:

$$\bar{x}_1 - \bar{x}_2 \sim N\left(\mu_1 - \mu_2, \frac{\sigma_1^2}{n_1} + \frac{\sigma_2^2}{n_2}\right) \tag{5.7}$$

The population variances, σ_1^2 and σ_2^2, may be replaced by their sample estimates, s_1^2 and s_2^2, if the former are unknown, as here. The hypothesis test is therefore as follows.

(1) H_0: $\mu_1 - \mu_2 = 0$
H_1: $\mu_1 - \mu_2 \neq 0$

The null hypothesis posits no real difference between the factories. This is a two-tail test since there is no *a priori* reason to believe one factory is better than the other, apart from the sample evidence.

(2) Significance level: $\alpha = 1\%$. This is chosen since the management does not want to interfere unless it is really confident of some difference between the factories. In order to favour the null hypothesis, a lower significance level than the conventional 5% is set.

(3) The critical value of the test is $z^* = 2.57$. This cuts off 0.5% in each tail of the standard Normal distribution.

(4) The test statistic is

$$z = \frac{(\bar{x}_1 - \bar{x}_2) - (\mu_1 - \mu_2)}{\sqrt{\dfrac{s_1^2}{n_1} + \dfrac{s_2^2}{n_2}}} = \frac{(420 - 408) - 0}{\sqrt{\dfrac{25^2}{30} + \dfrac{20^2}{30}}} = 2.05$$

Note that this is of the same form as in the single-sample cases. The hypothesised value of the difference (zero in this case) is subtracted from the sample difference and this is divided by the standard error of the random variable.

(5) Decision rule: $z < z^*$ so the test statistic falls into the non-rejection region. There does not appear to be a significant difference between the two factories (or, better expressed, there is not significant evidence of a difference between factories).

A number of remarks about this example should be made. First, it is not necessary for the two sample sizes to be equal (although they are in the example); 45 days' output from factory 1 and 35 days' from factory 2, for example, could have been sampled. Second, the values of s_1^2 and s_2^2 do not have to be equal. They are, respectively, estimates of σ_1^2 and σ_2^2, and although the null hypothesis asserts that $\mu_1 = \mu_2$ it does not assert that the variances are equal. Management wants to know if the *average* levels of output are the same; it is not concerned about daily fluctuations in output (although it might be). A test of the hypothesis of equal variances is set out in Chapter 6.

The final point to consider is whether all the necessary conditions for the correct application of this test have been met. The example noted that the 30 days were chosen at random. If the 30 days sampled were consecutive, we might doubt whether the observations were truly independent. Low output on one day (due to a mechanical breakdown, for example) might influence the following day's output (if a special effort were made to catch up on lost production, for example).

Testing the difference of two proportions

The general method should by now be familiar, so we will proceed by example for this case. Suppose that, in a comparison of two holiday companies' customers, of the 75 who went with Happy Days Tours, 45 said they were satisfied, while 48 of the 90 who went with Fly by Night Holidays were satisfied. Is there a significant difference between the companies?

This problem can be handled by a hypothesis test on the difference of two sample proportions. The procedure is as follows. The sample evidence is

$$p_1 = 45/75 = 0.6 \qquad n_1 = 75$$
$$p_2 = 48/90 = 0.533 \qquad n_2 = 90$$

The hypothesis test is carried out as follows

(1) H_0: $\pi_1 - \pi_2 = 0$
H_1: $\pi_1 - \pi_2 \neq 0$
(2) Significance level: $\alpha = 5\%$.
(3) Critical value: $z^* = 1.96$.
(4) Test statistic: The distribution of $p_1 - p_2$ is

$$p_1 - p_2 \sim N\left(\pi_1 - \pi_2, \frac{\pi_1(1 - \pi_1)}{n_1} + \frac{\pi_2(1 - \pi_2)}{n_2}\right)$$

so the test statistic is

$$z = \frac{(p_1 - p_2) - (\pi_1 - \pi_2)}{\sqrt{\dfrac{\pi_1(1 - \pi_1)}{n_1} + \dfrac{\pi_2(1 - \pi_2)}{n_2}}} \tag{5.8}$$

However, π_1 and π_2 in the denominator of equation (5.8) have to be replaced by estimates from the samples. They cannot simply be replaced by p_1 and p_2 because these are unequal; to do so would contradict the null hypothesis that they *are* equal. Since the null hypothesis is assumed to be true (for the moment), it makes no sense to use a test statistic which explicitly supposes the null hypothesis to be false. Therefore, π_1 and π_2 are replaced by an estimate of their common value which is denoted $\hat{\pi}$ and whose formula is

$$\hat{\pi} = \frac{n_1 p_1 + n_2 p_2}{n_1 + n_2} \tag{5.9}$$

i.e. a weighted average of the two sample proportions. This yields

$$\hat{\pi} = \frac{75 \times 0.6 + 90 \times 0.533}{75 + 90} = 0.564$$

This, in fact, is just the proportion of all customers who were satisfied, 93 out of 165. The test statistic therefore becomes

$$z = \frac{0.6 - 0.533 - 0}{\sqrt{\dfrac{0.564 \times (1 - 0.564)}{75} + \dfrac{0.564 \times (1 - 0.564)}{90}}} = 0.86$$

(5) The test statistic is less than the critical value so the null hypothesis cannot be rejected with 95% confidence. There is not sufficient evidence to demonstrate a difference between the two companies' performance.

STATISTICS IN PRACTICE

Are women better at multi-tasking?

The conventional wisdom is 'yes'. However, the concept of multi-tasking originated in computing and, in that domain, it appears men are more likely to multi-task. Oxford Internet Surveys (http://www.oii.ox.ac.uk/microsites/oxis/) asked a sample of 1578 people if they multi-tasked while on-line (e.g. listening to music, using the phone); 69% of men said they did, 57% of women did. Is this difference statistically significant?

The published survey does not give precise numbers of men and women respondents for this question, so we will assume equal numbers (the answer is not very sensitive to this assumption). We therefore have the test statistic:

$$z = \frac{0.69 - 0.57 - 0}{\sqrt{\frac{0.63 \times (1 - 0.63)}{789} + \frac{0.63 \times (1 - 0.63)}{789}}} = 4.94$$

(0.63 is the overall proportion of multi-taskers). The evidence is significant and clearly suggests this is a genuine difference: men are the multi-taskers.

Exercise 5.8

A survey of 80 voters finds that 65% are in favour of a particular policy. Test the hypothesis that the true proportion is 50%, against the alternative that a majority is in favour.

?

Exercise 5.9

A survey of 50 teenage girls found that on average they spent 3.6 hours per week chatting with friends over the internet. The standard deviation was 1.2 hours. A similar survey of 90 teenage boys found an average of 3.9 hours, with standard deviation 2.1 hours. Test if there is any difference between boys' and girls' behaviour.

?

Exercise 5.10

One gambler on horse racing won on 23 of his 75 bets. Another won on 34 out of 95. Is the second person a better judge of horses, or just luckier?

?

Hypothesis tests with small samples

As with estimation, slightly different methods have to be employed when the sample size is small ($n < 25$) and the population variance is unknown. When both of these conditions are satisfied, the t distribution must be used rather than the Normal, so a t test is conducted rather than a z test. This means consulting tables of the t distribution to obtain the critical value of a test, but otherwise the methods are similar. These methods will be applied to hypotheses about sample means only, since they are inappropriate for tests of a sample proportion, as was the case in estimation.

Testing the sample mean

A large chain of supermarkets sells 5000 packets of cereal in each of its stores each month. It decides to test-market a different brand of cereal in 15 of its stores. After a month the 15 stores have sold an average of 5200 packets each, with a standard deviation of 500 packets. Should all supermarkets switch to selling the new brand?

The sample information is

$$\bar{x} = 5200, s = 500, n = 15$$

From Chapter 4 the distribution of the sample mean from a small sample when the population variance is unknown is based upon

$$\frac{\bar{x} - \mu}{\sqrt{s^2/n}} \sim t_v \qquad (5.10)$$

with $\nu = n - 1$ degrees of freedom. The hypothesis test is based on this formula and is conducted as follows:

(1) H_0: $\mu = 5000$
    H_1: $\mu > 5000$
    (Only an improvement in sales is relevant.)
(2) Significance level: $\alpha = 1\%$ (chosen because the cost of changing brands is high).
(3) The critical value of the t distribution for a one-tail test at the 1% significance level with $\nu = -1 = 14$ degrees of freedom is $t^* = 2.62$.
(4) The test statistic is

$$t = \frac{\bar{x} - \mu}{\sqrt{s^2/n}} = \frac{5200 - 5000}{\sqrt{500^2/15}} = 1.55$$

(5) The null hypothesis is not rejected since the test statistic, 1.55, is less than the critical value, 2.62. It would probably be unwise to switch over to the new brand of cereal.

Testing the difference of two means

A survey of 20 British companies found an average annual expenditure on research and development of £3.7m with a standard deviation of £0.6m. A survey of 15 similar German companies found an average expenditure on research and development of £4.2m with standard deviation £0.9m. Does this evidence lend support to the view often expressed that Britain does not invest enough in research and development?

This is a hypothesis about the difference of two means, based on small sample sizes. The test statistic is again based on the t distribution, i.e.

$$\frac{(\bar{x}_1 - \bar{x}_2) - (\mu_1 - \mu_2)}{\sqrt{\dfrac{S^2}{n_1} + \dfrac{S^2}{n_2}}} \sim t_\nu \qquad (5.11)$$

where S^2 is the pooled variance (as given in equation (4.23)) and the degrees of freedom are given by $\nu = n_1 + n_2 - 2$.

The hypothesis test procedure is as follows:

(1) H_0: $\mu_1 - \mu_2 = 0$
    H_1: $\mu_1 - \mu_2 < 0$
    (a one-tail test because the concern is with Britain spending less than Germany.)
(2) Significance level: $\alpha = 5\%$.
(3) The critical value of the t distribution at the 5% significance level for a one-tail test with $\nu = n_1 + n_2 - 2 = 33$ degrees of freedom is approximately $t^* = 1.70$.
(4) The test statistic is based on equation (5.11):

$$t = \frac{(\bar{x}_1 - \bar{x}_2) - (\mu_1 - \mu_2)}{\sqrt{\dfrac{S^2}{n_1} + \dfrac{S^2}{n_2}}} = \frac{3.7 - 4.2 - 0}{\sqrt{\dfrac{0.55}{20} + \dfrac{0.55}{15}}} = -1.97$$

where S^2 is the pooled variance, calculated by

$$S^2 = \frac{(n_1 - 1)s_1^2 + (n_2 - 1)s_2^2}{n_1 + n_2 - 2} = \frac{19 \times 0.6^2 + 14 \times 0.9^2}{33} = 0.55$$

(5) The test statistic falls in the rejection region, $t < -t^*$, so the null hypothesis is rejected. The data do support the view that Britain spends less on R&D than Germany.

Exercise 5.11

It is asserted that parents spend, on average, £540 p.a. on toys for each child. A survey of 24 parents finds expenditure of £490, with standard deviation £150. Does this evidence contradict the assertion?

Exercise 5.12

A sample of 15 final-year students were found to spend on average 15 hours per week in the university library, with standard deviation 3 hours. A sample of 20 freshers found they spend on average 9 hours per week in the library, with standard deviation 5 hours. Is this sufficient evidence to conclude that finalists spend more time in the library?

Are the test procedures valid?

A variety of assumptions underlie each of the tests which we have applied above, and it is worth considering in a little more detail whether these assumptions are justified. This will demonstrate that one should not rely upon the statistical tests alone; it is important to retain one's sense of judgement.

The first test concerned the weekly turnover of a series of franchise operations. To justify the use of the Normal distribution underlying the test, the sample observations must be independently drawn. If, for example, all the sample franchises were taken from vibrant and growing cities and avoided those in less fortunate parts of the country, then in some sense the observations would not be independent, and furthermore the sample would not be representative of the whole. The answer to this would be to ensure the sample was properly stratified, representing different parts of the country. This type of sampling issue is covered in Chapter 9.

If one were using time-series data, as in the car factory comparison, similar issues arise. Do the 30 days represent independent observations or might there be an autocorrelation problem (e.g. if the sample days were close together in time)? Suppose that factory 2 suffered a breakdown of some kind which took three days to fix. Output would be reduced on three successive days and factory 2 would almost inevitably appear less efficient than factory 1. A look at the individual sample observations might be worthwhile, therefore, to see if there are any irregular patterns. It would have been altogether better if the samples had been collected on randomly chosen days over a longer time period to reduce the danger of this type of problem.

If the two factories both obtain their supplies from a common, but limited, source, then the output of one factory might not be independent of the output of the other. A high output of one factory would tend to be associated with a low output from the other, which has little to do with their relative efficiencies. This might leave the average difference in output unchanged but might increase the variance substantially (either a very high positive value of $\bar{x}_1 - \bar{x}_2$ or a very high negative value is obtained). This would lead to a low value of the test statistic and the conclusion of no difference in output. Any real difference in efficiency is masked by the common supplier problem. If the two samples are not independent, then the distribution of $\bar{x}_1 - \bar{x}_2$ may not be Normal.

Hypothesis tests and confidence intervals

Formally, two-tail hypothesis tests and confidence intervals are equivalent. Any value which lies within the 95% confidence interval around the sample mean cannot be rejected as the 'true' value using the 5% significance level in a hypothesis test using the same sample data. For example, our by now familiar accountant could construct a confidence interval for the firm's sales. This yields the 95% confidence interval

$$[4792, 5008] \tag{5.12}$$

Notice that the hypothesised value of 5000 is within this interval and that this value was not rejected by the hypothesis test carried out earlier. As long as the same confidence level is used for both procedures, they are equivalent.

Having said this, their interpretation is different. The hypothesis test forces us into the reject/do not reject dichotomy, which is rather a stark choice. We have already seen how it becomes more likely that a null hypothesis is rejected as the sample size increases. This problem does not occur with estimation. As the sample size increases the confidence interval gets narrower (around the unbiased point estimate) which is entirely beneficial. The estimation approach also tends to emphasise importance over significance in most people's minds. With a hypothesis test one might know that turnover is significantly different from 5000 without knowing how far from 5000 it actually is.

On some occasions a confidence interval is inferior to a hypothesis test, however. Consider the following case. In the United Kingdom only 72 out of 564 judges are women (12.8%). The Equal Opportunities Commission had earlier commented that since the appointment system is so secretive, it is impossible to tell if there is discrimination or not. What can the statistician say about this? No discrimination (in its broadest sense) would mean half of all judges would be women. Thus, the hypotheses are

H_0: $\pi = 0.5$ (no discrimination)
H_1: $\pi < 0.5$ (discrimination against women)

The sample data are $p = 0.128$, $n = 564$. The z score is

$$z = \frac{p - \pi}{\sqrt{\dfrac{\pi(1 - \pi)}{n}}} = \frac{0.128 - 0.5}{\sqrt{\dfrac{0.5 \times 0.5}{564}}} = -17.7$$

This is clearly significant (*and* 12.8% is a long way from 50%) so the null hypothesis is rejected. There is some form of discrimination somewhere against women (unless women choose not to be judges). But a confidence interval estimate of the 'true' proportion of female judges would be meaningless. To what population is this 'true' proportion related?

The lesson from all this is that differences exist between confidence intervals and hypothesis tests, despite their formal similarity. Which technique is more appropriate is a matter of judgement for the researcher. With hypothesis testing, the rejection of the null hypothesis at some significance level might actually mean a small (and unimportant) deviation from the hypothesised value. It should be remembered that the rejection of the null hypothesis based on a large sample of data is also consistent with the true value possibly being quite close to the hypothesised value.

Independent and dependent samples

The following example illustrates the differences between **independent samples** (as encountered so far) and **dependent samples** (also known as **matched** or **paired samples**) where slightly different methods of analysis are required. The example also illustrates how a particular problem can often be analysed by a variety of statistical methods.

Dependent samples occur, for example, when the same individuals are sampled twice, at two points in time. Alternatively, the observations in a first sample might be matched to or related in some way with the observations in the second sample. To ignore these facts in our analysis would be to ignore some potentially valuable information and hence not obtain the optimum results from the data.

To proceed via an example, suppose a company introduces a training programme to raise the productivity of its clerical workers, which is measured by the number of invoices processed per day. The company wants to know if the training programme is effective. How should it evaluate the programme? There is a variety of ways of going about the task, as follows:

- Take two (random) samples of workers, one trained and one not trained, and compare their productivity. This would comprise two independent samples.
- Take a sample of workers and compare their productivity before and after training. This would be a paired sample.
- Take two samples of workers, one to be trained and the other not. Compare the improvement of the trained workers with any change in the other group's performance over the same time period. This would consist of two independent samples but we are controlling for any time effects that are unrelated to the training.

We shall go through each method in turn, pointing out any possible difficulties.

Two independent samples

Suppose a group of 10 workers is trained and compared to a group of 10 non-trained workers, with the following data being relevant:

$$\bar{x}_T = 25.5 \qquad \bar{x}_N = 21.00$$
$$s_T = 2.55 \qquad s_N = 2.91$$
$$n_T = 10 \qquad n_N = 10$$

Thus, trained workers process 25.5 invoices per day compared to only 21 by non-trained workers. The question is whether this is significant, given that the sample sizes are quite small.

The appropriate test here is a t test of the difference of two sample means, as follows:

$$H_0: \mu_T - \mu_N = 0$$
$$H_1: \mu_T - \mu_N > 0$$

$$t = \frac{25.5 - 21.0}{\sqrt{\frac{7.49}{10} + \frac{7.49}{10}}} = 3.68$$

(7.49 is S^2, the pooled variance). The t statistic leads to rejection of the null hypothesis; the training programme does seem to be effective.

One problem with this test is that the two samples might have other differences apart from the effect of the training programme. This could be due either to simple random variation or to some selection factor. Poor workers might have been reluctant to take part in training, departmental managers might have selected better workers for training as some kind of reward, or better workers may have volunteered. In a well-designed experiment this should not be allowed to happen, of course, but we do not rule out the possibility. Hence we should consider ways of conducting a fairer test.

Paired samples

If we compare the same workers before and after training, then we are controlling for the inherent quality of the workers. We hence rule out this form of random variation which might otherwise weaken our test. We should therefore obtain a better idea of the true effect of the training programme. This is an example of paired or matched samples, where we can match up and compare the individual observations to each other, rather than just the overall averages. Suppose the sample data are as follows:

Worker	1	2	3	4	5	6	7	8	9	10
Before	21	24	23	25	28	17	24	22	24	27
After	23	27	24	28	29	21	24	25	26	28

In this case, the observations in the two samples are paired, and this has implications for the method of analysis. One *could* proceed by assuming these are two independent samples and conduct a t test. The summary data and results of such a test are:

$$\bar{x}_B = 23.50 \qquad \bar{x}_A = 25.5$$
$$s_B = 3.10 \qquad s_A = 2.55$$
$$n_B = 10 \qquad n_A = 10$$

The resulting test statistic is $t_{18} = 1.58$ which is not significant at the 5% level.

There are two problems with this test and its result. First, the two samples are not truly independent, since the before and after measurements refer to the same group of workers. Second, note that 9 out of 10 workers in the sample have shown an improvement, which is odd in view of the result found above, of no significant improvement. If the training programme really has no effect, then the probability of a single worker showing an improvement is $\frac{1}{2}$. The probability of nine or more workers showing an improvement is, by the Binomial method, $\left(\frac{1}{2}\right)^{10} \times 10C9 + \left(\frac{1}{2}\right)^{10}$, which is about one in a hundred. A very unlikely event seems to have occurred. Furthermore, the improvement is better measured as a proportion, which is 8.5% (25.5 versus 23.5), and any company would be pleased at such an improvement in productivity. Despite the lack of significance, it is worth investigating further.

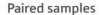

The t test used above is inappropriate because it does not make full use of the information in the sample. It does not reflect the fact, for example, that the before and after scores, 21 and 23, relate to the same worker. The Binomial calculation above does reflect this matching. A re-ordering of the data would not affect the t test result, but would affect the Binomial, since a different number of workers would now show an improvement. Of course, the Binomial does not use all the sample information either – it dispenses with the actual productivity data for each worker and replaces it with 'improvement' or 'no improvement'. It disregards the amount of improvement for each worker.

Better use of the sample data comes by measuring the improvement for each worker, as follows (if a worker had deteriorated, this would be reflected by a negative number):

Worker	1	2	3	4	5	6	7	8	9	10
Improvement	2	3	1	3	1	4	0	3	2	1

These new data can be treated by single sample methods, and account is taken both of the actual data values and of the fact that the original samples were dependent (re-ordering of the data would produce different, and incorrect, improvement figures). The summary statistics of the new data are as follows:

$$\bar{x} = 2.00, s = 1.247, n = 10$$

The null hypothesis of no improvement can now be tested as follows:

$$H_0: \mu = 0$$
$$H_1: \mu > 0$$
$$t = \frac{2.0 - 0}{\sqrt{\frac{1.247^2}{10}}} = 5.07$$

This is significant at the 5% level, so the null hypothesis of no improvement is rejected. The correct analysis of the sample data has thus reversed the previous conclusion. It is perhaps surprising that treating the same data in different ways leads to such a difference in the results. It does illustrate the importance of using the appropriate method.

Matters do not end here, however. Although we have discovered an improvement, this might be due to other factors apart from the training programme. For example, if the before and after measurements were taken on different days of the week (that Monday morning feeling . . .), or if one of the days were sunnier, making people feel happier and therefore more productive, this might bias the results. These may seem trivial examples but these effects do exist, for example the 'Friday afternoon car', which has more faults than the average, constructed when workers are thinking ahead to the weekend.

The way to solve this problem is to use a control group, so called because extraneous factors are controlled for, in order to isolate the effects of the factor under investigation. In this case, the productivity of the control group would be measured (twice) at the same times as that of the training group, although no training would be given to them. Ideally, the control group would be matched on other factors (e.g. age) to the treatment group to avoid other factors influencing the results. Suppose that the average improvement of the control group were 0.5 invoices per day with standard deviation 1.0 (again for a group of 10). This can be compared with the improvement of the training group via the two-sample t test, giving

$$t = \frac{2.0 - 0.5}{\sqrt{\frac{1.13^2}{10} + \frac{1.13^2}{10}}} = 2.97$$

(1.13^2 is the pooled variance). This adds more support to the finding that the training programme is of value.

Exercise 5.13

A group of students' marks on two tests, before and after instruction, were as follows:

Student	1	2	3	4	5	6	7	8	9	10	11	12
Before	14	16	11	8	20	19	6	11	13	16	9	13
After	15	18	15	11	19	18	9	12	16	16	12	13

Test the hypothesis that the instruction had no effect, using both the independent sample and paired sample methods. Compare the two results.

Issues with hypothesis testing

The above exposition has served to illustrate how to carry out a hypothesis test and the rationale behind it. However, the methodology has been subject to criticisms, some of which we have already discussed:

- The decision-making paradigm is problematic since we are not sure what we are choosing between (the alternative hypothesis is vague).
- The 5% significance level is just a convention.
- The focus on 'significance' leads to a neglect of the effect size.
- The experimental (i.e. alternative) hypothesis is never itself tested. This is a pity as it is often the one favoured by the researcher.
- The process is easily and often misunderstood. People tend to confuse the probability of observing the sample data assuming that the null is true with the probability that the null is true given the data. More succinctly, $\Pr(\text{data} \mid H_0)$ is confused with $\Pr(H_0 \mid \text{data})$. The significance level (P-value) relates to the former, not the latter.

There are other problems too which we have not yet discussed. It is common in research to be looking at several hypothesis tests rather than just one. Suppose we are trying to improve teaching of statistics to a group of students. We try five alternative approaches: smaller class sizes, regular assignments, online tests, etc. We test each of these with a conventional hypothesis test at the 5% significance level. What is our chance of a Type I error, assuming none of these innovations truly works? The chance that all five come up with 'no effect' is $0.95^5 = 0.77$. Hence the overall Type I error probability is $1 - 0.77 = 0.23$. There is a 23% chance that we (erroneously) find something significant. This reveals the danger of looking for things in the data – there is a good chance you will find something, but it will likely be a false positive.

This problem is pervasive, and even the researcher himself herself might be unaware of doing it. He/she investigates whether workers are less productive on a Friday but finds no significant effect. So he/she wonders whether there is a Monday effect and tests that. Or (even worse) he/she notices in the data that productivity looks low on a Wednesday, so he/she tests that. The results of these tests are largely meaningless and the true significance level (P-value) may be much higher than 5%.

Why is it, therefore, that hypothesis testing is so frequently used? One attraction is that it provides clear guidance on what to do, which does not require too much thought to apply. Follow the procedures and you will obtain a result. Moreover, this method will be generally accepted by others and is needed if the researcher wishes to get published.

What can be done to avoid some of these pitfalls? Some suggestions are as follows.

- If doing a hypothesis test, plan it in detail *before* obtaining the data, i.e. the null hypothesis (or hypotheses) to test, the significance level, how to measure the variables appropriately (e.g. look at wage rates or total earnings?), sample size and so on. Stick to these choices, do not alter them in the light of what you might observe in the data.
- Do not be overawed by significance. Look at the effect size as well. In fact, look at the effect size first. Your significant result could be unimportant.
- Calculating a confidence interval might be a better way of analysing your data than a hypothesis test. It gives more focus to the effect size while also telling you about the reliability of your finding.
- Do not rely only on a hypothesis test, there are lots of ways of gaining insight into a problem. Look at the data using descriptive statistics and charts (and present these results to the reader). Perhaps your significant result occurs because of a few outliers in the data.

If possible, validate your findings on new data. If the effect you have found is genuine, it ought to occur in a new sample. If your original data suggest a new hypothesis to you, you must get new data to test it, you cannot use the same data to test the hypothesis suggested by those data.

Exercise 5.14

Generally, to be published in an academic journal, a study needs to reject the null hypothesis at the 5% significance level. Of all studies published in journals, what proportion of them are likely to be Type I errors, i.e. false positives?

Exercise 5.15

Studies published in journals will usually have an effect size for the subject of study, e.g. smaller class sizes improve pupils' maths skills by 10% points. Would you expect a published effect size to be an underestimate, an accurate (unbiased) estimate, or an overestimate of the true effect size? Note that only 'significant' results get published.

Summary

- Hypothesis testing is the set of procedures for deciding whether a hypothesis is true or false. When conducting the test, we presume the hypothesis, termed the null hypothesis, is true until it is proved false on the basis of some sample evidence.

- If the null is proved false, it is rejected in favour of the alternative hypothesis. The procedure is conceptually similar to a court case, where the defendant is presumed innocent until the evidence proves otherwise.
- Not all decisions turn out to be correct, and there are two types of error that can be made. A Type I error is to reject the null hypothesis when it is in fact true. A Type II error is not to reject the null when it is false.
- Choosing the appropriate decision rule (for rejecting the null hypothesis) is a question of trading off Type I and Type II errors. Because the alternative hypothesis is imprecisely specified, the probability of a Type II error usually cannot be specified.
- The rejection region for a test is therefore chosen to give a 5% probability of making a Type I error (sometimes a 1% probability is chosen). The critical value of the test statistic (sometimes referred to as the critical value of the test) is the value which separates the acceptance and rejection regions.
- The decision is based upon the value of a test statistic, which is calculated from the sample evidence and from information in the null hypothesis.

$$\left(\text{e.g. } z = \frac{\bar{x} - \mu}{s/\sqrt{n}}\right)$$

- The null hypothesis is rejected if the test statistic falls into the rejection region for the test (i.e. it exceeds the critical value).
- For a two-tail test there are two rejection regions, corresponding to very high and very low values of the test statistic.
- Instead of comparing the test statistic to the critical value, an equivalent procedure is to compare the Prob-value of the test statistic with the significance level. The null is rejected if the Prob-value is less than the significance level.
- The power of a test is the probability of a test correctly rejecting the null hypothesis. Some tests have low power (e.g. when the sample size is small) and therefore are not very useful.

Key terms and concepts

alternative hypothesis
composite hypothesis
critical value
decision rule
dependent samples
effect size
independent samples
matched samples
non-rejection region
null or maintained hypothesis
one- and two-tail tests
paired samples
power
prior information
Prob-value
rejection region
significance level
test statistic
Type I and
Type II errors

References

Michael Oakes, *Statistical Inference: A Commentary for the Social and Behavioural Sciences*, Wiley, 1986.

McCloskey, D., and S. Ziliak, Size Matters: the Standard Error of Regressions in the *American Economic Review*, *Journal of Socio-Economics*, 33, 527–46, 2004.

Formulae used in this chapter

Formula	Description	Notes
$z = \dfrac{\bar{x} - \mu}{\sqrt{s^2/n}}$	Test statistic for H_0: mean $= \mu$	Large samples. For small samples, distributed as t with $\nu = n - 1$ degrees of freedom
$z = \dfrac{p - \pi}{\sqrt{\dfrac{\pi(1 - \pi)}{n}}}$	Test statistic for H_0: true proportion $= \pi$	Large samples
$z = \dfrac{(\bar{x}_1 - \bar{x}_2) - (\mu_1 - \mu_2)}{\sqrt{\dfrac{s_1^2}{n_1} + \dfrac{s_2^2}{n_2}}}$	Test statistic for H_0: $\mu_1 - \mu_2 = 0$	Large samples
$t = \dfrac{(\bar{x}_1 - \bar{x}_2) - (\mu_1 - \mu_2)}{\sqrt{\dfrac{S^2}{n_1} + \dfrac{S^2}{n_2}}}$	Test statistic for H_0: $\mu_1 - \mu_2 = 0$	Small samples. $S^2 = \dfrac{(n_1 - 1)s_1^2 + (n_2 - 1)s_2^2}{n_1 + n_2 - 2}$ Degrees of freedom $\nu = n_1 + n_2 - 2$
$z = \dfrac{(p_1 - p_2) - (\pi_1 - \pi_2)}{\sqrt{\dfrac{\pi_1(1 - \pi_1)}{n_1} + \dfrac{\pi_2(1 - \pi_2)}{n_2}}}$	Test statistic for H_0: $\pi_1 - \pi_2 = 0$	Large samples $\pi = \dfrac{n_1 p_1 + n_2 p_2}{n_1 + n_2}$

Problems

Some of the more challenging problems are indicated by highlighting the problem number in colour.

5.1 Answer true or false, with reasons if necessary.

(a) There is no way of reducing the probability of a Type I error without simultaneously increasing the probability of a Type II error.

(b) The probability of a Type I error is associated with an area under the distribution of $\bar{x}$ assuming the null hypothesis to be true.

(c) It is always desirable to minimise the probability of a Type I error.

(d) A larger sample, ***ceteris paribus***, will increase the power of a test.

(e) The significance level is the probability of a Type II error.

(f) The confidence level is the probability of a Type II error.

5.2 Consider the investor in the text, seeking out companies with weekly turnover of at least £5000. He or she applies a one-tail hypothesis test to each firm, using the 5% significance level. State whether each of the following statements is true or false (or not known) and explain why.

(a) 5% of his or her investments are in companies with less than £5000 turnover.

(b) 5% of the companies he ***fails*** to invest in have turnover greater than £5000 per week.

(c) He invests in 95% of all companies with turnover of £5000 or over.

5.3 A coin which is either fair or has two heads is to be tossed twice. You decide on the following decision rule: if two heads occur you will conclude it is a two-headed coin, otherwise you will presume it is fair. Write down the null and alternative hypotheses and calculate the probabilities of Type I and Type II errors.

5.4 In comparing two medical treatments for a disease, the null hypothesis is that the two treatments are equally effective. Why does making a Type I error not matter? What significance level for the test should be set as a result?

5.5 A firm receives components from a supplier, which it uses in its own production. The components are delivered in batches of 2000. The supplier claims that there are only 1% defective components on average from its production. However, production occasionally gets out of control and a batch is produced with 10% defective components. The firm wishes to intercept these low-quality batches, so a sample of size 50 is taken from each batch and tested. If two or more defectives are found in the sample, then the batch is rejected.

(a) Describe the two types of error the firm might make in assessing batches of components.

(b) Calculate the probability of each type of error given the data above.

(c) If, instead, samples of size 30 were taken and the batch rejected if one or more rejects were found, how would the error probabilities be altered?

(d) The firm can alter the two error probabilities by choice of sample size and rejection criteria. How should it set the relative sizes of the error probabilities

(i) if the product might affect consumer safety?

(ii) if there are many competitive suppliers of components?

(iii) if the costs of replacement under guarantee are high?

5.6 Computer diskettes (the precursor to USB drives) which do not meet the quality required for high-density diskettes are sold as low-density diskettes (storing less data) for 80 pence each. High-density diskettes are sold for £1.20 each. A firm samples 30 diskettes from each batch of 1000 and if any fail the quality test, the whole batch is sold as double-density diskettes. What are the types of error possible and what is the cost to the firm of a Type I error?

5.7 Testing the null hypothesis that $\mu = 10$ against $\mu > 10$, a researcher obtains a sample mean of 12 with standard deviation 6 from a sample of 30 observations. Calculate the z score and the associated Prob-value for this test.

5.8 Given the sample data $\bar{x} = 45$, $s = 16$, $n = 50$, at what level of confidence can you reject H_0: $\mu = 40$ against a two-sided alternative?

5.9 What is the power of the test carried out in Problem 5.3?

5.10 Given the two hypotheses

$$H_0: \mu = 400$$
$$H_1: \mu = 415$$

and $\sigma^2 = 1000$ (for both hypotheses):

(a) Draw the distribution of $\bar{x}$ under both hypotheses.

(b) If the decision rule is chosen to be: reject H_0 if $\bar{x} \geq 410$ from a sample of size 40, find the probability of a Type II error and the power of the test.

(c) What happens to these answers as the sample size is increased? Draw a diagram to illustrate.

5.11 Given the following sample data:

$$\bar{x} = 15 \qquad s^2 = 270 \qquad n = 30$$

test the null hypothesis that the true mean is equal to 12, against a two-sided alternative hypothesis. Draw the distribution of $\bar{x}$ under the null hypothesis and indicate the rejection regions for this test.

5.12 From experience it is known that a certain brand of tyre lasts, on average, 15 000 miles with standard deviation 1250. A new compound is tried and a sample of 120 tyres yields an average life of 15 150 miles, with the same standard deviation. Are the new tyres an improvement? Use the 5% significance level.

5.13 Test H_0: $\pi = 0.5$ against H_0: $\pi \neq 0.5$ using $p = 0.45$ from a sample of size $n = 35$.

5.14 Test the hypothesis that 10% of your class or lecture group are left-handed.

5.15 Given the following data from two independent samples:

$$\bar{x}_1 = 115 \qquad \bar{x}_2 = 105$$
$$s_1 = 21 \qquad s_2 = 23$$
$$n_1 = 49 \qquad n_2 = 63$$

test the hypothesis of no difference between the population means against the alternative that the mean of population 1 is greater than the mean of population 2.

5.16 A transport company wants to compare the fuel efficiencies of the two types of lorry it operates. It obtains data from samples of the two types of lorry, with the following results:

Type	Average mpg	Std devn	Sample size
A	31.0	7.6	33
B	32.2	5.8	40

Test the hypothesis that there is no difference in fuel efficiency, using the 99% confidence level.

5.17 (a) A random sample of 180 men who took the driving test found that 103 passed. A similar sample of 225 women found that 105 passed. Test whether pass rates are the same for men and women.

(b) If you test whether the group of people who passed the driving test contained the same proportion of men as the group of people who failed, what result would you expect to find? Carry out the test to check.

(c) Is your finding in part (b) inevitable or one that just arises with these data? Try to support your response with a proof.

5.18 (a) A pharmaceutical company testing a new type of pain reliever administered the drug to 30 volunteers experiencing pain. Sixteen of them said that it eased their pain. Does this evidence support the claim that the drug is effective in combating pain?

(b) A second group of 40 volunteers were given a placebo instead of the drug. Thirteen of them reported a reduction in pain. Does this new evidence cast doubt upon your previous conclusion?

5.19 (a) A random sample of 20 observations yielded a mean of 40 and standard deviation 10. Test the hypothesis that $\mu = 45$ against the alternative that it is not. Use the 5% significance level.

(b) What assumption are you implicitly making in carrying out this test?

5.20 A photo processing company sets a quality standard of no more than 10 complaints per week on average. A random sample of 8 weeks showed an average of 13.6 complaints, with standard deviation 5.3. Is the firm achieving its quality objective?

5.21 Two samples are drawn. The first has a mean of 150, variance 50 and sample size 12. The second has mean 130, variance 30 and sample size 15. Test the hypothesis that they are drawn from populations with the same mean.

5.22 (a) A consumer organisation is testing two different brands of battery. A sample of 15 of brand *A* shows an average useful life of 410 hours with a standard deviation of 20 hours. For brand *B*, a sample of 20 gave an average useful life of 391 hours with standard deviation 26 hours. Test whether there is any significant difference in battery life.

(b) What assumptions are being made about the populations in carrying out this test?

5.23 The output of a group of 11 workers before and after an improvement in the lighting in their factory is as follows:

Before	52	60	58	58	53	51	52	59	60	53	55
After	56	62	63	50	55	56	55	59	61	58	56

Test whether there is a significant improvement in performance

(a) assuming these are independent samples,

(b) assuming they are dependent.

5.24 Another group of workers were tested at the same times as those in Problem 5.23, although their department *also* introduced rest breaks into the working day.

Before	51	59	51	53	58	58	52	55	61	54	55
After	54	63	55	57	63	63	58	60	66	57	59

Does the introduction of rest days alone appear to improve performance?

5.25 Discuss in general terms how you might 'test' the following:

(a) astrology

(b) extra-sensory perception

(c) the proposition that company takeovers increase profits.

5.26 **(Project)** Can your class tell the difference between tap water and bottled water? Set up an experiment as follows: fill r glasses with tap water and $n - r$ glasses with bottled water. The subject has to guess which is which. If he or she gets more than p correct, you conclude he or she can tell the difference. Write up a report of the experiment including:

(a) a description of the experimental procedure

(b) your choice of n, r and p, with reasons

(c) the power of your test

(d) your conclusions.

5.27 **(Computer project)** Use the $= RAND()$ function in your spreadsheet to create 100 samples of size 25 (which are effectively all from the same population). Compute the mean and standard deviation of each sample. Calculate the *z* score for each sample, using a hypothesised mean of 0.5 (since the $= RAND()$ function chooses a random number in the range 0–1).

(a) How many of the *z* scores would you expect to exceed 1.96 in absolute value? Explain why.

(b) How many do exceed this? Is this in line with your prediction?

(c) Graph the sample means and comment upon the shape of the distribution. Shade in the area of the graph beyond $z = \pm 1.96$.

5.28 **(Project)** This is similar to Problem 5.26 but concerns digital music files. There is debate about whether listeners can tell the difference between high-quality WAV files and compressed MP3 files. Obtain the same song in both formats (most music players will convert a WAV file to MP3) and see if a listener can discern which is which. Some of your class colleagues might be better at this than others. You need to consider the same issues as in Problem 5.26.

The ABX comparison procedure would be interesting to follow (see http://wiki.hydrogenaudio.org/index.php?title=ABX) and you can download the WinABX program which automates much of the procedure (google 'WinABX' to find it).

Answers to exercises

Exercise 5.1

(a) H_0: crime is the same as last year, H_1: crime has increased.

(b) Type I error – concluding crime has risen, when in fact it has not. Type II – concluding it has not risen, when, in fact, it has. The cost of the former might be employing more police officers which are not in fact warranted; of the latter, not employing more police to counter the rising crime level. (The *Economist* (19 July 2003) reported that 33% of respondents to a survey in the United Kingdom felt that crime had risen in the previous two years, only 4% thought that it had fallen. In fact, crime had fallen slightly, by about 2%. A lot of people were making a Type I error, therefore.)

Exercise 5.2

(a) $z = (108 - 100)/\sqrt{36} = 1.33$. The area in the tail beyond 1.33 is 9.18%, which is the probability of a Type I error.

(b) $z = 1.64$ cuts off 5% in the upper tail of the distribution, hence we need the decision rule to be at $\bar{x} + 1.64 \times s/\sqrt{n} = 100 + 1.64 \times \sqrt{36} = 109.84$.

(c) Under H_1: $\mu = 112$, we can write $\bar{x} \sim N(112, 900/25)$. (We assume the same variance under both H_0 and H_1 in this case.) Hence $z = (108 - 112)/\sqrt{36} = -0.67$. This gives an area in the tail of 25.14%, which is the Type II error probability. Usually, however, we do not have a precise statement of the value of μ under H_1 so cannot do this kind of calculation.

Exercise 5.3

$\alpha = 0.05$ (significance level chosen), hence the critical value is $z^* = 1.96$ (two-tail test). The test statistic is $z = (530 - 500)/(90/\sqrt{30}) = 1.83 < 1.96$ so H_0 is not rejected at the 5% significance level.

Exercise 5.4

All of the statements are false. Any statement 'proving' or 'disproving' a hypothesis is wrong, as is one about the probability of a hypothesis being true. That deals with (a) to (d). (e) looks more plausible, but it asks the probability the hypothesis is true after you have rejected it. Again, this asks for a probability about a hypothesis. (f) is called the **replication fallacy**. It assumes the null is true but we do not know this from a sample.

It might help to recall that the P-value is the probability of obtaining such sample data, assuming the null is true. We may write this as $\Pr(\text{data} \mid H_0)$, like a conditional probability. The questions ask, in different ways, for $\Pr(H_0 \mid \text{data})$ which is something quite different.

Exercise 5.5

One wants to avoid making a Type I error if possible, i.e. rejecting H_0 when true. Hence, set a low significance level (1%) so that H_0 is rejected only by very strong evidence.

Exercise 5.6

(a) (i) Reject. The Prob-value should be halved, to 0.0385, which is less than 5%. Alternatively, think of comparing $1.77 > 1.64$, the one-tail critical value. (ii) Do not reject, the Prob-value is greater than 5%; equivalently $1.77 < 1.96$.

(b) In this case, the null is not rejected in both cases. In the one-tailed case, $0.0385 > 1\%$, so the null is not rejected.

Exercise 5.7

(a) We need to solve $z^* = 1.64 = \dfrac{\bar{x} - 160}{50/\sqrt{400}}$ which yields $\bar{x} = 164.1$ as the cutoff point.

(b) (i) The power of the test is obtained from the new z score: $z = \dfrac{164.1 - 163}{50/\sqrt{400}} = 0.44$, which cuts off 33% in the upper tail, and is the power of the test. (ii) The z score is now $z = \dfrac{164.1 - 166}{50/\sqrt{400}} = -0.76$, which cuts off 78% in the upper tail (note we have a negative z score yet want the right-hand tail, so need the complement of the value given in Table A2).

Exercise 5.8

$$z = \frac{0.65 - 0.5}{\sqrt{\dfrac{0.5 \times 0.5}{80}}} = 2.68$$

hence the null is decisively rejected. $z^* = 1.64$ (one-tailed test).

Exercise 5.9

We have the data: $\bar{x}_1 = 3.6$, $s_1 = 1.2$, $n_1 = 50$; $\bar{x}_2 = 3.9$, $s_2 = 2.1$, $n_2 = 90$. The null hypothesis is H_0: $\mu_1 = \mu_2$ versus H_1: $\mu_1 \neq \mu_2$. The test statistic is

$$z = \frac{(\bar{x}_1 - \bar{x}_2) - (\mu_1 - \mu_2)}{\sqrt{\dfrac{s_1^2}{n_1} + \dfrac{s_2^2}{n_2}}} = \frac{(3.6 - 3.9) - 0}{\sqrt{\dfrac{1.2^2}{50} + \dfrac{2.1^2}{90}}} = -1.08$$

The test statistic (absolute value) does not exceed the critical value of 1.96 so the null is not rejected at the 5% significance level.

Exercise 5.10

The evidence is $p_1 = 23/75$, $n_1 = 75$, $p_2 = 34/95$, $n_2 = 95$. The hypothesis to be tested is H_0: $\pi_1 - \pi_2 = 0$ versus H_1: $\pi_1 - \pi_2 < 0$. Before calculating the test statistic, we must calculate the pooled variance as

$$\hat{\pi} = \frac{n_1 p_1 + n_2 p_2}{n_1 + n_2} = \frac{75 \times 0.3067 + 95 \times 0.3579}{75 + 95} = 0.3353$$

The test statistic is then

$$z = \frac{0.3067 - 0.3579 - 0}{\sqrt{\dfrac{0.3353 \times (1 - 0.3353)}{75} + \dfrac{0.3353 \times (1 - 0.3353)}{95}}} = -0.70$$

This is less in absolute magnitude than 1.64, the critical value of a one tailed test, so the null is not rejected. The second gambler is just luckier than the first, we conclude. We have to be careful about our interpretation, however: one of the gamblers might prefer longer-odds bets, so wins less often but gets more money each time. Hence, this may not be a fair comparison.

Exercise 5.11

We shall treat this as a two-tailed test, although a one-tailed test might be justified if there were other evidence that spending had fallen. The hypothesis is H_0: $\mu = 540$ versus H_1: $\mu \neq 540$. Given the sample evidence, the test statistic is

$$t = \frac{\bar{x} - \mu}{\sqrt{s^2/n}} = \frac{490 - 540}{\sqrt{150^2/24}} = -1.63$$

The critical value of the t distribution for 23 degrees of freedom is 2.069, so the null is not rejected.

Exercise 5.12

The hypothesis to test is H_0: $\mu_F - \mu_N = 0$ versus H_1: $\mu_F - \mu_N > 0$ (F indexes finalists, N the new students). The pooled variance is calculated as

$$S^2 = \frac{(n_1 - 1)s_1^2 + (n_2 - 1)s_2^2}{n_1 + n_2 - 2} = \frac{14 \times 3^2 + 19 \times 5^2}{33} = 18.21$$

The test statistic is

$$t = \frac{(\bar{x}_1 - \bar{x}_2) - (\mu_1 - \mu_2)}{\sqrt{\frac{S^2}{n_1} + \frac{S^2}{n_2}}} = \frac{(15 - 9) - 0}{\sqrt{\frac{18.21}{15} + \frac{18.21}{20}}} = 4.12$$

The critical value of the t distribution with $15 + 20 - 2 = 33$ degrees of freedom is approximately 1.69 (5% significance level, for a one-tailed test). Thus the null is decisively rejected and we conclude finalists do spend more time in the library.

Exercise 5.13

For the case of independent samples, we obtain $\bar{x}_1 = 13$, $\bar{x}_2 = 14.5$, $s_1 = 4.29$, $s_2 = 3.12$, with $n = 12$ in both cases. The test statistic is therefore

$$t = \frac{(\bar{x}_1 - \bar{x}_2) - (\mu_1 - \mu_2)}{\sqrt{\frac{S^2}{n_1} + \frac{S^2}{n_2}}} = \frac{13 - 14.5 - 0}{\sqrt{\frac{14.05}{12} + \frac{14.05}{12}}} = -0.98$$

with pooled variance

$$S^2 = \frac{(n_1 - 1)s_1^2 + (n_2 - 1)s_2^2}{n_1 + n_2 - 2} = \frac{11 \times 4.29^2 + 11 \times 3.12^2}{22} = 14.05$$

The null of no effect is therefore accepted. By the method of paired samples, we have a set of improvements as follows:

Student	1	2	3	4	5	6	7	8	9	10	11	12
Improvement	1	2	4	3	−1	−1	3	1	3	0	3	0

The mean of these is 1.5 and the variance is 3. The t statistic is therefore

$$t = \frac{1.5 - 0}{\sqrt{3/12}} = 3$$

This now conclusively rejects the null hypothesis (critical value 1.8), in stark contrast to the former method. The difference arises because 10 out of 12 students have improved or done as well as before, only two have fallen back (slightly). The gain in marks is modest but applies consistently to nearly all candidates.

Another way to look at this question would be to ask, what is the probability of 8 (or more) improvements out of 12 students? This is the basis of **Wilcoxon's Sign Test.** For this we discard the two observations with zero improvement, leaving 10. If there were truly no improvement, we would expect 5 improvements out of 10, whereas we actually have 8. The probability of 8 or more improvements is given by the Binomial distribution with $n = 10$ and $P = 0.5$. Hence, $\Pr(8) + \Pr(9) + \Pr(10) = 0.5^{10} \times 10C8 + 0.5^{10} \times 10C9 + 0.5^{10} \times 10C10 = 0.055$ or 5.5%. This does not quite meet the criterion of 5% significance so we cannot reject the null hypothesis of no improvement. This test fails to find improvement because it discards information about the extent of improvement and hence is not a very powerful test.

Exercise 5.14

You might be tempted to answer 5% to this. However, the question asks for $\Pr(H_0 \text{ true} \mid H_0 \text{ rejected})$, or $\Pr(H_0 \mid \text{R})$ for short. H_0 indicates there is truly no effect, R indicates that H_0 is rejected and hence the study is published (with the wrong conclusion). Using Bayes' theorem:

$$\Pr(H_0 \mid R) = \frac{\Pr(R \mid H_0) \times \Pr(H_0)}{\Pr(R \mid H_0) \times \Pr(H_0) + \Pr(R \mid H_1) \times \Pr(H_1)}$$

We know that $\Pr(R \mid H_0) = 0.05$ but we have to assume some values for $\Pr(H_0)$ and $\Pr(R \mid H_1)$. If we let $\Pr(H_0) = 0.2$ (20% of hypotheses tested are in fact 'null') and $\Pr(R \mid H_1) = 0.6$ (this is the power of the test), then we obtain $\Pr(H_0 \mid R) = 0.020$, only 2% of published studies are Type I errors. However, if we set $\Pr(H_0) = 0.8$ (researchers test lots of crazy ideas) and have a low-powered test, $\Pr(R \mid H_1) = 0.4$, then we find that one-third of all published studies are false positives.

Exercise 5.15

One would expect published studies to overestimate the effect size. This is because any study's estimate of the effect is subject to random error. If the error is positive, the effect is still significant and so over-estimates will get published. If the error is negative, there will be an underestimate, and, if this means that the 5% significance threshold is not met, then the result is not published. Hence some under-estimates are not published, and published estimates will on average be an over-estimate of the true effect size.

6 The χ^2 and F distributions

Contents

Learning outcomes

By the end of this chapter you should be able to:

- **understand the uses of two new probability distributions: χ^2 and F**
- **construct confidence interval estimates for a variance**
- **perform hypothesis tests concerning variances**
- **analyse and draw inferences from data contained in contingency tables**
- **construct a simple analysis of variance table and interpret the results.**

Introduction

The final two distributions to be studied are the χ^2 (pronounced 'kye-squared') and F distributions. Both of these distributions have a variety of uses, the most common of which are illustrated in this chapter. These distributions allow us to extend some of the estimation and testing procedures covered in Chapters 4 and 5. The χ^2 distribution allows us to establish confidence interval estimates for a variance, just as the Normal and t distributions were used in the case of a mean. Further, just as the Binomial distribution was used to examine situations where the result of an experiment could be either 'success' or 'failure', the χ^2 distribution also allows us to analyse situations where there are more than two categories of outcome. The F distribution enables us to conduct hypothesis tests regarding the equality of two variances and also to make comparisons between the means of multiple samples, not just two. The F distribution also arises in Chapter 7 and 8 on regression analysis.

The χ^2 distribution

The χ^2 distribution has a number of uses. In this chapter we make use of it in three ways:

- To calculate a confidence interval estimate of the population variance.
- To compare actual observations on a variable with the (theoretically) expected values.
- To test for association between two variables in a contingency table.

The use of the distribution is in many ways similar to the Normal and t distributions already encountered. Once again, it is actually a family of distributions depending upon a single parameter, the degrees of freedom, in a similar fashion to the t distribution. The number of degrees of freedom can have slightly different interpretations, depending upon the particular problem, but is often related to sample size in some way. Some typical χ^2 distributions are drawn in Figure 6.1 for different values of the parameter. Note the distribution has the following characteristics:

- It is always non-negative.
- It is skewed to the right.
- It becomes more symmetric as the number of degrees of freedom increases.

Figure 6.1
The χ^2 distribution with different degrees of freedom

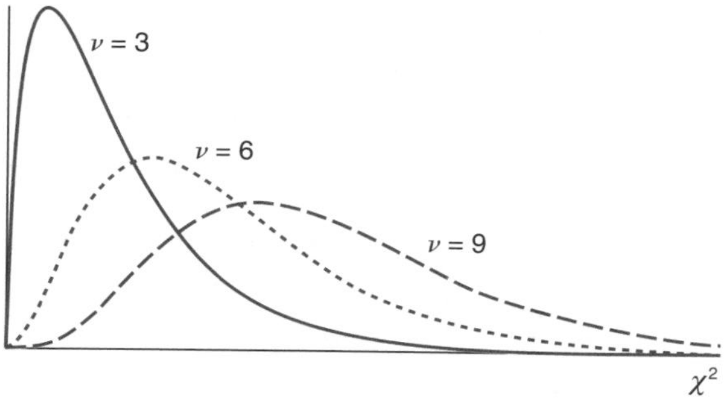

Using the χ^2 distribution to construct confidence intervals is done in the usual way, by using the critical values of the distribution (given in Table A4 (see page 452)) which cut off an area $\alpha/2$ in each tail of the distribution. For hypothesis tests, a rejection region is defined which cuts off an area α in either one or both tails of the distribution, whichever is appropriate. These principles should be familiar from previous chapters, so they are not repeated in detail. The following examples show how this works for the χ^2 distribution.

Estimating a variance

The sample variance is also a random variable like the mean; it takes on different values from sample to sample. We can therefore ask the usual question: given a sample variance, what can we infer about the true value?

To give an example, we use the data on spending by Labour boroughs in the example in Chapter 4 (see page 164). In that sample of 20 boroughs, the average spending on administration was £175 (per taxpayer), with standard deviation 25 (and hence variance of 625). What can we say about the true variance and standard deviation?

We work in terms of variances (this is more convenient when using the χ^2 distribution), taking the square root when we need to refer to the standard deviation. First of all, the sample variance is an unbiased estimator of the population variance[1], $E(s^2) = \sigma^2$, so we may use this as our point estimate, which is therefore 625. To construct the confidence interval around this we need to know about the distribution of s^2. Unfortunately, this does not have a convenient probability distribution, so we transform it to

$$\frac{(n-1)s^2}{\sigma^2} \tag{6.1}$$

which does have a χ^2 distribution, with $\nu = n - 1$ degrees of freedom. Again, we state this without a formal mathematical proof.

To construct the 95% confidence interval around the point estimate, we proceed in a similar fashion to the Normal or t distribution. First, we find the critical values of the χ^2 distribution which cut off 2.5% in each tail. These are no longer symmetric around zero as was the case with the standard Normal and t distributions. Table 6.1 shows an excerpt from the χ^2 table which is given in full in Table A4 in the Appendix at the end of the text (see page 452).

Like the t distribution, the first column gives the degrees of freedom, so we require the row corresponding to $\nu = n - 1 = 19$.

- For the *left-hand* critical value (cutting off 2.5% in the left-hand tail), we look at the column headed '0.975', representing 97.5% in the right-hand tail. This critical value is 8.91.
- For the *right-hand* critical value, we look up the column headed '0.025' (2.5% in the right-hand tail), giving 32.85.

The 95% confidence interval for $(n - 1)s^2/\sigma^2$, therefore, lies between these two values, i.e.

$$\left[8.91 \le \frac{(n-1)s^2}{\sigma^2} \le 32.85\right] \tag{6.2}$$

[1]This was stated, without proof, in Chapter 1, see page 38.

Table 6.1 Excerpt from Table A4 – the χ^2 distribution

ν	0.99	0.975	...	0.10	0.05	0.025	0.01
1	0.0002	0.0010	...	2.7055	3.8415	5.0239	6.6349
2	0.0201	0.0506	...	4.6052	5.9915	7.3778	9.2104
⋮	⋮	⋮	...	⋮	⋮	⋮	⋮
18	7.0149	8.2307	...	25.9894	28.8693	31.5264	34.8052
19	7.6327	8.9065	...	27.2036	30.1435	32.8523	36.1908
20	8.2604	9.5908	...	28.4120	31.4104	34.1696	37.5663

Note: The two critical values are found at the intersections of the shaded row and columns. Alternatively, you can use Excel. Since Excel 2010 the formula = *CHISQ.INV.RT* (0.975, 19) gives the left-hand critical value, 8.91; similarly, = *CHISQ.INV.RT* (0.025, 19) gives the answer 32.85, the right-hand critical value. In older versions of Excel, use = *CHIINV* (0.975, 19), etc.

We actually want an interval estimate for σ^2 so we need to rearrange equation (6.2) so that σ^2 lies between the two inequality signs. Rearranging yields

$$\left[\frac{(n-1)s^2}{32.85} \leq \sigma^2 \leq \frac{(n-1)s^2}{8.91}\right] \tag{6.3}$$

and evaluating this expression leads to the 95% confidence interval for σ^2 which is

$$\left[\frac{19 \times 625}{32.85} \leq \sigma^2 \leq \frac{19 \times 625}{8.91}\right] = [361.5, 1332.8]$$

Note that the point estimate, 625, is no longer at the centre of the interval but is closer to the lower limit. This is a consequence of the skewness of the χ^2 distribution.

Worked example 6.1

Given a sample of size $n = 51$ yielding a sample variance $s^2 = 81$, we may calculate the 95% confidence interval for the population variance as follows.

Since we are using the 95% confidence level, the critical values cutting off the extreme 5% of the distribution are 32.36 and 71.42, from Table A4. We can therefore use equation (6.3) to find the interval:

$$\left[\frac{(n-1) \times s^2}{71.42} \leq \sigma^2 \leq \frac{(n-1) \times s^2}{32.36}\right]$$

Substituting in the values gives

$$\left[\frac{(51-1) \times 81}{71.42} \leq \sigma^2 \leq \frac{(51-1) \times 81}{32.36}\right]$$

yielding a confidence interval of [56.71, 125.15].

Note that if we wished to find a 95% confidence interval for the standard deviation we can simply take the square root of the result to obtain [7.53, 11.19].

The 99% CI for the variance can be obtained by altering the critical values. The values cutting off 0.5% in each tail of the distribution are (again from Table A4) 27.99 and 79.49. Using these critical values results in an interval [50.95, 144.69]. Note that, as expected, the 99% CI is wider than the 95% interval.

Exercise 6.1

?

(a) Given a sample variance of 65 from a sample of size $n = 30$, calculate the 95% confidence interval for the variance of the population from which the sample was drawn.

(b) Calculate the 95% CI for the standard deviation.

(c) Calculate the 99% interval estimate of the variance.

Comparing actual and expected values of a variable

A second use of the χ^2 distribution provides a hypothesis test, allowing us to compare a set of observed values to expected values, the latter calculated on the basis of some null hypothesis to be tested. If the observed and expected values differ significantly, as judged by the χ^2 test (the test statistic falls into the rejection region of the χ^2 distribution), then the null hypothesis is rejected. Again, this is similar in principle to hypothesis testing using the Normal or t distributions, but allows a slightly different type of problem to be handled.

This can be illustrated with a very simple example. Suppose that throwing a die 72 times yields the following data:

Score on die	1	2	3	4	5	6
Frequency	6	15	15	7	15	14

Are these data consistent with the die being unbiased? Previously we might have investigated this problem by testing whether the proportion of (say) sixes is more or less than expected, using the Binomial distribution. One could still do this, but this does not make full use of the information in the sample, it only compares sixes against all other values together. The χ^2 test allows one to see if there is *any* bias in the die, for or against a particular number. It therefore answers a slightly different and more general question than if we made use of the Binomial distribution.

A crude examination of the data suggests a slight bias against 1 and 4, but is this truly bias or just a random fluctuation quite common in this type of experiment? First the null and alternative hypotheses are set up:

H_0: the die is unbiased
H_1: the die is biased

Note that the null hypothesis should be constructed in such a way as to permit the calculation of the expected outcomes of the experiment. Thus the null and alternative hypotheses could not be reversed in this case, since 'the die is biased' is a vague statement (exactly how biased, for example?) and would not permit the calculation of the expected outcomes of the experiment.

On the basis of the null hypothesis, the expected values are based on the **uniform distribution**, i.e. each number should come up an equal number of times. The expected values are therefore 12 ($= 72/6$) for each number on the die.

This gives the data shown in Table 6.2 with observed and expected frequencies in columns 2 and 3, respectively (ignore columns 4–6 for the moment). These are now compared using the χ^2 test statistic, constructed using the formula

$$\chi^2 = \sum \frac{(O - E)^2}{E} \tag{6.4}$$

Table 6.2 Calculation of the χ^2 statistic for the die problem

Score	Observed frequency (O)	Expected frequency (E)	$O - E$	$(O - E)^2$	$\frac{(O-E)^2}{E}$
1	6	12	−6	36	3.00
2	15	12	3	9	0.75
3	15	12	3	9	0.75
4	7	12	−5	25	2.08
5	15	12	3	9	0.75
6	14	12	2	4	0.33
Totals	72	72	0		7.66

which has a χ^2 distribution with $\nu = k - 1$ degrees of freedom (k is the number of different outcomes, here 6)[2]. O represents the observed frequencies and E the expected. If the value of this test statistic falls into the rejection region, i.e. the tail of the χ^2 distribution, then we conclude the die is biased, rejecting the null. The calculation of the test statistic is shown in columns 4–6 of Table 6.2, and is straightforward, yielding a value of the test statistic of $\chi^2 = 7.66$, to be compared to the critical value of the distribution, for $6 - 1 = 5$ degrees of freedom.

Trap!

In my experience many students misinterpret formula (6.4) and use

$$\chi^2 = \frac{\Sigma(O - E)^2}{\Sigma E}$$

instead. This is not the same as the correct formula and gives a wrong answer. Check that you recognise the difference between the two and that you always use the correct version.

Looking up the critical value for this test takes a little care as one needs first to consider if it is a one- or two-tailed test. Looking at the alternative hypothesis *suggests* a two-sided test, since the error could be in either direction. However, this intuition is wrong, for the following reason. Looking closely at equation (6.4) reveals that large discrepancies between observed and expected values (however occurring) can *only* lead to large values of the test statistic. Conversely, small values of the test statistic must mean that differences between O and E are small, so the die must be unbiased. Thus the null is *only* rejected by large values of the χ^2 statistic or, in other words, the rejection region is in the right-hand tail only of the χ^2 distribution. It is a one-tailed test. This is illustrated in Figure 6.2.

The critical value of the χ^2 distribution in this case ($\nu = 5$, 5% significance level) is 11.1, found from Table A4. Note that we require 5% of the distribution in the right-hand tail to establish the rejection region. Since the test statistic is less than the critical value ($7.66 < 11.1$) the null hypothesis is not rejected. The differences between scores are due to chance rather than to bias in the die. Alternatively, we could find the P-value associated with 7.66, which is 0.176 (use =*CHISQ.DIST.RT*(7.66,5) in Excel to obtain this). Since this is greater than 0.05, the null is not rejected.

[2]Note that, on this occasion, the degrees of freedom are not based on the sample size.

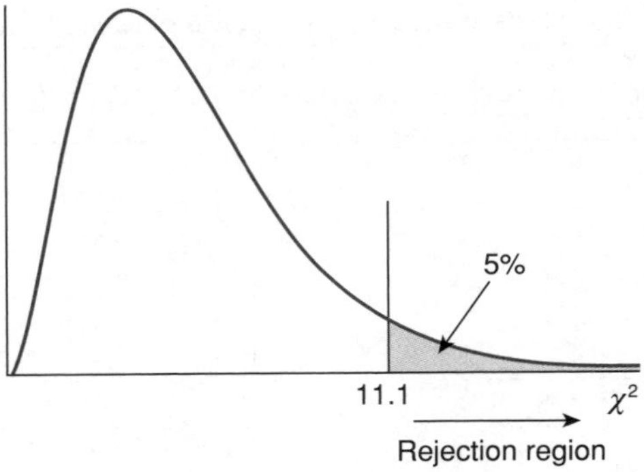

Figure 6.2
The rejection region for the χ^2 test

An important point to note is that the test should not be carried out on the *proportion* of occasions on which each number comes up (the expected values would all be $12/72 = 0.167$, and the observed values $8/72$, $13/72$, etc.), since information about the 'sample size' (number of rolls of the die, 72) would be lost. As with all sampling experiments, the inferences that can be drawn depend upon the sample size, with larger sample sizes giving more reliable results, so care must be taken to retain information about sample size in the calculations. If the test had been incorrectly conducted in terms of proportions, all O and E values would have been divided by 72, and this would have reduced the test statistic by a factor of 72 (check the formula to confirm this), reducing it to 0.14, nowhere near significance. It would be surprising if any data would yield significance given this degree of mistreatment. (See the "Oops!" box later in this chapter.)

A second, more realistic, example will now be examined to reinforce the message about the use of the χ^2 distribution and to show how the expected values might be generated in different ways. This example looks at road accident figures to see if there is any variation through the year. One might reasonably expect more accidents in the winter months due to weather conditions, poorer light, etc. Quarterly data on the number of fatal accidents on British roads are used, and the null hypothesis is that the number does not vary seasonally.

H_0: there is no difference in fatal accidents between quarters
H_1: there is some difference in fatal accidents between quarters

Such a study might be carried out by government, for example, to try to find the best means of reducing road accidents.

Table 6.3 shows data on road fatalities in 2014 by quarter in Great Britain, adapted from data taken from the UK government's transport data available at https://www.gov.uk/government/statistical-data-sets/ras30-reported-casualties-in-road-accidents. There does appear some evidence of more accidents in the final two quarters of the year, but is this convincing evidence or just random variation? Under the null hypothesis the total number of fatalities (1775) would be evenly split between the four quarters, yielding Table 6.4 and the χ^2 calculation that follows.

The calculated value of the test statistic is 22.42, given at the foot of the final column. The number of degrees of freedom is $\nu = k - 1 = 3$, so the critical value

Table 6.3 Road casualties in Great Britain, 2014

Quarter	I	II	III	IV	Total
Casualties	376	428	457	514	1775

Table 6.4 Calculation of the χ^2 statistic for road fatalities

Quarter	Observed	Expected	$O - E$	$(O - E)^2$	$\frac{(O-E)^2}{E}$
I	376	443.75	−67.75	4590.06	10.34
II	428	443.75	−15.75	248.06	0.56
III	457	443.75	13.25	175.56	0.40
IV	514	443.75	70.25	4935.06	11.12
Totals	1775	1775			22.42

Figure 6.3
The seasonal pattern of road casualties

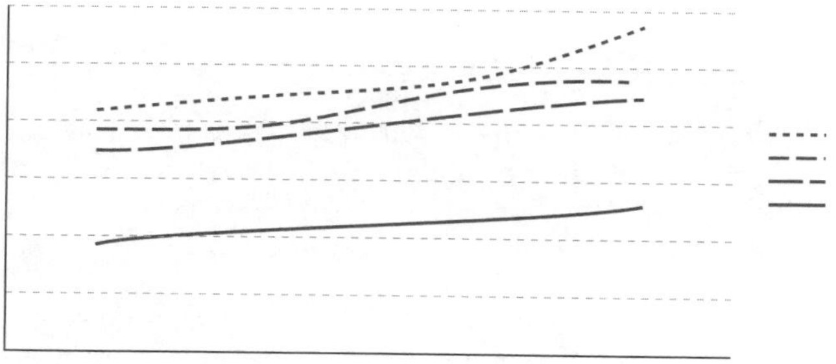

at the 5% significance level is 7.82. Since the test statistic exceeds this, the null hypothesis is rejected; there is a difference between seasons in the accident rate. Earlier editions of this text analysed data from earlier years and it is useful to compare across the different years. As so often, a graph is a useful way to do this, illustrated in Figure 6.3.

A couple of points are worth making about this chart. First, the quarterly pattern appears to be fairly consistent over time, casualties rising throughout the year. Second is the substantial progress made in reducing casualties since 2006, which is far bigger than seasonal differences.

The reason for the quarterly difference might be the increased hours of darkness during winter months, leading to more accidents. This particular hypothesis can be tested using the same data, but combining quarters I and IV (to represent winter) and quarters II and III (summer). The null hypothesis is of no difference between summer and winter, and the calculation is set out in Table 6.5. The χ^2 test statistic is now extremely small, and falls below the new critical value ($\nu = 1$, 5% significance level) of 3.84, so the null hypothesis is not rejected. Thus, the variation between quarters does not appear to be a straightforward summer/winter effect (providing, of course, that combining quarters I and IV to represent winter and II and III to represent summer is a valid way of combining the quarters).

Another point which the example brings out is that the data can be examined in a number of ways using the χ^2 technique. Some of the classes were combined to test a slightly different hypothesis from the original one. This is a quite acceptable technique but should be used with caution. In any set of data (even totally random

Table 6.5 Seasonal variation in road casualties

Season	Observed	Expected	$O - E$	$(O - E)^2$	$\frac{(O-E)^2}{E}$
Summer	885	887.5	–	6.25	0.007
Winter	890	887.5	2.5	6.25	0.007
Totals	1775	1775	0		0.014

data), there is bound to be *some* way of dividing it up such that there are significant differences between the divisions. The point, however, is whether there is any meaning to the division. In the above example the amalgamation of the quarters into summer and winter has some intuitive meaning, and we have good reason to believe that there might be differences between them. Driving during the hours of darkness might be more dangerous and might have had some relevance to accident prevention policy (e.g. an advertising campaign to persuade people to check that their lights work correctly). The hypothesis is led by some prior theorising and is worth testing.

STATISTICS IN PRACTICE

Road accidents and darkness

The question of the effect of darkness on road accidents has been extensively studied, particularly in relation to putting the clocks forward and back in spring and autumn. A study by H. Green in 1980 reported the following numbers of accidents (involving death or serious injury) on the five weekday evenings before and after the clocks changed:

	Spring		Autumn	
Year	Before	After	Before	After
1975	19	11	20	31
1976	14	9	23	36
1977	22	8	12	29

It is noticeable that accidents fell in spring after the hour change (when it becomes lighter) but increased in autumn (when it becomes darker). This is a better test than simply combining quarterly figures as in our example, so casts doubt upon our result. Evidence from other countries also supports the view that the light level has an important influence on accidents.

Source: H. Green, Some effects on accidents of changes in light conditions at the beginning and end of British Summer Time, *Supplementary Report 587*, Transport and Road Research Laboratory, 1980. For an update on research, see J. Boughton *et al.*, Influence of light level on the incidence of road casualties, *J. Royal Statistical Society*, Series A, 162 (2), 1999.

As discussed in Chapter 5, it is dangerous to look at the data and then formulate a hypothesis. From Table 6.4 there appears to be a large difference between the first and second halves of the year. If quarters I and II were combined, and III and IV combined, the χ^2 test statistic might be significant (in fact it is, $\chi^2 = 15.7$), but does this signify anything? It is extremely easy to look for a big difference *somewhere* in any set of data and then pronounce it 'significant' according to some test. The probability of making a Type I error (rejecting a correct null) is much greater than 5% in such a case. The point, as usual, is that it is no good looking at data in a vacuum and simply hoping that they will 'tell you something'.

A related warning is that we should be wary of testing one hypothesis and, on the basis of that result, formulating another hypothesis and testing it (as we have done by going on to compare summer and winter). Once again we are (indirectly) using the data to help formulate the hypothesis, and the true significance level of the test is likely to be different from 5% (even though we use the 5% critical value). We have therefore sinned, but is difficult to do research without sometimes resorting to this kind of behaviour. There are some formal methods for dealing with such situations, but they are beyond the scope of this text.

Combining classes

There is one further point to make about carrying out a χ^2 test, and this involves circumstances where classes *must* be combined. The theoretical χ^2 distribution from which the critical value is obtained is a continuous distribution, yet the calculation of the test statistic comes from data which are divided up into a discrete number of classes. The calculated test statistic is therefore only an approximation to a true χ^2 variable, but this approximation is good enough as long as each expected (not observed) value is greater than or equal to five. It does not matter what the observed values are. If this condition is not satisfied, the class (or classes) with expected values less than five must be combined with other classes until all expected values are at least five. An example of this will be given below.

In all cases of χ^2 testing the most important part of the analysis is the calculation of the expected values (the rest of the analysis is mechanical). Therefore, it is always worth devoting most of the time to this part of the problem. The expected values are, of course, calculated on the basis of the null hypothesis being true, so different null hypotheses will give different expected values. Consider again the case of road fatalities. Although the null hypothesis ('no differences in accidents between quarters') seems clear enough, it could mean different things. Here it was taken to mean an equal number in each quarter; but another interpretation is an equal number of casualties per car-kilometre travelled in each quarter; in other words, accidents might be higher in a given quarter simply because there are more journeys in that quarter (during holiday periods, for example). Table 6.6 gives an index of average daily traffic flows on British roads in each quarter of the year.

The pattern of accidents might follow the pattern of road usage – the first quarter of the year has the fewest casualties and also the least amount of travel. This may be tested by basing the expected values on the average traffic flow: the 1775 total casualties are allocated to the four quarters proportionally to the traffic. This is shown in Table 6.7, along with the calculation of the χ^2 statistic.

The χ^2 test statistic is 17.24, well in excess of the critical value, 7.82. This indicates that there are significant differences between the quarters, even after accounting for different amounts of traffic. In fact, the statistic is little changed from before, suggesting either that traffic flows do not affect accident probabilities much or that the flows do not actually vary very much. It is evident that the

Table 6.6 Index of road traffic flows, 2014

	Q1	Q2	Q3	Q4	Total
Index	73.3	78.7	81.5	77.4	310.9

Table 6.7 Calculation with alternative pattern of expected values

Quarter	Observed	Expected	$O - E$	$(O - E)^2$	$\frac{(O - E)^2}{E}$
I	376	418.5	−42.5	1805.12	4.31
II	428	449.3	−21.3	454.39	1.01
III	457	465.3	−8.3	68.93	0.15
IV	514	441.9	72.1	5199.20	11.77
Totals	1775	1775			17.24

Note: The first expected value is calculated as $1775 \times 95 \div 400 = 753.4$, the second as $1775 \times 102 \div 400 = 808.9$ and so on.

variation in traffic flows is much less than the variation in casualties. One possible explanation is that increased traffic means lower speed and hence a lower severity of accidents.

Worked example 6.2

One hundred volunteers each toss a coin twice and note the numbers of heads. The results of the experiment are as follows:

Heads	0	1	2	Total
Frequency	15	55	30	100

Can we reject the hypothesis that a fair coin (or strictly, coins) was used for the experiment?

On the basis of the Binomial distribution the probability of no heads is 0.25 ($= \frac{1}{2} \times \frac{1}{2}$), of one head is 0.5 and of two heads is again 0.25, as explained in Chapter 2. The expected frequencies are therefore 25, 50 and 25. The calculation of the test statistic is set out below:

Number of heads	O	E	$O - E$	$(O - E)^2$	$\frac{(O - E)^2}{E}$
0	15	25	−10	100	4
1	55	50	5	25	0.5
2	30	25	5	25	1
Totals	100	100			5.5

The test statistic of 5.5 compares to a critical value of 5.99 ($\nu = 2$) so we cannot reject the null hypothesis of a fair coin being used.

Note that we could also test this hypothesis via a z test, using the methods of Chapter 5. There have been a total of 200 tosses, of which 115 ($= 55 + 2 \times 30$) were heads, i.e. a ratio of 0.575 against the expected 0.5. We can therefore test H_0: $\pi = 0.5$ against H_1: $\pi \neq 0.5$ using the evidence $n = 200$ and $p = 0.575$. This yields the test statistic

$$z = \frac{0.575 - 0.5}{\sqrt{\frac{0.5 \times 0.5}{200}}} = 2.12$$

Interestingly, we now *reject* the null as the test statistic is greater than the critical value of 1.96. How can we reconcile these conflicting results?

Note that both results are close to the critical values, so narrowly reject or accept the null. The χ^2 and z distributions are both continuous ones and in this case are approximations to the underlying Binomial experiment. This is the cause of the problem. If we alter the data very slightly, to 16, 55, 29 observed frequencies of no heads, one head and two heads, then both methods accept the null hypothesis. Similarly, for frequencies 14, 55, 31 both methods reject the null.

The lesson of this example is to be cautious when the test statistic is close to the critical value. We cannot say decisively that the null has been accepted or rejected.

Exercise 6.2

The following data show the observed and expected frequencies of an experiment with four possible outcomes, A–D.

Outcome	A	B	C	D
Observed	40	60	75	90
Expected	35	55	75	100

Test the hypothesis that the results are in line with expectations using the 5% significance level.

Exercise 6.3

(a) Verify the claim in worked example 6.2, that both χ^2 and z statistic methods give the same qualitative (accept or reject) result when the observed frequencies are 16, 55, 29 and when they are 14, 55, 31.

(b) In each case, look up or calculate (using Excel) the Prob-values for the χ^2 and z test statistics and compare.

Contingency tables

Data are often presented in the form of a two-way classification as shown in Table 6.8, known as a **contingency table**, and this is another situation where the χ^2 distribution is useful. It provides a test of whether or not there is an association between the two variables represented in the table.

The table shows the voting intentions of a sample of 200 voters, cross-classified by social class. The interesting question that arises from these data is whether there is any association between people's voting behaviour and their social class. Are manual workers (social class C in the table) more likely to vote for the Labour Party than for the Conservative Party? The table would appear to indicate some support for this view, but is this truly the case for the whole population or is the evidence insufficient to draw this conclusion?

This sort of problem is amenable to analysis by a χ^2 test. The data presented in the table represent the observed values, so expected values need to be calculated and then compared to them using a χ^2 test statistic. The first task is to formulate a null hypothesis, on which to base the calculation of the expected values, and an alternative hypothesis. These are

H_0: there is no association between social class and voting behaviour
H_1: there is some association between social class and voting behaviour

As always, the null hypothesis has to be precise, so that expected values can be calculated. In this case it is the precise statement that there is no association between the two variables, they are independent.

Table 6.8 Data on voting intentions by social class

Social class	Labour	Conservative	Liberal Democrat	Total
A	10	15	15	40
B	40	35	25	100
C	30	20	10	60
Totals	80	70	50	200

Constructing the expected values

If H_0 is true and there is no association, we would expect the proportions voting Labour, Conservative and Liberal Democrat to be the same in each social class. Further, the parties would be identical in the proportions of their support coming from social classes A, B and C. This means that, since the whole sample of 200 splits 80:70:50 for the Labour, Conservative and Liberal Democrat parties (see the bottom row of the Table 6.8), each social class should split the same way. Thus of the 40 people of class A, 80/200 of them should vote Labour, 70/200 Conservative and 50/200 Liberal Democrat. This yields:

Split of social class A:	
Labour	$40 \times 80/200 = 16$
Conservative	$40 \times 70/200 = 14$
Liberal Democrat	$40 \times 50/200 = 10$
For class B:	
Labour	$100 \times 80/200 = 40$
Conservative	$100 \times 70/200 = 35$
Liberal Democrat	$100 \times 50/200 = 25$

And for C the 60 votes are split Labour 24, Conservative 21 and Liberal Democrat 15.

Both observed and expected values are presented in Table 6.9 (expected values are in brackets). Notice that both the observed and expected values sum to the appropriate row and column totals. It can be seen that, compared with the 'no association' position, Labour gets too few votes from Class A and the Liberal Democrats too many. However, Labour gets disproportionately many class C votes, the Liberal Democrats too few. The Conservatives' observed and expected values are nearly identical, indicating that the propensities to vote Conservative are the same in all social classes.

A quick way to calculate the expected value in any cell is to multiply the appropriate row total by column total and divide through by the grand total (200). For example, to get the expected value for the class A/Labour cell:

$$\textit{expected value} = \frac{\textit{row total} \times \textit{column total}}{\textit{grand total}} = \frac{40 \times 80}{200} = 16$$

In carrying out the analysis care should again be taken to ensure that information is retained about the sample size, i.e. the numbers in the table should be actual numbers and not percentages or proportions. This can be checked by ensuring that the grand total is always the same as the sample size.

As was the case before, the χ^2 test is only valid if the expected value in each cell is not less than five. In the event of one of the expected values being less than five, some of the rows or columns have to be combined. How to do this is a matter of choice and depends upon the aims of the research. Suppose for example that the

Table 6.9 Observed and expected values (latter in brackets)

Social class	Labour	Conservative	Liberal Democrat	Total
A	10 (16)	15 (14)	15 (10)	40
B	40 (40)	35 (35)	25 (25)	100
C	30 (24)	20 (21)	10 (15)	60
Totals	80	70	50	200

expected number of class C voting Liberal Democrat were less than five. There are four options open:

(1) Combine the Liberal Democrat column with the Labour column.
(2) Combine the Liberal Democrat column with the Conservative column.
(3) Combine the class C row with the class A row.
(4) Combine the class C row with the class B row.

Whether rows or columns are combined depends upon whether interest centres more upon differences between parties or differences between classes. If the main interest is the difference between class A and the others, option 4 should be chosen. If it is felt that the Liberal Democrat and Conservative parties are similar, option 2 would be preferred, and so on. If there are several expected values less than five, rows and columns must be combined until all are eliminated.

The χ^2 test on a contingency table is similar to the one carried out before, the formula being the same:

$$\chi^2 = \Sigma \frac{(O - E)^2}{E} \tag{6.5}$$

with the number of degrees of freedom this time given by $\nu = (r - 1) \times (c - 1)$ where r is the number of rows in the table and c is the number of columns. In this case $r = 3$ and $c = 3$, so

$$\nu = (3 - 1) \times (3 - 1) = 4$$

The reason why there are only four degrees of freedom is that once any four interior cells of the contingency table have been filled, the other five are constrained by the row and column totals. The number of 'free' cells can always be calculated as the number of rows less one, times the number of columns less one, as given above.

Calculation of the test statistic

The evaluation of the test statistic then proceeds as follows, cell by cell:

$$\frac{(10 - 16)^2}{16} + \frac{(15 - 14)^2}{14} + \frac{(15 - 10)^2}{10}$$
$$+ \frac{(40 - 40)^2}{40} + \frac{(35 - 35)^2}{35} + \frac{(25 - 25)^2}{25}$$
$$+ \frac{(30 - 24)^2}{24} + \frac{(20 - 21)^2}{21} + \frac{(10 - 15)^2}{15}$$
$$= 2.25 + 0.07 + 2.50 + 0 + 0 + 0 + 1.5 + 0.05 + 1.67$$
$$= 8.04$$

This must be compared with the critical value from the χ^2 distribution with four degrees of freedom. At the 5% significance level this is 9.50 (from Table A4).

Since $8.04 < 9.50$ the test statistic is smaller than the critical value, so the null hypothesis cannot be rejected. The evidence is not strong enough to support an association between social class and voting intention. We cannot reject the null of the lack of any association with 95% confidence. Note, however, that the test statistic is fairly close to the critical value, so there is some weak evidence of an association, but not enough to satisfy conventional statistical criteria.

Oops!

A leading firm of chartered accountants produced a report for the UK government on education funding. One question it asked of schools was: Is the school budget sufficient to provide help to pupils with special needs? This produced the following table:

	Primary schools	Secondary schools
Yes	34%	45%
No	63%	50%
No response	3%	5%
Totals	100%	100%
$n =$	137	159
$\chi^2 = 3.50$ n.s.		

Their analysis produces the conclusion that there is no significant difference between primary and secondary schools. But the χ^2 statistic is based on the percentage figures. Using frequencies (which can be calculated from the sample size figures) gives a correct χ^2 figure of 5.05. Fortunately for the accountants, this is still not significant.

Source: Adapted for *Local Management in School Report,1988* by Coopers and Lybrand for the UK government. Contains public sector information licensed under the Open Government Licence (OGL) v3.0. http://www.nationalarchives.gov.uk/doc/open-government-licence/open-government

Cohabitation

J. Ermisch and M. Francesconi examined the rise in cohabitation in the United Kingdom and asked whether it led on to marriage or not. One of their tables shows the relation between employment status and the outcome of living together. Their results, including the calculation of the χ^2 statistic for association between the variables, are shown in the figure.

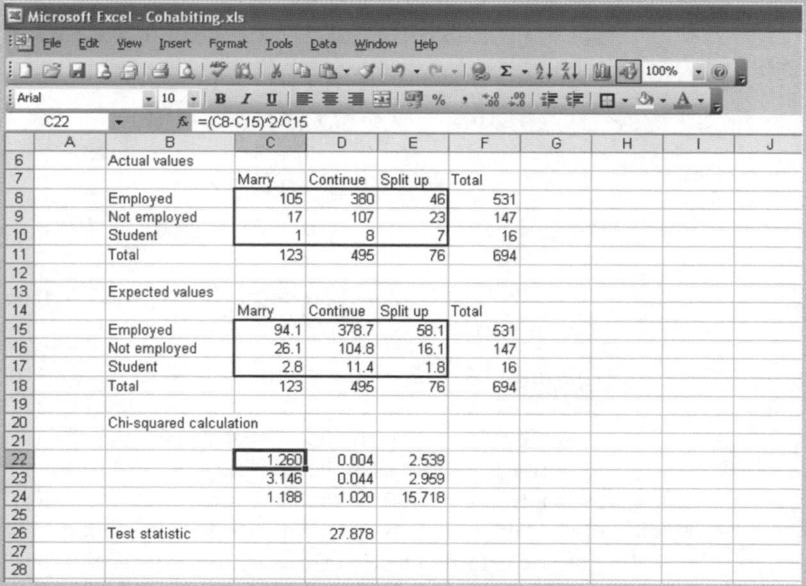
Microsoft Excel - Cohabiting.xls

C22 =(C8-C15)^2/C15

	A	B	C	D	E	F
6		Actual values				
7			Marry	Continue	Split up	Total
8		Employed	105	380	46	531
9		Not employed	17	107	23	147
10		Student	1	8	7	16
11		Total	123	495	76	694
12						
13		Expected values				
14			Marry	Continue	Split up	Total
15		Employed	94.1	378.7	58.1	531
16		Not employed	26.1	104.8	16.1	147
17		Student	2.8	11.4	1.8	16
18		Total	123	495	76	694
19						
20		Chi-squared calculation				
21						
22			1.260	0.004	2.539	
23			3.146	0.044	2.959	
24			1.188	1.020	15.718	
25						
26		Test statistic		27.878		
27						
28						

There were 694 cohabiting women in the sample. Of the 531 who were employed, 105 of them went on to marry their partner, 46 split up and 380 continued living together. Similar figures are shown for unemployed women and for students. The expected values for the contingency table then appear (based on the null hypothesis of no association), followed by the calculation of the χ^2 test statistic. You can see the formula for one of the elements of the calculation in the formula bar.

The test statistic is significant at the 5% level (critical value 9.49 for four degrees of freedom), so there is an association. The biggest contribution to the test statistic comes from the bottom right-hand cell, where the actual value is much higher than the expected. It appears that, unfortunately, those student romances often do not turn out to be permanent.

However, a reader of an earlier edition of this text pointed out that two of the expected values are less than five, so use of the χ^2 statistic is strictly inappropriate in this context.

Source: J. Ermisch and M. Francesconi, Cohabitation: not for long but here to stay, *J. Royal Statistical Society*, Series A, 163 (2), 2000.

Exercise 6.4

Suppose that the data on educational achievement and employment status in Chapter 1 were obtained from a sample of 999 people, as follows:

	Higher education	A-levels	Other qualification	No qualification	Total
In work	257	145	269	52	723
Unemployed	10	11	31	10	62
Inactive	33	38	87	56	214
Total	300	194	387	118	999

(These values reflect the proportions in the population data.) Test whether there is an association between education and employment status, using the 5% significance level for the test.

The *F* distribution

The second distribution we encounter in this chapter is the *F* distribution. It has a variety of uses in statistics; in this section we look at two of these: testing for the equality of two variances and conducting an **analysis of variance** (ANOVA) test. Both of these are variants on the hypothesis test procedures which should by now be familiar. The *F* distribution will also be encountered in later chapters on regression analysis.

The *F* family of distributions resembles the χ^2 distribution in shape: it is always non-negative and is skewed to the right. It has two sets of degrees of freedom (these are its parameters, labelled ν_1 and ν_2) and these determine its precise shape. Typical *F* distributions are shown in Figure 6.4. As usual, for a hypothesis test we define an area in one or both tails of the distribution to be the rejection region. If a test statistic falls into the rejection region, then the null hypothesis upon which the test statistic was based is rejected. Once again, examples will clarify the principles.

Testing the equality of two variances

Just as one can conduct a hypothesis test on a mean, so it is possible to test the variance. It is unusual to want to conduct a test of a specific value of a variance, since we usually have little intuitive idea of what the variance should be in most

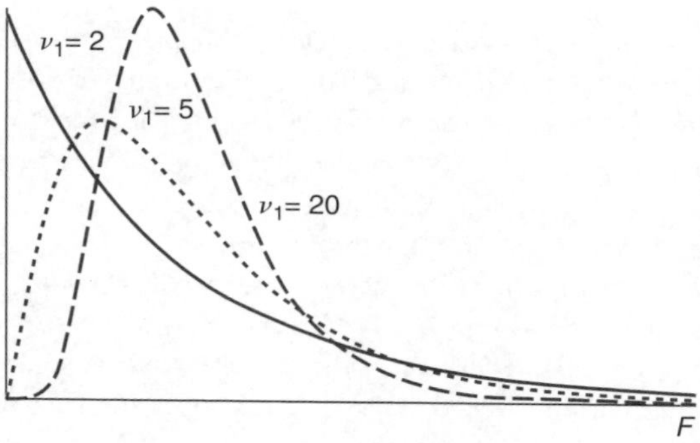

Figure 6.4
The *F* distribution, for different $\nu_1 (\nu_2 = 25)$

circumstances. A more likely circumstance is a test of the equality of two variances (across two samples). In Chapter 5 two car factories were tested for the equality of average daily output *levels*. One can also test whether the *variance* of output differs or not. A more consistent output (lower variance) from a factory might be beneficial to the firm, e.g. dealers can be reassured that they are more likely to be able to obtain models when they require them. In the example in Chapter 5, one factory had a standard deviation of daily output of 25, the second of 20, both from samples of size 30 (i.e. 30 days' output was sampled at each factory). We can now test whether the difference between these figures is significant or not.

Such a test is set up as follows. It is known as a **variance ratio test** for reasons which will become apparent.

The null and alternative hypotheses are

$$H_0: \sigma_1^2 = \sigma_2^2$$
$$H_1: \sigma_1^2 \neq \sigma_2^2$$

or, equivalently

$$H_0: \sigma_1^2/\sigma_2^2 = 1$$
$$H_1: \sigma_1^2/\sigma_2^2 \neq 1 \qquad (6.6)$$

It is appropriate to write the hypotheses in the form shown in (6.6) since the random variable and test statistic we shall use is in the form of the ratio of sample variances, s_1^2/s_2^2. This is a random variable which follows an F distribution with $\nu_1 = n_1 - 1$, $\nu_2 = n_2 - 1$ degrees of freedom. We require the assumption that the two samples are independent for the variance ratio to follow an F distribution. Thus we write:

$$\frac{s_1^2}{s_2^2} \sim F_{n_1-1, n_2-1} \qquad (6.7)$$

The F distribution thus has two parameters, the two sets of degrees of freedom, one (ν_1) associated with the numerator, the other (ν_2) associated with the denominator of the formula. In each case, the degrees of freedom are given by the relevant sample size minus one.

Note that s_2^2/s_1^2 is also an F distribution (i.e. it doesn't matter which variance goes into the numerator) but with the degrees of freedom reversed, $\nu_1 = n_2 - 1$, $\nu_2 = n_1 - 1$.

The sample data are:

$$s_1 = 25, s_2 = 20$$
$$n_1 = 30, n_2 = 30$$

Table 6.10 Excerpt from the *F* distribution: upper 2.5% points

ν_2 \ ν_1	1	2	3	...	20	24	30	40
1	647.7931	799.4822	864.1509	...	993.0809	997.2719	1001.4046	1005.5955
2	38.5062	39.0000	39.1656	...	39.4475	39.4566	39.4648	39.4730
3	17.4434	16.0442	15.4391	...	14.1674	14.1242	14.0806	14.0365
⋮	⋮	⋮	⋮	...	⋮	⋮	⋮	⋮
28	5.6096	4.2205	3.6264	...	2.2324	2.1735	2.1121	2.0477
29	5.5878	4.2006	3.6072	...	2.2131	2.1540	2.0923	2.0276
30	5.5675	4.1821	3.5893	...	2.1952	2.1359	2.0739	2.0089
40	5.4239	4.0510	3.4633	...	2.0677	2.0069	1.9429	1.8752

Note: The critical value lies at the intersection of the shaded row and column. Alternatively, use Excel or another computer package to give the answer. In *Excel* 2010 and later, the formula = *F.INV.RT*(0.025, 29, 29) will give the answer 2.09, the upper 2.5% critical value of the *F* distribution with $\nu_1 = 29, \nu_2 = 29$ degrees of freedom.

The test statistic is simply the ratio of sample variances. In testing it is less confusing if the larger of the two variances is made the numerator of the test statistic (you will see why soon). Therefore, we have the following test statistic:

$$F = \frac{25^2}{20^2} = 1.5625 \tag{6.8}$$

This must be compared to the critical value of the F distribution with $\nu_1 = 29$, $\nu_2 = 29$ degrees of freedom.

The rejection regions for the test are the two tails of the distribution, cutting off 2.5% in each tail. Since we have placed the larger variance in the denominator, only large values of F reject the null hypothesis so we need only consult the upper critical value of the F distribution, i.e. that value which cuts off the top 2.5% of the distribution. (This is the advantage of putting the larger variance in the numerator of the test statistic.)

Table 6.10 shows an excerpt from the F distribution. The degrees of freedom for the test are given along the top row (ν_1) and down the first column (ν_2). The numbers in the table give the critical values cutting off the top 2.5% of the distribution. The critical value in this case is 2.09, at the intersection of the row corresponding to $\nu_2 = 29$ and the column corresponding to $\nu_1 = 30$ ($\nu_1 = 29$ is not given so 30 is used instead; this gives a very close approximation to the correct critical value). Since the test statistic 1.56 does not exceed the critical value of 2.09, the null hypothesis of equal variances cannot be rejected with 95% confidence.

Exercise 6.5

?

Samples of 3-volt batteries from two manufacturers yielded the following outputs, measured in volts:

Brand A	3.1	3.2	2.9	3.3	2.8	3.1	3.2
Brand B	3.0	3.0	3.2	3.4	2.7	2.8	

Test whether there is any difference in the variance of output voltage of batteries from the two companies. Why might the variance be an important consideration for the manufacturer or for customers?

Analysis of variance

In Chapter 5 we learned how to test the hypothesis that the means of two samples are the same, using a z or t test, depending upon the sample size. This type of hypothesis test can be generalised to more than two samples using a technique called **analysis of variance** (ANOVA), based on the F distribution. Although it is called analysis of variance, it actually tests differences in means. The reason for this will be explained below. Using this technique, we can test the hypothesis that the means of *all* the samples are equal, versus the alternative hypothesis that at least one of them is different from the others. To illustrate the technique, we shall extend the example in Chapter 5 where different car factories' outputs were compared.

The assumptions underlying the analysis of variance technique are essentially the same as those used in the t test when comparing two different means. We assume that the samples are randomly and independently drawn from Normally distributed populations which have equal variances.

Suppose there are three factories, whose outputs have been sampled, with the results shown in Table 6.11. We wish to answer the question whether this is evidence of different outputs from the three factories, or simply random variation around a (common) average output level. The null and alternative hypotheses are therefore:

H_0: $\mu_1 = \mu_2 = \mu_3$
H_1: at least one mean is different from the others

This is the simplest type of ANOVA, known as **one-way analysis of variance.** In this case there is only one **factor** which affects output – the factory. The factor which may affect output is also known as the **independent variable**. In more complex designs, there can be two or more factors which influence output. The output from the factories is the **dependent or response variable** in this case.

Figure 6.5 presents a chart of the output from the three factories, which shows the greatest apparent difference between factories 2 and 3. Their ranges scarcely overlap, which does suggest some genuine difference between them, but as yet we cannot be sure that this is not just due to sampling variation. Factory 1 appears to be mid-way between the other two and this must also be included in the analysis.

Table 6.11 Samples of output from three factories

Observation	Factory 1	Factory 2	Factory 3
1	415	385	408
2	430	410	415
3	395	409	418
4	399	403	440
5	408	405	425
6	418	400	
7		399	

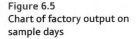

Figure 6.5
Chart of factory output on sample days

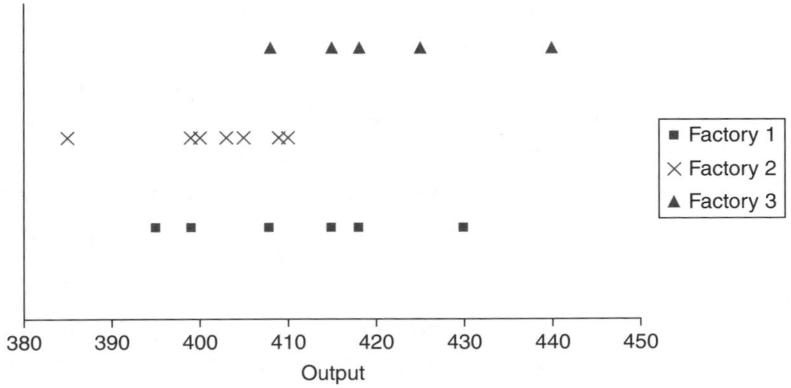

To decide whether or not to reject H_0, we compare the variance of output *within* factories to the variance of output *between* (the means of) the factories. Figure 6.6 provides an illustration. Where the variance between factories is large relative to the variance within each factory (Figure 6.6(a)), one is likely to reject H_0 and instead conclude there is a genuine difference. Alternatively, where the variance between factories is small relative to the within factory variance (Figure 6.6(b)), we are likely to accept H_0. The statistical test allows us to decide when the differences are large enough to warrant a particular conclusion and not just due to random variation.

Both *between* and *within* variance measures provide estimates of the overall true variance of output and, under the null hypothesis that factories make no difference, should provide similar estimates. The ratio of the variances should then be approximately unity. If the null is false however, the between-samples estimate will tend to be larger than the within-samples estimate and their ratio will exceed unity. This ratio has an F distribution and so if it is sufficiently large that it falls into the upper tail of the distribution, then H_0 is rejected.

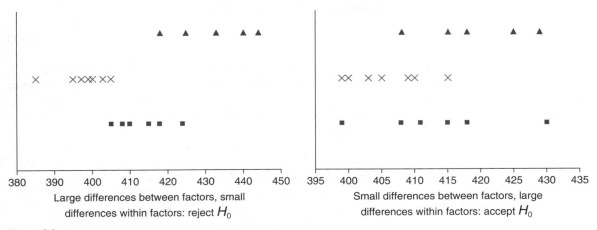

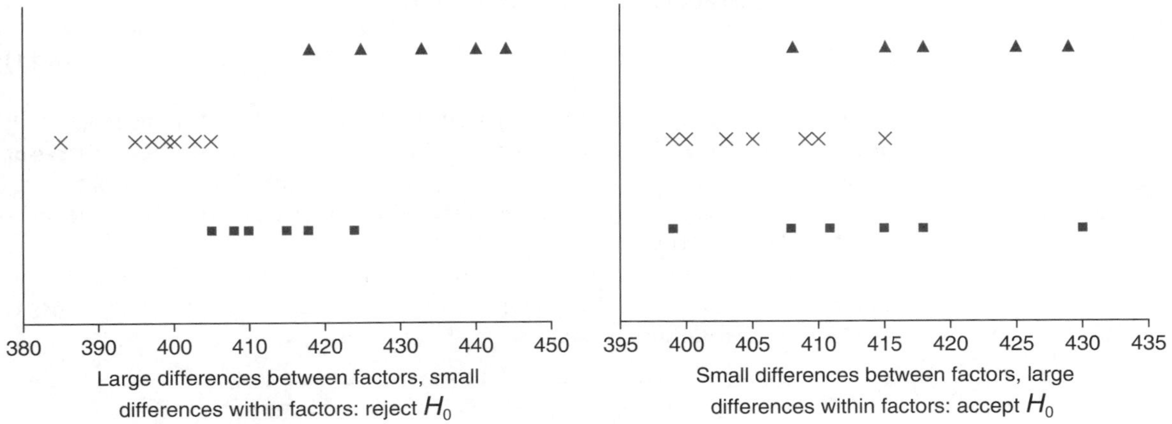

Figure 6.6
Illustration of when to reject H_0

To formally test the hypothesis, we break down the *total* variance of all the observations into

(1) the variance due to differences *between* factories, and
(2) the variance due to differences *within* factories (also known as the **error variance**).

Initially we work with **sums of squares** rather than variances. Recall from Chapter 1 that a sample variance is given by

$$s^2 = \frac{\Sigma(x - \bar{x})^2}{n - 1} \tag{6.9}$$

The numerator of the right-hand side of this expression, $\Sigma(x - \bar{x})^2$, gives the sum of squares, i.e. the sum of squared deviations from the mean.

Accordingly, we work with three sums of squares:

- The **total sum of squares** measures (squared) deviations from the overall or **grand average** using all the 18 observations. It ignores the existence of the different factors (factories).
- The **between sum of squares** measures how the three individual factor means vary around the grand average.
- The **within sum of squares** is based on squared deviations of observations from their own factor mean.

It can be shown that there is a relationship between these sums of squares, i.e.

$$\begin{matrix}\text{Time sum}\\ \text{of squares}\end{matrix} = \begin{matrix}\text{Between sum}\\ \text{of squares}\end{matrix} + \begin{matrix}\text{Within sum}\\ \text{of squares}\end{matrix} \tag{6.10}$$

The larger is the between sum of squares relative to the within sum of squares, the more likely it is that the null is false.

Because we have to sum over factors and over observations within those factors, the formulae look somewhat complicated, involving double summation signs. It is therefore important to follow the example showing how the calculations are actually done.

The **total sum of squares** is given by the formula:

$$\text{Total sum of squares} = \sum_{j=1}^{n_i}\sum_{i=1}^{k}(x_{ij} - \bar{x})^2 \tag{6.11}$$

where x_{ij} is the output from factory i on day j and $\bar{x}$ is the grand average of all observations. The index i runs from 1 to 3 in this case (there are three **classes or groups** for this factor) and the index j (indexing the observations) goes from 1 to 6, 7, or 5 (for factories 1, 2 and 3, respectively). Note that we do not require the same number of observations from each factory.

Although this looks complex, it simply means that we calculate the sum of squared deviations from the overall mean. The overall mean of the 18 values is 410.11 and the total sum of squares may be calculated as:

$$\text{Total sum of squares} = (415 - 410.11)^2 + (430 - 410.11)^2 + \cdots + (440 - 410.11)^2 + (425 - 410.11)^2 = 2977.778$$

An alternative formula for the total sum of squares is

$$\text{Total sum of squares} = \sum_{j=1}^{n_i}\sum_{i=1}^{k}x_{ij}^2 - n\bar{x}^2 \tag{6.12}$$

where n is the total number of observations. The sum of the squares of all the observations ($\sum x^2$) is $415^2 + 430^2 + \cdots + 425^2 = 3\,030\,418$ and the total sum of squares is then given by

$$\sum_{j=1}^{n_i}\sum_{i=1}^{k} x_{ij}^2 - n\bar{x}^2 = 3\,030\,418 - 18 \times 410.11^2 = 2977.778 \qquad (6.13)$$

as before.

The **between sum of squares** is calculated using the formula

$$\text{Between sum of squares} = \sum_j \sum_i (\bar{x}_i - \bar{x})^2 \qquad (6.14)$$

where $\bar{x}_i$ denotes the mean output of factor i. This part of the calculation effectively ignores the differences that exist *within* factors and compares the differences *between* them. It does this by replacing the observations within each factor by the mean for that factor. Hence, all the factor 1 observations are replaced by 410.83, for factor 2 they are replaced by the mean 401.57 and for factor 3 by 421.2[3]. We then calculate the sum of squared deviations of these values from the grand mean:

$$\begin{aligned}\text{Between sum of squares} &= 6 \times (410.83 - 410.11)^2 + 7 \times (401.57 - 410.11)^2 \\ &\quad + 5 \times (421.2 - 410.11)^2 = 1128.43\end{aligned}$$

Note that we take account of the number of observations within each factor in this calculation.

Once again there is an alternative formula which may be simpler for calculation purposes:

$$\text{Between sum of squares} = \sum_i n_i \bar{x}_i^2 - n\bar{x}^2 \qquad (6.15)$$

Evaluating this results in the same answer as above:

$$\begin{aligned}\sum_i n_i \bar{x}_i^2 - n\bar{x}^2 &= 6 \times 410.83^2 + 7 \times 401.57^2 + 5 \times 421.2^2 - 18 \times 410.10^2 \\ &= 1128.43 \qquad (6.16)\end{aligned}$$

We have arrived at the result that 37% (= 1128.43/2977.78) of the total variation (sum of squared deviations) is due to differences between factories and the remaining 63% is therefore due to variation (day to day) within factories. We can therefore immediately calculate the **within sum of squares** by straightforward subtraction as:

$$\begin{aligned}\text{Within sum of squares} &= \text{Total sum of squares} - \text{Between sum of squares} \\ &= 2977.778 - 1128.430 = 1849.348\end{aligned}$$

For completeness, the formula for the within sum of squares is

$$\text{Within sum of squares} = \sum_j \sum_i (x_{ij} - \bar{x}_i)^2 \qquad (6.17)$$

[3]Note that this is *not* the same as using the three factor means and calculating their variance.

The term $x_{ij} - \bar{x}_i$ measures the deviations of the observations from the factor mean and so the within sum of squares gives a measure of dispersion within the classes. Hence, it can be calculated as:

$$\begin{aligned}\text{Within sum of squares} &= (415 - 410.83)^2 + \cdots + (418 - 410.83)^2 \\ &\quad + (385 - 401.57)^2 + \cdots + (399 - 401.57)^2 \\ &\quad + (408 - 421.2)^2 + \cdots + (425 - 421.2)^2 \\ &= 1849.348\end{aligned}$$

This is the same value as obtained by subtraction.

The result of the hypothesis test

The F statistic is based upon comparing between and within sums of squares (*BSS* and *WSS*) but we must also take account of the degrees of freedom for the test. The degrees of freedom adjust for the number of observations and for the number of factors. Formally, the test statistic is

$$F = \frac{BSS/(k - 1)}{WSS/(n - k)}$$

which has $k - 1$ and $n - k$ degrees of freedom. k is the number of factors, 3 in this case, and n the overall number of observations, 18. We thus have

$$F = \frac{1128.43/(3 - 1)}{1849.348/(18 - 3)} = 4.576$$

The critical value of F for 2 and 15 degrees of freedom at the 5% significance level is 3.682. As the test statistic exceeds the critical value, we reject the null hypothesis of no difference between factories.

The analysis of variance table

ANOVA calculations are conventionally summarised in an **analysis of variance table**. Figure 6.7 shows such a table, as produced by Excel. Excel can produce the

Figure 6.7
One-way analysis of variance: Excel output

	A	B	C	D	E	F	G	H
13								
14								
15								
16	Anova: Single Factor							
17								
18	SUMMARY							
19	*Groups*	*Count*	*Sum*	*Average*	*Variance*			
20	Factory 1	6	2465	410.833	166.967			
21	Factory 2	7	2811	401.571	70.619			
22	Factory 3	5	2106	421.200	147.700			
23								
24								
25	ANOVA							
26	*Source of Variation*	*SS*	*df*	*MS*	*F*	*P-value*	*F crit*	
27	Between Groups	1128.430	2	564.215	4.576	0.028	3.682	
28	Within Groups	1849.348	15	123.290				
29								
30	Total	2977.778	17					
31								
32								

table automatically from data presented in the form shown in Table 6.11 and there is no need to do any of the calculations by hand. (In *Excel* you need to install the Analysis ToolPak in order to perform ANOVA. Other software packages, such as SPSS or Stata, also have routines to perform ANOVA.)

The first part of the table summarises the information for each factory, in the form of means and variances. Note that the means were used in the calculation of the *between sum of squares*. The ANOVA section of the output then follows, giving sums of squares and other information.

The column of the **ANOVA table** headed '*SS*' gives the sums of squares, which we calculated above. It can be seen that the between-group sum of squares makes up about 37% of the total, suggesting that the differences between factories (referred to as 'groups' by *Excel*) do make a substantial contribution to the total variation in output.

The '*df*' column gives the degrees of freedom associated with each sum of squares. These degrees of freedom are given by

Between sum of squares	$k - 1$
Within sum of squares	$n - k$
Total sum of squares	$n - 1$

The '*MS*' ('mean square') column divides the sums of squares by their degrees of freedom, and the *F* column gives the *F* statistic, which is the ratio of the two values in the *MS* column, i.e. $4.576 = 564.215/123.290$. This is the test statistic for the hypothesis test, which we calculated manually above. Excel helpfully gives the critical value of the test (at the 5% significance level) in the final column, 3.682. The P-value is given in the penultimate column and reveals that only 2.8% of the *F* distribution lies beyond the test statistic value of 4.576.

The test has found that the between sum of squares is 'large' relative to the within sum of squares, too large to be due simply to random variation, and this is why the null hypothesis of equal outputs is rejected. The rejection region for the test consists of the *upper* tail only of the *F* distribution; small values of the test statistic would indicate small differences between factories and hence non-rejection of H_0.

This simple example involves only three groups, but the extension to four or more follows the same principles, with different values of *k* in the formulae, and is fairly straightforward. Also, we have covered only the simplest type of ANOVA, with a one-way classification. More complex experimental designs are possible, with a two-way classification, for example, where there are two independent factors affecting the dependent variable. This is not covered in this text, although Chapter 8 on the subject of multiple regression does examine a method of modelling situations where two or more explanatory variables influence a dependent variable.

Worked example 6.3

ANOVA calculations are quite complex and are most easily handled by software which calculates all the results directly from the initial data. However, this is a kind of 'black box' approach to learning, so this example shows all the calculations mechanically.

→

Suppose we have six observations on each of three factors, as follows:

A	B	C
44	41	48
35	36	37
60	58	61
28	32	37
43	40	44
55	59	61

(These might be, for example, scores of different groups of pupils in a test.) We wish to examine whether there is a significant difference between the different groups. We need to see how the differences *between* the groups compare to those *within* groups.

First, we calculate the total sum of squares by ignoring the groupings and treating all 18 observations together. The overall mean is 45.5 so the squared deviations are $(44 - 45.5)^2$, $(41 - 45.5)^2$, etc. Summing these gives 2020.5 as the TSS.

For the between sum of squares we first calculate the means of each factor. These are 44.17, 44.33 and 48. We compare these to the grand average. The squared deviations are therefore $(44.17 - 45.5)^2$, $(44.33 - 45.5)^2$ and $(48 - 45.5)^2$. Rather than sum these, we must take account of the number of observations in each group which in this case is 6. Hence we obtain

$$\begin{aligned}\text{Between sum of squares} &= 6 \times (44.17 - 45.5)^2 + 6 \times (44.33 - 45.5)^2 \\ &\quad + 6 \times (48 - 45.5)^2 = 56.33\end{aligned}$$

The within sum of squares can be explicitly calculated as follows. For group A, the squared deviations from the group mean are $(44 - 44.17)^2$, $(35 - 44.17)^2$, etc. Summing these for group A gives 714.8. Similar calculations give 653.3 and 596 for groups B and C. These sum to 1964.2, which is the within sum of squares. As a check, we note:

$$2020.5 = 56.3 + 1964.2$$

The degrees of freedom are $k - 1 = 3 - 1 = 2$ for the between sum of squares, $n - k = 18 - 3 = 15$ for the within sum of squares and $n - 1 = 18 - 1 = 17$. The test statistic is therefore

$$F = \frac{56.33/2}{1964.2/15} = 0.22$$

The critical value at the 5% significance level is 3.68, so we cannot reject the null of no difference between the factors.

Exercise 6.6

?

The reaction times of three groups of sportsmen were measured on a particular task, with the following results (time in milliseconds):

Racing drivers	31	28	39	42	36	30	
Tennis players	41	35	41	48	44	39	38
Boxers	44	47	35	38	51		

Test whether there is a difference in reaction times between the three groups.

Summary

- The χ^2 and F distributions play important roles in statistics, particularly in problems relating to the goodness of fit of the data to that predicted by a null hypothesis.
- A random variable based on the sample variance, $(n - 1)s^2/\sigma^2$, has a χ^2 distribution with $n - 1$ degrees of freedom. Based on this fact, the χ^2 distribution may be used to construct confidence interval estimates for the variance σ^2. Since the χ^2 is not a symmetric distribution, the confidence interval is not symmetric around the (unbiased) point estimate s^2.
- The χ^2 distribution may also be used to compare actual and expected values of a variable and hence to test the hypothesis upon which the expected values were constructed.
- A two-way classification of observations is known as a contingency table. The independence or otherwise of the two variables may be tested using the χ^2 distribution, by comparing observed values with those expected under the null hypothesis of independence.
- The F distribution is used to test a hypothesis of the equality of two variances. The test statistic is the ratio of two sample variances which, under the null hypothesis, has an F distribution with $n_1 - 1, n_2 - 1$ degrees of freedom.
- The F distribution may also be used in an analysis of variance, which tests for the equality of means across several samples. The results are set out in an analysis of variance table, which compares the variation of the observations *within* each sample to the variation *between* samples.

Key terms and concepts

actual and expected values
analysis of variance
ANOVA table
between sum of squares
classes or groups
contingency table
dependent or response variable
error variance
factor
grand average
independent variable
one-way analysis of variance
sums of squares
total sum of squares
uniform distribution
variance ratio test
within sum of squares

Formulae used in this chapter

Formula	Description	Notes
$\left[\frac{(n-1)s^2}{U} \le \sigma^2 \le \frac{(n-1)s^2}{L}\right]$	Confidence interval for the variance	U and L are the upper and lower limits of the χ^2 distribution for the chosen confidence level, with $n - 1$ degrees of freedom
$\chi^2 = \sum \frac{(O-E)^2}{E}$	Test statistic for independence in a contingency table	$\nu = (r-1) \times (c-1)$, where r is the number of rows, c the number of columns
$F = \frac{s_1^2}{s_2^2}$	Test statistic for H_0: $\sigma_1^2 = \sigma_2^2$	$\nu = n_1 - 1, n_2 - 1$. Place larger sample variance in the numerator to ensure rejection region is in right-hand tail of the F distribution
$\sum_{j=1}^{n_i} \sum_{i=1}^{k} x_{ij}^2 - n\bar{x}^2$	Total sum of squares (ANOVA)	n is the total number of observations, k is the number of groups
$\sum_i n_i \bar{x}_i^2 - n\bar{x}^2$	Between sum of squares (ANOVA)	A n_i represents the number of observations in group i and $\bar{x}_i$ is the mean of the group
$\sum_j \sum_i (x_{ij} - \bar{x}_i)^2$	Within sum of squares (ANOVA)	

Problems

Some of the more challenging problems are indicated by highlighting the problem number in colour.

6.1 A sample of 40 observations has a standard deviation of 20. Estimate the 95% confidence interval for the standard deviation of the population.

6.2 Using the data $n = 70, s = 15$, construct a 99% confidence interval for the true standard deviation.

6.3 Use the data in Table 6.3 to see if there is a significant difference between road casualties in quarters I and III on the one hand and quarters II and IV on the other.

6.4 A survey of 64 families with five children found the following gender distribution:

Number of boys	0	1	2	3	4	5
Number of families	1	8	28	19	4	4

Test whether the distribution can be adequately modelled by the Binomial distribution.

6.5 Four different holiday firms which all carried equal numbers of holidaymakers reported the following numbers who expressed satisfaction with their holiday:

Firm	A	B	C	D
Number satisfied	576	558	580	546

Is there any significant difference between the firms? If told that the four firms carried 600 holiday-makers each, would you modify your conclusion? What do you conclude about your first answer?

6.6 A company wishes to see whether there are any differences between its departments in staff turnover. Looking at their records for the past year, the company finds the following data:

Department	Personnel	Marketing	Admin.	Accounts
Number in post at start of year	23	16	108	57
Number leaving	3	4	20	13

Do the data provide evidence of a difference in staff turnover between the various departments?

6.7 A survey of 100 firms found the following evidence regarding profitability and market share:

Profitability	Market share		
	<15%	15–30%	>30%
Low	18	7	8
Medium	13	11	8
High	8	12	15

Is there evidence that market share and profitability are associated?

6.8 The following data show the percentages of firms using computers in different aspects of their business:

Firm size	Computers used in			Total numbers of firms
	Admin.	Design	Manufacture	
Small	60%	24%	20%	450
Medium	65%	30%	28%	140
Large	90%	44%	50%	45

Is there an association between the size of firm and its use of computers?

6.9 (a) Do the accountants' job properly for them (see the *Oops!* box in the text (page 244)).

(b) It might be justifiable to omit the 'no responses' entirely from the calculation. What happens if you do this?

6.10 A roadside survey of the roadworthiness of vehicles obtained the following results:

	Roadworthy	Not roadworthy
Private cars	114	30
Company cars	84	24
Vans	36	12
Lorries	44	20
Buses	36	12

Is there any association between the type of vehicle and the likelihood of it being unfit for the road?

6.11 Given the following data on two sample variances, test whether there is any significant difference. Use the 1% significance level.

$$s_1^2 = 55 \qquad s_2^2 = 48$$
$$n_1 = 25 \qquad n_2 = 30$$

6.12 An example in Chapter 5 compared R&D expenditure in Britain and Germany. The sample data were:

$$\bar{x}_1 = 3.7 \quad \bar{x}_2 = 4.2$$
$$s_1 = 0.6 \quad s_2 = 0.9$$
$$n_1 = 20 \quad n_2 = 15$$

Is there evidence, at the 5% significance level, of difference in the variances of R&D expenditure between the two countries? What are the implications, if any, for the test carried out on the difference of the two means, in Chapter 4?

6.13 Groups of children from four different classes in a school were randomly selected and sat a test, with the following test scores:

Class	Pupil						
	1	2	3	4	5	6	7
A	42	63	73	55	66	48	59
B	39	47	47	61	44	50	52
C	71	65	33	49	61		
D	49	51	62	48	63	54	

(a) Test whether there is any difference between the classes, using the 95% confidence level for the test.

(b) How would you interpret a 'significant' result from such a test?

6.14 Lottery tickets are sold in different outlets: supermarkets, smaller shops and outdoor kiosks. Sales were sampled from several of each of these, with the following results:

Supermarkets	355	251	408	302
Small shops	288	257	225	299
Kiosks	155	352	240	

Does the evidence indicate a significant difference in sales? Use the 5% significance level.

6.15 **(Project)** Conduct a survey among fellow students to examine whether there is any association between:

(a) gender and political preference, or

(b) subject studied and political preference, or

(c) star sign and personality (introvert/extrovert – self-assessed: I am told that Aries, Cancer, Capricorn, Gemini, Leo and Scorpio are associated with an extrovert personality), or

(d) any other two categories of interest.

6.16 **(Computer project)** Use your spreadsheet or other computer program to generate 100 random integers in the range 0 to 9. Draw up a frequency table and use a χ^2 test to examine whether there is any bias towards any particular integer. Compare your results with those of others in your class.

Answers to exercises

Exercise 6.1

(a) The values cutting off the outer 2.5% in each tail of the χ^2 distribution ($df = 29$) are 16.05 and 45.72 (see Appendix Table A4). Using

$$\left[\frac{(n-1)\times s^2}{45.72} \leq \sigma^2 \leq \frac{(n-1)\times 65}{16.05}\right]$$

we therefore obtain

$$\left[\frac{(30-1)\times 65}{45.72} \leq \sigma^2 \leq \frac{(30-1)\times 65}{16.05}\right]$$

which yields the interval [41.2, 117.4]

(b) [6.4, 10.8], by taking the square roots of the answer to part (a).

(c) [36.0, 143.7], by similar methods but using critical values of 13.12 and 52.34

Exercise 6.2

The calculation of the test statistic is:

Outcome	Observed	Expected	$O - E$	$(O - E)^2$	$\frac{(O-E)^2}{E}$
A	40	35	5	25	0.714
B	60	55	5	25	0.455
C	75	75	0	0	0
D	90	100	−10	100	1
Total					2.169

This is smaller than the critical value of 7.81 (for three degrees of freedom) so the null is not rejected.

Exercise 6.3

The test statistics are for (16, 55, 29) $\chi^2 = 4.38$ (Prob-value $= 0.112$) and $z = 1.84$ (Prob-value $= 0.066$) and for (14, 55, 31) $\chi^2 = 6.78$ (Prob-value $= 0.038$) and $z = 2.40$ (Prob-value $= 0.016$). The two methods agree on the results (whether or not to reject the hypothesis), although the Prob-values are different.

Exercise 6.4

The expected values are:

	Higher education	A levels	Other qualifications	No qualifications	Total
In work	217	140	280	85	723
Unemployed	19	12	24	7	62
Inactive	64	42	83	25	214
Totals	300	194	387	118	999

(These are calculated by multiplying row and column totals and dividing by the grand total, e.g. $217 = 723 \times 3000/999$.)

The test statistic is:

$$7.3 + 0.2 + 0.4 + 13.1 + 4.0 + 0.1 + 2.0 + 1.0 + 15.2 + 0.3 + 0.2 + 37.3 = 81.1$$

This should be compared to a critical value of 12.59 ($\nu = (3 - 1) \times (4 - 1) = 6$), so the null is rejected.

Exercise 6.5

The two variances are $s_A^2 = 0.031$ and $s_B^2 = 0.066$. We therefore form the ratio $F = 0.066/0.031 = 2.09$, which has an F distribution with 5 and 6 degrees of freedom. The 5% critical value is therefore 4.39 and the null is not rejected. There appears no difference between manufacturers. The variance is important because consumers want a reliable product – they would not be happy if an incorrect voltage broke their MP3 player.

Exercise 6.6

The answer is summarised in this Excel table:

SUMMARY

Groups	Count	Sum	Average	Variance
Racing drivers	6	206	34.333	30.667
Tennis players	7	286	40.857	17.810
Boxers	5	215	43.000	42.500

ANOVA

Source of variation	*SS*	*df*	*MS*	*F*	P-value	*F* crit
Between groups	233.421	2	116.710	4.069	0.039	3.682
Within groups	430.190	15	28.679			
Totals	663.611	17				

The result shows that there is a difference between the three groups, with an F statistic of 4.069 (P-value 3.9%). The difference appears to be largely between racing drivers and the other two types.

Appendix Use of χ^2 and F distribution tables

Tables of the χ^2 distribution

Table A4 presents critical values of the χ^2 distribution for a selection of significance levels and for different degrees of freedom. As an example, to find the critical value of the χ^2 distribution at the 5% significance level, for $\nu = 20$ degrees of freedom, the cell entry in the column labelled '0.05' and the row labelled '20' are consulted. The critical value is 31.4. A test statistic greater than this value implies rejection of the null hypothesis at the 5% significance level.

An Excel function can alternatively be used to find this value. The formula '=CHIISQ.INV.RT(0.05, 20)' will give the result 31.4.

Tables of the F distribution

Table A5 (see page 454) presents critical values of the F distribution. Since there are two sets of degrees of freedom to be taken into account, a separate table is required for each significance level. Four sets of tables are provided, giving critical values cutting off the top 5%, 2.5%, 1% and 0.5% of the distribution (Tables A5(a), A5(b), A5(c) and A5(d) respectively). These allow both one- and two-tail tests at the 5% and 1% significance levels to be conducted. Its use is illustrated by example.

Two-tail test

To find the critical values of the F distribution at the 5% significance level for degrees of freedom ν_1 (numerator) $= 10$, $\nu_2 = 20$. The critical values in this case cut off the extreme 2.5% of the distribution in each tail, and are found in Table A5(b):

- Right-hand critical value: this is found from the cell of the table corresponding to the column $\nu_1 = 10$ and row $\nu_2 = 20$. Its value is 2.77.
- Left-hand critical value: this cannot be obtained directly from the tables, which only give right-hand values. However, it is obtained indirectly as follows:
  (a) Find the right-hand critical value for $\nu_1 = 20$, $\nu_2 = 10$ (note reversal of degrees of freedom). This gives 3.42.
  (b) Take the reciprocal to obtain the desired left-hand critical value. This gives $1/3.42 = 0.29$.

The rejection region thus consists of values of the test statistic less than 0.29 and greater than 2.77.

An Excel function can alternatively be used to find these values. The formula '=F.INV.RT(0.025, 10, 20)' will give the result 2.77. The formula '=F.INV.RT(0.975, 10, 20)' will give 0.29, the left hand value. Note that you do not need to reverse the degrees of freedom in the formula, Excel understands that the left-hand critical value is needed from the '0.975' figure.

One-tail test

To find the critical value at the 5% significance level for $\nu_1 = 15, \nu_2 = 25$. As long as the test statistic has been calculated with the larger variance in the numerator, the critical value is in the right-hand tail of the distribution and can be obtained directly from Table A5(a). For $\nu_1 = 15, \nu_2 = 25$ the value is 2.09. The null hypothesis is rejected, therefore, if the test statistic is greater than 2.09. Once again, in Excel, '=F.INV.RT(0.05, 15, 25)' = 2.09.

7 Correlation and regression

Contents

Learning outcomes

By the end of this chapter you should be able to:

- understand the principles underlying correlation and regression
- calculate and interpret a correlation coefficient and relate it to an XY graph of the two variables

→

- **calculate the line of best fit (regression line) and interpret the result**
- **recognise the statistical significance of the results, using confidence intervals and hypothesis tests**
- **recognise the importance of the units in which the variables are measured and of transformations to the data**
- **use computer software (Excel) to derive the regression line and interpret the computer output.**

Introduction

Correlation and regression are techniques for investigating the statistical relationship between two, or more, variables. In Chapter 1 we examined the relationship between investment and GDP using graphical methods (the *XY* chart). Although visually helpful, this did not provide any precise measurement of the strength of the relationship. In Chapter 6 the χ^2 test did provide a test of the significance of the association between two category-based variables, but this test cannot be applied to variables measured on a ratio scale. Correlation and regression fill in these gaps: the strength of the relationship between two (or more) ratio scale variables can be measured and the significance tested.

Correlation and regression are the techniques most often used by economists and forecasters. They can be used to answer such questions as

- Is there a link between the money supply and the price level?
- Do bigger firms produce at lower cost than smaller firms?
- Does spending more on advertising increase sales?

Each of these questions is about economics or business as much as about statistics. The statistical analysis is part of a wider investigation into the problem; it cannot provide a complete answer to the problem but, used sensibly, is a vital input. Correlation and regression techniques may be applied to time-series or cross-section data. The methods of analysis are similar in each case, although there are differences of approach and interpretation which are highlighted in this chapter and the next.

This chapter begins with the topic of correlation and simple (i.e. two variable) regression, using as an example the determinants of the birth rate in developing countries. In Chapter 8, multiple regression is examined, where a single dependent variable is explained by more than one explanatory variable. This is illustrated using time-series data pertaining to imports into the United Kingdom. This shows how a small research project can be undertaken, avoiding the many possible pitfalls along the way. Finally, a variety of useful additional techniques, tips and traps is set out, to help you understand and overcome a number of problems that can arise in regression analysis.

What determines the birth rate in developing countries?

This example follows the analysis in Michael Todaro's book, *Economic Development in the Third World* (3rd edn, pp. 197–200), where he tries to establish which of three variables (GNP per capita, the growth rate, or income inequality) is most important in determining a country's birth rate. (This analysis has been dropped from later editions of Todaro's book.)

The analysis is instructive as an example of correlation and regression techniques in a number of ways. First, the question is an important one; it was discussed at the UN International Conference on Population and Development in Cairo in 1995. It is felt by many that reducing the birth rate is a vital factor in economic development (birth rates in developed countries average around 12 per 1000 population, in developing countries around 30). Second, Todaro uses the statistical analysis to arrive at a questionable conclusion.

The data used by Todaro are shown in Table 7.1 using a sample of 12 developing countries. Two points need to be made initially. First, this is a very small sample which only includes developing countries, so the results will not give an all-embracing explanation of the birth rate. Different factors might be relevant to developed countries, for example. Second, there is the important question of why these particular countries were chosen as the sample and others ignored. The choice of country was limited by data availability, and one should ask whether countries with data available are likely to be representative of all countries. Data were, in fact, available for more than 12 countries, so Todaro was selective. You are asked to explore the implications of this in some of the problems at the end of the chapter.

The variables are defined as follows:

Birth rate:	the number of births per 1000 population in 1981.
GNP per capita:	1981 gross national product p.c., in US dollars.
Growth rate:	the growth rate of GNP p.c. p.a., 1961–81.
Income ratio:	the ratio of the income share of the richest 20% to that of the poorest 40%. A higher value of this ratio indicates greater inequality.

Table 7.1 Todaro's data on birth rate, GNP, growth and inequality

Country	Birth rate	1981 GNP p.c.	GNP growth	Income ratio
Brazil	30	2200	5.1	9.5
Colombia	29	1380	3.2	6.8
Costa Rica	30	1430	3.0	4.6
India	35	260	1.4	3.1
Mexico	36	2250	3.8	5.0
Peru	36	1170	1.0	8.7
Philippines	34	790	2.8	3.8
Senegal	48	430	−0.3	6.4
South Korea	24	1700	6.9	2.7
Sri Lanka	27	300	2.5	2.3
Taiwan	21	1170	6.2	3.8
Thailand	30	770	4.6	3.3

Figure 7.1
Graphs of the birth rate against (a) GNP, (b) growth and (c) income ratio

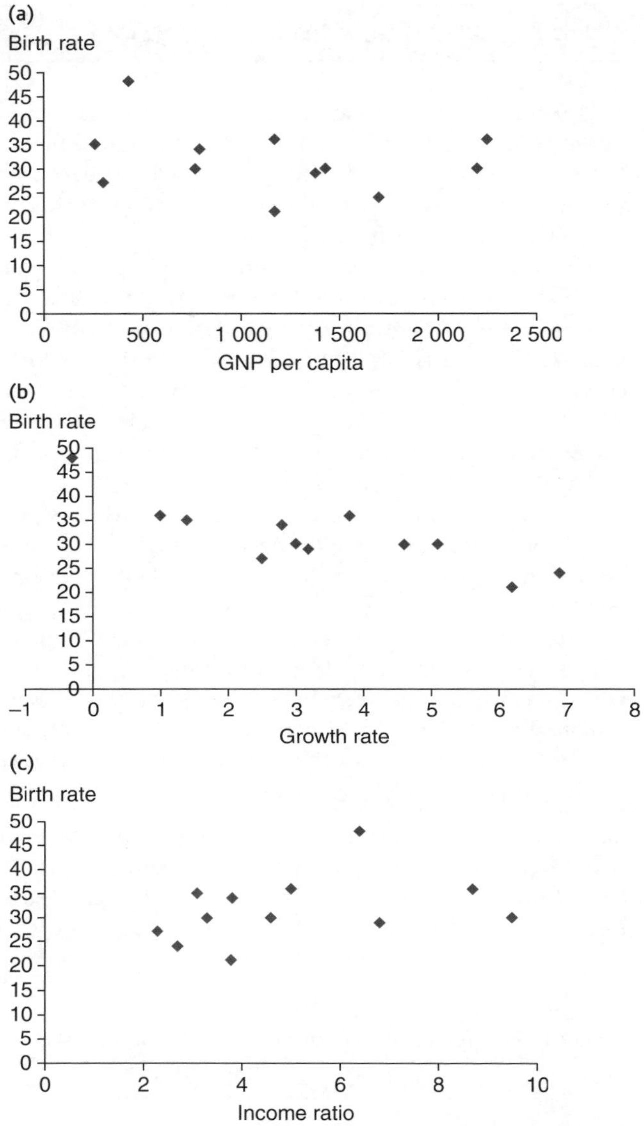

We leave aside the concerns about the sample until later and concentrate first on analysing the data. The first thing to do is to graph the variables to see if anything useful is revealed. *XY* graphs (see Chapter 1) are the most suitable in this case, and they are shown in Figure 7.1. From these we see a reasonably tidy relationship between the birth rate and the growth rate, with a negative slope; there is a looser relationship with the income ratio, with a positive slope; and there is little discernible pattern (apart from a flat line) in the graph of birth rate against GNP.

Todaro asserts that the best relationship is between the birth rate and income inequality. He rejects the growth rate as an important determinant of the birth rate because of the four countries near the top of the chart, which have very different growth rates, yet similar birth rates. In the following sections we shall see whether Todaro's conclusions are justified.

Correlation

The graphs are helpful, but it would be useful to have a simple numerical summary measure of each relationship. For this purpose, we use the **correlation coefficient** between any pair of variables. We illustrate this by calculating the correlation coefficient between the birth rate (B) and growth (G), although we also present the results for the other cases. Just as the mean is a number that summarises information about a single variable, so the correlation coefficient is a number which summarises the relationship between two variables.

The different types of possible relationship between any two variables, X and Y, may be summarised as follows:

- High values of X tend to be associated with low values of Y and vice versa. This is termed **negative correlation** and appears to be the case for B and G.
- High (low) values of X tend to be associated with high (low) values of Y. This is **positive correlation** and reflects (rather weakly) the relationship between B and the income ratio (IR).
- No relationship between X and Y exists. High (low) values of X are associated about equally with high and low values of Y. This is **zero**, or the absence of, **correlation**. There appears to be little correlation between the birth rate and per capita GNP.

It should be noted that positive correlation does not mean that high values of X are *always* associated with high values of Y, but usually they are. It is also the case that correlation only measures a *linear* relationship between the two variables. As a counter-example, consider the backward-bending labour supply curve, as suggested by economic theory (higher wages initially encourage extra work effort, but above a certain point the benefit of higher wage rates is taken in the form of more leisure). The relationship is non-linear and the measured degree of correlation between wages and hours of work is likely to be low, even though the former obviously influences the latter.

The sample correlation coefficient, r, is a numerical statistic which distinguishes between the types of cases shown in Figure 7.1. It has the following properties:

- It always lies between -1 and $+1$. This makes it relatively easy to judge the strength of an association.
- A positive value of r indicates positive correlation, a higher value indicating a stronger correlation between X and Y (i.e. the observations lie closer to a straight line). A value of $r = 1$ indicates perfect positive correlation and means that all the observations lie precisely on a straight line with positive slope, as Figure 7.2 illustrates.

Figure 7.2
Perfect positive correlation

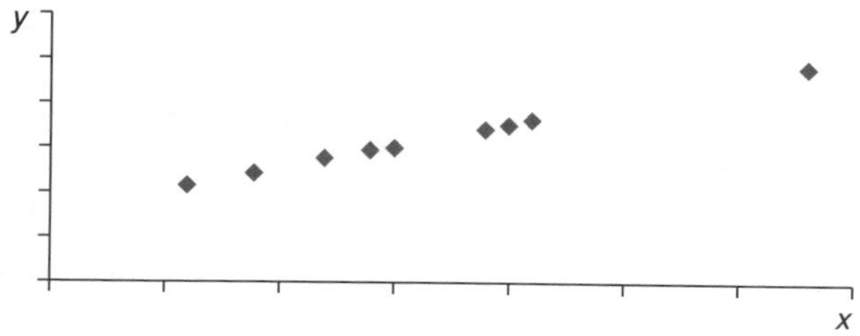

Table 7.2 Calculation of the correlation coefficient, *r*

Country	Birth rate *Y*	GNP growth *X*	Y^2	X^2	*XY*
Brazil	30	5.1	900	26.01	153.0
Colombia	29	3.2	841	10.24	92.8
Costa Rica	30	3.0	900	9.00	90.0
India	35	1.4	1 225	1.96	49.0
Mexico	36	3.8	1 296	14.44	136.8
Peru	36	1.0	1 296	1.00	36.0
Philippines	34	2.8	1 156	7.84	95.2
Senegal	48	−0.3	2 304	0.09	−14.4
South Korea	24	6.9	576	47.61	165.6
Sri Lanka	27	2.5	729	6.25	67.5
Taiwan	21	6.2	441	38.44	130.2
Thailand	30	4.6	900	21.16	138.0
Totals	380	40.2	12 564	184.04	1 139.7

Note: In addition to the *X* and *Y* variables in the first two columns, three other columns are needed, for X^2, Y^2 and *XY* values.

- A negative value of r indicates negative correlation. Similar to the above, a larger negative value indicates stronger negative correlation and $r = -1$ signifies perfect negative correlation.
- A value of $r = 0$ (or close to it) indicates a lack of correlation between X and Y.
- The relationship is symmetric, i.e. the correlation between X and Y is the same as between Y and X. It does not matter which variable is labelled Y and which is labelled X.

The formula[1] for calculating the correlation coefficient is given in equation (7.1):

$$r = \frac{n\sum XY - \sum X\sum Y}{\sqrt{(n\sum X^2 - (\sum X)^2)(n\sum Y^2 - (\sum Y)^2)}} \quad (7.1)$$

Although this looks rather complicated, it uses just six items which are easily calculated from the data: n, $\sum X$, $\sum Y$, $\sum X^2$, $\sum Y^2$ and $\sum XY$. These are shown in Table 7.2 for the relationship between birth rate (Y) and growth (X) variables, and r is then calculated in equation (7.2):

$$r = \frac{12 \times 1139.7 - 40.2 \times 380}{\sqrt{(12 \times 184.04 - 40.2^2)(12 \times 12564 - 380^2)}} = -0.824 \quad (7.2)$$

This result indicates a fairly strong negative correlation between the birth rate and growth, at least for this sample. Countries which have higher economic growth rates also tend to have lower birth rates. The result of calculating the correlation coefficient for the case of the birth rate and the income ratio is $r = 0.35$, which is positive as expected. Hence greater inequality (higher IR) is associated with a higher birth rate, although the degree of correlation is not particularly strong and less than the correlation with the growth rate. Between the birth rate and GNP per capita, the value of r is only -0.26, indicating only a modest degree of correlation. All of this begins to cast doubt upon Todaro's interpretation of the data.

[1]The formula for r can be written in a variety of different ways. The one given here is the most convenient for calculation.

Exercise 7.1

?

(a) Perform the required calculations to confirm that the correlation between the birth rate and the income ratio is 0.35.

(b) In Excel, use the =CORREL( ) function to confirm your calculations in part (a). (For example, the function =CORREL(A1:A12,B1:B12) would calculate the correlation between a variable X in cells A1:A12 and Y in cells B1:B12.)

(c) Calculate the correlation coefficient between the birth rate and the income ratio again, but expressing the birth rate per 100 population and dividing the income ratio by 100. Your calculation should confirm that changing the units of measurement leaves the correlation coefficient unchanged.

Are the results significant?

These results come from a (small) sample, one of many that could have been collected (either for different countries or different time periods). Once again we may ask the question, what can we infer about the population (of all developing countries) from the sample? *Assuming* the sample was drawn at random (which may not be justified) we can use the principles of hypothesis testing introduced in Chapter 5. As usual, there are two possibilities:

(1) The truth is that there is no correlation (in the population) and that our sample exhibits such a large (absolute) value by chance.
(2) There really is a correlation between the birth rate and the growth rate and the sample correctly reflects this.

By denoting the true but unknown population correlation coefficient by ρ (the Greek letter 'rho'), the possibilities can be expressed in terms of a hypothesis test:

$$H_0: \rho = 0$$
$$H_1: \rho \neq 0$$

The test statistic in this case is not r itself but a transformation of it:

$$t = \frac{r\sqrt{n-2}}{\sqrt{1-r^2}} \tag{7.3}$$

which has a t distribution with $n - 2$ degrees of freedom. The five steps of the test procedure are therefore:

(1) Write down the null and alternative hypotheses (shown above).
(2) Choose the significance level of the test: 5% by convention.
(3) Look up the critical value of the test for $n - 2 = 10$ degrees of freedom: $t^*_{10} = 2.228$ for a two-tail test.
(4) Calculate the test statistic using equation (7.3):

$$t = \frac{-0.824\sqrt{12-2}}{\sqrt{1-(-0.824)^2}} = -4.59$$

(5) Compare the test statistic with the critical value. In this case $t < -t^*_{10}$ so H_0 is rejected. There is a less than 5% chance of the sample evidence occurring if the null hypothesis were true, so the latter is rejected. There does appear to be a genuine association between the birth rate and the growth rate.

Performing similar calculations (see Exercise 7.2) for the income ratio and for GNP reveals that in both cases the null hypothesis cannot be rejected at the 5%

significance level. These observed associations could well have arisen by chance so the evidence is much less convincing.

Are significant results important?

Following the discussion in Chapter 5, we might ask if a certain value of the correlation coefficient is economically important as well as significant. We saw earlier that 'significant' results need not be important. The difficulty in this case is that we have little intuitive understanding of the correlation coefficient. Is $\rho = 0.5$ important, for example? Would it make much difference if it were only 0.4?

Our understanding may be helped if we look at some graphs of variables with different correlation coefficients (these data were generated artificially to illustrate the point). Three are shown in Figure 7.3. Panel (a) of the figure graphs two variables with a correlation coefficient of 0.2. Visually there seems little association between the variables, yet the correlation coefficient is (just) significant: $t = 2.06$ ($n = 100$ and the Prob-value is 0.046). This is a significant result which does not impress much.

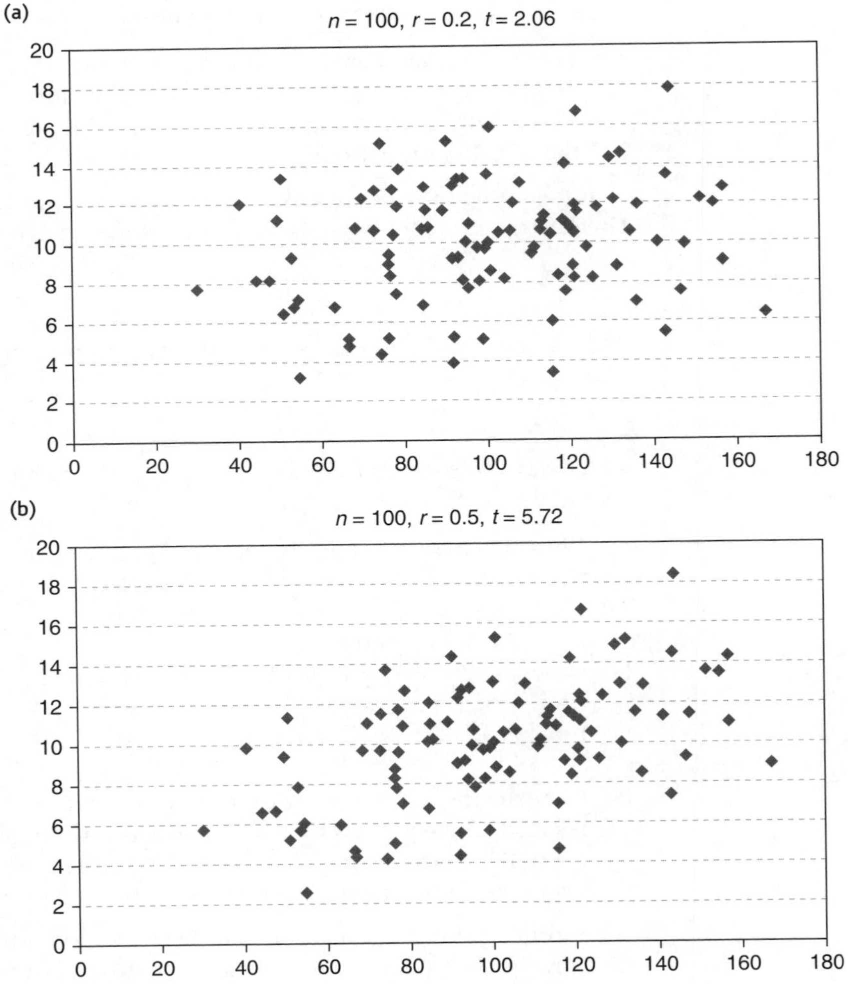

Figure 7.3
Variables with different degrees of correlation

Figure 7.3 (*cont'd*)

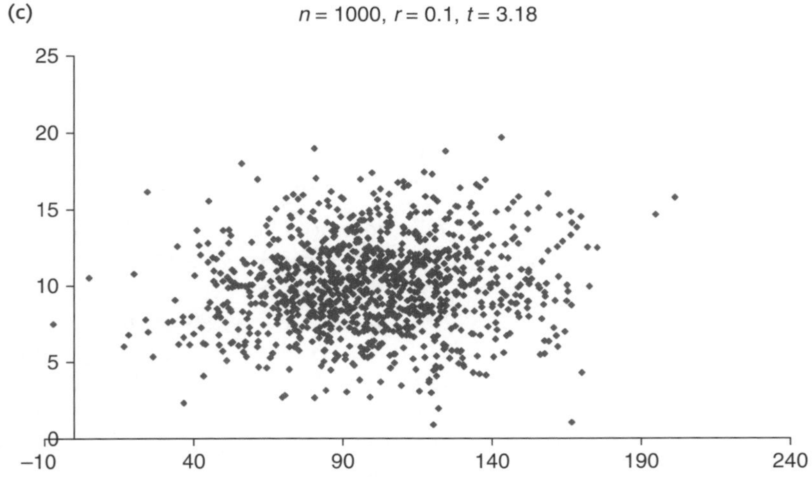

In panel (b) the correlation coefficient is 0.5 and the association seems a little stronger visually, although there is still a substantial scatter of the observations around a straight line. Yet the t statistic in this case is 5.72, highly significant (Prob-value 0.000).

Finally, panel (c) shows an example where $n = 1000$. To the eye this looks much like a random scatter, with no discernible pattern. Yet the correlation coefficient is 0.1 and the t statistic is 3.18, again highly significant (Prob-value = 0.002).

The lessons from this seem fairly clear. What looks like a random scatter on a chart may in fact reveal a relationship between variables which is statistically significant, especially if there are a large number of observations. On the other hand, a high t statistic and correlation coefficient can still indicate a lot of variation in the data, revealed by the chart. Panel (b) suggests, for example, that we are unlikely to get a very reliable prediction of the value of y, even if we know the value of x.

Exercise 7.2

?

(a) Test the hypothesis that there is no association between the birth rate and the income ratio.

(b) Look up the Prob-value associated with the test statistic and confirm that it does not reject the null hypothesis.

Correlation and causality

It is important to test the significance of any result because almost every pair of variables will have a non-zero correlation coefficient, even if they are totally unconnected (the chance of the sample correlation coefficient being *exactly* zero is very, very small). Therefore, it is important to distinguish between correlation coefficients which are significant and those which are not, using the t test just outlined. But even when the result is significant one should beware of the danger of 'spurious' correlation. Many variables which clearly cannot be related turn out to be 'significantly' correlated with each other. One now famous example is between the price level and cumulative rainfall. Since they both rise year after year, it is easy to see why they are correlated, yet it is hard to think of a plausible reason why they should be causally related to each other.

Apart from spurious correlation, there are four possible reasons for a non-zero value of r:

(1) X influences Y.
(2) Y influences X.
(3) X and Y jointly influence each other.
(4) Another variable, Z, influences both X and Y.

Correlation alone does not allow us to distinguish between these alternatives. For example, wages (X) and prices (Y) are highly correlated. Some people believe this is due to cost–push inflation, i.e. that wage rises lead to price rises. This is case (1) above. Others believe that wages rise to keep up with the cost of living (i.e. rising prices), which is (2). Perhaps a more convincing explanation is (3), a wage–price spiral where each feeds upon the other. Others would suggest that it is the growth of the money supply, Z, which allows both wages and prices to rise. To distinguish between these alternatives is important for the control of inflation, but correlation alone does not allow that distinction to be made.

Correlation is best used therefore as a suggestive and descriptive piece of analysis, rather than a technique which gives definitive answers. It is often a preparatory piece of analysis, which gives some clues to what the data might yield, to be followed by more sophisticated techniques such as regression.

The coefficient of rank correlation

On occasion it is inappropriate or impossible to calculate the correlation coefficient as described above and an alternative approach is helpful. Sometimes the original data are unavailable but the ranks are. For example, schools may be ranked in terms of their exam results, but the actual pass rates are not available. Similarly, they may be ranked in terms of spending per pupil, with actual spending levels unavailable. Although the original data are missing, one can still test for an association between spending and exam success by calculating the correlation between the ranks. If extra spending improves exam performance, schools ranked higher on spending should also be ranked higher on exam success, leading to a positive correlation.

Second, even if the raw data are available, they may be highly skewed and hence the correlation coefficient may be influenced heavily by a few outliers. In this case the hypothesis test for correlation may be misleading as it is based on the assumption of underlying Normal distributions for the data. In this case we could transform the values to ranks, and calculate the correlation of the ranks. In a similar manner to the median, described in Chapter 1, this can effectively deal with heavily skewed distributions.

Note the difference between the two cases. In the first, we would prefer to have the actual school pass rates and expenditures because our analysis would be better. We could actually see how much extra we have to spend in order to get better results. In the second case we actually prefer to use the ranks because the original data might mislead us, through the presence of outliers for example. **Non-parametric statistics** are those which are robust to the distribution of the data, such as the calculation of the median, rather than the mean which is a parametric measure. We do not cover many examples of the former in this text, but the **rank correlation coefficient** is one of them.

Spearman's coefficient of rank correlation is a measure that is robust to the underlying distribution of the data. It does not matter, for example, if the data are skewed. (The 'standard' correlation coefficient described above is more fully known as **Pearson's product-moment correlation coefficient**, to distinguish it.) The formula to be applied is the same as before, although there are a few tricks to be learned about constructing the ranks, and also the hypothesis test is conducted in a different manner.

Using the ranks is generally less efficient than using the original data, because one is effectively throwing away some of the information (e.g. by *how much* do countries' growth rates differ?). However, there is a trade-off: the rank correlation coefficient is more robust, i.e. it is less influenced by outliers or highly skewed distributions. If one suspects this is a risk, it may be better to use the ranks. This is similar to the situation where the median can prove superior to the mean as a measure of central tendency.

We will calculate the rank correlation coefficient for the data on birth and growth rates, to provide a comparison with the ordinary correlation coefficient calculated earlier. It is unlikely that the distributions of birth or of growth rates are particularly skewed (and we have too few observations to reliably tell), so the Pearson measure might generally be preferred, but we calculate the Spearman coefficient for comparison. Table 7.3 presents the data for birth and growth rates in the form of ranks. Calculating the ranks is fairly straightforward, although there are a couple of points to note.

The country with the highest birth rate has the rank of 1, the next highest 2, and so on. Similarly, the country with the highest growth rate ranks 1, etc. One could reverse a ranking, so the lowest birth rate ranks 1, for example; the direction of ranking is somewhat arbitrary. This would leave the rank correlation coefficient unchanged in value, but the sign would change, e.g. −0.691 would become +0.691. This could be confusing as we would now have a 'positive' correlation rather than a negative one (though the birth rate variable would now have to be redefined). It is better to use the 'natural' order of ranking for each variable, i.e. rank *both* variables in ascending order or *both* in descending order.

Table 7.3 Calculation of Spearman's rank correlation coefficient

Country	Birth rate Y	Growth rate X	Rank Y R_Y	Rank X R_X	R_Y^2	R_X^2	$R_X R_Y$
Brazil	30	5.1	7	3	49	9	21
Colombia	29	3.2	9	6	81	36	54
Costa Rica	30	3.0	7	7	49	49	49
India	35	1.4	4	10	16	100	40
Mexico	36	3.8	2.5	5	6.25	25	12.5
Peru	36	1.0	2.5	11	6.25	121	27.5
Philippines	34	2.8	5	8	25	64	40
Senegal	48	−0.3	1	12	1	144	12
South Korea	24	6.9	11	1	121	1	11
Sri Lanka	27	2.5	10	9	100	81	90
Taiwan	21	6.2	12	2	144	4	24
Thailand	30	4.6	7	4	49	16	28
Totals			78	78	647.5	650	409

Note: The country with the highest growth rate (South Korea) is ranked 1 for variable X; Taiwan, the next fastest growth nation, is ranked 2, etc. For the birth rate, Senegal is ranked 1, having the highest birth rate, 48. Taiwan has the lowest birth rate and so is ranked 12 for variable Y.

Confusion will usually follow if you rank one variable in ascending order, the other descending.

Where two or more observations are the same, as are the birth rates of Mexico and Peru, then they are given the same rank, which is the average of the relevant ranking values. For example, both countries are given the rank of 2.5, which is the average of 2 and 3. Similarly, Brazil, Costa Rica and Thailand are all given the rank of 7, which is the average of 6, 7 and 8. The next country, Colombia, is then given the rank of 9.

Excel warning

Microsoft Excel has a *rank.avg( )* function built in, which takes a variable and calculates a new variable consisting of the ranks, similar to the above table. This can obviously save a bit of work. A word of warning, however: the *rank( )* and *rank.eq( )* functions, also in *Excel*, will give incorrect answers if there are ties in the data. The *rank.avg( )* function is new from Excel 2010 onwards, earlier versions of Excel only have the unreliable functions.

We now apply formula (7.1) to the ranked data, giving:

$$r_s = \frac{n\sum XY - \sum X \sum Y}{\sqrt{(n\sum X^2 - (\sum X)^2)(n\sum Y^2 - (\sum Y)^2)}}$$

$$= \frac{12 \times 409 - 78 \times 78}{\sqrt{(12 \times 650 - 78^2)(12 \times 647.5 - 78^2)}} = -0.691$$

This indicates a negative rank correlation between the two variables, as with the standard correlation coefficient ($r = -0.824$), but with a slightly smaller absolute value.

To test the significance of the result a hypothesis test can be performed on the value of ρ_s, the corresponding population parameter:

$H_0: \rho_s = 0$
$H_1: \rho_s \neq 0$

This time the t distribution cannot be used (because we are no longer relying on the parent distribution being Normal), but prepared tables of the critical values for ρ_s itself may be consulted; these are given in Table A6 (see page 462), and an excerpt is given in Table 7.4.

The critical value at the 5% significance level, for $n = 12$, is 0.591. Hence the null hypothesis is rejected if the rank correlation coefficient falls outside the range

Table 7.4 Excerpt from Table A6: Critical values of the rank correlation coefficient

n	10%	5%	2%	1%
5	0.900			
6	0.829	0.886	0.943	
⋮	⋮	⋮	⋮	⋮
11	0.523	0.623	0.763	0.794
12	0.497	0.591	0.703	0.780
13	0.475	0.566	0.673	0.746

Note: The critical value is given at the intersection of the shaded row and column.

[−0.591, 0.591], which it does in this case. Thus the null can be rejected with 95% confidence; the data do support the hypothesis of a relationship between the birth rate and growth. This critical value shown in the table is for a two-tail test. For a one-tail test, the significance level given in the top row of the table should be halved, so we could reject the null at the 2.5% significance level or 97.5% confidence level in this case.

Exercise 7.3

(a) Rank the observations for the income ratio across countries (highest = 1) and calculate the coefficient of rank correlation with the birth rate.

(b) Test the hypothesis that $\rho_s = 0$.

(c) Reverse the rankings for both variables and confirm that this does not affect the calculated test statistic.

(d) Reverse the rankings of just the income ratio variable. How would you expect this to affect the value of the rank correlation coefficient?

Worked example 7.1

To illustrate all the calculations and bring them together without distracting explanation, we work through a simple example with the following data on X and Y:

Y	17	18	19	20	27	18
X	3	4	7	6	8	5

An XY graph of the data reveals the following picture, which suggests positive correlation:

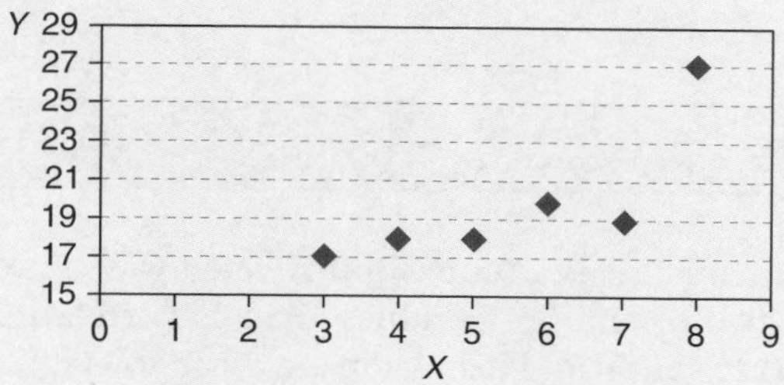

Note that one point appears to be something of an outlier. All the calculations for correlations may be based on the following table:

Obs	Y	X	Y^2	X^2	XY	Rank Y R_Y	Rank X R_X	R_Y^2	R_X^2	$R_X R_Y$
1	17	3	289	9	51	6	6	36	36	36
2	18	4	324	16	72	4.5	5	20.25	25	22.5
3	19	7	361	49	133	3	2	9	4	6
4	20	6	400	36	120	2	3	4	9	6
5	27	8	729	64	216	1	1	1	1	1
6	18	5	324	25	90	4.5	4	20.25	16	18
Totals	119	33	2427	199	682	21	21	90.5	91	89.5

The (Pearson) correlation coefficient r is therefore:

$$r = \frac{n\sum XY - \sum X \sum Y}{\sqrt{(n\sum X^2 - (\sum X)^2)(n\sum Y^2 - (\sum Y)^2)}}$$

$$= \frac{6 \times 682 - 33 \times 119}{\sqrt{(6 \times 199 - 33^2)(6 \times 2427 - 119^2)}} = 0.804$$

The hypothesis $H_0: \rho = 0$ versus $H_1: \rho \neq 0$ can be tested using the t test statistic:

$$t = \frac{r\sqrt{n-2}}{\sqrt{1-r^2}} = \frac{0.804 \times \sqrt{6-2}}{\sqrt{1-0.804^2}} = 2.7$$

which is compared to a critical value of 2.776, so the null hypothesis is not rejected, narrowly. This is largely due to the small number of observations, and anyway, it may be unwise to use the t distribution on such a small sample. The rank correlation coefficient is calculated as (using the R_X, R_Y etc., values)

$$r = \frac{n\sum XY - \sum X \sum Y}{\sqrt{(n\sum X^2 - (\sum X)^2)(n\sum Y^2 - (\sum Y)^2)}}$$

$$= \frac{6 \times 89.5 - 21 \times 21}{\sqrt{(6 \times 91 - 21^2)(6 \times 90.5 - 21^2)}} = 0.928$$

The critical value at the 5% significance level is 0.886, so the rank correlation coefficient *is* significant, in contrast to the previous result. Not too much should be read into this, however; with few observations the ranking process can easily alter the result substantially.

Regression analysis

Regression analysis is a more sophisticated way of examining the relationship between two (or more) variables than is correlation. The major differences between correlation and regression are the following:

- Regression can investigate the relationships between two *or more* variables.
- A *direction* of causality is asserted, from the explanatory variable (or variables) to the dependent variable.
- The *influence* of each explanatory variable upon the dependent variable is measured.
- The *significance* of each explanatory variable's influence can be ascertained.

Thus regression permits answers to such questions as:

- Does the growth rate influence a country's birth rate?
- If the growth rate increases, by how much might a country's birth rate be expected to fall?
- Are other variables important in determining the birth rate?

Figure 7.4
The line of best fit

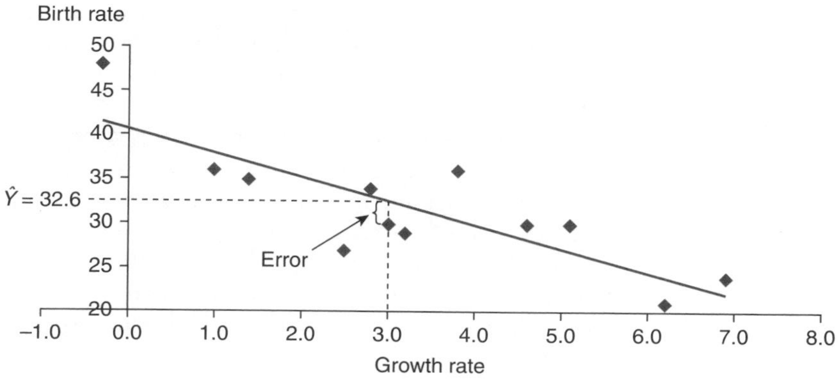

In this example we assert that the direction of causality is from the growth rate (X) to the birth rate (Y) and not vice versa. The growth rate is therefore the **explanatory variable** (also referred to as the **independent** or **exogenous variable**) and the birth rate is the **dependent variable** (also called the **explained** or **endogenous variable**).

Regression analysis describes this causal relationship by fitting a straight line drawn through the data, which best summarises them. It is sometimes called 'the line of best fit' for this reason. This is illustrated in Figure 7.4 for the birth rate and growth rate data. Note that (by convention) the explanatory variable is placed on the horizontal axis, the explained on the vertical axis. This regression line is downward sloping (its derivation will be explained shortly) for the same reason that the correlation coefficient is negative, i.e. high values of Y are generally associated with low values of X and vice versa.

Since the regression line summarises knowledge of the relationship between X and Y, it can be used to predict the value of Y given any particular value of X. In Figure 7.4 the value of $X = 3$ (the observation for Costa Rica) is related via the regression line to a value of Y (denoted[2] by $\hat{Y}$) of 32.6. This predicted value is close (but not identical) to the actual birth rate of 30. The difference reflects the absence of perfect correlation between the two variables.

The difference between the actual value, Y, and the predicted value, $\hat{Y}$, is called the **error term** or **residual**. It is labelled 'Error', e, in Figure 7.4[3]. Why should such errors occur? The relationship is never going to be an exact one for a variety of reasons. There are bound to be other factors besides growth which might affect the birth rate (e.g. the education of women) and these effects are all subsumed into the error term. There might additionally be simple measurement error (of Y) and, of course, people do act in a somewhat random fashion rather than follow rigid rules of behaviour.

All of these factors fall into the error term, and this means that the observations lie around the regression line rather than on it. If there are many of these factors, none of which is predominant, and they are independent of each other, then these errors may be assumed to be Normally distributed about the regression line.

[2]A 'hat' (^) over a symbol is often used to indicate the estimate of that variable.

[3]The italic e denoting the error term should not be confused with the use of the same letter as the base for natural logarithms (see the Appendix to Chapter 1, page 91). The correct interpretation should be clear from the context.

Why not include these factors explicitly? On the face of it this would seem to be an improvement, making the model more realistic. However, the costs of doing this are that the model becomes more complex, calculation becomes more difficult (not so important now with computers) and it is generally more difficult for the reader (or researcher) to interpret what is going on. If these other factors have only small effects upon the dependent variable, then it might be better to ignore them, adopt a simple model and focus upon the main relationship of interest. There is a virtue in simplicity, as long as the simplified model still gives an undistorted view of the relationship. In Chapter 10 on multiple regression the trade-off between simplicity and realism will be further discussed, particularly with reference to the problems which can arise if relevant explanatory variables are omitted from the analysis.

Calculation of the regression line

The equation of the sample **regression line** may be written

$$\hat{Y}_i = a + bX_i \tag{7.6}$$

where $\hat{Y}_i$ is the predicted value of Y for observation (country) i, X_i is the value of the explanatory variable for observation i, and a and b are fixed coefficients to be estimated; a measures the intercept of the regression line on the Y-axis, b measures its slope. This is illustrated in Figure 7.5.

The first task of regression analysis is to find the values of a and b so that the regression line may be drawn. To do this we proceed as follows. The difference between the actual value, Y_i, and its predicted value, $\hat{Y}_i$, is e_i, the error. Thus

$$Y_i = \hat{Y}_i + e_i \tag{7.7}$$

Substituting equation (7.6) into equation (7.7), the regression equation can be written as

$$Y_i = a + bX_i + e_i \tag{7.8}$$

Equation (7.8) shows that observed birth rates are made up of two components:

(1) that part explained by the growth rate, $a + bX_i$, and
(2) an error component, e_i.

In a good model, part (1) should be large relative to part (2) and the regression line is based upon this principle. The line of best fit is therefore found by finding

Figure 7.5
Intercept and slope of the regression line

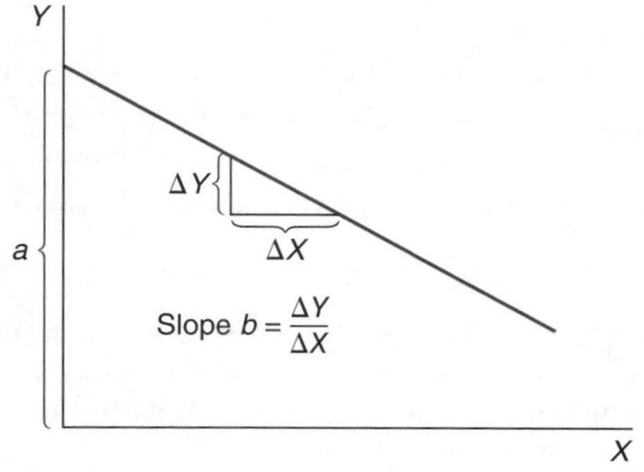

the values of a and b which *minimise the sum of squared errors* ($\sum e_i^2$) from the regression line. For this reason, this method is known as 'the method of least squares' or simply 'ordinary least squares' (OLS). The use of this criterion will be justified later on, but it can be said in passing that the sum of the errors is not minimised because that would not lead to a unique answer for the values a and b. In fact, there is an infinite number of possible lines through the data which all yield a sum of errors equal to zero. Minimising the sum of *squared* errors does yield a unique answer.

The task is therefore to

$$\text{minimise} \sum e_i^2 \tag{7.9}$$

by choice of a and b.

Rearranging equation (7.8), the error is given by

$$e_i = Y_i - a - bX_i \tag{7.10}$$

so equation (7.9) becomes

$$\text{minimise} \sum (Y_i - a - bX_i)^2 \tag{7.11}$$

by choice of a and b.

Finding the solution to (7.11) requires the use of differential calculus, and is not presented here. The resulting formulae for a and b are

$$b = \frac{n\sum XY - \sum X \sum Y}{n\sum X^2 - (\sum X)^2} \tag{7.12}$$

and

$$a = \bar{Y} - b\bar{X} \tag{7.13}$$

where $\bar{X}$ and $\bar{Y}$ are the mean values of X and Y, respectively. The values necessary to evaluate equations (7.12) and(7.13) can be obtained from Table 7.2 which was used to calculate the correlation coefficient. These values are repeated for convenience:

$$\begin{aligned} \sum Y &= 380 & \sum Y^2 &= 12\,564 \\ \sum X &= 40.2 & \sum X^2 &= 184.04 \\ \sum XY &= 1139.70 & n &= 12 \end{aligned}$$

Using these values, we obtain

$$b = \frac{12 \times 1139.70 - 40.2 \times 380}{12 \times 184.04 - 40.2^2} = -2.700$$

and

$$a = \frac{380}{12} - (-2.700) \times \frac{40.2}{12} = 40.711$$

Thus the regression equation can be written, to two decimal places for clarity, as

$$Y_i = 40.71 - 2.70X_i + e_i$$

Interpretation of the slope and intercept

The most important part of the result is the **slope** coefficient $b = -0.27$ since it measures the effect of X upon Y. This result implies that a unit increase in the growth rate (e.g. from 2% to 3% p.a.) would lower the birth rate by 2.7, e.g. from 30 births per 1000 population to 27.3. Given that the growth data refer to a 20-year period (1961–81), this increase in the growth rate might need to be sustained over such a time, not an easy task. It is unlikely that an increase in the growth rate in one year would have such an immediate effect on the birth rate. How big is the effect upon the birth rate? The average birth rate in the sample is 31.67, so a reduction of 2.7 for an average country would be a fall of 8.5% ($2.7/31.67 \times 100$). This is reasonably substantial (though not enough to bring the birth rate down to developed country levels) but would need a considerable, sustained increase in the growth rate to bring it about.

The value of a, the **intercept**, may be interpreted as the predicted birth rate of a country with zero growth (since $\hat{Y}_i = a$ at $X = 0$). This value of 40.71 is fairly close to that of Senegal, which actually had negative growth over the period and whose birth rate was 48, a little higher than the intercept value. Although a has a sensible interpretation in this case, this is not always so. For example, in a regression of the demand for a good on its price, a would represent demand at zero price, which is unlikely ever to be observed.

Exercise 7.4

(a) Calculate the regression line relating the birth rate to the income ratio.

(b) Interpret the coefficients of this equation.

Measuring the goodness of fit of the regression line

Having calculated the regression line, we now ask whether it provides a good fit for the data, i.e. do the observations tend to lie close to, or far away from, the line? Even though we have fitted a regression line, by itself this tells us nothing about the closeness of the fit. If the fit is poor, perhaps the effect of X upon Y is not so strong after all. Note that even if X has *no* true effect upon Y, we can still calculate a regression line and its slope coefficient b. Although b is likely to be small, it is unlikely to be exactly zero. Measuring the goodness of fit of the data to the line helps us to distinguish between good and bad regressions.

We proceed by comparing the three competing models explaining the birth rate. Which of them fits the data best? Using the income ratio and the GNP variable gives the following regressions (calculations not shown) to compare with our original model:

for the income ratio (IR):	$B = 26.44 + 1.045 \times \text{IR} + e$
for GNP:	$B = 34.72 - 0.003 \times \text{GNP} + e$
for growth:	$B = 40.71 - 2.70 \times \text{GROWTH} + e$

How can we decide which of these three is 'best' on the basis of the regression equations alone? From Figure 7.1 it is evident that some relationships appear stronger than others, yet this is not revealed by examining the regression equation alone. More information is needed. (You cannot choose the best equation simply by looking at the size of the coefficients. Consider why that is so.)

The goodness of fit is calculated by comparing two lines: the regression line and the 'mean line' (i.e. a horizontal line drawn at the mean value of Y). The regression line *must* fit the data better (if the mean line were the best fit, that is

Figure 7.6
The calculation of R^2

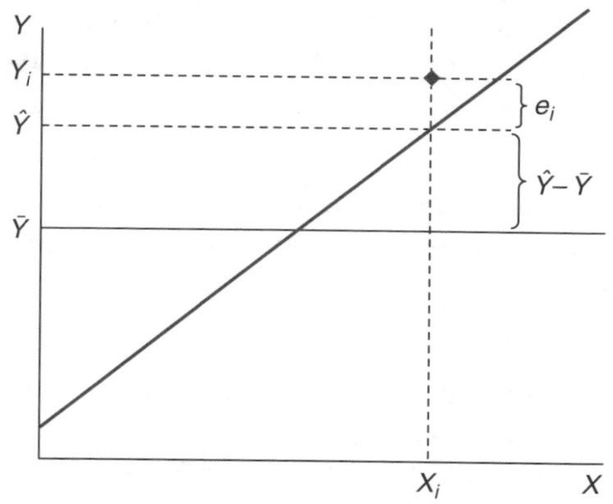

also where the regression line would be) but the question is how much better? This is illustrated in Figure 7.6, which demonstrates the principle behind the calculation of the **coefficient of determination**, denoted by R^2 and usually more simply referred to as 'R squared'.

The figure shows the mean value of Y, the calculated sample regression line and an arbitrarily chosen sample observation (X_i, Y_i). The difference between Y_i and $\overline{Y}$ (length $Y_i - \overline{Y}$) can be divided up into:

(1) That part 'explained' by the regression line, $\hat{Y}_i - \overline{Y}$ (i.e. explained by the value of X_i).
(2) The error term $e_i = Y_i - \hat{Y}_i$.

In algebraic terms,

$$Y_i - \overline{Y} = (Y - \hat{Y}_i) + (\hat{Y}_i - \overline{Y}) \tag{7.14}$$

A good regression model should 'explain' a large part of the differences between the Y_i values and $\overline{Y}$, i.e. the length $(\hat{Y}_i - \overline{Y})$ should be large relative to $Y_i - \overline{Y}$. A measure of fit could therefore be $(\hat{Y}_i - \overline{Y})/(Y_i - \overline{Y})$. We need to apply this to all observations rather than just a single one; hence we could sum this expression over all the sample observations. A problem with this is that some of the terms would take a negative value and offset the positive terms. To measure the goodness of fit, we do not want the positive and negative terms to cancel each other out. Hence, to get round this problem, we square each of the terms in equation (7.14) to make them all positive, and then sum over the observations. This gives

$\sum(Y_i - \overline{Y})^2$, known as the **total sum of squares** (TSS)
$\sum(\hat{Y}_i - \overline{Y})^2$, the **regression sum of squares** (RSS), and
$\sum(Y_i - \hat{Y}_i)^2$, the **error sum of squares** (ESS)

The measure of goodness of fit, R^2, is then defined as the ratio of the regression sum of squares to the total sum of squares, i.e.

$$R^2 = \frac{\text{RSS}}{\text{TSS}} \tag{7.15}$$

The better the divergences between Y_i and $\overline{Y}$ are explained by the regression line, the better the goodness of fit, and the higher the calculated value of R^2. Further, it is true that

$$\text{TSS} = \text{RSS} + \text{ESS} \tag{7.16}$$

From equations (7.15) and (7.16) we can then see that R^2 must lie between 0 and 1 (note that since each term in equation (7.16) is a sum of squares, none of them can be negative). Thus

$$0 \leq R^2 \leq 1$$

A value of $R^2 = 1$ (and hence ESS = 0) indicates that all the sample observations lie exactly on the regression line (equivalent to perfect correlation). If $R^2 = 0$, then the regression line is of no use at all $-X$ does not influence Y (linearly) at all, and to try to predict a value of Y one might as well use the mean $\overline{Y}$ rather than the value X_i inserted into the sample regression equation.

To calculate R^2, alternative formulae to those above make the task easier. Instead we use:

$$\text{TSS} = \Sigma(Y_i - \overline{Y})^2 = \Sigma Y_i^2 - n\overline{Y}^2 = 12564 - 12 \times 31.67^2 = 530.667 \tag{7.17a}$$

$$\begin{aligned}\text{ESS} &= \Sigma(Y_i - \hat{Y})^2 = \Sigma Y_i^2 - a\Sigma Y_i - b\Sigma X_i Y_i \\ &= 12564 - 40.711 \times 380 - (-2.7) \times 1139.70 = 170.754\end{aligned} \tag{7.17b}$$

$$\text{RSS} = \text{TSS} - \text{ESS} = 530.667 - 170.754 = 359.913 \tag{7.17c}$$

This gives the result

$$R^2 = \frac{\text{RSS}}{\text{TSS}} = \frac{359.913}{530.667} = 0.678$$

This is interpreted as follows. Countries' birth rates vary around the overall mean value of 31.67 and 67.8% of this variation is explained by variation in countries' growth rates. This is quite a respectable figure to obtain, leaving only 32.8% of the variation in Y left to be explained by other factors (or pure random variation). The regression seems to make a worthwhile contribution to explaining why birth rates differ. However, it does not explain the *mechanism* by which higher growth leads to a lower birth rate.

It turns out that in simple regression (i.e. where there is only one explanatory variable), R^2 is simply the square of the correlation coefficient between X and Y. Thus, for the income ratio and for GNP, we have:

for IR: $R^2 = 0.35^2 = 0.13$
for GNP: $R^2 = -0.26^2 = 0.07$

This shows, once again, that these other variables are not terribly useful in explaining why birth rates differ. Each of them explains only a small proportion of the variation in Y.

It should be emphasised at this point that R^2 is not the only criterion (or even an adequate one in all cases) for judging the quality of a regression equation and that other statistical measures, set out below, are also helpful.

Exercise 7.5

(a) Calculate the R^2 value for the regression of the birth rate on the income ratio, calculated in Exercise 7.4.

(b) Confirm that this result is the same as the square of the correlation coefficient between these two variables, calculated in Exercise 7.1.

Inference in the regression model

So far, regression has been used as a descriptive technique, to measure the relationship between the two variables. We now go on to draw inferences from the analysis about what the *true* regression line might look like. As with correlation, the estimated relationship is in fact a *sample* regression line, based upon data for 12 countries. The estimated coefficients a and b are random variables, since they would differ from sample to sample. What can be inferred about the true (but unknown) regression equation?

The question is best approached by first writing down a true or population regression equation, in a form similar to the sample regression equation:

$$Y_i = \alpha + \beta X_i + \varepsilon_i \tag{7.18}$$

As usual, Greek letters denote true, or population, values. α and β are thus the population *parameters*, of which a and b are (point) estimates, using the method of least squares. ε is the population error term. If we could observe the individual error terms ε_i, then we would be able to get exact values of α and β (even from a sample), rather than just estimates.

Given that a and b are estimates, we can ask about their properties: whether they are unbiased and how precise they are, compared to alternative estimators. Under reasonable assumptions (see, for example, Maddala and Lahiri (2009), Chapter 3) it can be shown that the OLS estimates of the coefficients are unbiased. Thus, OLS provides useful point estimates of the parameters (the true values α and β). This is one reason for using the least squares method. It can also be shown that, among the class of linear unbiased estimators, OLS has the minimum variance, i.e. the method provides the most precise estimates. This is another, powerful justification for the use of OLS.

Analysis of the errors

So far, we have found point estimates and we have learnt they are unbiased. However, just because they are unbiased (correct on average) does not mean that we might not get an estimate which is a long way from the truth. For some insight into this we need, as usual, a confidence interval estimate. To find the confidence intervals for α and β, we need to know which statistical distribution we should be using, i.e. the distributions of a and b. These can be derived, based on the assumptions that the error term ε in equation (7.18) is Normally distributed and that the errors are statistically independent of each other. Since we are using cross-section data from countries which are different geographically, politically and socially, it seems reasonable to assume the errors are independent.

To check the Normality assumption, we can graph the residuals calculated from the sample regression line. If the true errors are Normal, it seems likely that these residuals should be approximately Normal also. The residuals are calculated according to equation (7.10). For example, to calculate the residual for Brazil, we subtract the fitted value from the actual value. The fitted value is calculated by substituting the growth rate into the estimated regression equation, yielding $\hat{Y} = 40.712 - 2.7 \times 5.1 = 26.9$. Subtracting this from the actual value gives

Table 7.5 Calculation of residuals

	Actual birth rate	Fitted values	Residuals
Brazil	30	26.9	3.1
Colombia	29	32.1	−3.1
Costa Rica	30	32.6	−2.6
⋮	⋮	⋮	⋮
Sri Lanka	27	34.0	−7.0
Taiwan	21	24.0	−3.0
Thailand	30	28.3	1.7

$Y_i - \hat{Y} = 30 - 26.9 = 3.1$ Other countries' residuals are calculated in similar manner, yielding the results shown in Table 7.5.

These residuals may then be gathered together in a frequency table (as in Chapter 1) and graphed. This is shown in Figure 7.7.

Although the number of observations is small (and therefore the graph is not a smooth curve), the chart does have the greater weight of frequencies in the centre as one would expect, with less weight as one moves into the tails of the distribution. The assumption that the true error term is Normally distributed does not seem unreasonable.

If the residuals from the sample regression equation appeared distinctly non-Normal (heavily skewed, for example), then one should be wary of constructing confidence intervals using the formulae below, based on a small sample. Instead, one might consider transforming the data (see below) before continuing, since such a data transformation might make the new residuals Normally distributed. There are more formal tests for Normality of the residuals[4], but they are beyond the scope of this text. Drawing a graph is an informal alternative which can be useful, but remember that graphical methods can be misinterpreted.

If one were using time-series data one should also check the residuals for **autocorrelation** at this point. This occurs when the error in period t is dependent in

Figure 7.7
Bar chart of residuals from the regression equation

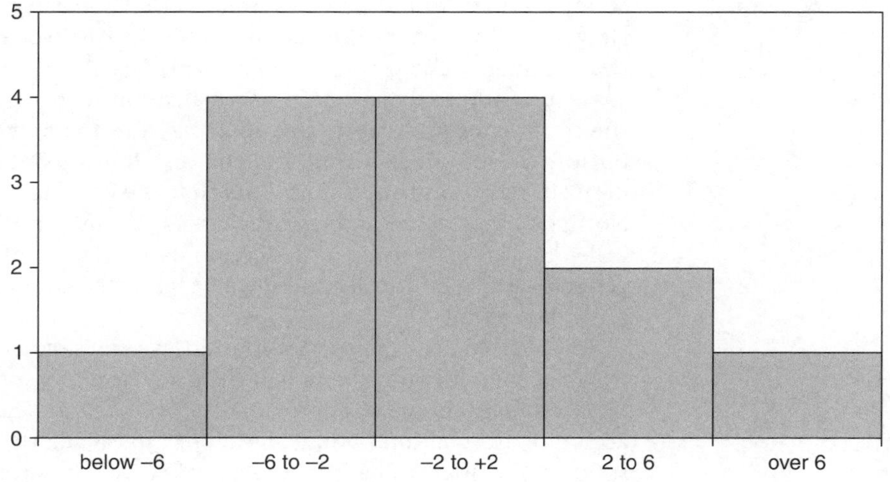

[4]One example is the Jarque-Bera test.

some way on the error in the previous period(s) and implies that the method of least squares may not be the best way of estimating the relationship. In this example we have cross-section data, so it is not appropriate to check for autocorrelation, since the ordering of the data does not matter. The next chapter, on multiple regression, covers this topic.

Confidence interval estimates of α and β

Having checked that the residuals appear reasonably Normal, we can proceed with inference. This means finding interval estimates of the parameters α and β and, later on, conducting hypothesis tests. As usual, the 95% confidence interval is obtained by adding and subtracting approximately two standard errors from the point estimate. We therefore need to calculate the **standard errors** of a and of b and we also need to look up tables to find the precise number of standard errors to add and subtract. The principle is just the same as for the confidence interval estimate of the sample mean, covered in Chapter 4.

The estimated sampling variance of b, the slope coefficient, is given by

$$s_b^2 = \frac{s_e^2}{\sum (X_i - \bar{X})^2} \tag{7.19}$$

where

$$s_e^2 = \frac{\sum e_i^2}{n - 2} = \frac{\text{ESS}}{n - 2} \tag{7.20}$$

is the **estimated variance of the error term**, ε.

The sampling variance of b measures the uncertainty associated with the estimate. Note that the uncertainty is greater (i) the larger the error variance s_e^2 (i.e. the more scattered are the points around the regression line) and (ii) the lower the dispersion of the X observations. When X does not vary much (less spread out along the x-axis) it is then more difficult to measure the effect of changes in X upon Y, and this is reflected in the formula.

First we need to calculate s_e^2. The value of this is

$$s_e^2 = \frac{170.754}{10} = 17.0754 \tag{7.21}$$

and so the estimated variance of b is

$$s_b^2 = \frac{17.0754}{49.37} = 0.346 \tag{7.22}$$

(Use $\sum (X_i - \bar{X})^2 = \sum X_i^2 - n\bar{X}^2$ in calculating (7.22) – it makes the calculation easier.) The estimated standard error of b is the square root of (7.22),

$$s_b = \sqrt{0.346} = 0.588 \tag{7.23}$$

To construct the confidence interval around the point estimate, $b = -2.7$, the **t distribution** is used (in regression this applies to all sample sizes, not just small ones). The 95% confidence interval is thus given by

$$b \pm t_\nu s_b \tag{7.24}$$

where t_ν is the (two-tail) critical value of the t distribution at the appropriate significance level (5% in this case), with $\nu = n - 2$ degrees of freedom. The critical value is 2.228. Thus the confidence interval evaluates to:

$$-2.7 \pm 2.228 \times 0.588 = [-4.01, -1.39]$$

Thus we can be 95% confident that the true value of β lies within this range. Note that the interval only includes negative values: we can rule out an upward-sloping regression line.

For the intercept a, the estimate of the variance is given by

$$s_a^2 = s_e^2 \times \left(\frac{1}{n} + \frac{\overline{X}^2}{\sum(X_i - \overline{X})^2}\right) = 17.0754 \times \left(\frac{1}{12} + \frac{3.35^2}{49.37}\right) = 5.304 \quad (7.25)$$

and the estimated standard error of a is the square root of this, 2.303. The 95% confidence interval for α, again using the t distribution, is

$$40.71 \pm 2.228 \times 2.303 = [35.57, 45.84]$$

The results so far can be summarised as follows:

$$\begin{aligned} &Y_i = 40.711 - 2.70X_i + e_i \\ &\text{s.e.} \quad (2.30) \quad (0.59) \\ &R^2 = 0.678 \quad n = 12 \end{aligned}$$

This conveys, at a glance, all the necessary information to the reader, who can then draw the inferences deemed appropriate. Any desired confidence interval (not just the 95% one) can be quickly calculated with the aid of a set of t tables.

Testing hypotheses about the coefficients

As well as calculating confidence intervals, one can use hypothesis tests as the basis for statistical inference in the regression model. These tests are quickly and easily explained given the information already assembled. Consider the following hypothesis:

$$H_0: \beta = 0$$
$$H_1: \beta \neq 0$$

This null hypothesis is interesting because it implies no influence of X upon Y at all (i.e. the slope of the true regression line is flat and Y_i can be equally well predicted by $\overline{Y}$). The alternative hypothesis asserts that X does in fact influence Y.

The procedure is in principle the same as in Chapter 5 on hypothesis testing. We measure how many standard errors separate the observed value of b from the hypothesised value. If this is greater than an appropriate critical value, we reject the null hypothesis. The test statistic is calculated using the formula:

$$t = \frac{b - \beta}{s_b} = \frac{-2.7 - 0}{0.588} = -4.59 \quad (7.26)$$

Thus the sample slope coefficient b differs by 4.59 standard errors from its hypothesised value $\beta = 0$. This is compared to the critical value of the t distribution, using $n - 2$ degrees of freedom. Since $t < -t_{10}^{*}(= -2.228)$, in this case, the null hypothesis is rejected with 95% confidence. X does have some influence on Y. Similar tests using the income ratio and GDP to attempt to explain the birth rate show that in neither case is the slope coefficient significantly different from zero, i.e. neither of these variables appears to influence the birth rate.

Rule of thumb for hypothesis tests

A quick and reasonably accurate method for establishing whether a coefficient is significantly different from zero is to see if it is at least twice its standard error. If so, it is significant. This works because the critical value (at 95%) of the *t* distribution for reasonable sample sizes is about 2.

Sometimes regression results are presented with the *t* statistic (as calculated above), rather than the standard error, below each coefficient. This implicitly assumes that the hypothesis of interest is that the coefficient is zero. This is not always appropriate: in the consumption function a test for the marginal propensity to consume being equal to 1 might be of greater relevance, for example. In a demand equation, one might want to test for unit elasticity. For this reason, it is better to present the standard errors rather than the *t* statistics.

Note that the test statistic $t = -4.59$ is exactly the same result as in the case of testing the correlation coefficient. This is no accident, for the two tests are equivalent. A non-zero slope coefficient means there is a relationship between X and Y which also means the correlation coefficient is non-zero. Both null hypotheses are rejected.

Testing the significance of R^2: the F test

Another check of the quality of the regression equation is to test whether the R^2 value, calculated earlier, is significantly greater than zero. This is a test using the F distribution and turns out once again to be equivalent to the two previous tests H_0: $\beta = 0$ and H_0: $\rho = 0$, conducted in previous sections, using the *t* distribution.

The null hypothesis for the test is H_0: $R^2 = 0$, implying once again that X does not influence Y (hence equivalent to $\beta = 0$). The test statistic is

$$F = \frac{R^2/1}{(1 - R^2)/(n - 2)} \tag{7.27}$$

or equivalently

$$F = \frac{\text{RSS}/1}{\text{ESS}/(n - 2)} \tag{7.28}$$

The F statistic is therefore the ratio of the regression sum of squares to the error sum of squares, each divided by their degrees of freedom (for the RSS there is one degree of freedom because of the one explanatory variable, for the ESS there are $n - 2$ degrees of freedom). A high value of the F statistic (i.e. RSS is large relative to ESS) rejects H_0 in favour of the alternative hypothesis, H_1: $R^2 > 0$. Evaluating (7.27) gives

$$F = \frac{0.678/1}{(1 - 0.678)/10} = 21.078 \tag{7.29}$$

The critical value of the F distribution at the 5% significance level, with $\nu_1 = 1$ and $\nu_2 = 10$, is $F^*_{1,10} = 4.96$. The test statistic exceeds this, so the regression as a whole is significant. It is better to use the regression model to explain the birth rate than to use the simpler model which assumes all countries have the same birth rate (the sample average).

As stated before, this test is equivalent to those carried out before using the t distribution. The F statistic is, in fact, the square of the t statistic calculated earlier ($(-4.59^2 = 21.078)$) and reflects the fact that, in general,

$$F_{1,n-2} = t_{n-2}^2$$

The Prob-value associated with both statistics is the same (approximately 0.001 in this case), so both tests reject the null at the same level of significance. However, in multiple regression with more than one explanatory variable, the relationship no longer holds and the tests do fulfil different roles, as we shall see in the next chapter.

Exercise 7.6

?

(a) For the regression of the birth rate on the income ratio, calculate the standard errors of the coefficients and hence construct 95% confidence intervals for both.

(b) Test the hypothesis that the slope coefficient is zero against the alternative that it is not zero.

(c) Test the hypothesis H_0: $R^2 = 0$.

Interpreting computer output

Having shown how to use the appropriate formulae to derive estimates of the parameters, their standard errors and to test hypotheses, we now present all these results as they would be generated by a computer software package, in this case Excel. This removes all the effort of calculation and allows us to concentrate on more important issues such as the interpretation of the results. Table 7.6 shows the computer output.

The table presents all the results we have already derived, plus a few more.

- The regression coefficients, standard errors and t ratios are given at the bottom of the table, suitably labelled. The column headed 'P-value' gives some additional information – it shows the significance level of the t statistic. For example, the slope coefficient is significant at the level of 0.1%[5], i.e. there is this probability of getting such a sample estimate by chance. This is much less than our usual 5% criterion, so we conclude that the sample evidence did not arise by chance.
- The program helpfully calculates the 95% confidence interval for the coefficients also, which were derived above in equation (7.24).
- Moving up the table, there is a section headed 'ANOVA'. This is similar to the ANOVA covered in Chapter 6. This table provides the sums of squares values (RSS, ESS and TSS, in that order) and their associated degrees of freedom in the '*df*' column. The '*MS*' ('mean square') column calculates the sums of squares each divided by their degrees of freedom, whose ratio gives the F statistic in the next column. This is the value calculated in equation (7.29). The '*Significance F*' value is similar to the *P-value* discussed previously: it shows the level at which the F statistic is significant (0.1% in this case) and saves us looking up the F tables.

[5]This is the area in *both* tails, so it is for a two-tail test.

Table 7.6 Regression output from Excel

	A	B	C	D	E	F	G	H
25								
26								
27								
28		*Regression Statistics*						
29		Multiple R	0.824					
30		R Square	0.678					
31		Adjusted R Square	0.646					
32		Standard Error	4.132					
33		Observations	12					
34								
35		ANOVA						
36			*df*	*SS*	*MS*	*F*	*Significance F*	
37		Regression	1	359.913	359.913	21.078	0.001	
38		Residual	10	170.754	17.075			
39		Total	11	530.667				
40								
41			*Coefficients*	*Standard Error*	*t Stat*	*P-value*	*Lower 95%*	*Upper 95%*
42		Intercept	40.71	2.30	17.68	7.15E-09	35.58	45.84
43		GR	-2.70	0.59	-4.59	0.001	-4.01	-1.39
44								

- At the top of the table is given the R^2 value and the standard error of the error term, s_e, labelled 'Standard Error', which we have already come across. 'Multiple R' is simply the square root of R^2; 'Adjusted R^2' (sometimes called '*R*-bar squared' and written $\bar{R}^2$) adjusts the R^2 value for the degrees of freedom. This is an alternative measure of fit, which is not affected by the number of explanatory variables, unlike R^2. See Maddala and Lahiri (2009, Chapter 4) for a more detailed explanation.

Prediction

Earlier we showed that the regression line could be used for **prediction**, using the figures for Costa Rica. The point estimate of Costa Rica's birth rate is calculated simply by putting its growth rate into the regression equation and assuming a zero value for the error, i.e.

$$\hat{Y} = a + bX + 0 = 40.711 - 2.7 \times 3 + 0 = 32.6$$

This is a point estimate, which is unbiased, around which we can build a confidence interval. There are, in fact, two confidence intervals we can construct, the first for the position of the *regression line* at $X = 3$, the second for an *individual observation* (on *Y*) at $X = 3$. Using the 95% confidence level, the first interval is given by the formula

$$\hat{Y} \pm t_{n-2} \times s_e \sqrt{\frac{1}{n} + \frac{(X_P - \bar{X})^2}{\sum(X - \bar{X})^2}} \tag{7.30}$$

where X_P is the value of X for which the prediction is made. t_{n-2} denotes the critical value of the *t* distribution at the 5% significance level (for a two-tail test) with $n - 2$ degrees of freedom. This evaluates to

$$32.6 \pm 2.228 \times 4.132\sqrt{\frac{1}{12} + \frac{(3 - 3.35)^2}{49.37}}$$
$$= [29.90, 35.30]$$

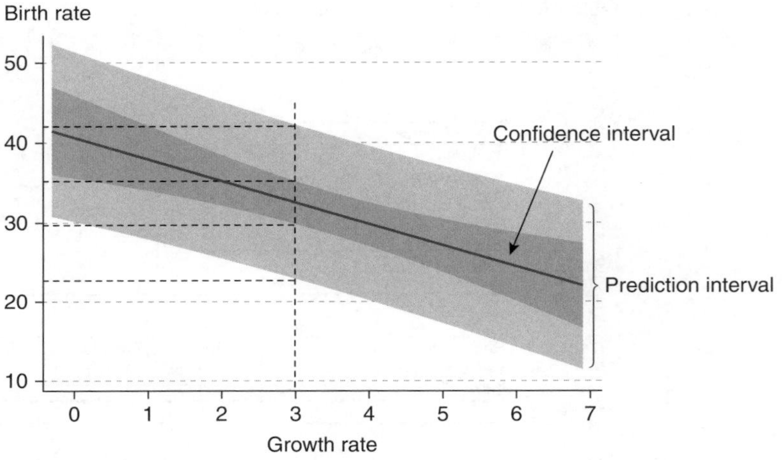

Figure 7.8
Confidence and prediction intervals

This means that we predict with 95% confidence that the *average* birth rate of all countries growing at 3% p.a. is between 29.9 and 35.3.

The second type of interval, for the value of Y itself at $X_P = 3$, is somewhat wider, because there is an additional element of uncertainty: individual countries do not lie on the regression line, but around it. This is referred to as the 95% **prediction interval**. The formula for this interval is

$$\hat{Y} \pm t_{n-2} \times s_e \sqrt{1 + \frac{1}{n} + \frac{(X_P - \overline{X})^2}{\sum(X - \overline{X})^2}} \tag{7.31}$$

Note the extra '1' inside the square root sign. When evaluated, this gives a 95% prediction interval of [23.01, 42.19]. Thus we are 95% confident that an individual country growing at 3% p.a. will have a birth rate within this range.

The two intervals are illustrated in Figure 7.8. The smaller confidence interval is shown in a darker shade, with the wider prediction interval being about twice as big. Note from the formulae that the prediction is more precise (the interval is smaller):

- the closer the sample observations lie to the regression line (smaller s_e),
- the greater the spread of sample X values (larger $\sum(X - \overline{X})^2$),
- the larger the sample size (larger n),
- the closer to the mean of X the prediction is made (smaller $X_P - \overline{X}$).

This last characteristic is evident in the diagram, where the intervals are narrower towards the centre of the diagram.

There is an additional danger of predicting far outside the range of sample X values, if the true regression line is not linear as we have assumed. The linear sample regression line might be close to the true line within the range of sample X values but diverge substantially outside. Figure 7.9 illustrates this point.

In the birth rate sample, we have a fairly wide range of X values; few countries grow more slowly than Senegal or faster than Korea.

Exercise 7.7

Use Excel's regression tool (or other software) to confirm your answers to Exercises 7.4 to 7.6.

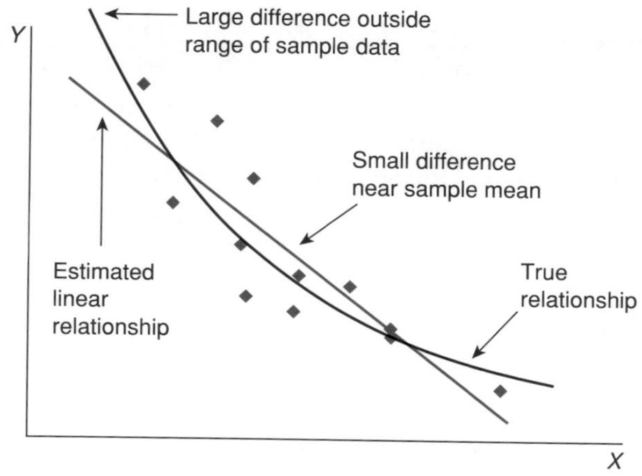

Figure 7.9
The danger of prediction outside the range of sample data

Exercise 7.8

(a) Predict (point estimate) the birth rate for a country with an income ratio of 10.

(b) Find the 95% confidence interval prediction for a typical country with IR = 10.

(c) Find the 95% confidence interval prediction for an individual country with IR = 10.

Route map of calculations

By now we have been through a lot of formulae and calculations, and the reader might be a little bewildered. Hence we summarise here in Figure 7.10, and in the worked example below, the order in which these calculations are carried out, a kind of route map. It is possible to vary parts of the ordering slightly, which is not important.

Figure 7.10
Route map of regression calculations

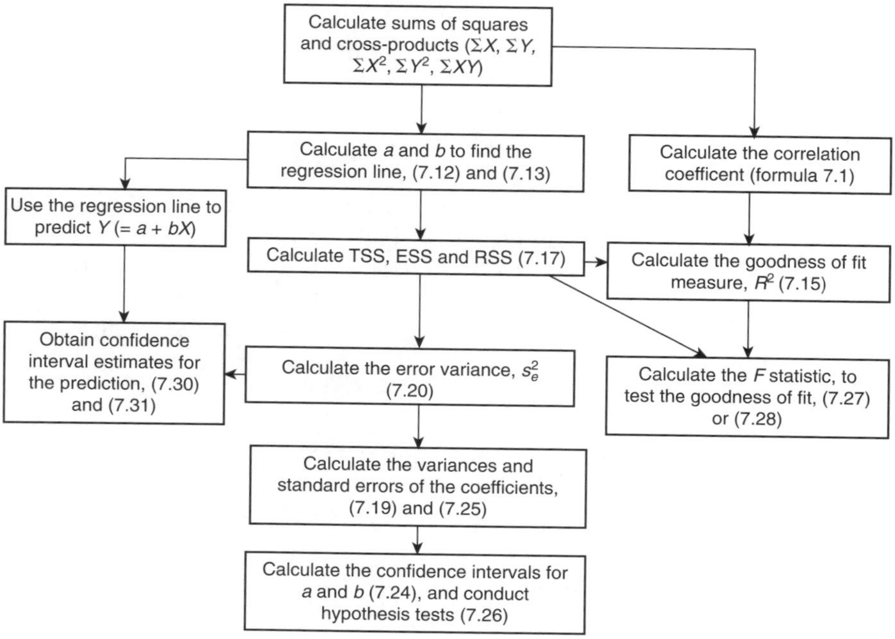

Worked example 7.2

We continue the previous worked example, completing the calculations needed for regression. The previous table contains most of the preliminary calculations. To find the regression line we use:

$$b = \frac{n\sum XY - \sum X \sum Y}{n\sum X^2 - \left(\sum X\right)^2} = \frac{6 \times 682 - 33 \times 119}{6 \times 199 - 33^2} = 1.57$$

and

$$a = 19.83 - 1.57 \times 5.5 = 11.19$$

Hence we obtain the equation:

$$Y_i = 11.19 + 1.57X_i + e_i$$

For inference, we start with the sums of squares:

$$\begin{aligned} \text{TSS} &= \sum (Y_i - \overline{Y})^2 = \sum Y_i^2 - n\overline{Y}^2 \\ &= 2427 - 6 \times 19.83^2 = 66.83 \\ \text{ESS} &= \sum (Y_i - \hat{Y})^2 = \sum Y_i^2 - a\sum Y_i - b\sum X_i Y_i \\ &= 2427 - 11.19 \times 119 - 1.57 \times 682 = 23.62 \\ \text{RSS} &= \text{TSS} - \text{ESS} = 66.83 - 23.62 = 43.21 \end{aligned}$$

We then get $R^2 = \text{RSS}/\text{TSS} = 43.21/66.83 = 0.647$ or 64.7% of the variation in Y explained by variation in X.

To obtain the standard errors of the coefficients, we first calculate the error variance as $s_e^2 = \text{ESS}/(n - 2) = 23.62/4 = 5.905$ and the estimated variance of the slope coefficient is:

$$s_b^2 = \frac{s_e^2}{\sum (X - \overline{X})^2} = \frac{5.905}{17.50} = 0.338$$

and the standard error of b is therefore $\sqrt{0.338} = 0.581$.

Similarly, for a we obtain:

$$s_a^2 = s_e^2 \times \left(\frac{1}{n} + \frac{\overline{X}^2}{\sum (X - \overline{X})^2}\right) = 5.905 \times \left(\frac{1}{6} + \frac{5.5^2}{17.50}\right) = 11.19$$

and the standard error of a is therefore 3.34.

Confidence intervals for a and b can be constructed using the critical value of the t distribution, $2.776(5\%, \nu = 4)$, yielding $1.57 \pm 2.776 \times 0.581 = [-0.04, 3.16]$ for b and $[1.90, 20.47]$ for a. Note that zero is inside the confidence interval for b. This is also reflected in the test of $H_0\colon \beta = 0$ which is

$$t = \frac{1.57 - 0}{0.581} = 2.71$$

which falls just short of the two-tailed critical value, 2.776. Hence H_0 cannot be rejected.

The F statistic to test $H_0\colon R^2 = 0$ is:

$$F = \frac{\text{RSS}/1}{\text{ESS}/(n - 2)} = \frac{43.21/1}{23.62/(6 - 2)} = 7.32$$

which compares to a critical value of $F(1,4)$ of 7.71. So, again, the null cannot be rejected (remember that this and the t test on the slope coefficient are equivalent in simple regression).

We shall predict the value of Y for a value of $X = 10$, yielding $\hat{Y} = 11.19 + 1.57 \times 10 = 26.90$. The 95% confidence interval for this prediction is calculated using (7.30) which gives

$$26.90 - 2.776 \times 2.43\sqrt{\frac{1}{6} + \frac{(10 - 5.5)^2}{17.50}} = [19.14, 34.66].$$

The 95% prediction interval for an actual observation at $X = 10$ is given by (7.31), resulting in

$$26.90 \pm 2.776 \times 2.43\sqrt{1 + \frac{1}{6} + \frac{(10 - 5.5)^2}{17.50}} = [16.62, 37.18].$$

Some other issues in regression

We have now completed our review of the main calculations and interpretation of the regression results. We now look at a few additional issues which often arise, and it is useful to be aware of them.

Units of measurement

The measurement and interpretation of the regression coefficients depend upon the units in which the variables are measured. For example, suppose we had measured the birth rate in births per *hundred* (not *thousand*) of population; what would be the implications? Obviously nothing fundamental is changed; we ought to get the same qualitative result, with the same interpretation. However, the regression coefficients cannot remain the same: if the slope coefficient remained $b = -2.7$, this would mean that an increase in the growth rate of one percentage point reduces the birth rate by 2.7 births *per hundred*, which is clearly wrong. The right answer should be 0.27 births per hundred (equivalent to 2.7 per thousand) so the coefficient should change to $b = -0.27$. Thus, in general, the sizes of the coefficients depend upon the units in which the variables are measured. This is why one cannot judge the importance of a regression equation from the size of the coefficients alone.

It is easiest to understand this in graphical terms. A graph of the data will look exactly the same, except that the scale on the Y-axis will change; it will be divided by 10. The intercept of the regression line will therefore change to $a = 4.0711$ and the slope to $b = -0.27$ (i.e. both are divided by 10). Thus the regression equation becomes

$$Y_i = 4.0711 - 0.27X_i + e_i'$$
$$(e_i' = e_i/10)$$

Since nothing fundamental has altered, any hypothesis test must yield the same result and, in particular, the same test statistic. Thus, t and F statistics are unaltered by changes in the units of measurement; nor is R^2 altered. However, standard errors will be divided by 10 (they have to be to preserve the t statistics; see equation (7.26) for example). Table 7.7 sets out the effects of changes in the

Table 7.7 The effects of data transformations

Factor (k) multiplying...		Effect upon			
Y	X	a	s_a	b	s_b
k	1	all multiplied by k			
1	k	unchanged		divided by k	
k	k	multiplied by k		Unchanged	

units of measurement (of either the X or Y variable) upon the coefficients and standard errors. In the table it is assumed that the variables have been multiplied by a constant k; in the above case $k = 1/10$ was used.

It is important to be aware of the units in which the variables are measured. If not, it is impossible to know how large is the effect of X upon Y. It may be statistically significant, but we have no idea of how important it is. This may occur if, for instance, one of the variables is presented as an index number (see Chapter 10) rather than in the original units.

How to avoid measurement problems: calculating the elasticity

A neat way to avoid the problems of measurement is to calculate the **elasticity**, i.e. the *proportionate* change in Y divided by the *proportionate* change in X. The proportionate changes are the same whatever units the variables are measured in. The proportionate change in X is given by $\Delta X/X$, where ΔX indicates the *change* in X. Thus if X changes from 100 to 110, the proportionate change is $\Delta X/X = 10/100 = 0.1$ or 10%. The elasticity, η, is therefore given by

$$\eta = \frac{\Delta Y/Y}{\Delta X/X} \text{ or equivalently } \eta = \frac{\Delta Y}{\Delta X} \times \frac{X}{Y} \tag{7.32}$$

The second form of the equation is more useful, since $\Delta Y/\Delta X$ is simply the slope coefficient b. We simply need to multiply this by the ratio X/Y, therefore. But what values should be used for X and Y? The convention is to use the means, so we obtain the following formula for the elasticity, from a linear regression equation:

$$\eta = b \times \frac{\overline{X}}{\overline{Y}} \tag{7.33}$$

This evaluates to $-2.7 \times 3.35/31.67 = -0.29$. This is interpreted as follows: a 1% increase in the growth rate would lead to a 0.29% decrease in the birth rate. Equivalently, and perhaps a little more usefully, a 10% rise in growth (from say 3% to 3.3% p.a.) would lead to a 2.9% decline in the birth rate (e.g. from 30 to 29.13). This result is the same whatever units the variables X and Y are measured in.

Note that this elasticity is measured at the means; it would have a different value at different points along the regression line. Later on we show an alternative method for estimating the elasticity, in this case the elasticity of demand which is familiar in economics.

Non-linear transformations

So far only *linear* regression has been dealt with, that is fitting a straight line to the data. This can sometimes be restrictive, especially when there is good reason to believe that the true relationship is non-linear (e.g. the labour supply curve). Poor results would be obtained by fitting a straight line through the data in Figure 7.11, yet the shape of the relationship seems clear at a glance.

Figure 7.11
Graph of *Y* against *X*

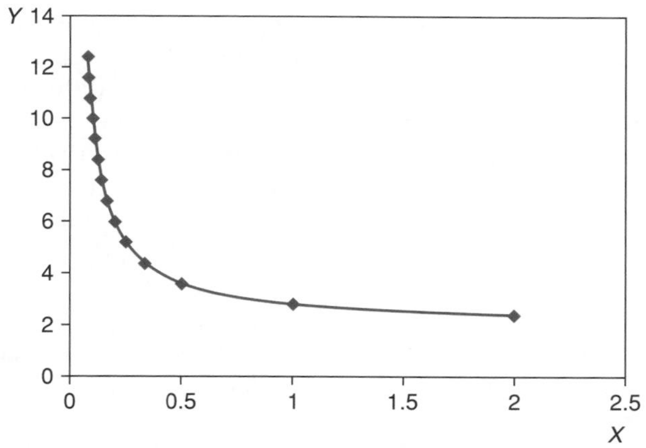

Fortunately, this problem can be solved by transforming the data, so that when graphed a linear relationship between the two variables appears. Then a straight line can be fitted to these transformed data. This is equivalent to fitting a curved line to the original data. All that is needed is to find a suitable transformation to 'straighten out' the data. Given the data represented in Figure 7.11, if Y were graphed against $1/X$, the relationship shown in Figure 7.12 would appear.

Thus, if the regression line

$$Y_i = a + b\frac{1}{X_i} = e_i \tag{7.34}$$

were fitted, this would provide a good representation of the data in Figure 7.11. The procedure is straightforward. First, calculate the reciprocal of each of the X values and then use these new values (together with the original data for Y), using exactly the same methods as before. This transformation appears inappropriate for the birth rate data (see Figure 7.1) but serves as an illustration. The transformed X values are 0.196 (=1/5.1) for Brazil, 0.3125 (=1/3.2) for Colombia, etc. The resulting regression equation is

$$Y_i = 31.92 - 3.96\frac{1}{X_i} + e_i \tag{7.35}$$

s.e. (1.64) (1.56)

$R^2 = 0.39$, $F = 6.44$, $n = 12$

Figure 7.12
Transformation of Figure 7.11: *Y* against 1/*X*

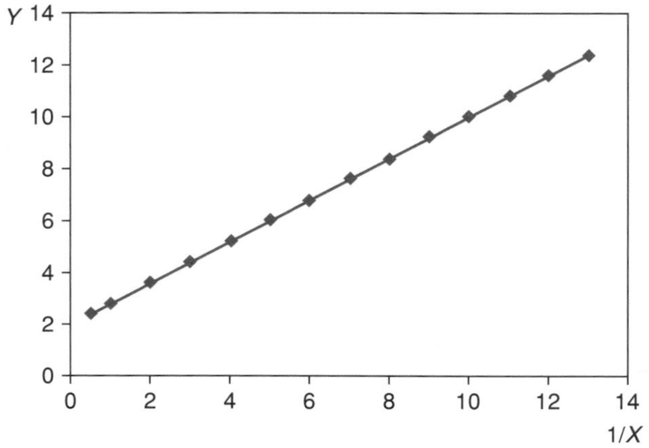

This appears worse than the original specification (the R^2 is low and the slope coefficient is not significantly different from zero), so the transformation does not appear to be a good one. Note also that it is difficult to calculate the effect of X upon Y in this equation. We can see that a unit increase in $1/X$ reduces the birth rate by 3.96, but we do not have an intuitive feel for the inverse of the growth rate. This latest result also implies that a *fall* in the growth rate (hence, $1/X$ rises) lowers the birth rate – the converse of our previous result. In the next chapter, we deal with a different example where a non-linear transformation does improve matters.

Table 7.8 presents a number of possible shapes for data, with suggested data transformations which will allow the relationship to be estimated using linear regression. In each case, once the data have been transformed, the methods and formulae used above can be applied.

It is sometimes difficult to know which transformation (if any) to apply. A graph of the data is unlikely to be as tidy as the diagrams in Table 7.8. Economic theory rarely suggests the form which a relationship should follow, and there are no simple statistical tests for choosing alternative formulations. The choice can sometimes be made after visual inspection of the data, or on the basis of convenience. The double log transformation is often used in economics as it has some very convenient properties. Unfortunately, it cannot be used with the growth rate data here because Senegal's growth rate was negative. It is impossible to take the logarithm of a negative number. We therefore postpone the use of the log transformation in regression until the next chapter.

Table 7.8 Data transformations

Name	Graph of relationship	Original relationship	Transformed relationship	Regression
Double log	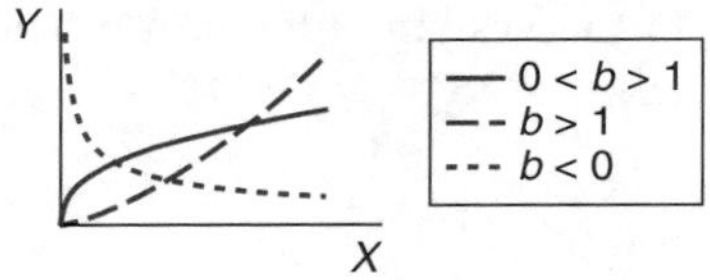	$Y = aX^b e$	$\ln Y - \ln a + b\ln X + \ln e$	$\ln Y$ on $\ln X$
Reciprocal	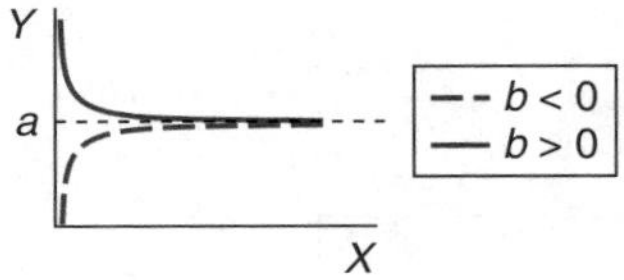	$Y = a + b/X + e$	$Y = a + b\frac{1}{X} + e$	Y on $\frac{1}{X}$
Semi-log	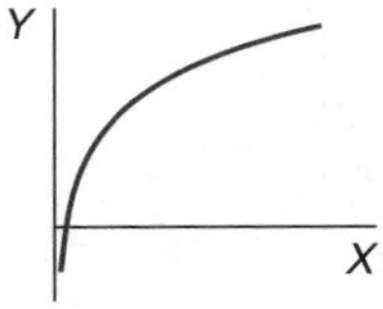	$e^Y = aX^b e$	$Y = \ln a + b\ln X + \ln e$	Y on $\ln X$
Exponential	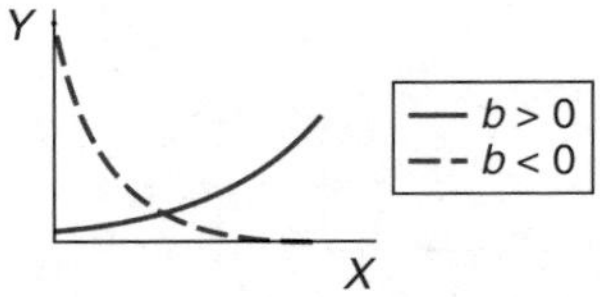	$Y = e^{a+bX+e}$	$\ln Y = a + bX + e$	$\ln Y$ on X

Gapminder again

It is instructive to return to Gapminder again (encountered in Chapter 1) to see what it can reveal with similar data but using more countries, in this case 92. This is a mixture of developing (mainly) and developed countries. First, we note that (as is often the case) it is difficult to replicate our earlier relationship between the birth rate and growth because (i) Gapminder has birth rate data for 1983, not 1981, and (ii) Gapminder only has 'Growth over the next 10 years' readily prepared for graphing, i.e. we would have to use the birth rate in 1983 and the growth rate over 1983 to 1993. Fortunately, one can download the underlying data and do the analysis[6]. This reveals a regression equation:

$$\begin{aligned} &BR = 41.8 - 4.08\,\text{growth rate} + \text{error} \\ &\text{s.e.}\ (2.03) \qquad (0.66) \\ &R^2 = 0.30, F_{1,90} = 38.4, n = 92. \end{aligned}$$

The slope coefficient uncovered earlier, using just 12 countries, was −2.70 (s.e. 0.59). Thus, the larger sample has found a much steeper relationship between the variables. The Gapminder data are slightly different (even for the same countries) from Todaro's, so we can also use the Gapminder data for the original 12 countries, revealing a slope coefficient of −3.24 (s.e. 0.38) which is not so different from Todaro.

It is easier to examine the birth rate/income relationship, since Gapminder can readily do this. It gives the following graph.

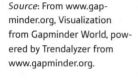

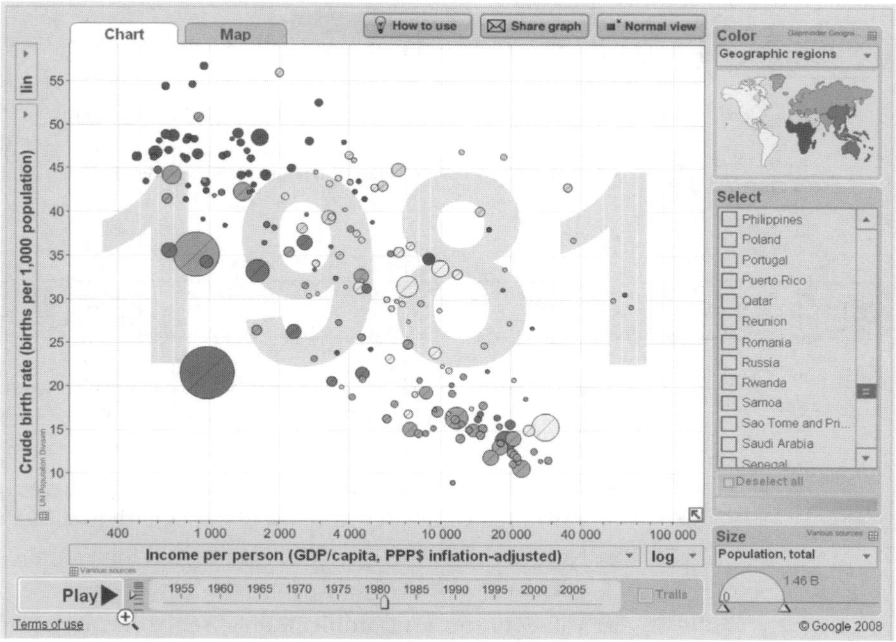

Source: From www.gapminder.org, Visualization from Gapminder World, powered by Trendalyzer from www.gapminder.org.

[6]Gapminder only produces graphs, it does not perform regression analysis. For this, you need to download the data and do it yourself.

Note that Gapminder, by default, uses the log of income, rather than income itself. This makes it a little difficult to compare with Figure 7.1(b), showing Todaro's data, which used income. Is it better to use income or log income? From looking at the Gapminder data there appears to be a curvilinear relationship between the birth rate and income, and a linear regression would not fit well. Also, income is highly skewed with some very high values (e.g. Kuwait) which will tend to have a large influence on the regression line's position. Taking the log of income addresses this issue. The birth rate data are not so skewed so there is less need to transform them.

Comparing regression results, we have:

	Income coefficient (s.e.)	R^2	N
Todaro	−0.003 (0.003)	0.07	12
Gapminder	−0.0006 (0.0001)	0.26	155
Gapminder (log income)	−7.81 (0.56)	0.56	155

The results using log income appear best, with the highest R^2 value (note the dependent variable is the same in all cases) and with a significant slope coefficient. The only tricky part is interpreting the coefficient – what does it mean? This is a useful exercise to work through. The initial interpretation is that a unit increase in log income leads to a 7.81 fall in the birth rate. But what is a unit increase in the (natural) log? The answer is that it corresponds to a 172% increase in income (because $\ln(2.72) = 1$). This is not the easiest of results to digest. Easier would be to ask, what is the effect of doubling a country's income? In this case, $\ln(2) = 0.69$, and this would be the increase in the log. This would imply a $0.69 \times 7.81 = 5.4$ fall in the birth rate. Thus, a doubling of a country's income reduces the birth rate by about five births per 1000 population (approximately a 15% reduction at the mean).

Exercise 7.9

?

(a) Calculate the elasticity of the birth rate with respect to the income ratio, using the results of previous exercises.

(b) Give a brief interpretation of the meaning of this figure.

Exercise 7.10

?

Calculate a regression relating the birth rate to the inverse of the income ratio $1/IR$.

Summary

- Correlation refers to the extent of association between two variables. The (sample) correlation coefficient is a measure of this association, extending from $r = -1$ to $r = +1$.
- Positive correlation ($r > 0$) exists when high values of X tend to be associated with high values of Y and low X values with low Y values.
- Negative correlation ($r < 0$) exists when high values of X tend to be associated with low values of Y and vice versa.
- Values of r around 0 indicate an absence of correlation.
- As the sample correlation coefficient is a random variable, we can test for its significance, i.e. test whether the true value is zero or not. This test is based upon the t distribution.

- The existence of correlation (even if 'significant') does not necessarily imply causality. There can be other reasons for the observed association.
- Regression analysis extends correlation by asserting a causality from X to Y and then measuring the relationship between the variables via the regression line, the 'line of best fit'.
- The regression line $Y = a + bX$ is defined by the intercept a and slope coefficient b. Their values are found by minimising the sum of squared errors around the regression line.
- The slope coefficient b measures the responsiveness of Y to changes in X.
- A measure of how well the regression line fits the data is given by the coefficient of determination, R^2, varying between 0 (very poor fit) and 1 (perfect fit).
- The coefficients a and b are unbiased point estimates of the true values of the parameters. Confidence interval estimates can be obtained, based on the t distribution. Hypothesis tests on the parameters can also be carried out using the t distribution.
- A test of the hypothesis $R^2 = 0$ (implying the regression is no better at predicting Y than simply using the mean of Y) can be carried out using the F distribution.
- The regression line may be used to predict Y for any value of X by assuming the residual to be zero for that observation.
- The measured response of Y to X (given by b) depends upon the units of measurement of X and Y. A better measure is often the elasticity, which is the proportionate response of Y to a proportionate change in X.
- Data are often transformed prior to regression (e.g. by taking logs) for a variety of reasons (e.g. to fit a curve to the original data).

Key terms and concepts

autocorrelation
coefficient of determination (R^2)
correlation coefficient
dependent (endogenous) variable
elasticity
error sum of squares
error term (or residual)
estimated variance of the error term
explained (endogenous) variable
explanatory variable
independent (exogenous) variable
intercept
negative correlation
non-parametric statistics
Pearson's product-moment correlation coefficient
positive correlation
prediction
prediction interval
rank correlation coefficient
regression line or equation
regression sum of squares
slope
Spearman's coefficient of rank correlation
standard error
t distribution
total sum of squares
zero correlation

References

Maddala, G.S., and K. Lahiri, *Introduction to Econometrics*, 4th edn, Wiley, 2009.
Todaro, M.P., *Economic Development in the Third World*, 3rd edn, Longman, 1985.

Formulae used in this chapter

Formula	Description	Notes
$r = \dfrac{n\sum XY - \sum X \sum Y}{\sqrt{(n\sum X^2 - (\sum X)^2)(n\sum Y^2 - (\sum Y)^2)}}$	Correlation coefficient	$-1 \leq r \leq 1$
$t = \dfrac{r\sqrt{n-2}}{\sqrt{1-r^2}}$	Test statistic for H_0: $\rho = 0$	$\nu = n - 2$
$r_s = 1 - \dfrac{6\sum d^2}{n(n^2-1)}$	Spearman's rank correlation coefficient	$-1 \leq r_s \leq 1$. d is the difference in ranks between the two variables. Only works if there are no tied ranks. Otherwise use standard correlation formula.
$b = \dfrac{n\sum XY - \sum X \sum Y}{n\sum X^2 - (\sum X)^2}$	Slope of the regression line (simple regression)	
$a = \bar{Y} - b\bar{X}$	Intercept (simple regression)	
$\text{TSS} = \sum Y^2 - n\bar{Y}^2$	Total sum of squares	
$\text{ESS} = \sum Y^2 - a\sum Y - b\sum XY$	Error sum of squares	
$\text{RSS} = \text{TSS} - \text{ESS}$	Regression sum of squares	
$R^2 = \dfrac{\text{RSS}}{\text{TSS}}$	Coefficient of determination	
$s_e^2 = \dfrac{\text{ESS}}{n-2}$	Variance of the error term in regression	Replace $n - 2$ by $n - k - 1$ in multiple regression
$s_b^2 = \dfrac{s_e^2}{\sum(X-\bar{X})^2}$	Variance of the slope coefficient in simple regression	
$s_a^2 = s_e^2 = \sqrt{\dfrac{1}{n} + \dfrac{\bar{X}^2}{\sum(X-\bar{X})^2}}$	Variance of the intercept in simple regression	
$b \pm t_\nu \times s_b$	Confidence interval estimate for b in simple regression	t_ν is the critical value of the t distribution with $\nu = n - 2$ degrees of freedom
$t = \dfrac{b - \beta}{s_b}$	Test statistic for H_0: $\beta = 0$	$\nu = n - 2$ in simple regression, $n - k - 1$ in multiple regression
$F = \dfrac{\text{RSS}/1}{\text{ESS}/(n-2)}$	Test statistic for H_0: $R^2 = 0$	$\nu = k, n - k - 1$ in multiple regression
$\hat{Y} \pm t_\nu \times s_e\sqrt{\dfrac{1}{n} + \dfrac{(X_P - \bar{X})^2}{\sum(X-\bar{X})^2}}$	Confidence interval for a prediction (simple regression) at $X = X_P$	$\nu = n - 2$
$\hat{Y} \pm t_\nu \times s_e\sqrt{1 + \dfrac{1}{n} + \dfrac{(X_P - \bar{X})^2}{\sum(X-\bar{X})^2}}$	Confidence interval for an observation on Y at $X = X_P$	$\nu = n - 2$

Problems

Some of the more challenging problems are indicated by highlighting the problem number in colour.

7.1 The other data which Todaro might have used to analyse the birth rate were:

Country	Birth rate	GNP	Growth	Income ratio
Bangladesh	47	140	0.3	2.3
Tanzania	47	280	1.9	3.2
Sierra Leone	46	320	0.4	3.3
Sudan	47	380	−1.3	3.9
Kenya	55	420	2.9	6.8
Indonesia	35	530	4.1	3.4
Panama	30	1910	3.1	8.6
Chile	25	2560	0.7	3.8
Venezuela	35	4220	2.4	5.2
Turkey	33	1540	3.5	4.9
Malaysia	31	1840	4.3	5.0
Nepal	44	150	0.0	4.7
Malawi	56	200	2.7	2.4
Argentina	20	2560	1.9	3.6

For *one* of the three possible explanatory variables (in class, different groups could examine each of the variables):

(a) Draw an *XY* chart of the data above and comment upon the result.

(b) Would you expect a line of best fit to have a positive or negative slope? Roughly, what would you expect the slope to be?

(c) What would you expect the correlation coefficient to be?

(d) Calculate the correlation coefficient, and comment.

(e) Test to see if the correlation coefficient is different from zero. Use the 95% confidence level.

(Analysis of this problem continues in Problem 7.5.)

7.2 The data below show alcohol expenditure and income (both in £s per week) for a sample of 17 families.

Family	Alcohol expenditure	Income	Family	Alcohol expenditure	Income
1	26.17	487	10	13.32	370
2	19.49	574	11	9.24	299
3	17.87	439	12	47.35	531
4	16.90	367	13	26.80	506
5	4.21	299	14	33.44	613
6	32.08	743	15	21.41	472
7	30.19	433	16	16.06	253
8	22.62	547	17	24.98	374

(a) Draw an *XY* plot of the data and comment.

(b) From the chart, would you expect the line of best fit to slope up or down? *In theory*, which way should it slope?

(c) What would you expect the correlation coefficient to be, approximately?

(d) Calculate the correlation coefficient between alcohol spending and income.

(e) Is the coefficient significantly different from zero? What is the implication of the result?

(The following totals will reduce the burden of calculation: $\sum Y = 137.990$; $\sum X = 7610$; $\sum Y^2 = 9\,918.455$; $\sum X^2 = 3\,680\,748$; $\sum XY = 181\,911.250$; *Y* is consumption, *X* is income. If you wish, you could calculate a logarithmic correlation. The relevant totals are: $\sum y = 50.192$; $\sum x = 103.079$; $\sum y^2 = 153.567$; $\sum x^2 = 626.414$; $\sum xy = 306.339$, where $y = \ln Y$ and $x = \ln X$.)

(Analysis of this problem continues in Problem 7.6.)

7.3 What would you expect to be the correlation coefficient between the following variables? Should the variables be measured contemporaneously or might there be a lag in the effect of one upon the other?

(a) Nominal consumption and nominal income.

(b) GDP and the imports/GDP ratio.

(c) Investment and the interest rate.

7.4 As Problem 7.3, for

(a) real consumption and real income;

(b) individuals' alcohol and cigarette consumption;

(c) UK and US interest rates.

7.5 Using the data from Problem 7.1, calculate the rank correlation coefficient between the variables and test its significance. How does it compare with the ordinary correlation coefficient?

7.6 (a) Calculate the rank correlation coefficient between income and quantity for the data in Problem 7.2. How does it compare to the ordinary correlation coefficient?

(b) Is there significant evidence that the ranks are correlated?

7.7 (a) For the data in Problem 7.1, find the estimated regression line and calculate the R^2 statistic. Comment upon the result. How does it compare with Todaro's findings?

(b) Calculate the standard error of the estimate and the standard errors of the coefficients. Is the slope coefficient significantly different from zero? Comment upon the result.

(c) Test the overall significance of the regression equation and comment.

(d) Taking your own results and Todaro's, how confident do you feel that you understand the determinants of the birth rate?

(e) What do you think will be the result of estimating your equation using all 26 countries' data? Try it. What do you conclude?

7.8 (a) For the data given in Problem 7.2, estimate the sample regression line and calculate the R^2 statistic. Comment upon the results.

(b) Calculate the standard error of the estimate and the standard errors of the coefficients. Is the slope coefficient significantly different from zero?

(c) Test the overall significance of the regression and comment upon your result.

7.9 From your results for the birth rate model, predict the birth rate for a country with *either* (a) GNP equal to \$3000, (b) a growth rate of 3% p.a., *or* (c) an income ratio of 7. How does your prediction compare with one using Todaro's results? Comment.

7.10 Predict alcohol consumption given an income of £700. Use the 99% confidence level for the interval estimate.

7.11 **(Project)** Update Todaro's study using more recent data.

7.12 Try to build a model of the determinants of infant mortality. You should use cross-section data for 20 countries or more and should include both developing and developed countries in the sample.

Write up your findings in a report which includes the following sections: discussion of the problem; data gathering and transformations; estimation of the model; interpretation of results. Useful data may be found in the *Human Development Report* (use Google to find it online) or on the World Bank website.

Answers to exercises

Exercise 7.1

(a) The calculation is:

	Birth rate Y	Income ratio X	Y^2	X^2	XY
Brazil	30	9.5	900	90.25	285
Colombia	29	6.8	841	46.24	197.2
Costa Rica	30	4.6	900	21.16	138
India	35	3.1	1 225	9.61	108.5
Mexico	36	5	1 296	25	180
Peru	36	8.7	1 296	75.69	313.2
Philippines	34	3.8	1 156	14.44	129.2
Senegal	48	6.4	2 304	40.96	307.2
South Korea	24	2.7	576	7.29	64.8
Sri Lanka	27	2.3	729	5.29	62.1
Taiwan	21	3.8	441	14.44	79.8
Thailand	30	3.3	900	10.89	99
Totals	380	60	12 564	361.26	1 964

$$r = \frac{12 \times 1964 - 60 \times 380}{\sqrt{(12 \times 136.26 - 60^2)(12 \times 12\,564 - 380^2)}} = 0.355$$

(b) As for (a) except $\Sigma X = 0.6$, $\Sigma Y = 38$, $\Sigma X^2 = 0.036\,126$, $\Sigma Y^2 = 125.64$, $\Sigma XY = 1.964$. Hence

$$r = \frac{12 \times 1.964 - 0.6 \times 38}{\sqrt{(12 \times 0.036\,126 - 0.6^2)(12 \times 125.64 - 38^2)}} = 0.355$$

Exercise 7.2

(a) $t = \dfrac{0.355\sqrt{12 - 2}}{\sqrt{1 - (0.355)^2}} = 1.20 < 2.228 = t^*$ (5% significance level, two-tailed test).

(b) The Prob-value, for a two-tailed test, is 0.257 or 25%. This is greater than 5% so we do not reject the null of no correlation.

Exercise 7.3

(a) The calculation is:

	Birth rate Y	Income ratio X	Rank of Y	Rank of X	Y^2	X^2	XY
Brazil	30	9.5	7	1	49	1	7
Colombia	29	6.8	9	3	81	9	27
Costa Rica	30	4.6	7	6	49	36	42
India	35	3.1	4	10	16	100	40
Mexico	36	5	2.5	5	6.25	25	12.5
Peru	36	8.7	2.5	2	6.25	4	5
Philippines	34	3.8	5	7.5	25	56.25	37.5
Senegal	48	6.4	1	4	1	16	4
South Korea	24	2.7	11	11	121	121	121
Sri Lanka	27	2.3	10	12	100	144	120
Taiwan	21	3.8	12	7.5	144	56.25	90
Thailand	30	3.3	7	9	49	81	63
Totals			78	78	647.5	649.5	569

$$r_s = \frac{12 \times 569 - 78^2}{\sqrt{(12 \times 649.5 - 78^2)(12 \times 647.5 - 78^2)}} = 0.438$$

(b) This is less than the critical value of 0.591 so the null of no rank correlation cannot be rejected.

(c) Reversing the rankings does not alter the result of the calculation.

(d) Reversing just the income ratio ranking changes the sign of the rank correlation coefficient but preserves the absolute value.

Exercise 7.4

(a) Using the data and calculations in the answer to Exercise 7.1 we obtain:

$$b = \frac{12 \times 1964 - 60 \times 380}{12 \times 361.26 - 60^2} = 1.045$$

$$a = \frac{380}{12} - (1.045) \times \frac{60}{12} = 26.443$$

(b) A unit increase in the measure of inequality (e.g. from four to five) leads to approximately one additional birth per 1000 mothers. This is not a very helpful interpretation as it is hard to envisage such a change in the ratio and how it would impact on families. The constant has no useful interpretation. The income ratio cannot be zero (in fact, it cannot be less than 0.5).

Exercise 7.5

(a) $\text{TSS} = \Sigma(Y_i - \bar{Y})^2 = \Sigma Y_i^2 - n\bar{Y}^2$

$= 12\,564 - 12 \times 31.67^2 = 530.667$

$\text{ESS} = \Sigma(Y_i - \hat{Y})^2 = \Sigma Y_i^2 - a\Sigma Y_i - b\Sigma X_i Y_i$

$= 12\,564 - 26.443 \times 380 - 1.045 \times 1139.70 = 463.804$

$\text{RSS} = \text{TSS} - \text{ESS} = 530.667 - 463.804 = 66.863$

$R^2 = 0.126.$

Hence only 12.6% of the variation in the birth rate is explained by variation in the income ratio.

(b) This is the square of the correlation coefficient, calculated earlier as 0.355.

Exercise 7.6

(a) $$s_e^2 = \frac{463.804}{10} = 46.3804$$

and so

$$s_b^2 = \frac{46.3804}{61.26} = 0.757$$

and

$$s_b = \sqrt{0.757} = 0.870$$

For a the estimated variance is

$$s_a^2 = s_e^2 \times \left(\frac{1}{n} + \frac{\bar{X}^2}{\Sigma(X_i - \bar{X})^2}\right) = 46.3804 \times \left(\frac{1}{12} + \frac{5^2}{61.26}\right) = 22.793$$

and hence $s_a = 4.774$. The 95% CIs are therefore $1.045 \pm 2.228 \times 0.87 = [-0.864, 2.983]$ for b and $26.443 \pm 2.228 \times 4.774 = [15.806, 37.081]$ for a

(b) $t = \dfrac{1.045 - 0}{0.870} = 1.201$

Not significant. The critical value of the t statistic is again 2.228.

(c) $F = \dfrac{\text{RSS}/1}{\text{ESS}/(n-2)} = \dfrac{66.863/1}{463.804/(12-2)} = 1.44 < 4.96 = F^*$

(5% significance level, with 1 and 10 degrees of freedom).

Exercise 7.7

Excel should give the same answers.

Exercise 7.8

(a) $\hat{BR} = 26.44 + 1.045 \times 10 = 36.9$.

(b) $\left[36.9 - 2.228 \times 6.81\sqrt{\dfrac{1}{12} + \dfrac{(10-5)^2}{61.26}}, 36.9 + 2.228 \times 6.81\sqrt{\dfrac{1}{12} + \dfrac{(10-5)^2}{61.26}}\right]$

$= [26.3, 47.5]$

(c) $\left[36.9 - 2.228 \times 6.81\sqrt{1 + \dfrac{1}{12} + \dfrac{(10-5)^2}{61.26}},\right.$

$\left.36.9 + 2.228 \times 6.81\sqrt{1\dfrac{1}{12} + \dfrac{(10-5)^2}{61.26}}\right]$

$= [18.4, 55.4]$

Exercise 7.9

(a) $e = 1.045 \times \dfrac{5}{31.67} = 0.165,$

where 5 and 31.67 are the means of X and Y, respectively.

(b) A 10% rise in the inequality measure (e.g. from 4 to 4.4) raises the birth rate by 1.65% (e.g. from 30 to 30.49).

Exercise 7.10

$$BR = 38.82 - 29.61 \times \frac{1}{IR} + e$$

s.e. (19.0)

$R^2 = 0.19, F(1.10) = 2.43.$

Note that taking the inverse is a non-linear transformation, so there is no simple relationship between the coefficient in the original regression (1.045) and the coefficient of this regression. For example, we do not simply get the inverse of the coefficient (which would be $1/1.045 = 0.957$) but a very different figure. The sign of the coefficient does change however, which we would expect. In general, this regression is rather poor and the F statistic is not significant.

8 Multiple regression

Contents

Learning outcomes

By the end of this chapter you should be able to:

- **understand the extension of simple regression to multiple regression, with more than one explanatory variable**
- **use computer software to calculate a multiple regression equation and interpret its output**
- **recognise the role of (economic) theory in deriving an appropriate regression equation**
- **interpret the effect of each explanatory variable on the dependent variable**
- **understand the statistical significance of the results**
- **judge the adequacy of the model and know how to improve it.**

Introduction

In the previous chapter we analysed the 'simple' regression model with just one explanatory variable. However, simple regression is rather limited, as it assumes that there is only the one explanatory factor affecting the dependent variable, which is unlikely to be true in most situations. Price *and* income affect demand, for example. Multiple regression, the subject of this chapter, overcomes this problem by allowing several explanatory variables (though still only one dependent variable) in a model. The techniques are an extension of those used in simple, or bivariate, regression. Multivariate regression allows more general and more helpful models to be estimated, although this does involve new problems as well as advantages.

The regression relationship now becomes

$$Y = b_0 + b_1X_1 + b_2X_2 + \cdots + b_kX_k + e \tag{8.1}$$

where there are now *k* explanatory variables. The principles used in multiple regression are basically the same as in the two-variable case: the coefficients $b_0, \ldots, b_k$ are found by minimising the sum of squared errors; a standard error can be calculated for each coefficient; R^2, *t* ratios, etc., can be calculated and hypothesis tests performed. However, there are a number of additional issues which arise, and these are dealt with in this chapter.

The formulae for calculating coefficients, standard errors, etc., become very complicated in multiple regression and are time-consuming (and error-prone) when done by hand. For this reason, these calculations are invariably done by computer nowadays. Therefore, the formulae are not given in this text: instead we present the results of computer calculations (which you can replicate) and concentrate on understanding and interpreting the results. This is as it should be; the calculations themselves are the means to an end, not the end in itself.

STATISTICS IN PRACTICE

Using spreadsheet packages

Standard spreadsheet packages such as Excel can perform multiple regression analysis and are sufficient for most straightforward tasks. A regression equation can be calculated via menus and dialogue boxes and no knowledge of the formulae is required. However, when problems such as autocorrelation (see below) are present, specialised packages such as *Stata* or *EViews* are much easier to use and provide more comprehensive results.

We also introduce a new example in this chapter, estimating a demand equation for imports into the United Kingdom over the period 1973 to 2005. There are a number of reasons for this switch, for we could have continued with the birth rate example (you are asked to do this in the exercises). First, it allows us to work through a small 'research project' from beginning to end, including the gathering of data, data transformations, interpretation of results, etc. Second, the example uses time-series data, and this allows us to bring out some of the particular issues that arise in such cases, in contrast to the cross-section data used in the previous chapter. Time-series data do not generally constitute a random sample of observations such as we have dealt with in the rest of this text. This is because the observations are constrained to follow one another in time rather than being randomly chosen. The proper analysis of time-series data goes far beyond the scope of this text; however, students often want or need to analyse such data using elementary techniques. This chapter therefore also emphasises the checking of the adequacy of the regression equation for such data. For a fuller treatment of the issues, the reader should consult a more advanced text such as Maddala and Lahiri (2009).

Principles of multiple regression

We illustrate some of the principles involved in multiple regression using two explanatory variables, X_1 and X_2. Since we are using time-series data, we replace the subscript i with a subscript t to denote the individual observations. This is not essential but reminds us that t represents time.

The sample regression equation now becomes

$$Y_t = b_0 + b_1X_{1t} + b_2X_{2t} + e_t \qquad t = 1, \ldots, T \tag{8.2}$$

with three coefficients, b_0, b_1 and b_2, to be estimated. Note that b_0 now signifies the constant. Rather than fitting a line through the data, the task is now to fit a *plane* to the data, in three dimensions, as shown in Figure 8.1.

The plane is drawn sloping down in the direction of X_1 and up in the direction of X_2. The observations are now points dotted about in three-dimensional space (with coordinates X_{1t}, X_{2t} and Y_t) and the task of regression analysis is to find the equation of the plane so as to minimise the sum of squares of vertical distances from each point to the plane. The principle is the same as in simple regression and the regression plane is the one that best summarises the data.

The coefficient b_0 gives the intercept on the Y-axis, b_1 is the slope of the plane in the direction of the X_1-axis and b_2 is the slope in the direction of the X_2-axis. Thus, b_1 gives the effect upon Y of a unit change in X_1 *assuming X_2 remains constant*[1]. Similarly, b_2 gives the response of Y to a unit change in X_2, *assuming no*

Figure 8.1
The regression plane in three dimensions

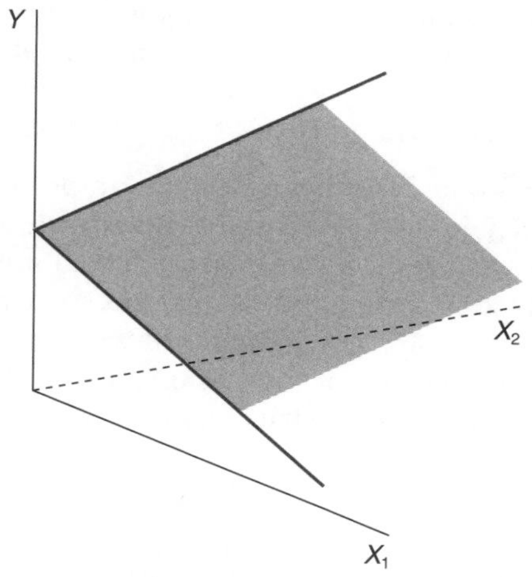

[1]Of course X_2 does not generally remain constant, so it might be better to say that b_1 shows the effect of a unit change in X_1, adjusting for the effect on Y of any contemporaneous change in X_2.

change in X_1. If X_1 and X_2 both change by 1, then the effect on Y is $b_1 + b_2$. Note that b_1 and b_2 are estimates of the (unknown) true parameters β_1 and β_2 and so standard errors and confidence intervals can be calculated, implying that we are not absolutely certain about the true position of the plane. In general, the smaller these standard errors, the better, since it implies less uncertainty about the true relationship between Y and the X variables.

When there are more than two explanatory variables, more than three dimensions are needed to draw a picture of the data. The reader will understand that this is a difficult (if not impossible) task; however, it is possible to estimate such a model and interpret the results in a similar manner to that set out below, for the two explanatory variable case.

What determines imports into the United Kingdom?

To illustrate multiple regression, we suppose that we have the job of finding out what determines the volume of imports into the United Kingdom and whether there are any policy implications of the result. We are given this very open-ended task, which we have to carry through from start to finish. We end up by estimating a demand equation for imports, so the analysis serves as a model for any demand estimation, for example, a firm trying to find out about the demand for its product.

How should we set about this task? The project can be broken down into the following steps:

(1) Theoretical considerations: what can economic theory tell us about the problem and how will this affect our estimation procedures?
(2) Data gathering: what data do we need? Are there any definitional problems, for example?
(3) Data transformation: are the data suitable for the task? We might want to transform one or more variables before estimation.
(4) Estimation: this is mainly done automatically, by the computer, although sometimes we have to choose the method of estimation.
(5) Interpretation of the results: what do the results tell us? Do they appear satisfactory? Do we need to improve the model? Are there any policy conclusions?

Although this appears reasonably clear-cut, in practice these steps are often mixed up. A researcher might gather the data, estimate a model and then not be happy with the results, realising he has overlooked some factors. He therefore goes back and gets some different data, perhaps some new variables, or maybe tries a different method of investigation until 'satisfactory' results are obtained. There is usually some element of data 'fishing' involved. These methodological issues are examined in more detail later on.

Theoretical issues

What does economic theory tell us about imports? Like any market, the quantity transacted depends upon supply and demand. Strictly, therefore, we should estimate a **simultaneous equation model** of both the demand and supply equations. Since this is beyond the scope of this text (see Maddala and Lahiri (2009),

Chapter 9, for analyses of such models) we simplify by assuming that, as the United Kingdom is a small economy in the world market, we can buy any quantity of imports that we demand (at the prevailing price). In other words, foreign supply is never a constraint, and the UK demand never influences the world price. This assumption, which seems reasonable, means that we can concentrate on estimating the demand equation alone.

Second, economic theory suggests that demand depends upon income and relative prices, particularly the prices of close substitutes and complements. Furthermore, rational consumers do not suffer from money illusion, so real variables should be used throughout.

Economic theory does *not* tell us some things, however. It does not tell us whether the relationship is linear or not. Nor does it tell us whether demand responds *immediately* to price or income changes, or whether there is a lag. For these questions, the data are more likely to give us the answer.

Data

The raw data are presented in Table 8.1, obtained from official UK statistics. Note that there is some slight rounding of the figures: imports are measured to the nearest £0.1bn (£100m) so there is a possible (rounding) error of up to about 0.1%. This is unlikely to substantially affect our estimates.

The variables are defined as follows:

- *Imports* (variable M): imports of goods and services into the United Kingdom, at current prices, in £bn.
- *Income* (GDP): UK gross domestic product at factor cost, at current prices, in £bn.
- *The GDP deflator* (P_{GDP}): an index of the ratio of nominal to real GDP, 1985 = 100. This is an index of general prices and may be used to transform nominal GDP to real GDP.
- *The price of imports* (P_M): the unit value index of imports, 1990 = 100.
- *The price of competing products* (P): the retail price index (RPI), 1985 = 100.

These variables were chosen from a wide range of possibilities. To take income as an example, we could use personal disposable income or GDP. Since firms as well as consumers import goods, the wider measure is used here. Then there is the question of whether to use GDP or GNP, and whether to measure them at factor cost or market prices. Because there is little difference between these different magnitudes, this is not an important decision in this case. However, in a research project one might have to consider such issues in more detail.

Data transformations

Before calculating the regression equation, we must transform the data in Table 8.1. This is because the expenditures on imports and GDP have not been adjusted for price changes (inflation). Part of the observed increase in the imports series is due to prices increasing over time, not increased consumption of imported goods. It is the latter we are trying to explain.

Since expenditure on any good (including imports) can be expressed as the quantity purchased multiplied by the price, to obtain the quantity of imports

Table 8.1 Original data for study of imports

Year	Imports	GDP	GDP deflator	Price of imports	RPI all items
1973	18.8	74.0	24.6	21.5	25.1
1974	27.0	83.8	28.7	31.3	29.1
1975	28.7	105.9	35.7	35.6	36.1
1976	36.5	125.2	41.4	43.6	42.1
1977	42.3	145.7	47.0	50.5	48.8
1978	45.2	167.9	52.5	52.4	52.8
1979	54.2	197.4	60.6	55.8	59.9
1980	57.4	230.8	71.5	65.5	70.7
1981	60.2	253.2	79.7	71.3	79.1
1982	67.6	277.2	85.8	77.3	85.9
1983	77.4	303.0	90.3	84.2	89.8
1984	92.6	324.6	94.9	91.8	94.3
1985	98.7	355.3	100.0	96.4	100.0
1986	100.9	381.8	103.8	91.9	103.4
1987	111.4	420.2	109.0	94.7	107.7
1988	124.7	469.0	116.3	93.7	113.0
1989	142.7	514.9	124.6	97.8	121.8
1990	148.3	558.2	134.1	100.0	133.3
1991	142.1	587.1	142.9	101.3	141.1
1992	151.7	612.0	148.5	102.1	146.4
1993	170.1	642.7	152.5	112.4	148.7
1994	185.4	681.0	155.3	116.1	152.4
1995	207.2	719.7	159.4	123.6	157.6
1996	227.7	765.2	164.6	123.4	161.4
1997	232.3	811.2	169.6	115.2	166.5
1998	239.2	860.8	174.1	109.3	172.2
1999	255.2	906.6	177.8	107.6	174.8
2000	287.0	953.2	180.6	111.2	180.0
2001	299.9	997.0	184.5	110.2	183.2
2002	307.4	1048.8	189.9	107.5	186.3
2003	314.8	1110.3	195.6	106.7	191.7
2004	333.7	1176.5	201.0	106.2	197.4
2005	366.5	1224.7	205.4	110.7	202.9

('real' imports) we must divide the expenditure by the price of imports. In algebraic terms:

$$\text{expenditure} = \text{price} \times \text{quantity, hence}$$

$$\text{quantity} = \frac{\text{expenditure}}{\text{price}}$$

We therefore adjust both imports and GDP for the effect of price changes in this way. P_M is used to deflate the imports series and P_{GDP} used to adjust GDP. This process is covered in more detail in Chapter 10 on index numbers (you may wish to read that before proceeding with this chapter, although it is not essential for understanding the rest of this chapter).

We also need to adjust the import price series, which influences the demand for imports. People make their spending decisions by looking at the price of an imported good *relative to* prices generally. Hence, we divide the price of imports by the retail price index to give the relative, or real, price of imports.

In summary, the transformed variables are derived as follows:

- *Real imports* (M/P_M): this series is obtained by dividing the nominal series for imports by the unit value index (i.e. the import price index). The series gives imports at 1990 prices (in £bn). (Note that the nominal and real series are identical in 1990.)
- *Real income* (GDP/P_{GDP}): this is the nominal GDP series divided by the GDP deflator to give GDP at 1990 prices (in £bn).
- *Real import prices* (P_M/P): the unit value index is divided by the RPI to give this series. It is an index number series, with its value set to 100 in 1990. It shows the price of imports relative to the price of all goods. The higher this price ratio, the less attractive imports would be relative to domestically produced goods.

The transformed variables are shown in Table 8.2. Do not worry if you have not fully understood the process of transforming to real terms. You can simply begin

Table 8.2 Transformed data

Year	Real imports	Real GDP	Real import prices
1973	87.4	403.4	114.2
1974	86.3	391.6	143.4
1975	80.6	397.8	131.5
1976	83.7	405.5	138.0
1977	83.8	415.7	137.9
1978	86.3	428.9	132.3
1979	97.1	436.8	124.2
1980	87.6	432.9	123.5
1981	84.4	426.0	120.2
1982	87.5	433.2	120.0
1983	91.9	450.0	125.0
1984	100.9	458.7	129.8
1985	102.4	476.5	128.5
1986	109.8	493.3	118.5
1987	117.6	517.0	117.2
1988	133.1	540.8	110.5
1989	145.9	554.2	107.0
1990	148.3	558.2	100.0
1991	140.3	550.9	95.7
1992	148.6	552.7	93.0
1993	151.3	565.2	100.8
1994	159.7	588.0	101.5
1995	167.6	605.5	104.5
1996	184.5	623.4	101.9
1997	201.6	641.4	92.2
1998	218.8	663.0	84.6
1999	237.2	683.8	82.1
2000	258.1	707.8	82.3
2001	272.1	724.6	80.2
2002	286.0	740.6	76.9
2003	295.0	761.2	74.2
2004	314.2	784.9	71.7
2005	331.1	799.6	72.7

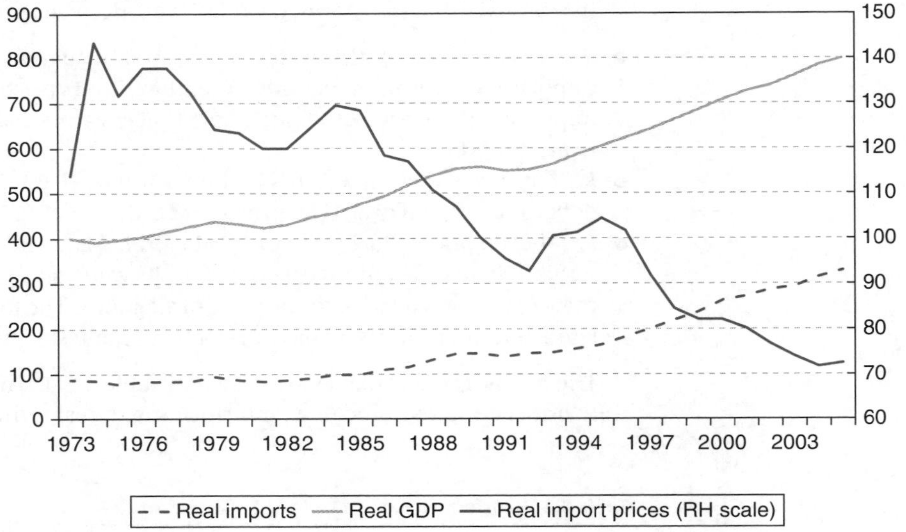

Figure 8.2
Time-series plot of imports, GDP and import prices (real terms)

with the data in Table 8.2, recognising them as the quantity of imports demanded, the level of real income or output and the price of imports relative to all goods.

We should now 'eyeball' the data using appropriate graphical techniques. This will give a broad overview of the characteristics of the data and any unusual or erroneous observations which may be spotted. This is an important step in the analysis.

Figure 8.2 shows a time-series plot of the three variables. The graph shows that both imports and GDP increase smoothly over the period, and that there appears to be a fairly close relationship between them. This is confirmed by the *XY* plot of imports and GDP in Figure 8.3, which shows an approximately linear relationship. One should take care in interpreting this however, since it shows only the *partial* relationship between two of the three variables. However, it does appear to be fairly strong.

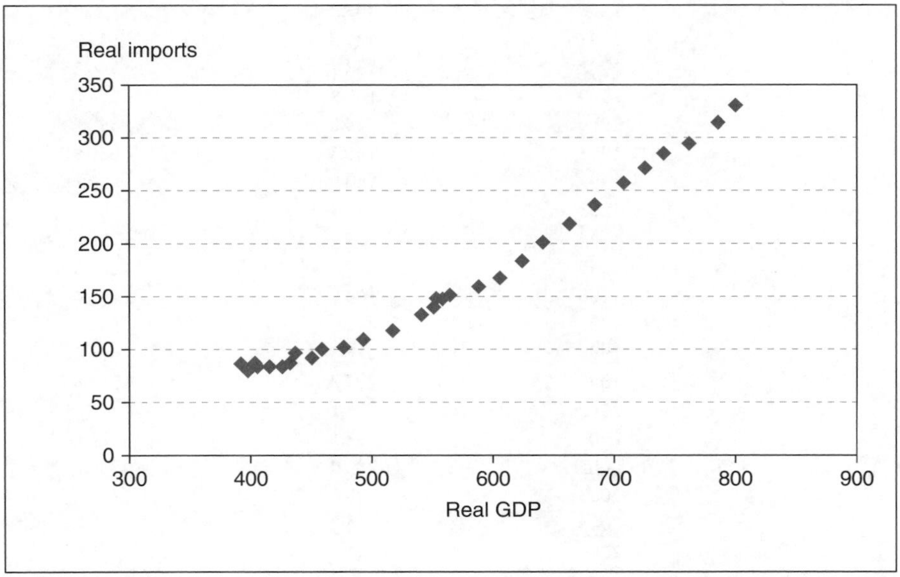

Figure 8.3
XY chart of imports against GDP

Figure 8.4
XY chart of imports against import prices

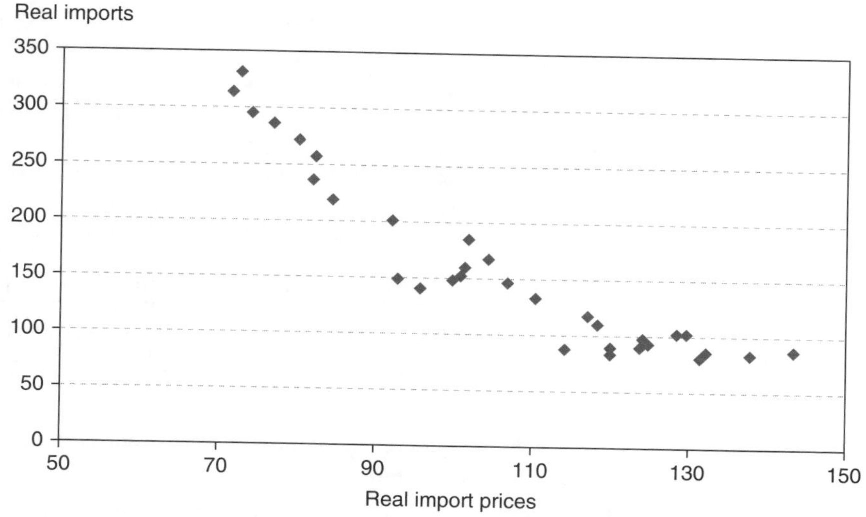

The price of imports (measured on the right-hand scale in the chart) has declined by about 35% over the period (this is relative to all goods generally), so this might also have contributed to the rise in imports. Figure 8.4 provides an *XY* chart of these two variables. There appears to be a clear negative relationship between imports and their price. On the basis of the graphs we might expect a positive relationship between imports and GDP, and a negative one between imports and their price. Both of these expectations are in line with what economic theory would predict.

Note that one does not always (or even often) get such neat graphs in line with expectations. In multivariate analyses the relationships between the variables can be complex and are not revealed by simple bivariate graphs. One needs to do a multiple regression to uncover more precise detail.

Exercise 8.1

?

For the exercises in this chapter we will be looking at the determinants of travel by car in the United Kingdom, which has obviously been increasing steadily and causes concern because of issues such as pollution and congestion. Data for these exercises are as follows:

Year	Car travel (billions of passenger-kilometres)	Real price of car travel	Real price of rail travel	Real price of bus travel	Real personal disposable income
1980	388	107.0	76.2	78.9	54.2
1981	394	107.1	77.8	79.3	54.0
1982	406	104.2	82.3	84.6	53.8
1983	411	106.4	83.4	85.5	54.9
1984	432	103.8	79.8	83.3	57.0
1985	441	101.7	80.4	81.6	58.9
1986	465	97.4	82.7	87.1	61.3
1987	500	99.5	84.1	88.4	63.6
1988	536	98.4	85.4	88.4	67.0
1989	581	95.9	85.7	88.6	70.2
1990	588	93.3	86.3	88.9	72.6
1991	582	96.4	89.9	92.4	74.1
1992	583	98.3	93.5	94.7	76.2
1993	584	101.6	97.6	97.3	78.3

→

Year	Car travel (billions of passenger-kilometres)	Real price of car travel	Real price of rail travel	Real price of bus travel	Real personal disposable income
1994	591	101.3	99.2	99.2	79.4
1995	596	99.7	100.4	100.5	81.3
1996	606	101.4	101.1	103.1	83.3
1997	614	102.7	100.6	105.1	86.6
1998	618	102.1	101.5	106.6	86.9
1999	613	103.9	103.2	109.3	89.8
2000	618	103.7	102.2	110.0	95.3
2001	624	101.2	102.9	112.0	100.0

(a) Draw time-series graphs of car travel and its price and comment on the main features.

(b) Draw *XY* plots of car travel against (i) price and (ii) income. Comment upon the major features of the graphs.

(c) In a multiple regression of car travel on its price and on income, what would you expect the signs of the two slope coefficients to be? Explain your answer.

(d) If the prices of bus and rail travel are added as further explanatory variables, what would you expect the signs on their coefficients to be? Justify your answer.

Estimation

The model to be estimated is therefore

$$\left(\frac{M}{P_M}\right)_t = b_0 + b_1\left(\frac{\text{GDP}}{P_{\text{GDP}}}\right)_t + b_2\left(\frac{P_M}{P}\right)_t + e_t \tag{8.3}$$

expressed in terms of the original variables. To simplify notation, we rewrite this in terms of the transformed variables, as

$$m_t = b_0 + b_1\text{GDP}_t + b_2 pm_t + e_t \tag{8.4}$$

The results of estimating this equation are shown in Table 8.3, which shows the output using Excel. We have used the data in years 1973 to 2003 for estimation purposes, ignoring the observations for 2004 and 2005. Later on we will use the results to predict imports in 2004 and 2005.

The print-out gives all the results we need, which may be summarised as

$$\begin{aligned} m_t &= -172.61 + \underset{(0.06)}{0.59}\text{GDP}_t + \underset{(0.37)}{0.05}pm_t + e_t \\ R^2 &= 0.96, F_{2,26} = 368.23, n = 31 \end{aligned} \tag{8.5}$$

How do we judge and interpret these results? As expected, we obtain a positive coefficient on income but, surprisingly, a positive one on price too. Note that it is difficult to give a sensible interpretation to the constant. The coefficients should be judged in two ways: in terms of their *size* and their *significance*.

Size

As noted earlier, the size of a coefficient depends upon the units of measurement. How 'big' is the coefficient 0.59, for income? This is the marginal propensity to import. It tells us that a rise in GDP, measured in 1990 prices, of £1bn would raise imports, also measured in 1990 prices, by £0.59bn. This is a bit cumbersome. It is better to interpret everything in *proportionate* terms, and calculate the *elasticity* of

Table 8.3 Regression results using Excel

SUMMARY OUTPUT

Regression Statistics	
Multiple *R*	0.98
R square	0.96
Adjusted *R* square	0.96
Standard error	13.24
Observations	31

ANOVA

	df	*SS*	*MS*	*F*	Significance *F*
Regression	2	129 031.05	64 515.52	368.23	7.82E−21
Residual	28	4 905.70	175.20		
Total	30	133 936.75			

	Coefficients	Standard Error	*t* Stat	P-value	Lower 95%	Upper 95%
Intercept	−172.61	73.33	−2.35	0.03	−322.83	−22.39
Real GDP	0.59	0.06	9.12	0.00	0.45	0.72
Real import prices	0.05	0.37	0.13	0.90	−0.70	0.79

imports with respect to income. This is the proportionate change in imports divided by the proportionate change in income:

$$\eta_{\mathrm{GDP}} = \frac{\Delta m/m}{\Delta \mathrm{GDP}/\mathrm{GDP}} \tag{8.6}$$

which can be evaluated (see equation (7.33)) as:

$$\eta_{\mathrm{GDP}} = b_1 \times \frac{\overline{\mathrm{GDP}}}{\overline{m}} = 0.59 \times \frac{536.4}{146.3} = 2.16 \tag{8.7}$$

which shows that imports are highly responsive to income. A 3% rise in real GDP (a fairly typical annual figure) leads to an approximate 6% rise in imports, as long as prices do not change at the same time. Thus as income rises, imports rise substantially faster. More generally we would interpret the result as showing that a 1% rise in GDP leads to a 2.16% rise in imports.

This does seem a large response and we might consider whether this estimate could be true in the long run. If GDP rises by 3% p.a. while imports rise by 6% p.a., then this is not sustainable. Imports will grow and grow as a percentage of GDP and, unless exports grow similarly quickly, this is likely to lead to a balance of payments crisis. Hence we might question our statistical finding because it is inconsistent with long run equilibrium of the economy. We are beginning to see that good modelling should fit with theoretical insights as well as with the data.

A similar calculation for the price variable yields

$$\eta_{pm} = 0.05 \times \frac{109.4}{146.3} = 0.04 \tag{8.8}$$

This yields the 'wrong' sign for the elasticity: a 10% price rise (relative to domestic prices) would *raise* import demand by 0.4%. This is an extremely small effect and for practical purposes can be regarded as zero.

Significance

We can test whether each coefficient is significantly different from zero, i.e. whether the variable truly affects imports or not, using a conventional hypothesis test. For income we have the test statistic

$$t = \frac{0.59 - 0}{0.06} = 9.12$$

as shown in Table 8.3. This has a t distribution with $n - k - 1 = 31 - 2 - 1 = 28$ degrees of freedom (k is the number of explanatory variables excluding the constant, so $k = 2$). The critical value for a two-tail test at the 95% confidence level is 2.048. Since the test statistic comfortably exceeds this we reject $H_0\colon \beta_1 = 0$ in favour of $H_1\colon \beta_1 \neq 0$. Hence, income does indeed affect imports; the sample data are unlikely to have arisen purely by chance. Note that this t ratio is given on the Excel print-out.

For price, the test statistic is

$$t = \frac{0.05 - 0}{0.37} = 0.13$$

which is smaller than 1.701 (the critical value for a one-tail test), so does not fall into the rejection region. $H_0\colon \beta_2 = 0$ cannot be rejected, therefore. We use a one-tailed test in this case since it is reasonable to expect a demand curve to slope downwards, on theoretical grounds. So not only is the coefficient on price quantitatively small, it is insignificantly different from zero, i.e. there is a reasonable probability of this result arising simply by chance. The fact that we had a positive coefficient is thus revealed as unimportant; it was just a small random fluctuation around zero. This result arises despite the fact that the graph of imports against price seemed to show a strong negative relationship. That graph was in fact somewhat misleading. The regression tells us that the more important relationship is with income and, once that is accounted for, price provides little additional explanation of imports. Well, that is the story so far.

The significance of the regression as a whole

We can test the overall significance via an F test as we did for simple regression. This is a test of the hypothesis that *all* the slope coefficients are simultaneously zero (equivalent to the hypothesis that $R^2 = 0$):

$$H_0\colon \beta_1 = \beta_2 = 0$$
$$H_1\colon \beta_1 \neq \beta_2 \neq 0$$

This tests whether *either* income *or* price (or both) affects demand. Since we have already found that income is a significant explanatory variable, via the t test, it would be surprising if this null hypothesis were not rejected. The test statistic is similar[2] to equation (7.28):

$$F = \frac{\text{RSS}/k}{\text{ESS}/(n - k - 1)} \tag{8.9}$$

[2]Note that the formula now contains $n - k - 1$ in the denominator, rather than $n - 2$, to reflect the fact that we now have k explanatory variables.

which has an F distribution with k and $n - k - 1$ degrees of freedom. Substituting in the appropriate values gives

$$F = \frac{129\,031.05/2}{4905.70/(31 - 2 - 1)} = 368.23$$

which is in excess of the critical value for the $F_{2,28}$ distribution of 3.34 (at 5% significance), so the null hypothesis is rejected, as expected. The actual significance level is given by Excel as '7.82E–21', i.e. 7.82×10^{-21}, effectively zero and certainly less than 5%.

STATISTICS IN PRACTICE

Does corruption harm investment?

The World Bank examined this question in its 1997 World Development Report, using regression methods. There is a concern that levels of corruption in many countries harm investment and hence also economic growth.

The study looked at the relationship between investment (measured as a percentage of GDP) and the following variables: the level of corruption, the predictability of corruption, the level of secondary school enrolment, GDP per capita and a measure of 'policy distortion'. Both the level and predictability of corruption were based upon replies to surveys of businesses in the 39 countries studied, which asked questions such as 'Do you have to make additional payments to get things done?' The policy distortion variable measures how badly economic policy is run, based on openness to trade, the exchange rate, etc. Higher values of the index indicate poorer economic management.

The regression obtained was

$$\underset{(\text{s.e.})}{\frac{Inv}{GDP}} = \underset{(13.5)}{19.5} - \underset{(2.2)}{5.8\,CORR} + \underset{(2.6)}{6.3\,PRED\_CORR} + \underset{(2.2)}{2.0\,SCHOOL} - \underset{(1.9)}{1.1\,GDP} - \underset{(1.5)}{2.0\,DISTORT}$$

$$R^2 = 0.24$$

Thus only the corruption variables prove significant at the 5% level. A rise in the level of corruption lowers investment (note the negative coefficient, −5.8) as expected, but a rise in the *predictability* of corruption raises it. This is presumably because people learn how to live with corruption. Unfortunately, units of measurement are not given, so it is impossible to tell just how important are the sizes of the coefficients and, in particular, to find the trade-off between corruption and its predictability.

Adapted from: *World Development Report*, 1997.

Exercise 8.2

(a) Using the data from Exercise 8.1, calculate a regression explaining the level of car travel, using price and income as explanatory variables. Use only the observations from 1980 to 1999. As well as calculating the coefficients you should calculate standard errors and t ratios, R^2 and the F statistic.

(b) Interpret the results. You should evaluate the size of the effect of the explanatory variables as well as their significance and evaluate the goodness of fit of the model.

Are the results satisfactory?

The results so far *appear* reasonably satisfactory: we have found one significant coefficient, the R^2 value is quite high at 96% (although R^2 values tend to be high in time-series regressions, sometimes artificially so) and the result of the F test proves the regression is worthwhile. Nevertheless, it is perhaps surprising to find no effect from the price variable; we might as well drop it from the equation and just regress imports on GDP.

A more stringent test is to use the equation for forecasting, since this uses out-of-sample information for the test. So far, the diagnostic tests such as the F test are based on the same data that were used for estimation. A more suitable test might be to see if the equation can forecast imports to within (say) 4% of the correct value. Since real imports increased by about 4.1% p.a. on average between 1973 and 2003, a simple forecasting rule would be to increase the current year's figure by 4.1%. The regression model might be compared to this standard.

Forecasts for 2004 and 2005[3] are obtained by inserting the values of the explanatory variables for these years into the regression equation, giving

$$2004: \hat{m} = -172.61 + 0.59 \times 784.9 + 0.05 \times 71.7 = 290.0$$
$$2005: \hat{m} = -172.61 + 0.59 \times 799.6 + 0.05 \times 72.7 = 298.6$$

Table 8.4 summarises the actual and forecast values, with the error between them. The percentage error is about 8% in 2004, 11% in 2005. This is not very good; both years are under-predicted by a large amount. The simple growth rule would have given predictions of $295.0 \times 1.04 = 306.8$ and $295.0 \times 1.04^2 = 319.1$ which are much closer. More work needs to be done.

Improving the model – using logarithms

There are various ways in which we might improve our model. We might try to find additional variables to improve the fit (although since we already have $R^2 = 0.96$, this might be difficult), or we might try lagged variables (e.g. the previous year's price) as explanatory variables, on the grounds that the effects do not work through instantaneously. Alternatively, we might try a different functional form for the equation. We have presumed that the regression should be a straight line, although we made no justification for this. Indeed, the graph of imports against income showed some degree of curvature (see Figure 8.3). Hence, we might try a non-linear transformation of the data, as briefly discussed at the end of Chapter 7.

We shall re-estimate the regression equation, having transformed all the data using (natural) logarithms. Remember (from Chapter 1) that logarithms are useful for representing multiplicative processes, such as we have here (where one year's figure tends to be a multiple of the previous year's). Not only does this method fit a curve to the data but has the additional advantage of giving more direct estimates of the elasticities, as we shall see. Because of such advantages, estimating a

Table 8.4 Actual, forecast and error values

Year	Actual	Forecast	Error
2004	314.2	290.0	24.2
2005	331.1	298.6	32.5

[3]Remember that data from 2004 and 2005 were not used to estimate the regression equation.

Table 8.5 Data in natural logarithm form

Year	Real imports	ln *m*	Real GDP	ln *GDP*	Real import prices	ln *pm*
1973	87.4	4.47	403.4	6.00	114.2	4.74
1974	86.3	4.46	391.6	5.97	143.4	4.97
1975	80.6	4.39	397.8	5.99	131.5	4.88
1976	83.7	4.43	405.5	6.01	138.0	4.93
⋮	⋮	⋮	⋮	⋮	⋮	⋮
2001	272.1	5.61	724.6	6.59	80.2	4.38
2002	286.0	5.66	740.6	6.61	76.9	4.34
2003	295.0	5.69	761.2	6.63	74.2	4.31
2004	314.2	5.75	784.9	6.67	71.7	4.27
2005	331.1	5.80	799.6	6.68	72.7	4.29

Note: You can obtain the natural logarithm by using the 'ln' key on your calculator or the 'ln' function in Excel (or other software). Thus we have ln (87.4) = 4.47, etc.

regression equation in logs is extremely common in economics and analysts often start with the logarithmic form in preference to the linear form.

We will therefore estimate the equation

$$\ln m_t = b_0 + b_1 \ln GDP_t + b_2 \ln pm_t + e_t$$

where $\ln m_t$ indicates the logarithm of imports in period t, etc. We therefore need to transform our three variables into logarithms, as shown in Table 8.5 (selected years only).

We now use the new data for the regression, with ln *m* as the dependent variable, ln *GDP* and ln *pm* as the explanatory variables. We also use exactly the same formulae as before, applied to this new data. This gives the following results:

SUMMARY OUTPUT

Regression Statistics	
Multiple *R*	0.99
R square	0.98
Adjusted *R* square	0.98
Standard error	0.05
Observations	31

ANOVA

	df	*SS*	*MS*	*F*	Significance *F*
Regression	2	5.309	2.655	901.43	3.83E−26
Residual	28	0.082	0.003		
Total	30	5.391			

	Coefficients	Standard Error	*t* Stat	P-value	Lower 95%	Upper 95%
Intercept	−3.60	1.65	−2.17	0.04	−6.98	−0.21
ln GDP	1.66	0.15	11.31	0.00	1.36	1.97
ln import prices	−0.41	0.16	−2.56	0.02	−0.74	−0.08

The regression equation we have is therefore

$$\mathrm{lm}\, m_t = -3.60 + 1.66 \ln GDP_t - 0.41 \ln pm_t$$

Because we have transformed the variables the slope coefficients are very different from the values we had before, from the linear equation. However, the interpretation of the log regression equation is different. A big advantage of this formulation is that the coefficients give direct estimates of the elasticities; there is no need to multiply by the ratio of the means, as with the linear form (see equation (8.7).

Hence the income elasticity of demand is estimated as 1.66 and the price elasticity is -0.41. These contrast with the values calculated from the linear equation, of 2.16 and 0.04, respectively. The contrast with the previous estimate of the price elasticity is particularly stark. We have gone from an estimate which was positive (though very small and statistically insignificant) to one which is negative and significant.

It is difficult to say which is the 'right' answer, both are estimates of the unknown, true values. One advantage of the log model is that the elasticity does not vary along the demand curve, as it does with the linear model. With the latter we had to calculate the elasticity at the means of the variables, but the value inevitably varies along the curve. For example, taking 2003 values for imports and income we obtain an elasticity of

$$\eta_{\mathrm{GDP}} = 0.59 \times \frac{761.2}{295.0} = 1.52$$

This is quite different from the value at the mean, 2.16. A convenient mathematical property of the log formulation is that the elasticity does not change along the curve. Hence, we can talk about 'the' elasticity, which is very convenient.

Graphing the regression coefficients

Regression results are usually presented in the form of a table, as above. However, this can result in complex and dense arrays of numbers which are not easy to read or interpret. In Chapter 1 we found that graphs were often a useful way to present complex information, and we will illustrate this again using our regression results.

The results of the first, linear model can be shown as follows:

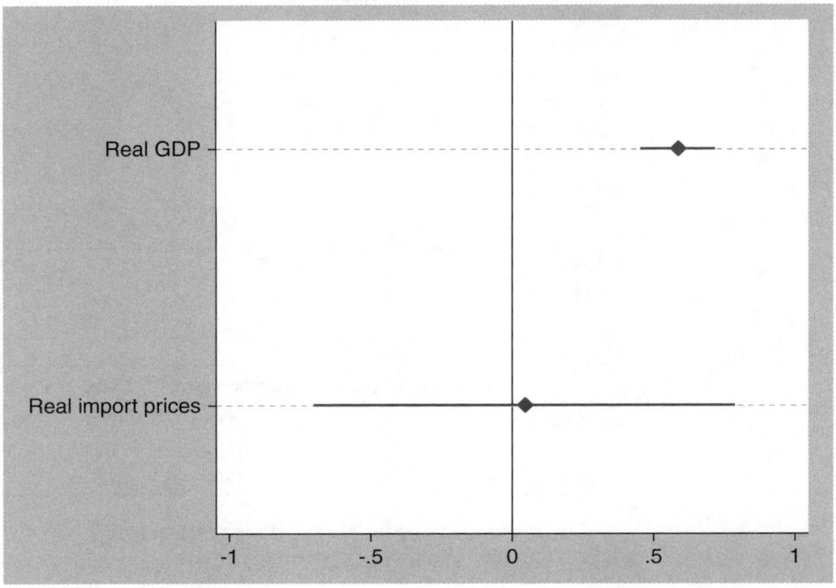

There is a horizontal line for each variable, showing the width of the confidence interval, with the point estimate of the coefficient marked at its centre. The vertical line is drawn at $\beta = 0$, indicating no effect of that variable.

The implications are immediately clear. The GDP coefficient is positive, measured fairly precisely and is statistically significant (since zero is not in the confidence interval). By contrast, the price effect is unclear (wide CI) and the point estimate is close to zero.

If we draw a similar diagram for the model with variables in logs, we obtain

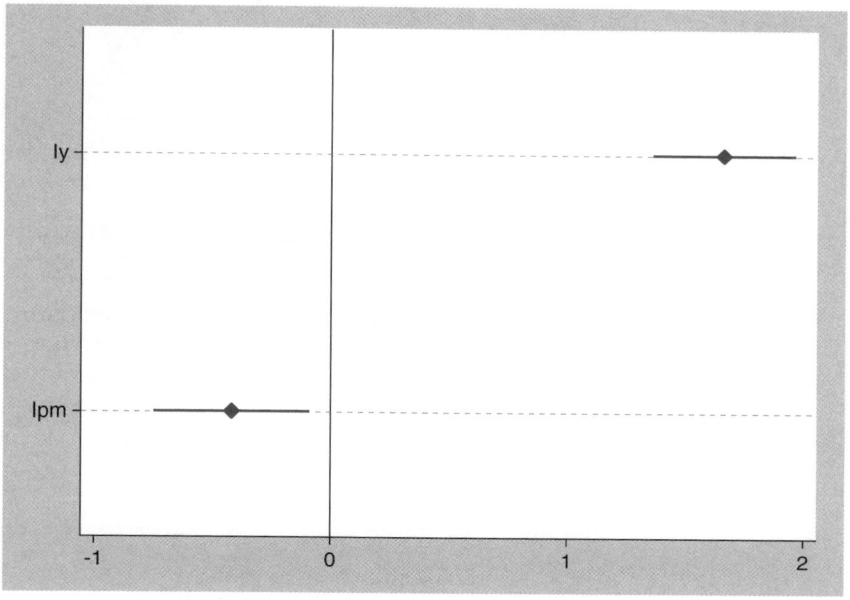

Qualitatively, little has changed for the income variable. It is still positive, statistically significant and with a small confidence interval. The price effect is different, however. The confidence interval is smaller than before, the coefficient is negative and significant. The chart might still lead us to be cautious about the price effect, since zero is not far outside the confidence interval.

This type of chart can therefore be useful. If making several estimates of the same coefficient (e.g. using different models, functional forms, etc.), one could draw a bar for each estimate and it would be easy to compare them, much easier than comparing across different tables (or columns of a table), with each estimate having a different standard error needed for the confidence interval.

Comparison of models and predictions

We can compare the linear and log models further to judge which is preferable. The log model has a higher price elasticity and is 'significant' ($t = -2.56$), so we can now reject the hypothesis that price has no effect upon import demand. This is more in line with what economic theory would predict. The R^2 value is also higher (0.98 versus 0.96), but this is a misleading comparison. R^2 tells us how much of the variation in the dependent variable is explained by the explanatory variables. However, we have a different dependent variable now: the log of imports

Figure 8.5
Chart of predictions up to 2010

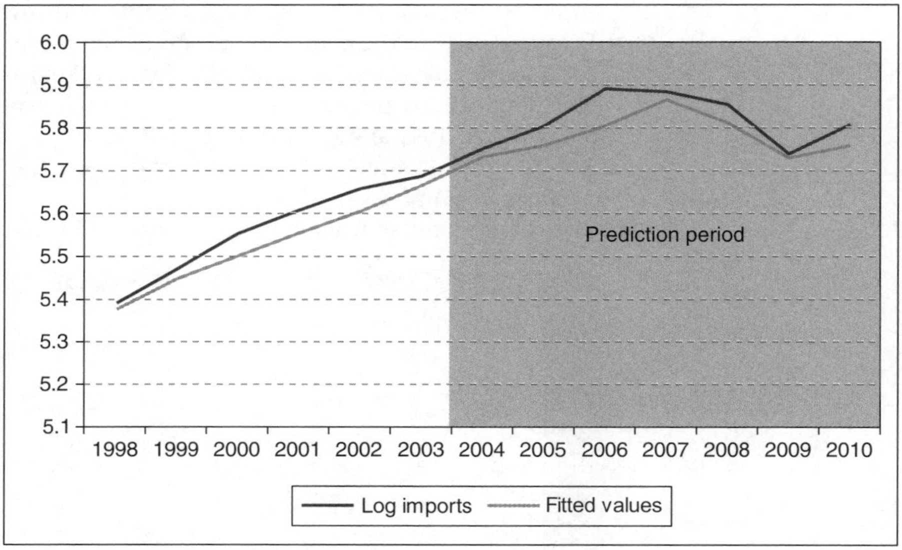

rather than imports. Although they are both measuring imports, they are different variables, making direct comparison of R^2 invalid.

We can also compare the predictive abilities of the two models. For the log model we have the following predictions:

$$2004\text{: } \ln \hat{m} = -3.60 + 1.66 \times 6.67 - 0.41 \times 4.27 = 5.73$$
$$2005\text{: } \ln \hat{m} = -3.60 + 1.66 \times 6.68 - 0.41 \times 4.29 = 5.76$$

These are log values, so we need to take anti-logs to get back to the original units:

$$e^{5.73} = 308.2 \text{ and } e^{5.76} = 316.0$$

These predictions are substantially better than from the linear equation, as we see below:

Year	Actual	Fitted	Error	% error
2004	314.2	308.2	6.0	1.9
2005	331.1	316.0	15.1	4.8

The errors are less than half the size they were in the linear formulation and, overall, the log regression is beginning to look the better. If we forecast further ahead, up to 2010, we get the results as illustrated in Figure 8.5.

The predictions track the actual values quite well, even up to seven years ahead. Note that this is not a true forecast – to obtain the prediction for 2010, for example, we have used the values of income and prices in 2010, which would not be known back in 2003. This picture shows that the relationship remains fairly stable over time even when imports turn down around 2006 and the economy is then hit by a severe financial crisis.

STATISTICS IN PRACTICE

The log transformation: a subtle trap

By using the log regression, we find the expected value of ln Y for a given value of ln X, i.e. $E(\ln Y \mid \ln X)$, by setting the error term to zero. For 2004 this prediction turned out to be 5.73. Of course, we really want the expected value of Y. How do we obtain $E(Y)$ for a given value of X? We simply raised e to the power ln Y: $e^{5.73} = 308.2$, but this is not strictly correct.

The problem lies in the fact that the logarithm is a *non-linear* transformation and hence $e^{E(\ln Y)} \neq E(Y)$. The correct formula to use is:

$$E(Y) = e^{E(\ln Y)+s^2/2}$$

where s^2 is the error variance in the regression. In this example it makes very little difference, since $s = 0.05$ and we obtain

$$E(Y) = e^{5.73+0.05^2/2} = 308.35$$

which is a trivial difference. The small difference is due the very small standard error of the regression or equivalently the very good fit. If the regression fits less well, with a standard error of, say, 0.5, the estimate of $E(Y)$ would be underestimated by about 13%.

Matters do not end here. The transformation $e^{E(\ln Y)}$ could be defended on the grounds that it provides an estimate of the *median* level of Y and this might give a better idea of a 'typical' value of Y if the original data are heavily skewed (refer back to the wealth data of Chapter 1). Fuller discussion is beyond our scope but does suggest that if your regression standard error is more than 0.5 or so, you should consider the implications of this when transforming back from logs.

Choosing between alternative models is a matter of judgement. The criteria are convenience, conformity with economic theory and the general statistical 'fit' of the model to the data. In this case the log model seems superior on all counts. It is more convenient as we get direct estimates of the elasticities. It is more in accord with economic theory as it suggests a significant price effect and also because the variables are growing over time, which is usually better represented by the log transformation. Finally, the model seems to fit the data better and, in particular, it gives better forecasts. There are more formal statistical methods for choosing between different models, but they are beyond the scope of this text.

The rest of this chapter looks at more advanced topics relating to the regression model. These are not essential as far as estimation of the regression model goes but are useful 'diagnostic tools' which allow us to check the quality of the estimates in more depth.

Testing the accuracy of the forecasts: the Chow test

There is a formal test for the accuracy of the forecasts (which can be applied to both linear and log forms of the equation), based on the F distribution. This is the **Chow test** (named after its inventor). The null hypothesis is that the true prediction errors are all equal to zero, so the errors we do observe are just random variation from the regression line. Alternatively, we can interpret the hypothesis as asserting that the same regression line applies to both estimation and prediction periods. If the predictions lie too far from the estimated regression line, then the null is rejected. The alternative hypothesis is that the model has changed in some way and that a different regression line should be applied to the prediction period.

The test procedure is as follows:

(1) Use the first n_1 observations for estimation, the last n_2 observations for the forecast. In this case we have $n_1 = 31$, $n_2 = 2$.

(2) Estimate the regression equation using the first n_1 observations, as above, and obtain the error sum of squares, ESS_1.
(3) Re-estimate the equation using all $n_1 + n_2$ observations, and obtain the pooled error sum of squares, ESS_P.
(4) Calculate the F statistic:

$$F = \frac{(ESS_P - ESS_1)/n_2}{ESS_1/(n_1 - k - 1)}$$

We then compare this test statistic with the critical value of the F distribution with $n_2, n_1 - k - 1$ degrees of freedom. If the test statistic exceeds the critical value, the model fails the prediction test. A large value of the test statistic indicates a large divergence between ESS_P and ESS_1 (adjusted for the different sample sizes), suggesting that the model does not fit the two periods equally well. The bigger the prediction errors, the more ESS_P will exceed ESS_1, leading to a large F statistic.

Evaluating the test (for the log regression), we have $ESS_1 = 0.08246$ (the *Excel* printout rounded this to 0.08). Estimating over the whole sample, 1973 to 2005, gives:

$$\ln m_t = -3.54 + 1.67 \ln GDP_t - 0.42 \ln pm_t$$
$$R^2 = 0.99, F_{2,30} = 1202.52, ESS_P = 0.08444$$

so the test statistic is

$$F = \frac{(0.08444 - 0.08246)/2}{0.08246/28} = 0.34$$

The critical value of the F distribution for 2, 28 degrees of freedom is 3.34, so the equation passes the test, i.e. the same regression line may be considered valid for both sub-periods and the errors in the forecasts are just random errors around the regression line. Repeating the calculation for the forecast period up till 2010 reveals a similar result with $F = 0.43$ (7 and 28 degrees of freedom).

It is noticeable that the predictions are always too low (for all the models); the errors in both years are positive. This suggests a slight 'boom' in imports relative to what one might expect (despite the result of the Chow test). Perhaps we have omitted an explanatory variable which has changed markedly in 2004 to 2005, or perhaps the errors are not truly random. Alternatively, we still could have the wrong functional form for the model. Since we already have an R^2 value of 0.98, we are unlikely to find another variable which adds significantly to the explanatory power of the model. We have already tried two functional forms. Therefore, we shall examine the errors in the model to see if they appear random.

Exercise 8.3

?

(a) Use the regression equation from Exercise 8.2 to forecast the level of car travel in 2000 and 2001. How accurate are your forecasts? Is this a satisfactory result?

(b) Convert the variables to (natural) logarithms and repeat the regression calculation. Interpret your result and compare to the linear equation.

(c) Calculate price and income elasticities from the linear model and compare to those obtained from the log model.

(d) Forecast car travel in 2000 and 2001 using the log model and compare the results to those from the linear model. (Use the function e^x to convert the forecasts in logs back to the original units.)

(e) Use a Chow test to test whether the forecasts are accurate. Is there any difference between linear and log models?

Analysis of the errors

Why analyse the errors, as surely they are just random? In setting out our model (equation 8.2) we *asserted* the error is random, but this does depend upon our formulating the correct model. Hence, if we study the errors and find they are not random, in some way, this suggests the model is not correct and hence could be improved. This is another important part of the checking procedure, to see if the model is adequate or whether it is mis-specified (e.g. has the wrong functional form, or a missing explanatory variable). If the model is a good one, then the error term should be random and ideally should be unpredictable. If there are any predictable elements to it, then we could use this information to improve our model and forecasts. Unlike forecasting, this is a within-sample procedure. Furthermore, we expect the observed errors to be approximately Normally distributed, since this assumption underlies the t and F distributions used for inference. If the errors are not Normal, this would cast doubt on our use of t and F statistics for inference purposes.

A complete, formal, treatment of these issues is beyond the scope of this text (see, for example, Maddala and Lahiri (2009), Chapters 5, 6 and 12). Instead, we give an outline of how to detect the problems and some simple procedures which might overcome them. At least, if you are aware of the problem, you will know that you should consult a more advanced text.

First, we can quickly deal with the issue of Normality of the errors. In this example we have only 31 observations, which is not really sufficient to check for a Normal distribution. Drawing a histogram of the errors (left as an exercise) does not give a nice, smooth distribution because of the few observations and it is hard to tell if it looks Normal or not. More formal methods also require more observations to be reliable, so we will have to take the assumption of Normality on trust in this case.

Second, we can examine the error term for evidence of **autocorrelation**. This was introduced briefly in Chapter 1. To recapitulate: autocorrelation occurs when one error observation is correlated with an earlier (often the previous) one. It only occurs with time-series data (in cross-section, the ordering of the observations does not matter, so there is not a natural 'preceding' observation). Autocorrelation often occurs in time-series data: if inflation is 'high' this month, it is likely to be high next month also; if low, it is likely to be low next month also. Many economic variables are 'sticky' in this way. Imports are likely to behave this way too, as the factors affecting imports (mainly GDP) change slowly.

This characteristic has not been incorporated into our model. If it were, we might improve our forecasts: noting that the actual value of imports in 2003 is above the predicted value (a positive error), we might expect another positive error in 2004. However, our forecast was made by setting the error for 2004 to zero (i.e. using the fitted value from the regression line). In light of this, perhaps we should not be surprised that the predicted value is below the actual value.

Table 8.6 Calculation of residuals

Observation	Actual	Predicted	Residual
1973	4.47	4.43	0.04
1974	4.46	4.29	0.17
1975	4.39	4.35	0.04
1976	4.43	4.36	0.07
⋮	⋮	⋮	⋮
2000	5.55	5.50	0.05
2001	5.61	5.55	0.05
2002	5.66	5.61	0.05
2003	5.69	5.67	0.02

Note: in logs, the residual is approximately the percentage error. So, for example, the first residual 0.04 indicates the error is of the order of 4%.

One should therefore check for this possibility before making forecasts by examining the errors (up to 2003) for the presence of autocorrelation. Poor forecasting is not the only consequence of autocorrelation – the estimated standard errors can also be affected (often biased downwards in practice) leading to incorrect inferences being drawn.

Checking for autocorrelation

The errors to be examined are obtained by subtracting the fitted values from the actual observations. Using time-series data, we have:

$$e_t = Y_t - \hat{Y}_t = Y_t - b_0 - b_1 X_{1t} - b_2 X_{2t} \tag{8.10}$$

The errors obtained from the import demand equation (for the logarithmic model of import demand) are shown in Table 8.6 and are graphed in Figure 8.6. The graph suggests a definite pattern, that of *positive* errors initially, followed by a series of negative errors, followed in turn by more positive errors. This is surely not a random pattern: a positive error is likely to be followed by a positive error, a negative error by another negative error. From this graph we might reasonably predict that the two errors for 2004 to 2005 will be positive (as in fact they are). This means our regression equation is inadequate in some way – we are *expecting* it to under-predict. If so, we ought to be able to improve it.

The phenomenon we have uncovered (positive errors usually following positive, negative following negative) is known as **positive autocorrelation**. In other words, there appears to be a positive correlation between successive errors e_t and e_{t-1}. A

Figure 8.6
Time-series graph of the errors from the import demand equation

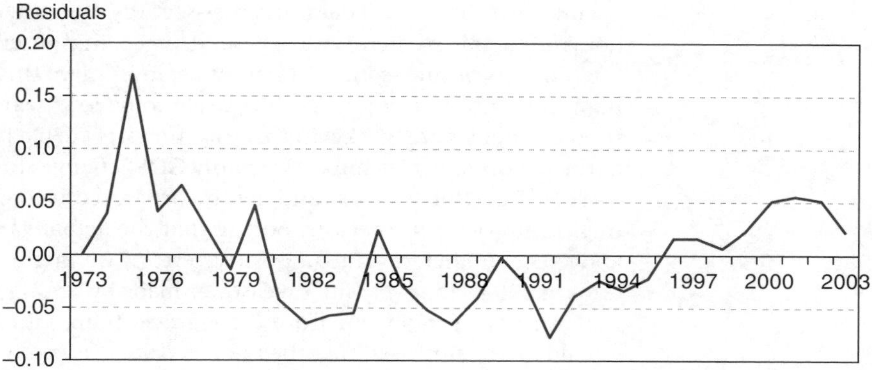

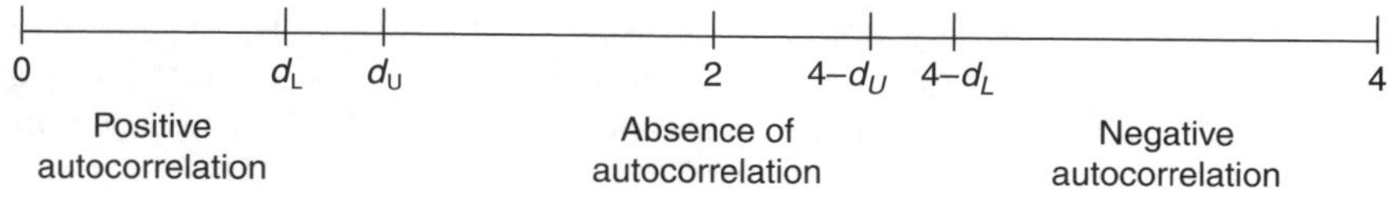

Figure 8.7
The Durbin–Watson test statistic

truly random series would have a low or zero correlation. Less common in economic models is **negative autocorrelation**, where positive errors tend to follow negative ones, negative follow positive. We will concentrate on positive autocorrelation.

This non-randomness can be summarised and tested numerically by the **Durbin–Watson** (DW) **statistic** (named after its two inventors). This is routinely printed out by specialist software packages but, unfortunately, not by spreadsheet programs. The statistic is a one-tailed test of the null hypothesis of no autocorrelation against the alternative of positive, or of negative, autocorrelation. The test statistic always lies in the range 0–4 and is compared to critical values d_L and d_U (given in Appendix Table A7, see page 463). The decision rule is best presented graphically, as in Figure 8.7.

Low values of DW (below d_L) suggest positive autocorrelation, high values (above $4 - d_L$) suggest negative autocorrelation and a value near 2 (between d_U and $4 - d_U$) suggests the problem is absent. There are also two regions where the test is, unfortunately, inconclusive (between the d_L and d_U values).

The test statistic can be calculated by the formula[4]

$$\text{DW} = \frac{\sum_{t=2}^{n}(e_t - e_{t-1})^2}{\sum_{t=1}^{n} e_t^2} \tag{8.11}$$

This is relatively straightforward to calculate using a spreadsheet program. Table 8.7 shows part of the calculation.

Hence we obtain:

$$\text{DW} = \frac{0.0705}{0.0825} = 0.855$$

Table 8.7 Calculation of the DW statistic

Year	e_t	e_{t-1}	$e_t - e_{t-1}$	$(e_t - e_{t-1})^2$	e_t^2
1973	0.0396				0.0016
1974	0.1703	0.0396	0.1308	0.0171	0.0290
1975	0.0401	0.1703	−0.1302	0.0170	0.0016
1976	0.0658	0.0401	0.0258	0.0007	0.0043
⋮	⋮	⋮	⋮	⋮	⋮
2000	0.0517	0.0236	0.0281	0.0008	0.0027
2001	0.0548	0.0517	0.0031	0.0000	0.0030
2002	0.0509	0.0548	−0.0039	0.0000	0.0026
2003	0.0215	0.0509	−0.0294	0.0009	0.0005
Totals				0.0705	0.0825

[4]The DW statistic can also be approximated using the correlation coefficient r between e_t and e_{t-1}, and then $\text{DW} \approx 2 \times (1 - r)$. The closer the approximation, the larger the sample size. It should be reasonably accurate if you have 20 observations or more.

The result suggests positive autocorrelation[5] of the errors. For $n = 30$ (close enough to $n = 31$) the critical values are $d_L = 1.284$ and $d_U = 1.567$ (using the 95% confidence level, see Table A7), so we clearly reject the null of no autocorrelation.

Consequences of autocorrelation

The presence of autocorrelation in this example causes our forecasts to be too low. If we took account of the pattern of errors over time, we could improve the forecasting performance of the model. A second general consequence of autocorrelation is that the standard errors are often under-estimated, resulting in excessive t and F statistics. This leads us to think the estimates are 'significant' when they might not, in fact, be so. We may have what is sometimes known as a **spurious regression** – it looks good but is misleading. The bias in the standard errors and t statistics can be large, and this is potentially a serious problem.

This danger occurs particularly when the variables used in the analysis are trended (as many economic variables are) over time. Variables trending over time *appear* to be correlated with each other, but there may be no true underlying relationship. The now-famous study by Hendry[6] noted a strong correlation between cumulative rainfall and the price level (both increase over time but are unlikely to be related). It has been suggested that a low value of the DW statistic (typically, less than the R^2 value) can be a symptom of such a problem. The fact that economic theory supports the idea of a causal relationship between demand, prices and income should make us a little more confident that we have found a valid economic relationship rather than a spurious one in this case.

This topic goes well beyond the scope of this text (once again), but it is raised because it is important to be aware of the potential shortcomings of simple models. If you estimate a time-series regression equation, check the DW statistic to test for autocorrelation. If present, you may want to seek further advice rather than accept the results as they are, even if they appear to be good. The cause of the autocorrelation is often (though not always) the omission of lagged variables in the model, i.e. a failure to recognise that it may take time for the effect of the independent variables to work through to the dependent variable.

Exercise 8.4

(a) Using the log model explaining car travel, calculate the residuals from the regression equation and draw a line graph of them. Do they appear to be random or is some time-dependence apparent?

(b) Calculate the Durbin–Watson statistic and interpret the result.

(c) If autocorrelation is present, what are the implications for your estimates?

Finding the right model

How do you know that you have found the 'right' model for the data? Can you be confident that another researcher, using the same data, would arrive at the same results? How can you be sure there isn't a relevant explanatory variable out there that you have omitted from your model? Without trying them all it is difficult to

[5]The correlation between e_t and e_{t-1} is, in fact, 0.494.

[6]Hendry, D.F., Econometrics – Alchemy or Science? in *Economica*, 47 (1980), 387–406.

be sure. Good modelling is based on theoretical considerations (e.g. models that are consistent with economic or business principles) and statistical ones (e.g. significant *t* ratios). One can identify two different approaches to modelling:

- *General to specific:* this starts off with a comprehensive model, including all the likely explanatory variables, then simplifies it.
- *Specific to general:* this begins with a simple model that is easy to understand, then explanatory variables are added to improve the model's explanatory power.

There is something to be said for both approaches, but it is not guaranteed that the two will end up with the same model. The former approach is usually favoured nowadays; it suffers less from the problem of **omitted variable bias** (discussed below), and the simplifying procedure is usually less ad hoc than that of generalising a simple model. A very general model will almost certainly initially contain a number of irrelevant explanatory variables. However, this is not much of a problem (and less serious than omitted variable bias): standard errors on the coefficients tend to be higher than otherwise, but this is remedied once the irrelevant variables are excluded.

It is rare for either of these approaches to be adopted in its pure, ideal form. For example, in the import demand equation we should have started out with several lags on the price variable, since we cannot be sure how long imports take to adjust to price changes. Therefore, we might have started with (assuming a maximum lag of one year is 'reasonable'):

$$m_t = b_0 + b_1\text{GDP}_t + b_2\text{GDP}_{t-1} + b_3 pm_t + b_4 pm_{t-1} + b_5 m_{t-1} + e_t \qquad (8.12)$$

If b_4 proved to be not significantly different from zero, we would then re-estimate the equation without pm_{t-1} and obtain new coefficient estimates. If the new b_2 proved to be not significant, we would omit GDP_{t-1} and re-estimate. This process would continue until all the remaining coefficients had significant *t* ratios. We would then have the final, simplified model. At each stage we would omit the variable with the least significant coefficient. Having found the right model, we could then test it on new data, to see if it can explain the new observations.

STATISTICS IN PRACTICE

Uncertainty regarding the correct model

The remarks about finding the right model apply to many of the other techniques used in this text. For example, we might employ the Poisson distribution to model manufacturing faults in televisions, but we are *assuming* this is the correct distribution to use. In the example of railway accidents recounted in Chapter 4, it was found that the Poisson distribution did not fit the data precisely – the real world betrayed less variation than predicted by the model.

Our estimates of parameters, and the associated confidence intervals, are based on the assumption that we are using the correct model. To our uncertainty about the estimates we should ideally add the uncertainty about the correct model, but unfortunately this is difficult to measure. It may be that if we used a different model we would obtain a different conclusion. If possible, therefore, it is a good idea to try out different models to see if the results are robust, and also to inform the reader about alternative methods that have been tried but not reported.

In practice the procedure is not as mechanical (nor as pure) as this, and more judgement should be exercised. You may not want to exclude all the price variables from a demand equation even though the *t* ratios are small. A coefficient may be large in *size* even though it is not *significant*. 'Not significant' does not mean the same as 'not important', rather that there is a lot of uncertainty about its true value. In modelling imports, we used the 2004 and 2005 observations to test the model's forecasts. When it failed, we revised the model and applied the forecast test again. But this is no longer a strictly independent test, since we used the 2004 to 2005 observations to decide upon revision to the model.

STATISTICS IN PRACTICE

Interpreting the R^2 coefficient

Students often worry that their R^2 value is not large enough, hence their regression is a poor one. However, interpretation of the statistic is tricky. It is easier to 'explain' aggregate data than it is to explain individual observations, hence a regression using country level data typically has a higher R^2 value than a similar one using individual data (the values might vary between 0.6 for the former but 0.1 for the latter). Also, time-series data typically have much higher R^2 values (e.g. 0.95) than cross-section data. Estimating a regression using first differences of the data (i.e. explaining the change in *Y* by changes in the *X* values) typically lowers the R^2 value substantially, even though the underlying phenomena are the same. The size of the R^2 statistic needs to be considered in the light of the type of data used in the model.

To briefly sum up a complex and contentious debate, a good model should be:

- *consistent with theory*: an estimated demand curve should not slope upwards, for example.
- *statistically satisfactory*: there should be good explanatory power (e.g. R^2, *F* statistics), the coefficients should be statistically significant (*t* ratios) and the errors should be random. It should also predict well, using new data (i.e. data not used in the estimation procedure).
- *simple*: although a very complicated model predicts better, it might be difficult for the reader to understand and interpret.

Sometimes these criteria conflict and then the researcher must use his or her judgement and experience to decide between them.

Testing compound hypotheses

Simplifying a general model is largely based on hypothesis testing. Usually this means a hypothesis of the form H_0: $\beta = 0$ using a *t* test. Accepting this hypothesis would mean we can simplify by dropping that variable from the equation. Sometimes, however, the hypothesis is more complex, as in the following examples:

- You want to test the equality of two coefficients, H_0: $\beta_1 = \beta_2$.
- You want to test if a *group* of coefficients are all zero, H_0: $\beta_1 = \beta_2 = 0$.

A general method for testing these compound hypotheses is to use an *F* test, comparing the general (unrestricted) model with a more restricted version. An unrestricted model will always fit the data better (higher R^2) than a restricted version. However, if the restricted version fits almost as well, we conclude that the restriction is valid and that the simpler restricted model is the more appropriate one to use.

We illustrate this by examining whether consumers suffer from money illusion in the import demand equation. We assumed, in line with economic theory, that only relative prices matter and used the real price of imports P_M/P as an explanatory variable. But suppose consumers actually respond differently to changes in P_M and in P? In that case we should enter P_M and P as separate explanatory variables and they would have different coefficients. In other words, we should estimate (using the log form[7]):

$$\ln m_t = c_0 + c_1 \ln GDP_t + c_2 \ln P_{Mt} + c_3 \ln P_t + e_t \tag{8.13}$$

rather than

$$\ln m_t = b_0 + b_1 \ln GDP_t + b_2 \ln pm_t + e_t \tag{8.14}$$

where P_M is the nominal price of imports and P is the nominal price level. We would expect $c_2 < 0$ and $c_3 > 0$. Note that (8.14) is a *restricted* form of (8.13), with the restriction $c_2 = -c_3$ imposed. A lack of money illusion implies that this restriction should be valid and that (8.14) is the correct form of model. The hypothesis to test is therefore H_0: $c_2 = -c_3$ (or alternatively H_0: $c_2 + c_3 = 0$).

If the restriction is valid, (8.13) and (8.14) should fit equally well and thus have similar error sums of squares. Conversely, if they have very different ESS values, then we would reject the validity of the restriction. To carry out the test, we therefore do the following:

- Estimate the *unrestricted* model (8.13) and obtain the *unrestricted* ESS from it (ESS_U).
- Estimate the *restricted* model (8.14) and obtain the *restricted* ESS (ESS_R). Note that by definition, $ESS_U \leq ESS_R$.
- Compare the ESS values using the test statistic

$$F = \frac{(ESS_R - ESS_U)/q}{ESS_U/(n - k - 1)} \tag{8.15}$$

  where q is the number of restrictions (1 in this case) and k is the number of explanatory variables in the *unrestricted* model.
- Compare the test statistic with the critical value of the F distribution with q and $n - k - 1$ degrees of freedom. If the test statistic exceeds the critical value, reject the restricted model in favour of the unrestricted one.

We have already estimated the restricted model (equation (8.14)) and from that we obtain $ESS_R = 0.08246$. Estimating the unrestricted model gives

$$\ln m_t = -8.77 + 2.31 \ln GDP_t - 0.20 \ln P_{Mt-1} + 0.02 \ln P_{t-1+et} \tag{8.16}$$

with $ESS_U = 0.02720$. The test statistic is therefore

$$F = \frac{(0.08246 - 0.02720)/1}{0.02720/(31 - 3 - 1)} = 54.85 \tag{8.17}$$

[7]Note that it is much easier to test the restriction in log form, since P_M and P are entered additively. It would be much harder to do this in levels form.

The critical value[8] at the 95% confidence level is 4.21, so the restriction is rejected. Consumers do not use relative prices alone in making decisions, but are somehow influenced by the general rate of inflation as well. This is contrary to what one would expect from economic theory. Interestingly, the equation using nominal prices does not suffer from autocorrelation, so imposing the restriction (estimating with the real price of imports) induces autocorrelation, another indication that the restriction might be inappropriate.

To our earlier finding we might therefore add that consumers appear to take account of nominal prices. We do not have space to investigate this issue in more detail, but further analysis of these nominal effects would be worthwhile. There may be a theoretical reason for nominal prices to have an influence. Alternatively, there could be measurement problems with the data or inadequacies in the model which mask the truth that it is, after all, relative prices that matter.

Whatever the results, this method of hypothesis testing is quite general: it is possible to test any number of (linear) restrictions by estimating the restricted and unrestricted forms of the equation and comparing how well they fit the data. The unrestricted will always fit the data better but if the restricted model fits almost as well, it is preferred on the grounds of simplicity. The F test is the criterion by which we compare the fit of the two models, using error sums of squares.

Omitted variable bias

Omitting a relevant explanatory variable from a regression equation can lead to serious problems. Not only is the model inadequate because there is no information about the effect of the omitted variable but, in addition, the coefficients on the variables which *are* included are usually biased. This is called **omitted variable bias** (OVB).

We encountered an example of this in the model of import demand. Notice how the coefficient on income changed from 1.66 to 2.31 when nominal prices were included. This is a substantial change and shows that the original equation with only the real price of imports included may be misleading with respect to the effect of income upon imports. The coefficient on income was biased downwards.

The direction of OVB depends upon two things: the correlation between the omitted and included explanatory variables and the sign of the coefficient on the omitted variable. Thus, if you have to omit what you believe is a relevant explanatory variable (because the observations are unavailable, for example) you might be able to infer the direction of bias on the included variable(s). Table 8.8 summarises the possibilities, where the true model is $Y = b_0 + b_1X_1 + b_2X_2 + e$ but the estimated model omits the X_2 variable. Table 8.8 only applies to a single omitted variable; when there are several, matters are more complicated (see Maddala and Lahiri (2009), Chapter 4).

In addition to coefficients being biased, their standard errors are biased upwards as well, so that inferences and confidence intervals will be incorrect. The best advice, therefore, is to ensure you don't omit a relevant variable.

[8]Because only large values of the F statistic reject H_0, we use the critical value cutting off the upper tail of the distribution.

Table 8.8 The effects of omitted variable bias

Sign of omitted coefficient, b_2	Correlation between X_1 and X_2	Direction of bias of b_1	Example values of b_1	
			True	Estimated
>0	>0	upwards	0.5	0.9
			-0.5	-0.1
>0	<0	downwards	0.5	0.1
			-0.5	-0.9
<0	>0	downwards	0.5	0.1
			-0.5	-0.9
<0	<0	upwards	0.5	0.9
			-0.5	-0.1

Exercise 8.5

(a) Calculate the simple correlation coefficients between price, income, the price of rail travel and the price of bus travel.

(b) The prices of rail and bus travel may well influence the demand for car travel. If so, the models calculated in previous exercises are mis-specified. What are the possible consequences of this? How might the correlations calculated in part (a) help?

(c) Extend the regression equation to include these two extra prices. (Estimate in logs, using 1980–99.) Does this change any of your conclusions?

(d) One might expect the bus and rail price variables to have similar coefficients, as they are both substitutes for car travel. Test the hypothesis $H_0: \beta_{rail} - \beta_{bus} = 0$ by comparing error sums of squares from restricted and unrestricted regressions.

Dummy variables and trends

These are types of artificial variable which can be very useful in regression. A **dummy variable** is one that takes on a restricted range of values, usually just 0 and 1. Despite this simplicity, it can be useful in a number of situations. For example, suppose we suspect that the United Kingdom's import demand function shifted after the rise in oil prices in 1979. Ideally, we might include oil prices in our model, but suppose these data are unavailable. How could we then explore this possibility empirically?

One answer is to construct a variable, D_t, which takes the value 0 for the years 1973–79, and 1 thereafter (i.e. 0, 0, . . . , 0, 1, 1, . . . , 1, the switch occurring after 1979). We then estimate:

$$\ln m_t = b_0 + b_1 \ln GDP_t + b_2 \ln pm_t + b_3 D_t + e_t \quad (8.18)$$

The coefficient b_3 gives the size of the shift in 1979. The constant in this equation is now given by $b_0 + b_3 D_t$, which evaluates to b_0 for 1973–79, when $D_t = 0$, and to $b_0 + b_3$ thereafter, when $D_t = 1$. The sign of b_3 shows the direction of any shift, and one can also test its significance, via the *t* ratio. If it turns out not to be significant, then there was probably no shift in the relationship.

Note that we do not use the log of D – this would be impossible as ln 0 is not defined. In any case, a dummy variable only needs to have two different values, it does not matter what they are (although 0, 1 is convenient for interpretation). Note also that b_3 will give the change in ln m, which is approximately[9] the percentage change in m.

[9]This approximation is reasonably accurate for values of b between -0.3 and $+0.3$. It is more accurate to calculate $e^b - 1$ to obtain the percentage change, especially outside this range of values.

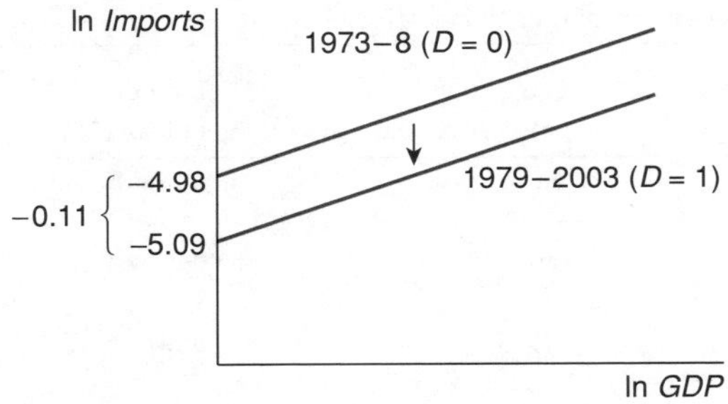

Figure 8.8
The dummy variable effect

Estimating equation (8.18) yields the following result:

$$\ln m_t = -4.98 + 1.85 \ln GDP_t - 0.35 \ln pm_t - 0.11 D_t + e_t \qquad \textbf{(8.19)}$$
$$\text{s.e.} \qquad (0.12) \qquad (0.12) \qquad (0.02)$$
$$R^2 = 0.99 \quad F(3, 27) = 1029.1 \quad n = 31$$

We note that the dummy variable has a significant coefficient and that after 1979 imports were 11% lower than before, after taking account of any price and income effects. We presume the oil shock has caused this, but in fact it could be due to anything that changed in 1979. Figure 8.8 shows the effect of introducing such a dummy variable, and from the figure we can see that the effect of the dummy variable is to shift the regression line downwards for the years from 1979 onwards.

Trap!

There were, in fact, two oil shocks – in 1973 and 1979. With a longer series of data you might therefore be tempted to use a dummy variable {0, 0, . . . , 0, 1, . . . , 1, 2, . . . , 2}, with the first switch in 1973, the second in 1979 (this assumes you have some pre-1973 observations). This is incorrect because it implicitly assumes that the two shocks had the same effect upon the dependent variable. The correct technique is to use two dummies, both using only zeros and ones. The first dummy would switch from 0 to 1 in 1973, the second would switch in 1979. Their individual coefficients would then measure the size of each shock.

A **time trend** is another useful type of dummy variable used with time-series data. It takes the values {1, 2, 3, 4, . . . , T} where there are T observations. It is used as a proxy for a variable which we cannot measure and which we believe increases in a linear fashion. For example, suppose we are trying to model petrol consumption of cars. Price and income would obviously be relevant explanatory variables; but in addition, technical progress has made cars more fuel-efficient over time. It is difficult to measure this accurately, so we use a time trend as an additional regressor. In this case it should have a negative coefficient which would measure the annual reduction in consumption due to more fuel-efficient cars. Remember also that if the dependent variable is in logs, the coefficient on the time trend shows the percentage change per annum (or per time period), e.g. a coefficient of −0.05 would indicate a 5% p.a. fall in the dependent variable, independent of movements in other explanatory variables.

Exercise 8.6

(a) The graph of car travel suggests a possible break in 1990. Test whether this break is significant or not using a dummy variable with a value of 0 up to (and including) 1990, 1 thereafter. (Estimate in logs using all three prices and income, 1980–99.)

(b) The quality of cars has improved steadily over time, perhaps leading to increased travel by car. Add a time trend to the regression equation in part (a) and re-estimate. Is there evidence to support this idea?

Multicollinearity

Sometimes some or all of the explanatory variables are highly correlated (in the sample data), which means that it is difficult to tell *which* of them is influencing the dependent variable. This is known as **multicollinearity**. Since all variables are correlated to some degree, multicollinearity is a problem of degree also. For example, if GDP and import prices both rise over time, it may be difficult to tell which of them influences imports. There has to be some independent movement of the explanatory variables for us to be able to disentangle their separate influences.

The symptoms of multicollinearity are:

- high correlation between two or more of the explanatory variables
- high standard errors of the coefficients leading to low *t* ratios
- a high value of R^2 (and significant F statistic) in spite of the insignificance of the individual coefficients.

In this situation, one might conclude that a variable is insignificant because of a large standard error, when, in fact, multicollinearity is to blame. It may be useful, therefore, to examine the correlations between all the explanatory variables to see if such a problem is apparent. For example, the correlation between nominal import prices and the retail price index is 0.97. Hence, it may be difficult to disentangle their individual effects.

The best cure is to obtain more data which might exhibit more independent variation of the explanatory variables. This is not always possible, however, for example if a sample survey has already been completed. An alternative is to drop one of the correlated variables from the regression equation, although the choice of which to exclude is somewhat arbitrary. Another procedure is to obtain alternative estimates of the effects of one of the collinear variables (for example, from another study). These effects can then be allowed for when estimates of the remaining coefficients are made.

Measurement error

The variables in a regression equation are rarely measured without error. Chapter 9 has an example showing that the measured balance of payments (exports minus imports) for 1970 varied considerably as better information became available in later years. The question therefore arises whether **measurement error** in the data is a problem for our estimates.

Measurement error could either affect the standard errors of the regression coefficients or, worse, it could cause bias. If the measurement error is systematic rather than random, then biased estimates can arise. For example, if transport costs are left out of the measured price of imported goods, and these costs have declined over time, then there is systematic measurement error in the price variable. The actual price fall (as experienced by the consumer) is less than the measured fall. The estimated price elasticity of imports might therefore have a bias away from zero (consumers have reacted to a price fall but we attribute this to a smaller price change, hence we estimate that price changes have a bigger effect than they really do).

Even when the measurement error is random, this can bias coefficient estimates. Such error in the explanatory variable generally biases the coefficient towards zero and lowers the t statistic. Hence, one might mistakenly conclude that X does not affect Y when in fact it does. The more severe the problem is, the greater is the variance of the measurement error relative to the variance of X. For example, if the variance of measurement error is 25% the size of the variance of X, then the coefficient of X will be biased downward by about 20%.

Exercise 8.7

?

We noted in Exercise 8.5 that rail and bus prices were highly correlated. This may be why they both appear to be 'insignificant' in the regression equation. It could be the case that either of them could be influencing demand, but we cannot tell which. We can examine this by testing the hypothesis H_0: $\beta_{rail} = \beta_{bus} = 0$. The restricted regression therefore excludes these two variables, the unrestricted regression includes them. One can then use equation (8.15) with $q = 2$ restrictions to test the hypothesis. What is the result? (Do not include dummy or trend in the equation.)

Some final advice on regression

- As always, large samples are better than small. Reasonable results were obtained above with only 31 observations, but this is rather a small sample size on which to base solid conclusions.
- Check the data carefully before calculation. This is especially true if a computer is used to analyse the data. If the data are entered incorrectly, *every* subsequent result will be wrong. A substantial part of any research project should be devoted to verifying the data, checking the definitions of variables, etc. The work is tedious, but important.
- Do not go fishing. This is searching through the data hoping something will turn up. Some idea of what the data are expected to reveal, and why, allows the search to be conducted more effectively. It is easy to see imaginary patterns in data if an aimless search is being conducted. Try looking at the table of random numbers (Table A1, see page 448), which will probably soon reveal something 'significant', like your telephone number or your credit card number.
- Do not be afraid to start with fairly simple techniques. Draw a graph of demand against price to see what it looks like, if it looks linear or log linear, if there are any outliers (a data error?), etc. This will give an overview of the problem which can be kept in mind when more refined techniques are used.

Summary

- Multiple regression extends the principles of simple regression to models using several explanatory variables to explain variation in Y.
- The multiple regression equation is derived by minimising the sum of squared residuals, as in simple regression. This principle leads to the formulae for slope coefficients, standard errors, etc.

- The significance of the individual slope coefficients can be tested using the *t* distribution and the overall significance of the model is based on the *F* distribution.
- It is important to check the adequacy of the model. This can be done in various ways, including examining the accuracy of predictions and checking that the residuals appear random.
- One important form of non-randomness is termed autocorrelation, where the error in one period is correlated with earlier errors (this can occur in time-series data). This can lead to incorrect inferences being drawn.
- The Durbin–Watson statistic is one diagnostic test for autocorrelation. If there is a problem of autocorrelation, it can often be eliminated by including lagged regressors.
- A good model should be (i) consistent with economic (or some other) theory, (ii) statistically satisfactory and (iii) simple. Sometimes there is a trade-off between these different criteria.
- Complex hypothesis tests can often be performed by comparing restricted and unrestricted forms of the model. If the former fits the data almost as well as the latter, then the (simplifying) restrictions specified in the null hypothesis are accepted.
- Omitting relevant explanatory variables from the model is likely to cause bias to the estimated coefficients. This suggests it is often best to start off with a fairly general model and simplify it.
- Regression analysis can become very complicated (well beyond the scope of this text), involving issues such as multicollinearity and simultaneous equations. However, the methods given in this chapter can provide helpful insights into a range of problems, especially if the potential shortcomings of the model are appreciated.

Key terms and concepts

autocorrelation
Chow test
dummy variables
Durbin–Watson statistic
measurement error
multicollinearity
negative autocorrelation
omitted variable bias
positive autocorrelation
regression coefficients
simultaneous equation model
spurious regression
time trend

Reference

Maddala, G.S., and K. Lahiri, *Introduction to Econometrics*, 4th edn, Wiley, 2009.

Formulae used in this chapter

Formula	Description	Notes
$F = \dfrac{(ESS_P - ESS_1)/n_2}{ESS_1/(n_1 - k - 1)}$	Chow test for a prediction	First n_1 observations used for estimation, last n_2 for prediction
$DW = \dfrac{\sum (e_t - e_{t-1})^2}{\sum e_t^2}$	Durbin–Watson statistic for testing autocorrelation	
$F = \dfrac{(ESS_R - ESS_U)/q}{ESS_U/(n - k - 1)}$	Test statistic for testing q restrictions in the regression model	$v = q, n - k - 1$

Problems

Some of the more challenging problems are indicated by highlighting the problem number in colour.

8.1 (a) Using the data in Problem 7.1 (page 301), estimate a multiple regression model of the birth rate explained by GNP, the growth rate and the income ratio. Comment upon:

(i) the sizes and signs of the coefficients,

(ii) the significance of the coefficients,

(iii) the overall significance of the regression.

(b) How would you simplify the model?

(c) Test for the *joint* significance of the coefficients on growth and the income ratio.

(d) Repeat the above steps for all 26 observations. Comment.

(e) Do you feel your understanding of the birth rate is improved after estimating the multiple regression equation?

(f) What other possible explanatory variables do you think it might be worth investigating?

8.2 The following data show the number of adults in each of 17 households and whether or not the family contains at least one person who smokes, to supplement the data in Problem 7.2 on alcohol spending (see page 301).

Family	Adults	Smoker	Family	Adults	Smoker
1	2	0	10	2	1
2	2	0	11	1	1
3	1	1	12	4	1
4	2	0	13	2	1
5	1	0	14	2	0
6	2	1	15	3	0
7	2	0	16	1	1
8	2	0	17	2	0
9	1	0			

(a) Estimate a multiple regression model of expenditure on alcohol, using income, the number of adults and whether there is a smoker in the household as the three explanatory variables. Do the coefficients have the expected signs? Interpret your results.

(b) Test the significance of the individual coefficients and of the regression as a whole.

(c) Should the model be simplified?

(d) Calculate the elasticity of alcohol expenditure with respect to income. Do this first using the linear model and then one based on logarithms of these two variables (but keep 'Adults' and 'Smokers' in linear form).

(e) Estimate the effect of one additional adult in the family on their spending on alcohol.

(f) To compare the alcohol expenditure of families with a smoker and families without, one could try two different methods:

(i) use a two sample *t* test to compare mean expenditures, or

(ii) run a regression of alcohol expenditure on the 'Smoker' variable, using all 17 observations.

Try both of these methods. What do you conclude?

8.3 Using the results from Problem 8.1, forecast the birth rate of a country with the following characteristics: GNP equal to $3000, a growth rate of 3% p.a. and an income ratio of 7. (Construct the point estimate only).

8.4 (This problem continues the analysis from Problem 8.2.) Given the following data for a family:

Family	Income	Adults	Smoker
18	700	2	1

(a) Predict the level of alcohol expenditure for this family.

(b) If their actual expenditure turned out to be 32.50, how accurate would you judge the prediction?

8.5 How would you most appropriately measure the following variables?

(a) social class in a model of alcohol consumption

(b) crime

(c) central bank independence from political interference.

8.6 As Problem 8.5, for

(a) the output of a car firm, in a production function equation,

(b) potential trade union influence in wage bargaining,

(c) the performance of a school.

8.7 Would it be better to use time-series or cross-section data in the following models?

(a) the relationship between the exchange rate and the money supply

(b) the determinants of divorce

(c) the determinants of hospital costs.

Explain your reasoning.

8.8 As Problem 8.7, for

(a) measurement of economies of scale in the production of books,

(b) the determinants of cinema attendances,

(c) the determinants of the consumption of perfume.

8.9 How would you estimate a model explaining the following variables?

(a) airline efficiency,

(b) infant mortality,

(c) bank profits.

You should consider such issues as whether to use time-series or cross-section data; the explanatory variables to use and any measurement problems; any relevant data transformations; the expected results.

8.10 As Problem 8.9, for

(a) investment,

(b) the pattern of UK exports (i.e. which countries they go to),

(c) attendance at football matches.

8.11 R. Dornbusch and S. Fischer (in R.E. Caves and L.B. Krause, *Britain's Economic Performance*, Brookings, 1980) report the following equation for predicting the UK balance of payments:

$$B = 0.29 + 0.24U + 0.17 \ln Y - 0.004t - 0.10 \ln P - 0.24 \ln C$$
$$t \quad (.56) \quad (5.9) \quad (2.5) \quad (3.8) \quad (3.2) \quad (3.9)$$
$$R^2 = 0.76, s_e = 0.01, n = 36 \text{ (quarterly data 1970: 1–1978: 1)}$$

where

B: the current account of the balance of payments as a percentage of gross domestic product (a balance of payments deficit of 3% of GDP would be recorded as −3.0, for example)
U: the rate of unemployment
Y: the OECD index of industrial production
t: a time trend
P: the price of materials relative to the GDP deflator (price index)
C: an index of UK competitiveness (a lower value of the index implies greater competitiveness).

(ln indicates the natural logarithm of a variable)

(a) Explain why each variable is included in the regression. Do they all have the expected sign for the coefficient?

(b) Which of the following lead to a higher BOP deficit (relative to GDP): (i) higher unemployment; (ii) higher OECD industrial production; (iii) higher material prices and (iv) greater competitiveness?

(c) What is the implied shape of the relationship between B and (i) U, (ii) Y?

(d) Why cannot a double log equation be estimated for this data? What implications does this have for obtaining elasticity estimates? Why are elasticity estimates not very useful in this context?

(e) Given the following values of the explanatory variables, estimate the state of the current account (point estimate): unemployment rate = 10, OECD index = 110, time trend = 37, materials price index = 100, competitiveness index = 90.

8.12 In a cross-section study of the determinants of economic growth (National Bureau of Economic Research, *Macroeconomic Annual*, 1991), Stanley Fischer obtained the following regression equation:

$$GY = 1.38 - \underset{(-5.9)}{0.52RGDP70} + \underset{(2.69)}{2.51PRIM70} + \underset{(3.91)}{11.16INV} - \underset{(2.7)}{4.75INF} + \underset{(4.34)}{0.17SUR}$$
$$- \underset{(-0.79)}{0.33DEBT80} - \underset{(-3.71)}{2.02SSA} - \underset{(-3.76)}{1.98LAC}$$
$$R^2 = 0.60, n = 73$$

where

GY:	growth per capita, 1970–85
RGDP:	real GDP per capita, 1970
PRIM70:	primary school enrolment rate, 1970
INV:	investment/GNP ratio
INF:	inflation rate
SUR:	budget surplus/GNP ratio
DEBT80:	foreign debt/GNP ratio
SSA:	dummy for sub-Saharan Africa
LAC:	dummy for Latin America and the Caribbean.

(a) Explain why each variable is included. Does each have the expected sign on its coefficient? Are there any variables which are left out, in your view?

(b) If a country were to increase its investment ratio by 0.05, by how much would its estimated growth rate increase?

(c) Interpret the coefficient on the inflation variable.

(d) Calculate the *F* statistic for the overall significance of the regression equation. Is it significant?

(e) What do the *SSA* and *LAC* dummy variables tell us?

8.13 **(Project)** Build a suitable model to predict car sales in the United Kingdom. You should use time-series data (at least 20 annual observations). You should write a report in a similar manner to Problem 7.12.

Answers to exercises

Exercise 8.1

(a) Demand rises rapidly until around 1990, then rises more slowly

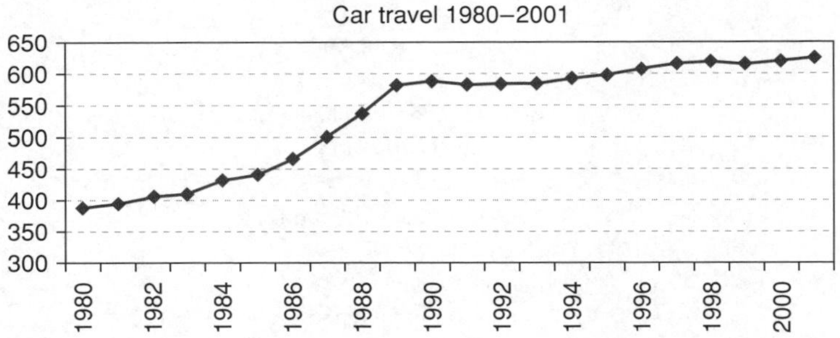

Price falls quite quickly until 1990, then rises. This may relate to the pattern of travel demand, above.

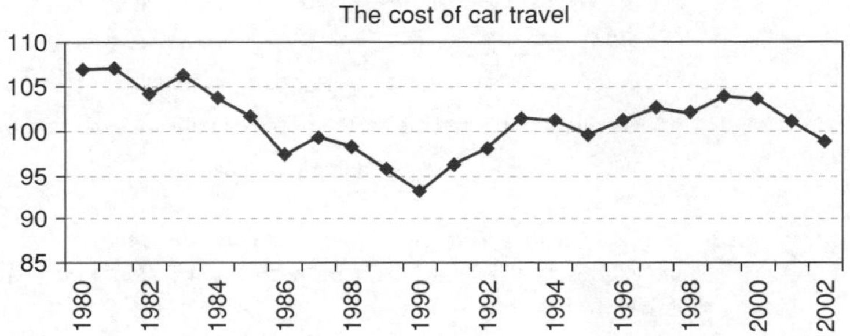

(b) The cross-plot of travel (vertical axis) against price is not clear-cut. There may be a slight negative relationship.

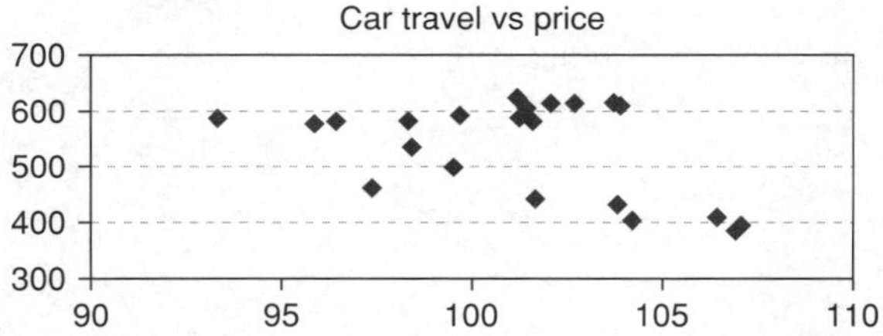

Again, there is not an obvious bivariate relationship between travel and income. For both graphs it looks as if a single line might not represent all of the data and therefore it might be that both explanatory variables are needed. The bivariate graphs are not very informative.

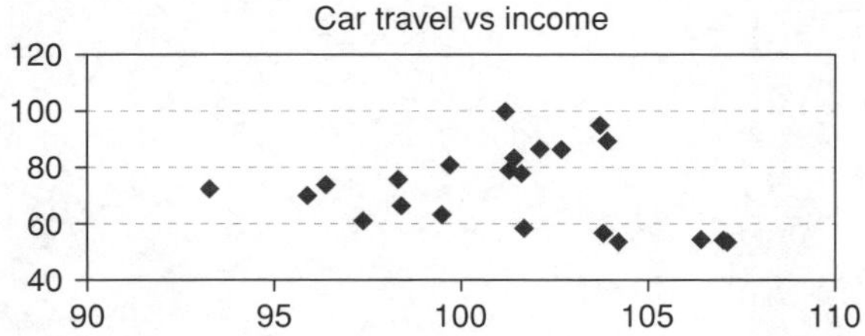

(c) Economic theory would suggest a negative price coefficient and a positive income coefficient.

(d) If bus and rail are substitutes for car travel, one would expect positive coefficients on their prices. However, they might be complements – commuters may drive to the station to catch the train.

Exercise 8.2

(a) The regression is:[10]

```
  Source |       SS      df           MS           Number of obs  =      20
---------+-----------------------------            F( 2, 17)      =  483.10
   Model |  138136.46     2       69068.23         Prob > F       =  0.0000
Residual |    2430.48    17         142.96         R-squared      =  0.9827
---------+-----------------------------            Adj R-squared  =  0.9807
   Total |  140566.95    19        7398.26         Root MSE       =   11.96
---------------------------------------------------------------------------
     car |      Coef.  Std. Err.       t    P>|t|   [95% Conf.  Interval]
---------+-----------------------------------------------------------------
   rpcar |      -6.39        .76   -8.37    0.000      -8.00        -4.77
    rpdi |       6.04        .23   25.85    0.000       5.55         6.54
   _cons |     748.11      83.85    8.92    0.000     571.18       925.03
---------+-----------------------------------------------------------------
```

(b) The signs of the coefficients are as expected. A unit increase in price lowers demand by 6.4 units; a unit rise in income raises demand by about 6 units. Without knowledge of the units of measurement it is hard to give a more precise interpretation. Both coefficients are highly significant, as is the *F* statistic. Ninety-eight per cent of the variation of car travel demand is explained by these two variables, a high figure.

Exercise 8.3

(a) The forecast values are 661.9 and 706.3 in 2000 and 2001. These compare with actual values of 618 and 624, so the errors are −6.6% and −11.7%. Assuming 2000 and 2001 would be the same as 1999 would actually give better results.

(b) In logs the results are:

```
  Source |       SS      df           MS           Number of obs  =      20
---------+-----------------------------            F( 2, 17)      =  599.39
   Model | .557417045     2     .278708523         Prob > F       =  0.0000
Residual | .007904751    17     .000464985         R-squared      =  0.9860
---------+-----------------------------            Adj R-squared  =  0.9844
   Total | .565321796    19     .029753779         Root MSE       =  .02156
---------------------------------------------------------------------------
    lcar |      Coef.  Std. Err.       t    P>|t|   [95% Conf.  Interval]
---------+-----------------------------------------------------------------
  lrpcar |      -1.19        .14   -8.45    0.000      -1.49         -.89
   lrpdi |        .84        .03   28.35    0.000        .78          .90
   _cons |       8.19        .71   11.61    0.000       6.70         9.68
---------+-----------------------------------------------------------------
```

Demand is elastic with respect to price ($e = -1.19$) and slightly less than elastic for income ($e = 0.84$). The coefficients are again highly significant.

[10] These results were produced using Stata. The layout is similar to that of Excel. Prob-values are indicated by 'Prob > F' and 'P > |t|'. 'rpcar' indicates the real price of car travel, 'rpdi' indicates real personal disposable income. Later on, an 'l' in front of a variable name indicates it is in log form.

(c) Price and income elasticities from the linear model are $-6.4 \times 101.1/526.5 = -1.23$ and $6.0 \times 70.2/526.5 = 0.8$. These are very similar to the log coefficients.

(d) The forecasts in logs are 6.492 and 6.561 which translate into 659.8 and 706.8. The predictions (and errors) are similar to the linear model.

(e) For the linear model the Chow test is

$$F = \frac{(ESS_P - ESS_1)/n_2}{ESS_1/(n_1 - k - 1)} = \frac{(7672.6 - 2430.5)/2}{2430.5/(20 - 2 - 1)} = 18.3$$

The critical value is $F(2, 17) = 3.59$, so there appears to be a change between estimation and forecast periods. A similar calculation for the log model yields an F statistic of 13.9 ($ESS_P = 0.0208$), also significant.

Exercise 8.4

(a) The residuals from the log regression are as follows:

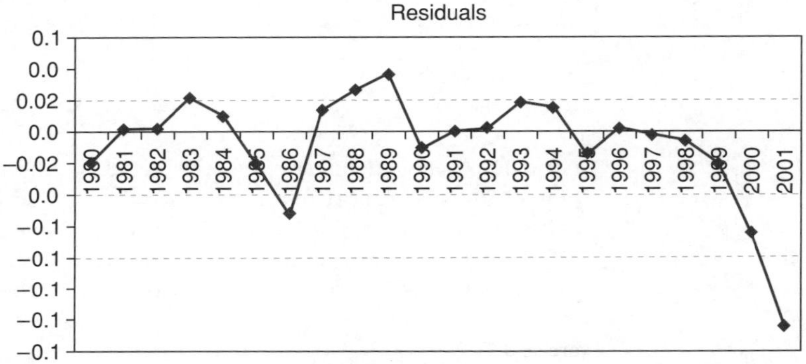

There is some evidence of positive autocorrelation and, in particular, the last two residuals (from the forecast period) are substantially larger than the rest.

(b) The Durbin–Watson statistic is DW = 1.52, against an upper critical value of $d_U = 1.54$. The test statistic (just) falls into the uncertainty region, but the evidence for autocorrelation is very mild.

(c) Autocorrelation would imply biased standard errors, so inference would be dubious, but the coefficients themselves are still unbiased.

Exercise 8.5

(a) The correlations are:

	rpcar	rpdi	rprail	rpbus
rpcar	1.0000			
rpdi	-0.3112	1.0000		
rprail	-0.1468	0.9593	1.0000	
rpbus	-0.1421	0.9632	0.9827	1.0000

The price of car travel has a low correlation with the other variables, which are all highly correlated with each other ($r > 0.95$).

(b) There may be omitted variable bias. Since the omitted variables are correlated with income, the income coefficient we have observed may be misleading. The car price variable is unlikely to be affected much, as it has a low correlation with the omitted variables.

(c) The results are:

Source	SS	df	MS		
Model	.557989155	4	.139497289	Number of obs =	20
Residual	.007332641	15	.000488843	F(4, 15) =	285.36
Total	.565321796	19	.029753779	Prob > F =	0.0000
				R-squared =	0.9870
				Adj R-squared =	0.9836
				Root MSE =	.02211

lcar	Coef.	Std. Err.	t	P>|t|	[95% Conf.	Interval]
lrpcar	-1.195793	.1918915	-6.23	0.000	-1.6048	-.786786
lrpdi	.8379483	.1372577	6.10	0.000	.5453904	1.130506
lrprail	.3104458	.3019337	1.03	0.320	-.3331106	.9540023
lrpbus	-.3085937	.3166891	-0.97	0.345	-.9836004	.3664131
_cons	8.22269	.7318088	11.24	0.000	6.662877	9.782503

The new price variables are not significant, so there is unlikely to have been a serious OVB problem. Neither car price nor income coefficients have changed. The simpler model seems to be preferred.

(d) The restricted equation is $y = \beta_1 + \beta_2 P_{car} + \beta_3 RPDI + \beta_4(P_{rail} + P_{bus}) + u$ (in logs) and estimating this yields $ESS_R = 0.007\,901$. The test statistic is therefore

$$F = \frac{(0.007\,901 - 0.007\,333)/1}{0.007\,333/(20 - 4 - 1)} = 1.16$$

This is not significant, so the hypothesis of equal coefficients is accepted.

Exercise 8.6

(a) The result is:

Source	SS	df	MS		
Model	.558588344	5	.111717669	Number of obs =	20
Residual	.006733452	14	.000480961	F(5, 14) =	232.28
Total	.565321796	19	.029753779	Prob > F =	0.0000
				R-squared =	0.9881
				Adj R-squared =	0.9838
				Root MSE =	.02193

lcar	Coef.	Std. Err.	t	P>|t|	[95% Conf.	Interval]
lrpcar	-1.107049	.2062769	-5.37	0.000	-1.549469	-.6646293
lrpdi	.8898566	.1438706	6.19	0.000	.581285	1.198428
lrprail	.5466294	.3667016	1.49	0.158	-.2398673	1.333126
lrpbus	-.4867887	.3523676	-1.38	0.189	-1.242542	.2689648
d1990	-.0314327	.0281614	-1.12	0.283	-.091833	.0289676
_cons	7.352081	1.065511	6.90	0.000	5.066787	9.637375

The new coefficient, −0.03, suggests car travel is 3% lower after 1990 than before, *ceteris paribus*. However, the coefficient is not significantly different from zero, so there is little evidence of structural break. The change in car usage appears due to changes in prices and income.

(b) The result is:

```
    Source |       SS       df       MS              Number of obs =      20
-----------+------------------------------           F( 6, 13)     =  191.34
     Model | .558991816      6  .093165303           Prob > F      =  0.0000
  Residual |  .00632998     13  .000486922           R-squared     =  0.9888
-----------+------------------------------           Adj R-squared =  0.9836
     Total | .565321796     19  .029753779           Root MSE      =  .02207
------------------------------------------------------------------------------
      lcar |      Coef.   Std. Err.       t     P>|t|    [95% Conf. Interval]
-----------+------------------------------------------------------------------
    lrpcar |  -1.116536   .2078126     -5.37    0.000   -1.565488   -.6675841
     lrpdi |   1.107112   .2791366      3.97    0.002    .5040736     1.71015
   lrprail |    .558322   .3691905      1.51    0.154   -.2392655    1.355909
    lrpbus |  -.2707759   .4266312     -0.63    0.537   -1.192457    .6509048
     d1990 |   -.036812   .0289451     -1.27    0.226    -.099344      .02572
     trend |  -.0099434   .0109234     -0.91    0.379    -.033542    .0136552
     _cons |   5.553859   2.247619      2.47    0.028    .6981737    10.40954
-----------+------------------------------------------------------------------
```

The trend is not significant. Note that the income coefficient has changed substantially. This is due to the high correlation between income and the trend ($r = 0.99$). It seems preferable to keep income and exclude the trend.

Exercise 8.7

The F statistic is

$$F = \frac{(0.007905 - 0.007333)/2}{0.007333/(20 - 4 - 1)} = 0.59$$

This is less than the critical value of $F(2, 15) = 3.68$, so the hypothesis that both coefficients are zero is accepted.

9 Data collection and sampling methods

Contents

Learning outcomes

By the end of this chapter you should be able to:

- **recognise the distinction between primary and secondary data sources**
- **avoid a variety of common pitfalls when using secondary data**
- **make use of electronic sources to gather data**

→

- recognise the main types of random sample and understand their relative merits
- appreciate how such data are collected
- conduct a small sample survey yourself.

Introduction

It may seem a little odd to look at data collection now, after several chapters covering the analysis of data. Collection of data logically comes first, but the fact is that most people's experience is as a user of data, which determines their priorities. Also, it is difficult to have the motivation for learning about data collection when one does not know what it is subsequently used for. Having spent considerable time learning how to analyse data, it is now time to look at its collection and preparation.

There are two reasons why you might find this chapter useful. First, it will help if you have to carry out some kind of survey yourself. Second, it will help you in your data analysis, even if you are using someone else's data. Knowing the issues involved in data collection can help your judgement of the quality of the data you are using. The material in this chapter has been reorganised a little for this edition to reflect the fact that most data are nowadays available in electronic format and that collecting numbers from dusty tomes of statistics is now a rarity.

When conducting statistical research, there are two ways of obtaining data:

(1) use **secondary data** sources, such as from the World Bank, or
(2) collect sample data personally, a **primary data** source.

Using secondary data sources sounds simple, but it is easy to waste valuable time by making elementary errors. The first part of this chapter provides some simple advice to help you avoid such mistakes.

Much of this text has been concerned with the analysis of sample evidence and the inferences that can be drawn from it. It has been stressed that this evidence must come from randomly drawn samples and, although the notion of randomness was discussed in Chapter 2, the practical details of random sampling have not been set out.

The second part of this chapter is therefore concerned with the problems of collecting sample survey data prior to its analysis. The decision to collect the data personally depends upon the type of problem faced, the current availability of data relating to the problem and the time and cost needed to conduct a survey. It should not be forgotten that the first question that needs answering is whether the answer obtained is worth the cost of finding it. It is probably not worthwhile for the government to spend £50 000 to find out how many biscuits people eat, on average (though it may be worth biscuit manufacturers doing this). The sampling procedure is always subject to some limit on cost, therefore, and the researcher is trying to obtain the best value for money. The emergence of online surveys has greatly altered the cost of collecting data and allowed very large samples to be gathered. However, there are particular issues around such data.

Using secondary data sources

Much of the research in economics and finance is based on secondary data sources, i.e. data which the researcher did not collect himself or herself. The data may be in the form of official statistics such as those published in *Economic Trends* or they may come from unofficial surveys. In either case one has to use the data as presented; there is no control over sampling procedures. Nevertheless, knowing how the data were collected is likely to influence how we analyse and interpret the data. For example, if we know that a particular group is under-represented in the sample, we might want to give more weight to those observations we do have.

It may seem easy enough to look up some figures in a publication or online, but there are a number of pitfalls for the unwary. The following advice comes from experience, some of it painful, and it may help you to avoid wasting time and effort. I have also learned much from the experiences of my students, whom I have also watched suffer.

Make sure you collect the right data

This may seem obvious, but most variables can be measured in a variety of different ways. Suppose you want to measure the cost of labour (over time) to firms. Should you use the wage rate or earnings? The latter includes payment for extra hours such as overtime payments and reflects general changes in the length of the working week. Is the wage measured per hour or per week? Does it include part-time workers? If so, a trend in the proportion of part-timers will bias the wage series. Does the series cover all workers, men only or women only? Again, changes in the composition might influence the wage series. What about tax and social security costs? Are they included? There are many questions one could ask.

One needs to have a clear idea, therefore, of the precise variable one needs to collect. This will presumably depend upon the issue in question. Economic theory might provide some guidance: for instance, theory suggests that firms care about *real* wage rates (i.e. after taking account of inflation, so related to the price of the goods the firm sells), so this is what one should measure. Check the definition of any series you collect (this is often in a separate supplement giving explanatory notes and definitions, or at the back of the printed publication). Make sure that the definition has not changed over the time period you require: the definition of unemployment used in the United Kingdom changed about 20 times in the 1980s, generally with the effect of reducing *measured* unemployment, even if actual unemployment was unaffected. In the United Kingdom the geographical coverage of data may vary: one series may relate to the United Kingdom, another to Great Britain and yet another to England and Wales. Care should obviously be taken if one is trying to compare such series.

To illustrate how the measurement of a variable can matter, Figure 9.1 shows two measures of inflation, the RPI and the CPI (the retail and the consumer price index, respectively). The major differences are that the RPI includes housing costs, whereas the CPI does not, and also that the former is calculated as an arithmetic mean of the separate items in the index whereas the CPI is a geometric mean. The UK has recently switched the indexation of certain benefits, including pensions,

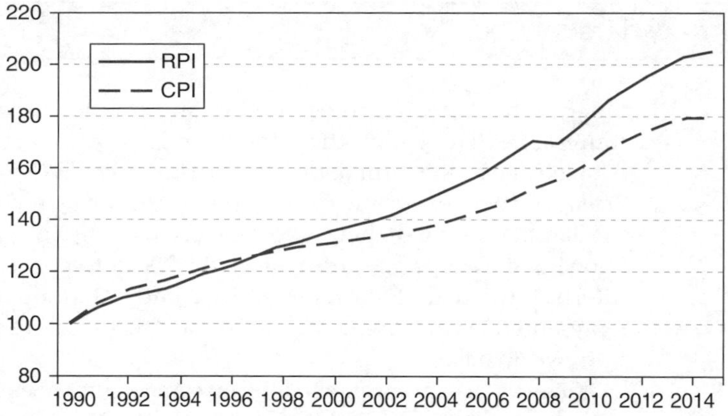

Figure 9.1
The CPI and RPI
Source: Adapted from Office of National Statistics (ONS) data.

from the RPI to the CPI. This means that a pension is worth about 11% less after 25 years, a substantial difference.

Get the most up-to-date figures

Many macroeconomic series are revised as more information becomes available. The balance of payments serves as a good example. The first edition of this text showed the balance of payments (current balance, in £m for the United Kingdom) for 1970, as published in successive years, as follows:

1971	1972	1973	1974	1975	1976	1977	1978	...	1986
579	681	692	707	735	733	695	731	...	795

The difference between the largest and smallest figures is of the order of 37%, a wide range. In the third edition of this text the figure was (from the 1999 edition of *Economic Trends Annual Supplement*) £911m, which is 57% higher than the initial estimate. The latest figure at the time of writing is £361m. Most series are better than this. The balance of payments is hard to measure because it is the small difference between two large numbers, exports and imports. A 5% increase in measured exports and a 5% decrease in measured imports could thus change the measured balance by 100% or more.

One should always try to get the most up-to-date figures, therefore, which often means working *backwards* through printed data publications, i.e. use the current issue first and get data back as far as is available, then get the previous issue to go back a little further, etc. This can be tedious but it will also give some idea of the reliability of the data from the size of data revisions. This should be less of a problem with electronic sources where you can download the latest data in one go.

Keep a record of your data sources

You should always keep *precise* details of where you obtained each item of data. If you need to go back to the original source (to check on the definition of a series, for example), you will then be able to find it easily. It is easy to spend hours (if not *days*) trying to find the source of some interesting numbers that you wish to update. For **online data** this means taking down the URL of the site you visit. Remember that some sites generate the page 'on demand', so the web address is not a permanent one and typing it in later on will not take you back to the same source. In these circumstances it may be better to note the 'root' part of the address (e.g. www.imf.org/data/) rather than the complete detail. You should also take a note of the date you accessed the site; this may be needed if you put the source into a bibliography.

STATISTICS IN PRACTICE

Keeping data in *Excel* or another spreadsheet

Spreadsheets are ideal for keeping your data. It is often a good idea to keep the data all together in one worksheet and extract portions of them as necessary and analyse them in another worksheet. Alternatively, it is usually quite easy to transfer data from the spreadsheet to another program (e.g. *SPSS* or *Stata*) for more sophisticated analysis. In most spreadsheets you can attach a comment to any cell, so you can use this to keep a record of the source of each observation, changes of definition, etc. Thus you can retain all the information about your data together in one place.

For printed sources you should retain the name of the publication, issue number or date, and table or page number. It also helps to keep the library reference number of the publication if it is obscure. It is best to take a photocopy of the data (but check copyright restrictions) rather than just copy it down, if possible.

STATISTICS IN PRACTICE

Tips on downloading data

- If you are downloading a spreadsheet, save it to your hard disk then include the URL of the source within the spreadsheet itself. You will always know where it came from. You can do the same with Word documents.
- You often cannot do this with PDF files, which are read-only, but if the document is not protected you can add a comment, containing the source address. Alternatively, you could save the file to your disk, including the URL within the file name. (Avoid putting extra full stops in the file name, that confuses the operating system. Replace them with hyphens.)
- You can use the 'Text select tool' within Acrobat to copy items of data from a PDF file and then paste them into a spreadsheet.
- Often, when pasting several columns of such data into Excel, all the numbers go into a single column. You can fix this using the Data, Text to Columns menu. Experimentation is required, but it works well.

Since there are now so many online sources (and they are constantly changing), a list of useful data sites rapidly becomes out of date. The following sites seem to have withstood the test of time so far and have a good chance of surviving throughout the life of this edition.

- The UK Office for National Statistics is at http://www.ons.gov.uk/ons/index.html and is a source of all official statistics. However, it is *hugely* frustrating to use (even after a recent re-design) and very difficult to find what you want. The 'Time series explorer' section of the site is a little better.

- It is easier to access time-series ONS data via the Econstats website (http://www.econstats.com/) which puts the data into some kind of manageable order. It also contains similar data for a range of other countries.
- Another alternative is the OECD site at http://stats.oecd.org/. This is a well-organised site which allows easy customisation of the data you want and then allows downloading the data in Excel or other formats.
- The IMF's World Economic Database is at http://www.imf.org/ (follow the link to Data). It has macroeconomic series for most countries going back many years.
- The Penn World Tables are at http://cid.econ.ucdavis.edu/pwt.html and provide national accounts data on a comparable basis for most countries.
- The World Bank provides a lot of information, particularly relating to developing countries, at http://www.worldbank.org. It is now very easy to download the data into a spreadsheet or specialist software such as Stata (using the 'wbopendata' add-in, see the WB website).
- Bill Goffe's Resources for Economists site (http://rfe.org) contains a data section which is a good starting-off point for data sources.
- Financial and business databases are often supplied by commercial enterprises and hence are not freely available. One useful free (or partially free) site is *Yahoo Finance* (http://finance.yahoo.com/).
- Google is possibly the most useful website of all. Intelligent use of this search tool is often the best way to find what you want.

With the continuing development of the web there is also the ability now to 'scrape' data from websites such as Amazon, Facebook, etc. Such data can be useful for both businesses and academic researchers. For a useful guide see Edelman (2012).

Check your data

Once you have collected your data, you must check it. Once you have done this, you must check it again. Better, get someone else to help with the second check. Note that if your data are wrong, then all your subsequent calculations could be incorrect and you will have wasted much time. I have known many students who have spent months or even years on a dissertation or thesis who have then found an error in the data they collected earlier.

Obtaining data electronically should avoid input errors and provide consistent, up-to-date figures. However, this is not always guaranteed. For example, the UK Office for National Statistics (ONS) online databank provides plenty of information, but some of the series clearly have breaks in them, and there is little warning of this in the on-screen documentation. The series for revenue per admission to cinemas (roughly the price of admission) was obtained as follows:

1963	1964	1965	1966	1967
37.00	40.30	45.30	20.60	21.80

which strongly suggests an artificial break in the series in 1966 (especially as admissions *fell* by 12% between 1965 and 1966). Later in the series, the observations appear to be divided by 100. The lesson is that, even with electronic data, you should check the numbers to ensure they are correct[1].

A useful first step to check the data is to graph it (e.g. a time-series plot). Obvious outliers or other unexpected features will show up and you can investigate them for possible errors. Do not just rely on the graphs, however, look through your data and check it against the original source. It can then be useful to calculate some summary descriptive statistics; for example, does the average level of the weekly earnings data you have downloaded look approximately correct? Does the rate of growth of a time-series variable look reasonable? If not, go back and check. Better that you find an error in the data than someone else finds the error in your results.

Collecting primary data

Primary data are data that you have collected yourself from original sources, often by means of a sample survey. This has the advantage that you can design the questionnaire to include the questions of interest to you and you have total control over all aspects of data collection. You can also choose the size of the sample (as long as you have sufficient funds available) so as to achieve the desired width of any confidence intervals.

Almost all surveys rely upon some method of sampling, whether random or not. The probability distributions and formulae which have been used in previous chapters as the basis of the techniques of estimation and hypothesis testing rely upon the samples having been drawn at random from the population. If this is not the case, then the formulae for confidence intervals, hypothesis tests, etc., are incorrect and not strictly applicable (they may be reasonable approximations, but it is difficult to know how reasonable). In addition, the results about the bias and precision of estimators will be incorrect. For example, suppose an estimate of the average expenditure on servicing by car owners were obtained from a sample survey. A poor estimate would arise if only Rolls-Royce owners were sampled, since they are not representative of the population as a whole. These are expensive cars, with commensurately high servicing costs, not representative of the majority of cars on the road.

Thus, some form of random sampling method is needed to be able to use the theory of the probability distributions of random variables. Nor should it be believed that the theory of random sampling can be ignored if a very large sample is taken, as the following cautionary tale shows. In 1936 the *Literary Digest* tried to predict the result of the forthcoming US election by sending out 10 million mail questionnaires. Two million were returned, but even with this enormous sample size Roosevelt's vote was incorrectly estimated by a margin of 19 percentage points, and the *Digest* predicted the wrong candidate winning the presidency. The problem here was two-fold: first, the questionnaires may have not been sent out to a representative sample of voters, and second, those who respond to questionnaires are not necessarily a random sample of those who receive them.

[1] I wrote this for a previous edition of this text. I can no longer find the same data on the ONS site, it seems to have disappeared into the ether.

Random sampling

The definition of random sampling is that every element of the population should have a known, non-zero probability of being included in the sample. The problem with the sample of cars used above was that Ford cars (for example) had a zero probability of being included. Many sampling procedures give an equal probability of being selected to each member of the population, but this is not an essential requirement. It is possible to adjust the sample data to take account of unequal probabilities of selection. If, for example, Rolls-Royce had a much greater chance of being included than Ford, then the estimate of the population mean (of servicing costs) would be calculated as a weighted average of the sample observations, with greater weight being given to the few 'Ford' observations than to relatively abundant 'Rolls-Royce' observations. A very simple illustration of this is given below. Suppose that for the population we have the following data:

	Rolls-Royce	Ford
Number in population	20 000	2 000 000
Annual servicing cost	£1000	£200

Then the true average repair bill is

$$\mu = \frac{20\,000 \times 1000 + 2\,000\,000 \times 200}{2\,020\,000} = 207.92$$

Suppose the sample data are as follows:

	Rolls-Royce	Ford
Number in sample	20	40
Probability of selection	1/1000	1/50 000
Servicing cost	£990	£205

To calculate the average cost from the sample data we use a weighted average, using the relative population sizes as weights, not the sample sizes:

$$\bar{x} = \frac{20\,000 \times 990 + 2\,000\,000 \times 205}{2\,020\,000} = 212.77$$

If the sample sizes were used as weights, the average would come out at £466.67, which is substantially incorrect.

As long as the probability of being in the sample is known (and hence the relative population sizes must be known), the weight can be derived; but if the probability is zero, this procedure breaks down.

Other theoretical assumptions necessary for deriving the probability distribution of the sample mean or proportion are that the population is of infinite size and that each observation is independently drawn. In practice the former condition is never satisfied, since no population is of infinite size, but most populations are large enough that it does not matter. For each observation to be independently drawn (i.e. the fact of one observation being drawn does not alter the probability of others in the sample being drawn) strictly requires that sampling be done with replacement, i.e. each observation drawn is returned to the population before the next observation is drawn. Again in practice this is often not the case, sampling being done without replacement, but again this is of negligible practical importance where the population is large relative to the sample.

On occasion, the population is quite small and the sample constitutes a substantial fraction of it. In these circumstances the **finite population correction (fpc)** should be applied to the formula for the variance of $\bar{x}$, the fpc being given by

$$\text{fpc} = (1 - n/N) \tag{9.1}$$

where N is the population size and n the sample size. The table below illustrates its usage:

Variance of $\bar{x}$ from infinite population	Variance of $\bar{x}$ from finite population	Example values of fpc			
		$n = 20$	25	50	100
		$N = 50$	100	1000	10 000
σ^2/n	$\sigma^2/n \times (1 - n/N)$	0.60	0.75	0.95	0.99

The finite population correction serves to narrow the confidence interval because a sample size of (say) 25 reveals more about a population of 100 than about a population of 100 000, so there is less uncertainty about population parameters. When the sample size constitutes only a small fraction of the population (e.g. 5% or less) the finite population correction can be ignored in practice. If the whole population is sampled ($n = N$) then the variance becomes zero and there is no uncertainty about the population mean.

A further important aspect of random sampling occurs when there are two samples to be analysed, when it is important that the two samples are independently drawn. This means that the drawing of the first sample does not influence the drawing of the second sample. This is a necessary condition for the derivation of the probability distribution of the difference between the sample means (or proportions).

Types of random sample

The meaning and importance of randomness in the context of sampling has been explained. However, there are various different types of sampling, all of them random, but which have different statistical properties. Some methods lead to greater precision of the estimates, while others can lead to considerable cost savings in the collection of the sample data, but at the cost of lower precision. The aim of sampling is usually to obtain the most precise estimates of the parameter in question, but the best method of sampling will depend on the circumstances of each case. If it is costly to sample individuals, a sampling method which lowers cost may allow a much larger sample size to be drawn and thus good (precise) estimates to be obtained, even if the method is inherently not very precise. These issues are investigated in more detail below, as we will look at a number of different methods:

- simple random sampling – the simplest type, on which our earlier formulae for confidence intervals, etc., were based
- stratified sampling – a method which ensures the sample is more representative of the population (ruling out some forms of unrepresentative samples)

- cluster sampling (a relatively cheap method of sampling, allowing a larger sample size)
- quota sampling (a non-random method, carried out because it is cheap)
- multi-stage sampling (a method which combines the advantages of some of the above types).

Simple random sampling

This type of sampling has the property that every possible sample that could be obtained from the population has an equal chance of being selected. This implies that each element of the population has an equal probability of being included in the sample, but this is not the defining characteristic of **simple random sampling**. As will be shown below, there are **sampling methods** where every member of the population has an equal chance of being selected, but some samples (i.e. certain combinations of population members) can never be selected.

The statistical methods in this text are based upon the assumption of simple random sampling from the population. It leads to the most straightforward formulae for estimation of the population parameters. Although many statistical surveys are not based upon simple random sampling, the use of statistical tests based on simple random sampling is justified since the sampling process is often hypothetical. For example, if one were to compare annual growth rates of two countries over a 30-year period, a z test on the difference of two sample means (i.e. the average annual growth rate in each country) would be conducted. In a sense the data are not a sample since they are the only possible data for those two countries over that time period. Why not just regard the data as constituting the whole population, therefore? Then it would just be a case of finding which country had the higher growth rate; there would be no uncertainty about it.

The alternative way of looking at the data would be to suppose that there exists some hypothetical population of annual growth rates and that the data for the two countries were drawn by (simple) random sampling from this population. Is this story consistent with the data available? In other words, could the data we have simply arise by chance? If the answer to this is no (i.e. the z score exceeds the critical value) then there is something causing a difference between the two countries (it may not be clear what that something is). In this case it is reasonable to assume that all possible samples have an equal chance of selection, i.e. that simple random sampling takes place. Since the population is hypothetical, one might as well suppose it to have an infinite number of members, again required by sampling theory.

Stratified sampling

Returning to the practical business of sampling, one problem with simple random sampling is that it is possible to collect 'bad' samples, i.e. those which are unrepresentative of the population. Suppose that we are investigating how often people buy a daily newspaper, where it is the case that older people tend to buy one more frequently (with younger people more likely to use the internet as a news source, for example). The situation can be illustrated in Figure 9.2.

Now suppose that we take a random sample from this population which, by bad luck or bad design, contains exclusively or predominately the young. It is

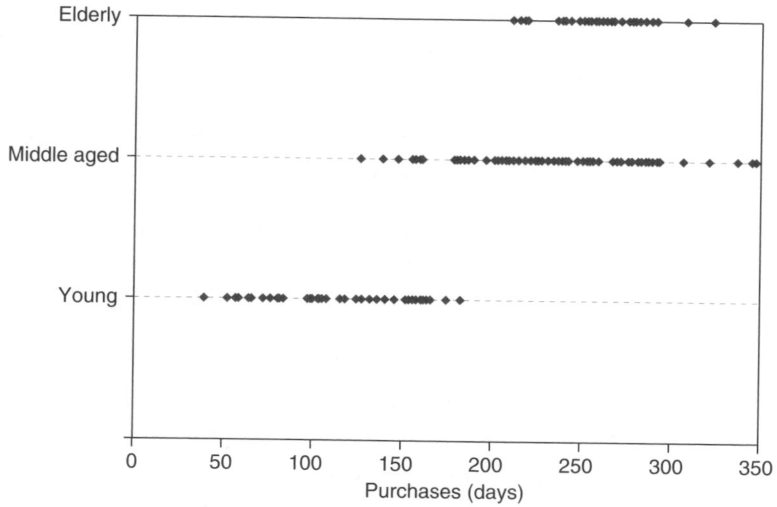

Figure 9.2
Differences in newspaper buying habits

clear that average frequency of purchase would be estimated at around 120 days per year and that this is an underestimate for the population as a whole.

To remedy this problem, we should ensure that each of the age groups is fairly represented in the sample. Hence, the proportion of younger people in the sample should reflect the proportion in the population, and the same goes for the other groups. The age groups are referred to as **strata** and each should be fairly represented in the sample. Suppose the strata are made up as follows:

Percentage of population in age group		
Elderly	*Middle aged*	*Young*
20%	50%	30%

Suppose a sample of size 100 is taken. With luck it would contain 20 older people, 50 who are middle-aged and 30 young people, and thus would be representative of the population as a whole. But if, by bad luck (or bad sample design), all 100 people in the sample were young, poor results will be obtained since newspaper buying differs between age groups.

To avoid this type of problem a stratified sample is taken, which ensures that all age groups are represented in the sample. This means that the survey would have to ask people about their age as well as their reading habits. The simplest form of stratified sampling is equiproportionate sampling, whereby a stratum which constitutes (say) 20% of the population also makes up 20% of the sample. For the example above the sample would be made up as follows:

Class	Elderly	Middle-aged	Young	Total
Number in sample	20	50	30	100

It should be clear why stratified sampling constitutes an improvement over simple random sampling, since it rules out 'bad' samples, i.e. those not representative of the population. It is simply impossible to get a sample consisting completely of young people. In fact, it is impossible to get a sample in anything but the proportions 20:50:30, as in the population; this is ensured by the method of collecting the sample.

It is easy to see when stratification leads to large improvements over simple random sampling. If there were no difference between strata (age groups) in buying habits, then there would be no gain from stratification. If behaviour were the same regardless of age group, there would be no point in dividing up the population according to that factor. On the other hand, if there were large differences between strata, but within strata reading habits were similar, then the gains from stratification would be large. (The fact that newspaper habits are similar within strata means that even a small sample from a single stratum should give an accurate picture of that stratum.)

Stratification is beneficial, therefore, when

- the between-strata differences are large, and
- the within-strata differences are small.

These benefits take the form of greater precision of the estimates, i.e. narrower confidence intervals[2]. The greater precision arises because stratified sampling makes use of supplementary information – i.e. the proportion of the population in each age group. Simple random sampling does not make use of this. Obviously, therefore, if those proportions of the population are unknown, stratified sampling cannot be carried out. However, even if the proportions are only known approximately, there could be a gain in precision.

In this example age is a **stratification factor**, i.e. a variable which is used to divide the population into strata. Other factors could, of course, be used, such as income or even height. A good stratification factor is one which is related to the subject of investigation. Income would probably be a good stratification factor, therefore, since it is likely to be related to reading habits, but height is not since there is probably little difference between tall and short people in newspaper purchases. What is a good stratification factor obviously depends upon the subject of study. A bed manufacturer might well find height to be a good stratification factor if conducting an enquiry into preferences about the size of beds. Although good stratification factors improve the precision of estimates, bad factors do not make them worse; there will simply be no gain over simple random sampling. It would be as if there were no differences between the age groups in reading habits, so that ensuring the right proportions in the sample is irrelevant, but it has no detrimental effects.

Proportional allocation of sample observations to the different strata (as done above) is the simplest method but is not necessarily the best. For the optimal allocation there should generally be a divergence from proportional allocation, and the sample should have more observations in a particular stratum (relative to proportional allocation):

- the more diverse the stratum, and
- the cheaper it is to sample the stratum.

[2]The formulae for calculating confidence intervals with stratified sampling are not given here, since they merit a whole book to themselves. The classic reference is C.A. Moser and G. Kalton, *Survey Methods in Social Investigation*, Heinemann, 1971. A more recent text is Barnett, *Sample Survey Principles and Methods*, Wiley, 2002.

Starting from the 20:50:30 proportional allocation derived earlier, suppose that older people are all very similar in their buying habits but that youngsters vary considerably in their behaviour. Then the representation of youngsters in the sample should be increased and that of older people reduced. If it were true that every elderly person bought a newspaper every day, then a single observation from that class would be sufficient to yield all there is to know about it. Furthermore, if it is cheaper to sample younger readers, perhaps because they are easier to contact than older people, then again the representation of youngsters in the sample should be increased. This is because, for a given budget, it will allow a larger total sample size.

Stratified sampling is also useful when it is desired to look in detail at some or all of the strata. For example, if one wanted to look at the variation in buying behaviour by the young, it would be inconvenient if one's sample contained (by chance) very few young people. Stratified sampling avoids this by ensuring that each stratum is adequately covered.

Surveying concert-goers

A colleague and I carried out a survey of people attending a concert in Brighton (by Jamiroquai – I hope they are still popular by the time you read this) to find out who they were, how much they spent in the town and how they travelled to the concert. The spreadsheet gives some of the results.

Microsoft Excel - Book1

	A	B	C	D	E	F
3						
4		Number	%	Cumulative %		
5	Work in Brighton	11	15.1	15.1		
6	Student in Brighton	19	26.0	41.1		
7	Work outside Brighton	42	57.5	98.6		
8	Not working	1	1.4	100.0		
9	Total	73	100.0			
10						
11	**Method of travelling to concert:**					
12		Number	%	Cumulative %		
13	Car	23	31.5	31.5		
14	Car share	24	32.9	64.4		
15	Bus/coach	4	5.5	69.9		
16	On foot	9	12.3	82.2		
17	Train	10	13.7	95.9		
18	Taxi	3	4.1	100.0		
19	Total	73	100.0			
20						

The data were collected by face-to-face interviews before the concert. We did not have a sampling frame, so the (student) interviewers simply had to choose the sample themselves on the night. The one important instruction about sampling we gave them was that they should not interview more than one person in any group. People in the same group are likely to be influenced by each other (e.g. travel together) so we would not get independent observations, reducing the effective sample size.

From the results you can see that 41.1% either worked or studied in Brighton and that only one person in the sample was neither working nor studying. The second half of the table shows that 64.4% travelled to the show in a car (obviously adding to congestion in the town), about half of whom shared a car ride. Perhaps surprisingly, Brighton residents were just as likely to use their car to travel as were those from out of town.

The average level of spending was £24.20, predominantly on food (£7.38), drink (£5.97) and shopping (£5.37). The last category had a high variance associated with it – many people spent nothing, one person spent £200 in the local shops.

Cluster sampling

A third form of sampling is **cluster sampling** which, although intrinsically inefficient, can be much cheaper than other forms of sampling, allowing a larger sample size to be collected. Drawing a simple or a stratified random sample of size 100 from the whole of Britain would be very expensive to collect since the sample observations would be geographically very spread out. Interviewers would have to make many long and expensive journeys simply to collect one or two observations. To avoid this, the population can be divided into 'clusters' (for example, regions or local authorities), and one or more of these clusters are then randomly chosen. Sampling takes place only within the selected clusters, is therefore geographically concentrated and the cost of sampling falls, allowing a larger sample to be collected for the same expenditure of money.

Within each cluster one can have either a 100% sample or a lower sampling fraction, which is called multi-stage sampling (this is explained further below). Cluster sampling gives unbiased estimates of population parameters but, for a given sample size, these are less precise than the results from simple or stratified sampling. This arises in particular when the clusters are very different from each other, but fairly homogeneous within themselves. In this case once a cluster is chosen, if it is unrepresentative of the population, a poor (inaccurate) estimate of the population parameter is inevitable. The ideal circumstances for cluster sampling are when all clusters are very similar, since in that case examining one cluster is almost as good as examining the whole population.

Dividing up the population into clusters and dividing it into strata are similar procedures, but the important difference is that sampling is from one or at most a few clusters, but from all strata. This is reflected in the characteristics which make for good sampling. In the case of stratified sampling, it is beneficial if the between-strata differences are large and the within-strata differences small. For cluster sampling this is reversed: it is desirable to have small between-cluster differences but heterogeneity within clusters. Cluster sampling is less efficient (precise) for a given sample size, but is cheaper and so can offset this disadvantage with a larger sample size. In general, cluster sampling needs a much larger sample to be effective, so is only worthwhile where there are significant gains in cost.

Multi-stage sampling

Multi-stage sampling was briefly referred to in the previous section and is commonly found in practice. It may consist of a mixture of simple, stratified and cluster sampling at the various stages of sampling. Consider the problem of selecting a random sample of 1000 people from a population of 25 million to find out about voting intentions. A nationwide simple random sample would be extremely expensive to collect, for the reasons given above, so an alternative method must be found. Suppose further that it is suspected that voting intentions differ according to whether one lives in the north or south of the country and whether one is a home owner or renter. How is the sample to be selected? The following would be one appropriate method.

First, the country is divided up into clusters of counties or regions, and a random sample of these taken, say one in five. This would be the first way of reducing the cost of selection, since only one-fifth of all counties now need to be visited. This one-in-five sample would be stratified to ensure that north and south were both appropriately represented. To ensure that each voter has an equal chance of being in the sample, the probability of a county being drawn should be proportional to its adult population. Thus, a county with twice the population of another should have twice the probability of being in the sample.

Having selected the counties, the second stage would be to select a random sample of local authorities within each selected county. This might be a 1-in-10 sample from each county and would be a simple random sample within each cluster. Finally, a selection of voters from within each local authority would be taken, stratified according to tenure. This might be a 1-in-500 sample. The sampling fractions would therefore be

$$\frac{1}{5} \times \frac{1}{10} \times \frac{1}{500} = \frac{1}{25\,000}$$

So from the population of 25 million voters a sample of 1000 would be collected. For different population sizes the sampling fractions could be adjusted so as to achieve the goal of a sample size of 1000.

The sampling procedure is a mixture of simple, stratified and cluster sampling. The two stages of cluster sampling allow the selection of 50 local authorities for study and so costs are reduced. The north and south of the country are both adequately represented and housing tenures are also correctly represented in the sample by the stratification at the final stage. The resulting confidence intervals will be complicated to calculate and may in fact be wider than for a simple random sample of the same size (see the discussion of the UK Time Use Survey), but this more complicated procedure will be much less costly to implement.

STATISTICS IN PRACTICE

The UK Time Use Survey

The UK Time Use Survey provides a useful example of the effects of multi-stage sampling. It uses a mixture of cluster and stratified sampling and the results are weighted to compensate for unequal probabilities of selection into the sample and for the effects of non-response. Together, these act to increase the size of standard errors, relative to those obtained from a simple random sample of the same size. This increase can be measured by

→

the **design factor**, defined as the ratio of the true standard error to the one arising from a simple random sample of the same size. For the time use survey, the design factor is typically 1.5 or more. Thus the standard errors are increased by 50% or more, but a simple random sample of the same size would be much more expensive to collect (for example, the clustering means that only a minority of geographical areas are sampled).

The following table shows the average amount of time spent sleeping by 16 to 24 year olds (in minutes per day):

	Mean	True s.e.	95% CI	Design factor	*n*	Effective sample size
Male	544.6	6.5	[531.9, 557.3]	1.63	1090	412
Female	545.7	4.2	[537.3, 554.0]	1.14	1371	1058

The true standard error, taking account of the sample design, is 6.5 minutes for men. The design factor is 1.63, meaning this standard error is 63% larger than for a similar sized ($n = 1090$) simple random sample. Equivalently, a simple random sample of size $n = 412(= 1090/1.63^2)$ would achieve the same precision (but at greater cost).

How the design factor is made up is shown in the following table:

Design factor (deft)	Deft due to stratification	Deft due to clustering	Deft due to weighting
1.63	1.00	1.17	1.26

It can be seen that stratification has no effect on the standard error, but both clustering and the post-sample weighting serve to increase the standard errors.

Source: The UK 2000 Time Use Survey, Technical Report 2003, HMSO.

Quota sampling

Quota sampling is a non-random method of sampling and therefore it is impossible to use sampling theory to calculate confidence intervals from the sample data, or to find whether or not the sample will give biased results. Quota sampling simply means obtaining the sample information as best one can, for example, by asking people in the street (hence it is sometimes called 'convenience sampling'). However, it is by far the cheapest method of sampling and so allows much larger sample sizes and can be carried out quickly. As shown above, large sample sizes can still give biased results if sampling is non-random; but in some cases the budget is too small to afford even the smallest properly conducted random sample, so a quota sample is the only alternative.

Even with quota sampling, where the interviewer is simply told to go out and obtain (say) 1000 observations, it is worth making some crude attempt at stratification. The problem with human interviewers is that they are notoriously non-random, so that when they are instructed to interview every 10th person they see (a reasonably random method), if that person turns out to be a shabbily dressed tramp slightly the worse for drink, they are quite likely to select the 11th person instead. Shabbily dressed tramps, slightly the worse for drink, are therefore under-represented in the sample. To combat this sort of problem the interviewers are given quotas to fulfil, for example, 20 men and 20 women, 10 old-age pensioners, one shabbily dressed tramp, etc., so that the sample will at least broadly reflect the population under study and give reasonable results.

It is difficult to know how accurate quota samples are, since it is rare for their results to be checked against proper random samples or against the population itself. Probably the most common quota samples relate to voting intentions and so can be checked against actual election results. The 1992 UK general election provides an interesting illustration. The opinion polls predicted a fairly substantial Labour victory, but the outcome was a narrow Conservative majority. An enquiry concluded that the erroneous forecast occurred because a substantial number of voters changed their minds at the last moment and that there was 'differential turn-out', i.e. Conservative supporters were more likely to vote than Labour ones. Since then, pollsters have tried to take this factor into account when trying to predict election outcomes.

Unfortunately, the pollsters were caught out again in 2015, this time predicting a tie between the two major parties when the election resulted in a clear majority for the Conservative Party. This time the blame was put on unrepresentative samples, with some groups under-represented (e.g. 45–64 year olds). Attempts to correct this via weighting of the sample results proved ineffective. Interestingly, a proper random sample (the British Election Survey) did get the right result, but this was one carried out *after* the election and so is difficult to compare impartially with the quota samples.

STATISTICS IN PRACTICE

Can you always believe surveys?

Many surveys are more interested in publicising something than in finding out the facts. One has to be wary of surveys finding that people enjoy high-rise living . . . when the survey is sponsored by an elevator company, for example. In July 2007 a survey of 1000 adults found that 'the average person attends 3.4 weddings each year'. This sounds suspiciously high to me. I have never attended three or more weddings in a year, nor have friends I have asked. Let us do some calculations. There were 283 730 weddings in the United Kingdom in 2005. There are about 45m adults, so if they each attend 3.4 weddings, that makes $45 \times 3.4 = 153$ million attendees. This means 540 per wedding. That seems excessively high (remember this excludes children) and probably means the sample design was poor, obtaining an unrepresentative result.

A good way to make a preliminary judgement on the likely accuracy of a survey is to ask 'who paid for this?'

Calculating the required sample size

Before collecting sample data, it is obviously necessary to know how large the sample size has to be. The required sample size will depend upon two factors:

- the desired level of precision of the estimate, and
- the funds available to carry out the survey.

The greater the precision required, the larger the sample size needs to be, other things being equal. But a larger sample will obviously cost more to collect, and this might conflict with a limited amount of funds being available. There is a

Figure 9.3
The desired width of the confidence interval

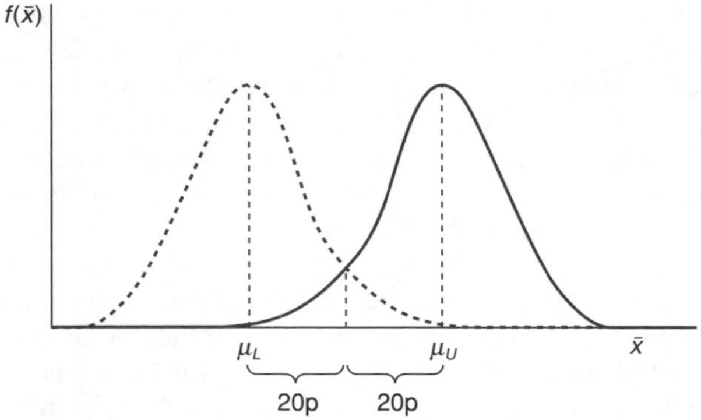

trade-off, therefore, between the two desirable objectives of high precision and low cost. The following example shows how these two objectives conflict.

A firm producing sweets wishes to find out the average amount of pocket money children receive per week. It wants to be 99% confident that the estimate is within 20 pence of the correct value. How large a sample is needed?

The problem is one of estimating a confidence interval, turned on its head. Instead of having the sample information $\bar{x}$, s and n, and calculating the confidence interval for μ, the desired width of the confidence interval is given and it is necessary to find the sample size n which will ensure this. The formula for the 99% confidence interval, assuming a Normal rather than t distribution (i.e. it is assumed that the required sample size will be large), is

$$\bar{x} \pm 2.58 \times \sqrt{s^2/n} \tag{9.2}$$

Diagrammatically this can be represented as in Figure 9.3.

The firm wants the distance between $\bar{x}$ and μ to be no more than 20 pence in either direction, which means that the confidence interval must be 40 pence wide. The value of n which makes the confidence interval 40 pence wide has to be found. This can be done by solving the equation

$$20 = 2.58 \times \sqrt{s^2/n}$$

and hence, by rearranging:

$$n = \frac{2.58^2 \times s^2}{20^2} \tag{9.3}$$

All that is now required to solve the problem is the value of s^2, the sample variance; but since the sample has not yet been taken, this is not available. There are a number of ways of trying to get round this problem:

- using the results of existing surveys if available,
- conducting a small, preliminary, survey, and
- guessing.

These may not seem very satisfactory (particularly the last), but something has to be done and some intelligent guesswork should give a reasonable estimate of s^2. Suppose, for example, that a survey of children's *spending* taken five years previously showed a standard deviation of 30p. It might be reasonable to expect that the standard deviation of spending would be similar to the standard deviation of income, so 30p (updated for inflation) can be used as an estimate of the standard deviation. Suppose that five years' inflation turns the 30p into 50p. Using $s = 50$, we obtain

$$n = \frac{2.58^2 \times 50^2}{20^2} = 41.6$$

giving a required sample size of 42 (the sample size has to be an integer). This is a large ($n \geq 25$) sample size, so the use of the Normal distribution was justified.

Is the firm willing to pay for such a large sample? Suppose it was willing to pay out £1000 in total for the survey, which costs £600 to set up and then £6 per person sampled. The total cost would be £600 + 42 × 6 = £852, which is within the firm's budget. If the firm wished to spend less than this, it would have to accept a smaller sample size and thus a lower precision or a lower level of confidence. For example, if only a 95% confidence level were required, the appropriate z score would be 1.96, yielding

$$n = \frac{1.96^2 \times 50^2}{20^2} = 24.01$$

A sample size of 24 would only cost £600 + 6 × 24 = £804. (At this sample size the assumption that $\bar{x}$ follows a Normal distribution becomes less tenable, so the results should be treated with caution. Use of the t distribution is tricky, because the appropriate t value depends upon the number of degrees of freedom which in turn depends on sample size, which is what is being looked for.)

The general formula for finding the required sample size is

$$n = \frac{z_\alpha^2 \times s^2}{p^2} \tag{9.4}$$

where z_α is the z score appropriate for the $(100 - \alpha)\%$ confidence level and p is the desired accuracy (20 pence in this case).

Caution should be used with this type of calculation as the result is true for a simple random sample only. If a different type of sampling is used (e.g. cluster sampling) then the sample might need to be bigger, perhaps double the size for simple random sample.

Collecting the sample

The sampling frame

We now move on to the fine detail of how to select the individual observations which make up the sample. In order to do this, it is necessary to have some sort of **sampling frame**, i.e. a list of all the members of the population from which the sample is to be drawn. This can be a problem if the population is extremely large, for example the population of a country, since it is difficult to manipulate so much information (cutting up 65 million pieces of paper to put into a hat for a random draw is a tedious business). Alternatively, the list might not even exist or, if it does, not be in one place convenient for consultation and use. In this case there is often an advantage to multi-stage sampling, for the selection of regions or even local authorities is fairly straightforward and not too time-consuming. Once at this lower level the sampling frame is more manageable (each local authority has an electoral register, for example) and individual observations can be relatively easily chosen. Thus it is not always necessary to have a complete sampling frame for the entire population in one place.

Choosing from the sampling frame

There is a variety of methods available for selecting a sample of (say) 1000 observations from a sampling frame of 25 000 names, varying from the manual to the electronic. The oldest method is to cut up 25 000 pieces of paper, put them in a (large) hat, shake it (to randomise) and pick out 1000. This is fairly time-consuming, however, and has some pitfalls – if the pieces are not all cut to the same size, is the probability of selection the same? It is much better if the population in the sampling frame is numbered in some way, for then one only has to select random numbers. This can be done by using a table of random numbers (see Table A1 on page 448, for example), or a computer. The use of random number tables for such purposes is an important feature of statistics, and in 1955 the Rand Corporation produced a book entitled *A Million Random Digits with 100 000 Normal Deviates*. This book, as the title promises, contained nothing but pages of random numbers which allowed researchers to collect random samples. Interestingly, the authors did not bother to fully proofread the text, since a few (random) errors here and there would not matter. These numbers were calculated electronically and nowadays every computer has a facility for rapidly choosing a set of random numbers. (It is an interesting question how a computer, which follows rigid rules of behaviour, can select random numbers which, by definition, are unpredictable by any rule.)

A further alternative, if a 1-in-25 sample is required, is to select a random starting point between 1 and 25 and then select every subsequent 25th observation (e.g. the 3rd, 28th, 53rd, etc.). This is a satisfactory procedure if the sampling frame is randomly sorted to start with, but otherwise there can be problems. For example, if the list is sorted by income (poorest first), a low starting value will almost certainly give an underestimate of the population mean. If all the numbers were randomly sorted, this 'error' in the starting value would not be important.

Interviewing techniques

Good training of interviewers is vitally important to the results of a survey. It is very easy to lead an interviewee into a particular answer to a question. Consider the following two sets of questions:

A

(1) Do you know how many people were killed by the atomic bomb at Hiroshima?
(2) Do you think nuclear weapons should be banned?

B

(1) Do you believe in nuclear deterrence?
(2) Do you think nuclear weapons should be banned?

*A*2 is almost certain to get a higher 'yes' response than *B*2. Even a different ordering of the questions can have an effect upon the answers (consider asking

$A2$ before $A1$). The construction of the questionnaire has to be done with care, therefore. The manner in which the questions are asked is also important, since it can often suggest the answer. Good interviewers are trained to avoid these problems by sticking precisely to the wording of the question and not to suggest an expected answer.

Telephone surveys

An article by M. Collins in the *Journal of the Royal Statistical Society* reveals some of the difficulties in conducting surveys by telephone. First, the sampling frame is incomplete since, although most people have a telephone, some are not listed in the directory. In the late 1980s this was believed to be around 12% of all numbers, but it has been growing since, to around 40%. (Part of this trend, of course, may be due to people getting fed up with being pestered by salespersons and 'market researchers'.) Researchers have responded with 'random digit dialling' which is made easier by modern computerised equipment.

Matters are unlikely to improve for researchers in the future. The answering machine is often used as a barrier to unwanted calls and some residential lines connect to fax machines. Increasing deregulation and mobile phone use mean it will probably become more and more difficult to obtain a decent sampling frame for a proper survey.

Source: M. Collins, Sampling for UK telephone surveys, *J. Royal Statistical Society*, Series A, 162 (1), 1999.

Even when these procedures are adhered to there can be various types of response bias. The first problem is of non-response, due to the subject not being at home when the interviewer calls. There might be a temptation to remove that person from the sample and call on someone else, but this should be resisted. There could well be important differences between those who are at home all day and those who are not, especially if the survey concerns employment or spending patterns, for example. Continued efforts should be made to contact the subject. One should be wary of surveys which have low response rates, particularly where it is suspected that the non-response is in some way systematic and related to the goal of the survey.

A second problem is that subjects may not answer the question truthfully for one reason or another, sometimes inadvertently. An interesting example of this occurred in the survey into sexual behaviour carried out in Britain in 1992 (see *Nature*, 3 December 1992). Amongst other things, this found the following:

- The average number of heterosexual partners during a woman's lifetime is 3.4.
- The average number of heterosexual partners during a man's lifetime is 9.9.

This may be in line with one's beliefs about behaviour, but, in fact, the figures must be wrong. The *total* number of partners of all women must by definition equal the *total* number for all men. Since there are approximately equal numbers of males and females in the United Kingdom, the averages must therefore be about the same. So how do the above figures come about?

It is too much to believe that international trade holds the answer. It seems unlikely that British men are so much more attractive to foreign women than British women are to foreign men. Nor is an unrepresentative sample likely. It was carefully chosen and quite large (around 20 000 respondents). The answer would appear to be that some people are lying. Either women are being excessively modest or (more likely?) men are boasting. Perhaps the answer is to divide by three whenever a man talks about his sexual exploits.

For an update on this story, see the article by J. Wadsworth *et al.*, What is a mean? An examination of the inconsistency between men and women in reporting sexual partnerships, *J. Royal Statistical Society*, 1996, Series A. 159 (1).

Case study: the UK Living Costs and Food Survey

Introduction

The Living Costs and Food Survey (LCF) is an example of a large government survey which examines households' expenditure patterns (with a particular focus on food expenditures) and income receipts. It is worth having a brief look at it, therefore, to see how the principles of sampling techniques outlined in this chapter are put into practice. The LCF succeeded the Expenditure and Food Survey in 2008, which used a similar design. The LCF is used for many different purposes, including the calculation of weights to be used in the UK Consumer Price Index and the assessment of the effects of changes in taxes and state benefits upon different households.

Choosing the sample

The sample design which follows is known as a **multi-stage, stratified random sample with clustering**. This is obviously quite a complex design so will be examined in a little detail.

Stage 1

The primary sampling unit (PSU) is the Postcode Sector, a small geographical area. There are 638 of these selected at random each year (from a total of around 10,600 sectors in the country), stratified according to

- region,
- socio-economic classification, and
- car ownership.

This ensures the chosen areas are representative of the country as a whole on these characteristics.

Stage 2

Eighteen households are drawn at random from each postcode sector, giving a prospective sample size of $638 \times 18 = 11\,484$ households. To capture spending patterns throughout the year, interviewing takes place in approximately 53 sectors each month. The response rate to the survey is approximately 50%, having fallen in recent years, hence the number of households providing information is around 5700.

The sampling frame

The Postcode Address File, a list of all postal delivery addresses, is used as the sampling frame. Previously the register of electors in each ward had been used but had some drawbacks: it was under-representative of those who have no permanent home or who move frequently (e.g. tramps, students, etc.). The fact that many people took themselves off the register in the early 1990s in order to avoid paying the Community Charge could also have affected the sample.

Collection of information

The data are collected by interview, and by asking participants to keep a diary in which they record everything they purchase over a two-week period. Highly skilled interviewers are required to ensure accuracy and compliance with the survey, and each participating family is visited several times. As an inducement to cooperate, each member of the family is paid a small sum of money – it is to be hoped that the anticipation of this does not distort their expenditure patterns.

Sampling errors

Given the complicated survey design, it is difficult to calculate sampling errors exactly. The multi-stage design of the sample actually tends to increase the sampling error relative to a simple random sample, but, of course, this is offset by cost savings which allow a greatly increased sample size. Overall, the results of the survey are of good quality and can be verified by comparison with other statistics, such as retail sales, for example.

Summary

- A primary data source is one where you obtain the data yourself, designed the questions and have access to all the original observations.
- A secondary data source is one collected by others, perhaps for a different purpose to your own. You may have access to all the original observations (typical with electronic sources) or a summary, usually in the form of tables.
- When collecting data always keep detailed notes of the sources of all information, how it was collected, precise definitions of the variables, etc.
- Most data can be obtained electronically, which saves having to type it into a computer, but the data still need to be checked for errors.
- There are various types of random sample, including simple, stratified and clustered random samples. The methods are sometimes combined in multi-stage samples.
- The type of sampling affects the size of the standard errors of the sample statistics. The most precise sampling method is not necessarily the best if it costs more to collect (since the overall sample size that can be afforded will be smaller).
- Quota sampling is a non-random method of sampling which has the advantage of being extremely cheap. It is often used for opinion polls and surveys.
- The sampling frame is the list (or lists) from which the sample is drawn. If it omits important elements of the population, its use could lead to biased results.
- Careful interviewing techniques are needed to ensure reliable answers are obtained from participants in a survey.

Key terms and concepts

cluster sampling
design factor
finite population correction
multi-stage sampling
online data
primary data
quota sampling
random sample
sampling frame
sampling methods
secondary data
simple random sampling
strata
stratification factor
stratified sampling

References

Barnett, V., *Sample Survey Principles and Methods*, Wiley, 2002.

Edelman, B., Using Internet Data for Economic Research, *Journal of Economic Perspectives*, 26(2), 189–206, 2012.

Moser, C.A., and G. Kalton, *Survey Methods in Social Investigations*, Heinemann, 1971.

Rand Corporation, *A Million Random Digits with 100 000 Normal Deviates*, The Glencoe Press, 1955.

Formulae used in this chapter

Formula	Description	Note
$\text{fpc} = 1 - n/N$	Finite population correction for the variance of $\bar{x}$	
$n = \dfrac{z_\alpha^2 \times s^2}{p^2}$	Required sample size to obtain desired confidence interval	p is the desired accuracy (half the width of the CI), Z_α is the critical value from the Normal distribution (depends on confidence level specified).

Problems

Some of the more challenging problems are indicated by highlighting the problem number in colour.

9.1 What issues of definition arise in trying to measure 'output'?

9.2 What issues of definition arise in trying to measure 'unemployment'?

9.3 Find the gross domestic product for both the United Kingdom and the United States for the period 1995 to 2003. Obtain both series in constant prices.

9.4 Find figures for the monetary aggregate M0 for the years 1995 to 2003 in the United Kingdom, in nominal terms.

9.5 A firm wishes to know the average weekly expenditure on food by households to within £2, with 95% confidence. If the variance of food expenditure is thought to be about 400, what sample size does the firm need to achieve its aim?

9.6 A firm has £10 000 to spend on a survey. It wishes to know the average expenditure on gas by businesses to within £30 with 99% confidence. The variance of expenditure is believed to be about 40 000. The survey costs £7000 to set up and then £15 to survey each firm. Can the firm achieve its aim with the budget available?

9.7 **(Project)** Visit your college library or online sources to collect data to answer the following question. Have females' earnings risen relative to men's over the past 10 years? You should write a short report on your findings. This should include a section describing the data collection process, including any problems encountered and decisions you had to make. Compare your results with those of other students. It might be interesting to compare your experiences of using online and offline sources of data.

9.8 **(Project)** Do a survey to find the average age of cars parked on your college campus. (A letter or digit denoting the registration year can be found on the number plate (applies to the United Kingdom, other countries will differ) – precise details can be obtained in various guides to used-car prices.) You might need stratified sampling (e.g. if administrators have newer cars than faculty and students, for example). You could extend the analysis by comparing the results with a public car park. You should write a brief report outlining your survey methods and the results you obtain. If several students do such a survey you could compare results.

10 Index numbers

Contents

Learning outcomes

By the end of this chapter you should be able to:

- **represent a set of data in index number form**
- **understand the role of index numbers in summarising or presenting data**

- recognise the relationship between price, quantity and expenditure index numbers
- turn a series measured at current prices into one at constant prices (or in volume terms)
- splice separate index number series together
- measure inequality using index numbers.

Introduction

'Consumer price index up 3.8%. Retail price index up 4.6%.' (UK, June 2008)

'Vietnam reports an inflation rate of 27.04%' (July 2008)

'Zimbabwe inflation at 2,200,000%' (July 2008)

The above headlines reveal startling differences between the inflation rates of three different countries. This chapter is concerned with how such measures are constructed and then interpreted. Index numbers are not restricted to measuring inflation, although that is one of the most common uses. There are also indexes of national output, of political support, of corruption in different countries of the world, and even of happiness (Danes are the happiest, it seems).

An **index number** is a descriptive statistic, in the same sense as the mean or standard deviation, which summarises a mass of information into some readily understood statistic. As such, it shares the advantages and disadvantages of other summary statistics: it provides a useful overview of the data but misses out the finer detail. The **retail price index** (RPI) referred to above is one example, which summarises information about the prices of different goods and services, aggregating them into a single number. We have used index numbers earlier in the text (for example, in the chapters on regression), without fully explaining their derivation or use. This will now be remedied.

Index numbers are most commonly used for following trends in data over time, such as the consumer price index (CPI) measuring the price level or the index of industrial production (IIP) measuring the output of industry. The CPI (or the similar retail price index, RPI) also allows calculation of the rate of inflation, which is simply the rate of change of the price index; and from the IIP it is easy to measure the rate of growth of output. Index numbers are also used with cross-section data, for example, an index of regional house prices would summarise information about the different levels of house prices in different regions of the country at a particular point in time. There are many other examples of index numbers in use, common ones being the Financial Times All Share index, the trade weighted exchange rate index, and the index of the value of retail sales.

This chapter will explain how index numbers are constructed from original data and the problems which arise in doing this. There is also a brief discussion of the CPI to illustrate some of these problems and to show how they are resolved in practice. Finally, a different set of index numbers is examined, which are used to measure inequality, such as inequality in the distribution of income, or in the market shares held by different firms competing in a market. Constructing index numbers is not the most glamorous part of statistics but it is important for any applied researcher to understand them.

A simple index number

We begin with the simplest case, where we wish to construct an index number series for a single commodity. In this case, we shall construct an index number series representing the price of coal. This is a series of numbers showing, in each year, the price of coal and how it changes over time. More precisely, we measure the cost of coal to industrial users, for the years 2006 to 2010 (as I wrote above, not the most glamorous part of statistics). Later in the chapter we will expand the analysis to include other fuels and thereby construct an index of the price of energy as a whole. The raw data for coal are given in Table 10.1 (adapted from the Digest of UK Energy Statistics, available on-line). We assume that the product itself has not changed from year to year, so that the index provides a fair representation of costs. This means, for example, that the quality of coal has not changed during the period.

To construct a price index from these data, we choose one year as the **reference year** (we will use 2006 in this case) and set the price index in that year equal to 100. The prices in the other years are then measured *relative* to the reference year figure of 100. The index, and its construction, are presented in Table 10.2.

All we have done so far is to change the form in which the information is presented. We have perhaps gained some degree of clarity (for example, it is easy to see that the price in 2010 is 62.5% higher than in 2006), but we have lost the original information about the actual *level* of prices. Since it is usually *relative* prices that are of interest, this loss of information about the actual price level is not too serious, and information about relative prices is retained by the price index. For example, using either the index or actual prices, we can see that the price of coal was 6.6% higher in 2007 than in 2006.

In terms of a formula we have calculated, for each year:

$$P^t = \frac{\text{price of coal in year } t}{\text{price of coal in 2006}} \times 100$$

where P^t represents the value of the index in year t.

Table 10.1 The price of coal, 2006–10

	2006	2007	2008	2009	2010
Price (£/tonne)	43.63	46.49	60.31	59.60	70.90

Table 10.2 The price index for coal, 2006 = 100

Year	Price	Index	
2006	43.63	100.0	$\left(= \frac{43.63}{43.63} \times 100\right)$
2007	46.49	106.6	$\left(= \frac{46.49}{43.63} \times 100\right)$
2008	60.31	138.2	$\left(= \frac{60.31}{43.63} \times 100\right)$
2009	59.60	136.6	etc.
2010	70.90	162.5	

Table 10.3 The price index for coal, 2008 = 100

Year	Price	Index	
2006	43.63	72.3	$\left(= \frac{43.63}{60.31} \times 100\right)$
2007	46.49	77.1	$\left(= \frac{46.49}{60.31} \times 100\right)$
2008	60.31	100.0	$\left(= \frac{60.31}{60.31} \times 100\right)$
2009	59.60	98.8	etc.
2010	70.90	117.6	

The choice of reference year is arbitrary and we can easily change it for a different year. If we choose 2008 to be the reference year, then we set the price in that year equal to 100 and again measure all other prices relative to it. The formula is thus

$$P^t = \frac{\text{price of coal in year } t}{\text{price of coal in 2008}} \times 100$$

This calculation is shown in Table 10.3 which can be derived from Table 10.2 or directly from the original data on prices. You should choose whichever reference year is most convenient for your purposes. Whichever year is chosen, the informational content is the same. The values of the index with 2008 as reference year are all 72.3% of the corresponding values of the index with 2006 as reference year.

Exercise 10.1

?

(a) Average house prices in the United Kingdom for 2000–4 were:

Year	2000	2001	2002	2003	2004
Price (£)	86 095	96 337	121 137	140 687	161 940

Turn this into an index with a reference year of 2000.

(b) Recalculate the index with reference year 2003.

(c) Check that the ratio of house prices in 2004 relative to 2000 is the same for both indexes.

A price index with more than one commodity

Constructing an index for a single commodity is a simple process but only of limited use, mainly for presentation purposes. Once there is more than a single commodity, index numbers become more useful but are a little more difficult to calculate. Industry uses other sources of energy as well as coal, such as gas, petroleum and electricity, and managers might wish to know the overall price of energy, which affects their costs. This is a more common requirement in reality, rather than the simple index number series calculated above. If the price of each fuel were rising at the same rate, say at 5% per year, then it is straightforward to say that the price of energy is also rising at 5% per year. But supposing, as is likely, that the prices are all rising at different rates, as shown in Table 10.4. Is it now possible to say how fast the price of energy is increasing? Several different prices now have

Table 10.4 Fuel prices to industry, 2006–10

Year	Coal (£/tonne)	Petroleum (£/tonne)	Electricity (£/MWh)	Gas (£/MWh)
2006	43.63	260.47	55.07	18.04
2007	46.49	269.68	54.49	14.74
2008	60.31	392.94	68.36	21.14
2009	59.60	383.22	72.70	19.06
2010	70.90	471.46	65.12	17.38

to be combined in order to construct an index number, a more complex process than the simple index number calculated above.

From the data presented in Table 10.4 we can calculate that the price of coal has risen by 63% over the five-year period, petroleum has risen by 81%, electricity by 18% and gas has fallen by 4%. It is fairly clear prices are volatile and increasing at very different rates. How do we combine these to measure the rise in the price of energy?

Using base-year weights: the Laspeyres index

We find the desired index by constructing a hypothetical 'shopping basket' of the fuels used by industry, and measure how the cost of this basket has risen (or fallen) over time. We therefore take a **weighted average** of the price changes of the individual fuels, the weights being derived from the quantities of each fuel used by the industry (the 'shopping basket'). Thus, if industry uses relatively more coal than petrol, more weight is given to the rise in the price of coal in the calculation.

Table 10.5 gives the quantities of each fuel consumed by industry in 2006 (again from the *Digest of UK Energy Statistics*) and it is this which forms the shopping basket. The year 2006 is referred to as the **base year** since the quantities consumed in this year are used to make up the shopping basket.

The cost of this basket is obtained by multiplying together the prices and quantities, as shown in Table 10.6 (using information from Tables 10.4 and 10.5).

The final column of the table shows the expenditure on each of the four energy inputs and the total cost of the basket is 10 457.93 (this is in £m, so altogether about £10.5bn was spent on energy by industry). This sum may be written algebraically as:

$$\sum_{i=1}^{4} p_{0i}q_{0i} = 10\,457.93$$

Table 10.5 Quantities of fuel used by industry, 2006

Coal (m. tonnes)	1.76
Petroleum (m. tonnes)	5.52
Electricity (m. MWh)	114.90
Gas (m. MWh)	145.00

Table 10.6 Cost of the energy basket, 2006

	Price	Quantity	Price × quantity
Coal (£/tonne)	43.63	1.76	76.79
Petroleum (£/tonne)	260.47	5.52	1 437.79
Electricity (£/MWh)	55.07	114.90	6 327.54
Gas (£/MWh)	18.04	145.00	2 615.80
Total			10 457.93

Table 10.7 The cost of the 2006 energy basket at 2007 prices

	2007 Price	2006 Quantity	Price × quantity
Coal (£/tonne)	46.49	1.76	81.82
Petroleum (£/tonne)	269.68	5.52	1488.63
Electricity (£/MWh)	54.49	114.90	6260.90
Gas (£/MWh)	14.740	145.00	2137.30
Total			9968.66

where the summation is calculated over all the four fuels. Here, p refers to prices, q to quantities. The first subscript (0) refers to the year, the second (i) to each energy source in turn. We refer to 2006 as year 0, 2007 as year 1, etc., for brevity of notation. Thus, for example, p_{01} means the price of coal in 2006, q_{12} the consumption of petroleum by industry in 2007.

We now need to find what the 2006 basket of energy would cost in each of the subsequent years, using the prices pertaining to those years. For example, for 2007 we value the same 2006 basket using the 2007 prices. This calculation is shown in Table 10.7 and yields a cost of £9968.66bn.

Firms would therefore have to spend £489m less (= 9969 − 10 458) in 2007 to buy the same quantities of energy as in 2006, mainly due to the fall in the price of gas. This amounts to a saving of 4.7% over the expenditure in 2006. The sum of £9969m may be expressed as $\sum p_{1i}q_{0i}$, since it is obtained by multiplying the prices in year 1 (2007) by quantities in year 0 (2006).

Similar calculations for subsequent years produce the costs of the 2006 basket as shown in Table 10.8.

It can be seen that *if* firms had purchased the same quantities of each energy source in the following years, they would have had to pay more in each subsequent year up until another small drop in 2010.

To obtain the energy price index from these numbers we simply calculate an index of the values in the final column of Table 10.8. In other words, we measure the cost of the basket in each year relative to its 2006 cost, i.e. we divide the cost of the basket in each successive year by $\sum p_{0i}q_{0i}$ and multiply by 100.

This index is given in Table 10.9 and is called the **Laspeyres price index**, after its inventor. We say that it uses **base-year weights** (i.e. quantities in the base year 2006 form the weights in the basket). We have set the value of the index to 100 in 2006, i.e. the reference year and the base year coincide; this is convenient although not essential.

Figure 10.1 charts the prices of the four individual fuels over the period as well as the Laspeyres index just calculated. One can see that the index tracks the electricity price fairly closely, with a small influence from the other fuel prices. Later

Table 10.8 The cost of the energy basket, 2006–10

	Formula	
2006	$\sum p_0q_0$	10 457.93
2007	$\sum p_1q_0$	9 968.66
2008	$\sum p_2q_0$	13 195.04
2009	$\sum p_3q_0$	13 337.20
2010	$\sum p_4q_0$	12 729.63

Note: For brevity, we have dropped the *i* subscript in the formula.

Table 10.9 The Laspeyres price index

Year	Formula	Index	
2006	$\frac{\sum p_0q_0}{\sum p_0q_0} \times 100$	100	$\left(= \frac{10\,457.93}{10\,457.93} \times 100\right)$
2007	$\frac{\sum p_1q_0}{\sum p_0q_0} \times 100$	95.32	$\left(= \frac{9968.66}{10\,457.93} \times 100\right)$
2008	$\frac{\sum p_2q_0}{\sum p_0q_0} \times 100$	126.17	etc.
2009	$\frac{\sum p_3q_0}{\sum p_0q_0} \times 100$	127.53	
2010	$\frac{\sum p_4q_0}{\sum p_0q_0} \times 100$	121.72	

Figure 10.1
Fuel prices and the Laspeyres index

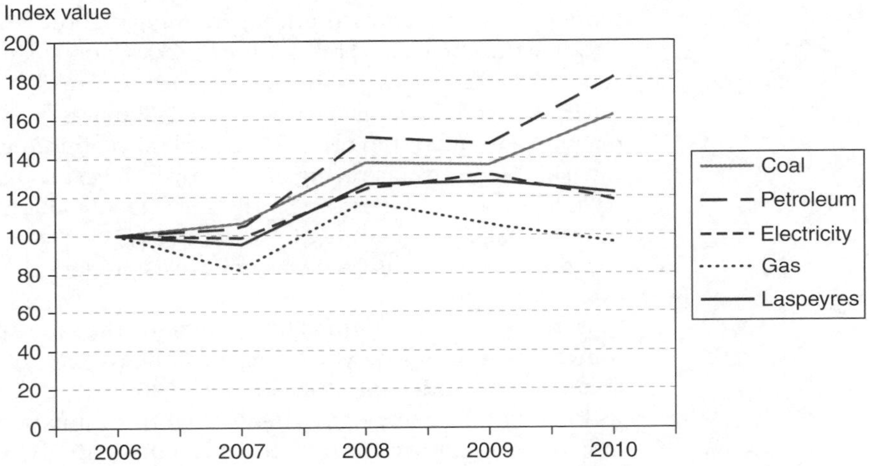

we will see that about 60% of firms' expenditure on energy goes on electricity, which is why the index closely tracks the electricity price.

In general, the Laspeyres index for year n with the base year as year 0 is given by the following formula:

$$P_L^n = \frac{\sum p_{ni}q_{0i}}{\sum p_{0i}q_{0i}} \times 100 \qquad \textbf{(10.1)}$$

(Henceforth we shall omit the i subscript on prices and quantities in the formulae for index numbers, for brevity.) The index shows that energy prices increased by 21.72% over the period. This might seem slightly surprising, given the large increases in coal and petroleum prices noted earlier. The reason for the modest overall increase is that industry uses little coal and petroleum, relative to electricity and gas. The rise amounts to an average increase of 5% p.a. in the cost of energy. During the same period, prices in general rose by 12.9% (or 3.1% p.a.), so in relative terms energy became more expensive.

The choice of 2006 as the base year for the index was an arbitrary one; any year will do. If we choose 2007 as the base year, then the cost of the 2007 basket is evaluated in each year (including 2006), and this will result in a slightly different

Table 10.10 The Laspeyres price index using the 2007 basket

Year	Cost of 2007 basket	Laspeyres index 2007 = 100	Laspeyres index 2006 = 100 (2007 basket)	Laspeyres index using 2006 basket
2006	10 189.58	104.60	100	100
2007	9 741.04	100	95.60	95.32
2008	12 878.54	132.21	126.39	126.17
2009	13 040.69	133.87	127.98	127.53
2010	12 464.08	127.95	122.32	121.72

Note: The quantities used in the 2007 basket can be found in Table 10.11, in the second row of data. These are multiplied by the relevant prices each year (see Table 10.4) to give the cost of the 2007 basket.

Laspeyres index. The calculations are in Table 10.10. The final two columns of the table compare the Laspeyres index constructed using the 2007 and 2006 baskets, respectively (the former adjusted to 2006 = 100). A very small difference can be seen, which is due to the fact that consumption patterns were very similar, although not identical, in 2006 and 2007. It would not be uncommon to get a larger difference between the series than in this instance.

The Laspeyres price index shows the increase in the price of energy for the 'average' firm, i.e. one which consumes energy in the same proportions as the 2006 basket overall. There are probably very few such firms: most would use perhaps only one or two energy sources. Individual firms may therefore experience price rises quite different from those shown here. For example, a firm depending upon coal alone would face a 63% price increase over the four years, significantly different from the figure of 21.7% suggested by the Laspeyres index.

Exercise 10.2

?

(a) The prices of fuels used by industry 2002–6 were:

Year	Coal (£/tonne)	Petroleum (£/tonne)	Electricity (£/MWh)	Gas (£/MWh)
2002	36.97	132.24	29.83	7.80
2003	34.03	152.53	28.68	8.09
2004	37.88	153.71	31.26	9.61
2005	44.57	204.28	42.37	13.87
2006	43.63	260.47	55.07	18.04

and quantities consumed by industry were:

Year	Coal (m. tonnes)	Petroleum (m. tonnes)	Electricity (m. MWh)	Gas (m. MWh)
2002	1.81	5.70	112.65	165.17

Calculate the Laspeyres price index of energy based on these data. Use 2002 as the reference year.

(b) Recalculate the index making 2004 the reference year.

(c) The quantities consumed in 2003 were:

	Coal (m tonnes)	Petroleum (m tonnes)	Electricity (m MWh)	Gas (m therms)
2003	1.86	6.27	113.36	166.22

Calculate the Laspeyres index using this basket and compare to the answer to part (a).

Using current-year weights: the Paasche index

Firms do not of course consume the same basket of energy every year. One would expect them to respond to changes in the relative prices of fuels and to other factors. Technological progress means that the efficiency with which the fuels can be used changes, causing fluctuations in demand. Table 10.11 shows the quantities consumed in 2006 and later years, indicating that firms did indeed alter their pattern of consumption.

Any of these annual patterns of consumption could be used as the 'shopping basket' for the purpose of constructing a Laspeyres index and each would give a slightly different price index, as we saw with the usage of the 2006 and 2007 baskets. One cannot say that one of these is more correct than the others. One further problem is that whichever basket is chosen remains the same over time and eventually becomes unrepresentative of the current pattern of consumption.

The **Paasche index** (denoted P_P^n to distinguish it from the Laspeyres index) overcomes these problems by using **current-year weights** to construct the index; in other words the basket is continually changing. Suppose 2006 is to be the reference year, so $P_P^0 = 100$. To construct the Paasche index value for 2007 we use the 2007 weights (or basket), for the 2008 value of the index we use the 2008 weights, and so on. An example will clarify matters.

The Paasche index for 2007 will be the cost of the 2007 basket at 2007 prices relative to its cost at 2006 prices, i.e.

$$P_P^1 = \frac{\sum p_1 q_1}{\sum p_0 q_1} \times 100$$

The cost of the 2007 basket at 2006 prices ($\sum p_0 q_1$) has already been calculated in Table 10.10, as 10 189.58. The numerator of the formula, $\sum p_1 q_1$, is calculated[1] as $46.49 \times 1.9 + 269.68 \times 5.53 + 54.49 \times 113.8 + 14.74 \times 133 = 9741.04$. The formula therefore evaluates to:

$$P_P^1 = \frac{9741.04}{10\,189.58} \times 100 = 95.60$$

This is the value of the Paasche index for 2007, with 2006 = 100 as the reference year.

The general formula for the Paasche index in year n is given in equation (10.2).

$$P_P^n = \frac{\sum p_n q_n}{\sum p_0 q_n} \times 100 \quad \textbf{(10.2)}$$

Table 10.11 Quantities of energy used, 2006–10

Year	Coal (m. tonnes)	Petroleum (m. tonnes)	Electricity (m. MWh)	Gas (m. MWh)
2006	1.76	5.52	114.90	145.00
2007	1.90	5.53	113.80	133.00
2008	1.94	5.08	114.51	139.00
2009	1.74	4.51	100.84	116.00
2010	1.72	4.53	104.50	122.00

[1]Prices from Table 10.4, quantities from Table 10.11.

Table 10.12 The Paasche price index

	Cost of basket:		
	At current prices: $\sum p_n q_n$	At 2006 prices $\sum p_0 q_n$	Index
2006	10 457.93	10 457.93	100
2007	9 741.04	10 189.58	95.60
2008	12 879.50	10 221.46	126.00
2009	11 374.05	8 896.53	127.85
2010	11 183.06	9 210.67	121.41

Table 10.12 shows the calculation of this index for all years.

The Paasche formula gives a slightly different result to the Laspeyres formula, as is usually the case. The Paasche should generally give a slower rate of increase than does the Laspeyres index. This is because one would expect profit-maximising firms to respond to changing relative prices by switching their consumption in the direction of the inputs which are becoming relatively cheaper. The Paasche index, by using the current weights, captures this change, but the Laspeyres, assuming fixed weights, does not. This may happen slowly, as it takes time for firms to switch to different fuels, even if technically possible. This is why the Paasche can increase faster than the Laspeyres in some years (e.g. 2007) although in the long run it should increase more slowly.

Is one of the indices more 'correct' than the other? The answer is that neither is definitively correct, they are different interpretations of the phrase 'the cost of energy'. It can be shown theoretically that the 'true' value lies somewhere between the two, but it is difficult to say exactly where. If all the items which make up the index increase in price at the same rate, then the Laspeyres and Paasche indices would give the same answer, so it is the change in *relative* prices and the resultant change in consumption patterns which causes problems.

Units of measurement

It is important that the units of measurement in the price and quantity tables be consistent. Note that in the example, the price of coal was measured in £/tonne and the consumption was measured in millions of tonnes. The other fuels were similarly treated (in the case of electricity, one MWh equals one million watt-hours). But suppose we had measured electricity consumption in kWh instead of MWh (1 MWh = 1000 kWh), but still measured its price in £ per MWh? We would then have 2006 data of 55.07 for price as before, but 114 900 for quantity. It is as if electricity consumption has been boosted 1000-fold, and this would seriously distort the results. The (Laspeyres) energy price index would be (by a similar calculation to the one above):

2006	2007	2008	2009	2010
100	98.94	124.14	132.01	118.26

This gives an incorrect result. By chance it is not a large distortion because electricity prices were rising at about the same rate as the average for all fuels. If we had mis-measured the coal data in a similar manner, there would have been a much larger discrepancy.

STATISTICS IN PRACTICE

The Human Development Index

One of the more interesting indices to appear in recent years is the Human Development Index (HDI), produced by the United Nations Development Programme (UNDP). The HDI aims to provide a more comprehensive socioeconomic measure of a country's progress than GDP (national output). Output is a measure of how well-off we are in material terms, but makes no allowance for the quality of life and other factors.

The HDI combines a measure of well-being (GDP per capita) with longevity (life expectancy) and knowledge (based on years of schooling). As a result, each country obtains a score, from 0 (poor) to 1 (good). Some selected values are given in the table.

Country	HDI 1980	HDI 2008	HDI 2013	Rank (HDI)	Rank (GDP)
New Zealand	0.793	0.899	0.910	7	30
United Kingdom	0.735	0.890	0.892	14	27
Hong Kong	0.698	0.877	0.891	15	10
Gabon	0.468	0.525	0.674	124	65
Senegal	0.443	0.474	0.485	163	159

One can see that there is an association between the HDI and GDP, but not a perfect one. New Zealand has the world's 30th highest GDP per capita but is 7th in the HDI rankings. In contrast, Gabon, some way up the GDP rankings (for a developing country), is much lower when the HDI is calculated.

So how is the HDI calculated from the initial data? How can we combine life expectancy (which can stretch from 0 to 80 years or more) with literacy (the proportion of the population who can read and write)? The answer is to score all of the variables on a scale from 0 to 100.

The HDI sets a range for (national average) life expectancy between 25 and 85 years. A country with a life expectancy of 52.9 (the case of Gabon) therefore scores 0.465, i.e. 52.9 is 46.5% of the way between 25 and 85.

Adult literacy can vary between 0% and 100% of the population, so it needs no adjustment. Gabon's figure is 0.625. The scale used for years of schooling is 0 to 15, so Gabon's very low average of 2.6 yields a score of 0.173. Literacy and schooling are then combined in a weighted average (with a $\frac{2}{3}$ weight on literacy) to give a score for knowledge of $\frac{2}{3} \times 0.625 + \frac{1}{3} \times 0.173 = 0.473$.

For income, Gabon's average of \$3498 is compared to the global average of \$5185 to give a score of 0.636. (Incomes above \$5185 are manipulated to avoid scores above 1.)

A simple average of 0.465, 0.473 and 0.636 then gives Gabon's final figure of 0.525 (2008 figures). One can see that its average income is brought down by the poorer scores in the two other categories, resulting in a poorer HDI ranking.

The construction of this index number shows how disparate information can be brought together into a single index number for comparative purposes. Further work by UNDP adjusts the HDI on the basis of gender and reveals the stark result that no country treats its women as well as it does its men.

Source: http://hdr.undp.org/en/content/table-2-human-development-index-trends-1980-2013.

It is possible to make some manipulations of the units of measurement (usually to make calculation easier) as long as all items are treated alike. If, for example, all prices were measured in pence rather than pounds (so all prices in Table 10.4 were multiplied by 100) then this would have no effect on the resultant index, as you would expect. Similarly, if all quantity figures were measured in thousands of tonnes and thousands of MWh, there would be no effects on the index, even if prices remained in £/tonne, etc. But if electricity were measured in pence per MWh, while all other fuels were in £/tonne or £/MWh, a wrong answer would again be obtained. Quantities consumed should also be measured over the same time period, e.g. millions of MWh *p.a.* It does not matter what the time period is (days, weeks, months or years) as long as all the items are treated similarly.

Exercise 10.3

?

The quantities of energy used in subsequent years were:

Year	Coal (m. tonnes)	Petroleum (m. tonnes)	Electricity (m. MWh)	Gas (m. MWh)
2004	1.85	6.45	115.84	153.95
2005	1.79	6.57	118.52	151.44
2006	1.71	6.55	116.31	144.54

Calculate the Paasche index for 2002–6 with 2002 as reference year. Compare this to the Laspeyres index result.

Using expenditures as weights

On occasion the quantities of each commodity consumed are not available, but expenditures are, and a price index can still be constructed using slightly modified formulae. It is often easier to find the expenditure on a good than to know the actual quantity consumed (think of housing as an example). We shall illustrate the method with a simplified example, using the data on energy prices and consumption for the years 2006 and 2007 only.

The data for consumption are assumed to be no longer available, but only the expenditure on each energy source as a percentage of total expenditure. Expenditure is derived as the product of price and quantity consumed. The calculation of expenditure shares for 2006 is shown in Table 10.13.

The formula for the Laspeyres index can be easily manipulated to suit the data as presented in Table 10.13. Note that it confirms the earlier claim that 60% of firms' energy expenditure goes on electricity, dominating the other fuel costs.

The Laspeyres index formula based on expenditure shares is given in equation (10.3):[2]

$$P_L^n = \sum \frac{p_n}{p_0} \times s_0 \times 100 \tag{10.3}$$

Table 10.13 Expenditure shares, 2006

	Prices	Quantity	Expenditure	Share
Coal (£/tonne)	43.63	1.76	76.79	0.7%
Petroleum (£/tonne)	260.47	5.52	1 437.79	13.7%
Electricity (£/MWh)	55.07	114.9	6 327.54	60.5%
Gas (£/MWh)	18.04	145	2 615.80	25.0%
Total			10 457.93	100.0%

Note: The 0.7% share of coal, for example, is calculated as (76.79/10 457.93) × 100. Other shares are calculated similarly.

[2] See the appendix to this chapter (page 419) for the derivation of this formula.

Equation (10.3) is made up of two component parts. The first, p_n/p_0, is simply the price in year n relative to the base-year price for each energy source. The second component, $s_0 (= p_0q_0/\Sigma p_0q_0)$, is the share or proportion of total expenditure spent on each energy source in the base year, the data for which are in Table 10.13. It should be easy to see that the sum of the s_0 values is 1, so that equation (10.3) calculates a weighted average of the individual price increases, the weights being the expenditure shares.

The calculation of the Laspeyres index for 2007 using 2006 as the base year is therefore:

$$P_L^1 = \frac{46.49}{43.63} \times 0.007 + \frac{269.68}{260.47} \times 0.137 + \frac{54.49}{55.07} \times 0.605 + \frac{14.74}{18.04} \times 0.250$$
$$= 0.9532$$

giving the value of the index as $0.9532 \times 100 = 95.32$, the same value as derived earlier using the more usual methods. Values of the index for subsequent years are calculated by appropriate application of equation (10.3). This is left as an exercise for the reader, who may use Table 10.9 to verify the answers.

The Paasche index may similarly be calculated from data on prices and expenditure shares, as long as these are available for each year for which the index is required. The formula for the Paasche index is

$$P_P^n = \frac{1}{\sum \frac{p_0}{p_n} s_n} \times 100 \qquad \textbf{(10.4)}$$

The calculation of the Paasche index using this formula is also left as an exercise.

Comparison of the Laspeyres and Paasche indices

The advantages of the Laspeyres index are that it is easy to calculate and that it has a fairly clear intuitive meaning, i.e. the cost each year of a particular basket of goods. The Paasche index involves more computation, and it is less easy to envisage what it refers to. As an example of this point, consider the following simple case. The Laspeyres index values for 2008 and 2009 are 126.17 and 127.53. The ratio of these two numbers, 1.011, would suggest that prices rose by 1.1% between these years. What does this figure actually represent? The 2009 Laspeyres index has been divided by the same index for 2008, i.e.

$$\frac{P_L^3}{P_L^2} = \frac{\sum p_3q_0}{\sum p_0q_0} \Big/ \frac{\sum p_2q_0}{\sum p_0q_0} = \frac{\sum p_3q_0}{\sum p_2q_0}$$

which is the ratio of the cost of the 2006 basket at 2009 prices to its cost at 2008 prices. This makes some intuitive sense. Note that it is not the same as the Laspeyres index for 2009 with 2008 as base year, which would require using q_2 (2008 quantities) in the calculation.

If the same is done with the Paasche index numbers, a rise of 1.5% is obtained between 2008 and 2009, a similar result. But the meaning of this is not so clear, since the relevant formula is:

$$\frac{P_P^3}{P_P^2} = \frac{\sum p_3q_3}{\sum p_0q_3} \Big/ \frac{\sum p_2q_2}{\sum p_0q_2}$$

which does not simplify further. This is a curious mixture of 2008 and 2009 quantities, and 2006, 2008 and 2009 prices.

The major advantage of the Paasche index, however, is that the weights are continuously updated, so that the basket of goods never becomes out of date. In the case of the Laspeyres index the basket remains unchanged over a period, becoming less and less representative of what is being bought by consumers. When revision is finally made (an updated shopping basket is used) there may therefore be a large change in the weighting scheme. The extra complexity of calculation involved in the Paasche index is less important now that computers do most of the work.

Exercise 10.4

?

(a) Calculate the share of expenditure going to each of the four fuel types in the previous exercises and use this result to re-calculate the Laspeyres and Paasche indexes using equations (10.3) and(10.4).

(b) Check that the results are the same as calculated in previous exercises.

The story so far – a brief summary

We have encountered quite a few different concepts and calculations thus far and it might be worthwhile to briefly summarise what we have covered before moving on. In order, we have examined:

- A simple index for a single commodity
- A Laspeyres price index, which uses base year weights
- A Paasche price index, which uses current year weights and is an alternative to the Laspeyres formulation
- The same Laspeyres and Paasche indices, but calculated using the data in a slightly different form, using expenditure shares rather than quantities.

We now move on to examine quantity and expenditure indices, then look at the relationship between them all.

Quantity and expenditure indices

Just as one can calculate price indices, it is also possible to calculate **quantity** and **value** (or **expenditure**) **indices**. We first concentrate on quantity indices, which provide a measure of the total quantity of energy consumed by industry each year. The problem again is that we cannot easily aggregate the different sources of energy. It makes no sense to add together tonnes of coal and petroleum, with megawatts of electricity and gas. Some means has to be found to put these different fuels on a comparable basis. To do this, we now reverse the roles of prices and quantities: the quantities of the different fuels are weighted by their different prices (prices represent the value to the firm, at the margin, of each different fuel). As with price indices, one can construct both Laspeyres and Paasche quantity indices.

The Laspeyres quantity index

The Laspeyres quantity index for year n is given by

$$Q_L^n = \frac{\sum q_n p_0}{\sum q_0 p_0} \times 100 \qquad (10.5)$$

i.e. it is the ratio of the cost of the year n basket to the cost of the year 0 basket, both valued at year 0 prices. Note that it is the same as equation (10.1) but with prices and quantities interchanged.

Using 2006 as the base year, the cost of the 2007 basket at 2006 prices is[3]:

$$\begin{aligned}\sum q_1 p_0 &= 1.90 \times 43.63 + 5.53 \times 260.47 + 113.80 \times 55.07 + 133.00 \times 18.04 \\ &= 10\,189.58\end{aligned}$$

and the cost of the 2006 basket at 2006 prices is 10 457.93 (calculated earlier). The value of the quantity index for 2007 is therefore

$$Q_L^1 = \frac{10\,189.58}{10\,457.93} \times 100 = 97.43$$

In other words, energy consumption fell by 2.57% between 2006 and 2007. If prices had remained at their 2006 levels firms would have spent 2.57% less on energy.

The value of the index for subsequent years is shown in Table 10.14, using the formula given in equation (10.5). Note that the quantity of energy consumed falls significantly, particularly in 2009. Much of this is due to the financial crisis of 2007 and subsequent recession.

The Paasche quantity index

Just as there are Laspeyres and Paasche versions of the price index, the same is true for the quantity index. The Paasche quantity index is given by

$$Q_P^n = \frac{\sum q_n p_n}{\sum q_0 p_n} \times 100 \qquad (10.6)$$

and is the analogue of equation (10.2) with prices and quantities swapped. The calculation of this index is shown in Table 10.15, which shows a similar trend to the Laspeyres index in Table 10.14. Normally one would expect the Paasche to show a slower increase than the Laspeyres quantity index: firms should switch to inputs whose relative prices fall; the Paasche gives lesser weight (current prices) to these quantities than does the Laspeyres (base-year prices) and thus shows a slower rate of increase.

Table 10.14 Calculation of the Laspeyres quantity index

	$\Sigma p_0 q_n$	Index	
2006	10 457.93	100	$\left(= \frac{10\,457.93}{10\,457.93} \times 100\right)$
2007	10 189.58	97.43	$\left(= \frac{10\,189.58}{10\,457.93} \times 100\right)$
2008	10 221.46	97.74	etc.
2009	8 896.53	85.07	
2010	9 210.67	88.07	

[3]The values for the calculation can be found in Tables 10.11 and 10.4, respectively.

Table 10.15 Calculation of the Paasche quantity index

	$\sum p_n q_n$	$\sum p_n q_0$	Index
2006	10 457.93	10 457.93	100
2007	9 741.04	9 968.66	97.72
2008	12 879.50	13 195.04	97.61
2009	11 374.05	13 337.20	85.28
2010	11 183.06	12 729.63	87.85

Note: The final column is calculated as the ratio of the previous two columns, then multiplied by 100.

In summary, quantity indices are similar to price indices but measure the changes in quantity over time, combining together several different commodities into one overall index.

Expenditure indices

The **expenditure** or **value index** is simply an index of the cost of the year n basket at year n prices and so it measures how expenditure changes over time. The formula for the index in year n is

$$E^n = \frac{\sum p_n q_n}{\sum p_0 q_0} \times 100 \tag{10.7}$$

There is obviously only one value index and one does not distinguish between Laspeyres and Paasche formulations. The index can be easily derived, as shown in Table 10.16. The expenditure index shows how industry's expenditure on energy is changing over time. Thus, expenditure in 2010 was 7% higher than in 2006, for example.

The increase in expenditure over time is a consequence of two effects: (i) changes in the prices of energy and (ii) changes in quantities purchased. It should therefore be possible to decompose the expenditure index into price and quantity effects. You may not be surprised to learn that these effects can be measured by the price and quantity indices we have already covered. We look at this decomposition in more detail in the next section.

Relationships between price, quantity and expenditure indices

Just as multiplying a price by a quantity gives total value, or expenditure, the same is true of index numbers. The value index can be decomposed as the product of a price index and a quantity index. In particular, it is the product of a Paasche quantity index and a Laspeyres price index, *or* the product of a Paasche price index and a

Table 10.16 The expenditure index

	$\sum p_n q_n$	Index
2006	10 457.93	100
2007	9 741.04	93.15
2008	12 879.50	123.16
2009	11 374.05	108.76
2010	11 183.06	106.93

Note: The expenditure index is a simple index of the expenditures in the previous column.

Laspeyres quantity index. This can be very simply demonstrated using Σ notation:

$$E^n = \frac{\sum p_n q_n}{\sum p_0 q_0} = \frac{\sum p_n q_n}{\sum p_n q_0} \times \frac{\sum p_n q_0}{\sum p_0 q_0} = Q_P^n \times P_L^n \tag{10.8}$$

(Paasche quantity times Laspeyres price index)

or

$$E^n = \frac{\sum p_n q_n}{\sum p_0 q_0} = \frac{\sum p_n q_n}{\sum p_0 q_n} \times \frac{\sum p_0 q_n}{\sum p_0 q_0} = P_P^n \times Q_L^n \tag{10.9}$$

(Paasche price times Laspeyres quantity index)

Thus increases in value or expenditure can be decomposed into price and quantity effects. Two decompositions are possible and give slightly different answers.

It is also evident that a quantity index can be constructed by dividing a value index by a price index, since by simple manipulation of (10.8) and (10.9) we obtain

$$Q_P^n = E^n / P_L^n \tag{10.10}$$

and

$$Q_L^n = E^n / P_P^n \tag{10.11}$$

Note that dividing the expenditure index by a Laspeyres price index gives a Paasche quantity index, and dividing by a Paasche price index gives a Laspeyres quantity index. In either case we go from a series of expenditures to one representing quantities, having taken out the effect of price changes. This is known as **deflating** a series and is a widely used and very useful technique. We shall illustrate this using our earlier data. Table 10.17 provides the detail. Column 2 of the table shows the expenditure on fuel at **current prices** or in **cash terms**. This is simply the total expenditures each year (see Table 10.16). Column 3 contains the Laspeyres price index repeated from Table 10.9 above. Deflating (dividing) column 2 by column 3 and multiplying by 100 yields column 4 which shows expenditure on fuel in **quantity** or **volume terms**. The final column turns the volume series in column 4 into an index with 2006 = 100.

This final index is identical to the Paasche quantity index, as illustrated by equation (10.10) and can be seen by comparison with Table 10.15.

Trap!

A common mistake is to believe that once a series has been turned into an index, it is inevitably in real (or volume) terms. This is *not* the case. One can have an index of a cash (or nominal) series (e.g. in Table 10.16) *or* of a real series (the final column of Table 10.17). An index number is really just a change of the units of measurement to something more useful for presentation purposes; it is not the same as deflating the series.

In the example above we used the energy price index to deflate the expenditure series. However, it is also possible to use a *general* price index (such as the consumer price index or the GDP deflator) to deflate. This gives a slightly different result, both in numerical terms and in its interpretation. Deflating by a general price index yields a series of expenditures in **constant prices** or in **real terms.** Deflating by a specific price index (e.g. of energy) results in a **quantity** or **volume series.**

Table 10.17 Deflating the expenditure series

Year	Expenditure at current prices	Laspeyres price index	Expenditure in volume terms	Index
2006	10 457.93	100	10 457.93	100
2007	9 741.04	95.32	10 219.14	97.72
2008	12 879.50	126.17	10 207.84	97.61
2009	11 374.05	127.53	8 918.59	85.28
2010	11 183.06	121.72	9 187.35	87.85

An example should clarify this (see Problem 10.11 for data). The government spends billions of pounds each year on the health service. If this cash expenditure series is deflated by a general price index (e.g. the GDP deflator), then we obtain expenditure on health services at constant prices, or real expenditure on the health service. If the NHS pay and prices index is used as a deflator, then the result is an index of the quantity or volume of health services provided. Since the NHS pay and prices index tends to rise more rapidly than the GDP deflator, the volume series rises more slowly than the series of expenditure at constant prices. This can lead to a vigorous, if pointless, political debate. The government claims it is spending more on the health service, in real terms, while the opposition claims that the health service is getting fewer resources. As we have seen, both can be right.

Exercise 10.5

(a) Use the data from earlier exercises to calculate the Laspeyres quantity index.

(b) Calculate the Paasche quantity index.

(c) Calculate the expenditure index.

(d) Check that dividing the expenditure index by the price index gives the quantity index (remember that there are two ways of doing this).

STATISTICS IN PRACTICE

The real rate of interest

Another example of 'deflating' is calculating the 'real' rate of interest. This adjusts the actual (sometimes called 'nominal') rate of interest for changes in the value of money, i.e. inflation. If you earn a 7% rate of interest on your money over a year, but the price level rises by 5% at the same time, you are clearly not 7% better off. The real rate of interest in this case would be given by

$$\text{real interest rate} = \frac{1 + 0.07}{1 + 0.05} - 1 = 0.019 = 1.9\% \qquad (10.12)$$

In general, if r is the interest rate and i is the inflation rate, the real rate of interest is given by

$$\text{real interest rate} = \frac{1 + r}{1 + i} - 1 \qquad (10.13)$$

A simpler method is often used in practice, which gives virtually identical results for small values of r and i. This is to subtract the inflation rate from the interest rate, giving 7% − 5% = 2% in this case.

Chain indices

Whenever an index number series over a long period of time is wanted, it is usually necessary to link together a number of separate, shorter indices, resulting in a **chain index**. Without access to the original raw data it is impossible to construct a proper Laspeyres or Paasche index for the complete time period, so the result will be a mixture of different types of index number; but it is the best that can be done in the circumstances.

Suppose that the following two index number series are available. Access to the original data is assumed to be impossible.

Laspeyres price index for energy, 2006–10 (from Table 10.9)

2006	2007	2008	2009	2010
100	95.32	126.17	127.53	121.72

Laspeyres price index for energy, 2002–6

2002	2003	2004	2005	2006
100	101.68	115.22	161.31	209.11

The two series have different reference years and use different shopping baskets of consumption. The first index measures the cost of the 2006 basket in each of the subsequent years. The second measures the price of the 2002 basket in subsequent years. There is an 'overlap' year which is 2006. How do we combine these into one continuous index covering the whole period?

The obvious method is to use the ratio of the costs of the two baskets in 2006, $209.11/100 = 2.0911$, to alter one of the series. To base the continuous series on $2002 = 100$ requires multiplying each of the post-2006 figures by 2.0911, as is demonstrated in Table 10.18. Alternatively, the continuous series could just as easily be based on $2006 = 100$ by dividing the pre-2006 numbers by 2.0911.

The continuous series is not a proper Laspeyres index number as can be seen if we examine the formulae used. We shall examine the 2010 figure, 254.53, by way of example. This figure is calculated as $254.53 = 2.0911 \times 121.72$ which in terms of our formulae is

$$\frac{\sum p_{10}q_{06}}{\sum p_{06}q_{06}} \times \frac{\sum p_{06}q_{02}}{\sum p_{02}q_{02}} \Big/ 100 \qquad (10.14)$$

Table 10.18 A chain index of energy prices, 2002–10

	'Old' index	'New' index	Chain index
2002	100		100
2003	101.68		101.68
2004	115.22		115.22
2005	161.31		161.31
2006	209.11	100	209.11
2007		95.32	199.33
2008		126.17	263.84
2009		127.53	266.68
2010		121.72	254.53

Note: After 2006, the chain index values are calculated by multiplying the 'new' index by 2.0911; e.g. $199.33 = 95.32 \times 2.0911$ for 2007.

The proper Laspeyres index for 2010 using 2002 weights is

$$\frac{\sum p_{10}q_{02}}{\sum p_{02}q_{02}} \times 100 \tag{10.15}$$

There is no way that this latter equation can be derived from equation (10.14), proving that the former is not a properly constructed Laspeyres index number. Although it is not a proper Laspeyres index number series, it should be a reasonable approximation and it does have the advantage of the weights being revised (once, in 2006) and therefore more up-to-date.

Similar problems arise when deriving a chain index from two Paasche index number series. Investigation of this is left to the reader; the method follows that outlined above for the Laspeyres case.

The Consumer Price Index

To consider a real-world example, we examine the UK **Consumer Price Index (CPI)**, which is one of the more sophisticated of index numbers, involving the recording of the prices of around 700 items each month and weighting them on the basis of households' expenditure patterns (derived from the National Accounts and based on surveys such as the Living Costs and Food Survey, explained in more detail in Chapter 9 on sampling methods). The principles involved in the calculation are similar to those set out earlier, with a number of small complications which we shall discuss below.

The CPI is compiled in accordance with international standards and replaces the older Retail Prices Index (RPI). The CPI is the inflation measure that the UK government monitors and targets and is also the measure used to uprate pensions and benefits. There is a variant of the CPI (called CPIH) which includes some additional measures of housing costs, which we won't consider further. Both CPI measures, and the older RPI, track each other closely over time; there is not a lot of difference between them, at least over a short time period.

The CPI is something of a compromise between a Laspeyres and a Paasche index. It is calculated monthly, and within each calendar year the weights used remain constant, so that it takes the form of a Laspeyres index. Each January, however, the weights are updated on the basis of household spending patterns, so that the index is in fact a set of chain-linked Laspeyres indices, the chaining taking place in January each year. Despite the formal appearance as a Laspeyres index, the CPI measured over a period of years has the characteristics of a Paasche index, due to the annual change in the weights.

Some further adjustments need to be made for the index to be accurate. A change in the quality of goods purchased can be problematic, as alluded to earlier. If a manufacturer improves the quality of a product and charges more, is it fair to say that the price has gone up? Sometimes it is possible to measure improvement (if the power of a vacuum cleaner is increased, for example), but other cases are more difficult, such as if the punctuality of a train service is improved. By how much has quality improved? In many circumstances the statistician has to make a judgement about the best procedure to adopt. The ONS does make explicit allowance for the increase in quality of personal computers, for example, taking account of such as factors as increased memory and processing speed.

STATISTICS IN PRACTICE

Prices in the long run

Table 10.19 shows how prices have changed over the longer term. The 'inflation-adjusted' column shows what the item would have cost if it had risen in line with the overall retail price index. It is clear that some relative prices have changed substantially and you can try to work out the reasons.

Table 10.19 Eighty years of prices: 1914–94

Item	1914 price	Inflation-adjusted price	1994 price
Car	£730	£36 971	£6995
London–Manchester 1st class rail fare	£2.45	£124.08	£130
Pint of beer	1p	53p	£1.38
Milk (quart)	1.5p	74p	70p
Bread	2.5p	£1.21	51p
Butter	6p	£3.06	68p
Double room at Savoy Hotel, London	£1.25	£63.31	£195

The Office for National Statistics has gone back even further and shown that, since 1750, prices have increased about 140 times. Most of this occurred after 1938: up till then prices had only risen by about three times (over two centuries, about half a percent per year on average), since then prices have risen 40-fold, or about 6% p.a.

Exercise 10.6

?

The Laspeyres index of energy prices for the years 1999–2003 was:

1999	2000	2001	2002	2003
100	101.01	109.49	99.40	99.22

Use these data to calculate a chain index from 1999 to 2010, setting 1999 = 100.

Discounting and present values

Deflating makes expenditures in different years comparable by correcting for the effect of inflation. The future sum is deflated (reduced) because of the increase in the general price level. **Discounting** is a similar procedure for comparing amounts across different years, correcting for **time preference**. For example, suppose that by investing £1000 today a firm can receive £1100 in a year's time. To decide if the investment is worthwhile, the two amounts need to be compared, to see which the firm prefers.

If the prevailing interest rate is 12%, then the firm could simply place its £1000 in the bank and earn £120 interest, giving it £1120 at the end of the year. Hence the firm should not invest in this particular project; it does better keeping money in the bank. The investment is not undertaken because

$$£1000 \times (1 + r) > £1100$$

where r is the interest rate, 12% or 0.12. Alternatively, this inequality may be expressed as

$$1000 > \frac{1100}{(1+r)} = \frac{1100}{1+0.12} = 982.14$$

The expression on the right-hand side of the inequality sign is the **present value** (*PV*) of £1100 received in one year's time, £982.14. Here, r is the rate of discount and is equal to the rate of interest in this example because this is the rate at which the firm can transform present into future income, and vice versa. In what follows, we use the terms interest rate and discount rate interchangeably. The term $1/(1+r)$ is known as the **discount factor**. Multiplying an amount by the discount factor results in the present value of the sum.

We can also express the inequality as follows (by subtracting £1000 from each side):

$$0 > -1000 + \frac{1100}{(1+r)} = -17.86$$

The right-hand side of this expression is known as the **net present value** (*NPV*) of the project. It represents the difference between the initial outlay and the present value of the return generated by the investment. Since this is negative, the investment is not worthwhile (the money would be better placed on deposit in a bank). The general rule is to invest if the *NPV* is positive.

Similarly, the present value of £1100 to be received in two years' time is

$$PV = \frac{£1100}{(1+r)^2} = \frac{£1100}{(1+0.12)^2} = £876.91$$

When $r = 12\%$. In general, the *PV* of a sum S to be received in t years is

$$PV = \frac{S}{(1+r)^t}$$

The *PV* may be interpreted as the amount a firm would be prepared to pay today to receive an amount S in t years' time. Thus a firm would not be prepared to make an outlay of more than £876.91 in order to receive £1100 in two years' time. It would gain more by putting the money on deposit and earning 12% interest p.a.

Most investment projects involve an initial outlay followed by a *series* of receipts over the following years, as illustrated by the figures in Table 10.20. In order to decide if the investment is worthwhile, the present value of the income stream needs to be compared to the initial outlay. The *PV* of the income stream is obtained by adding together the present value of each year's income. Thus we calculate[4]

$$PV = \frac{S_1}{(1+r)} + \frac{S_2}{(1+r)^2} + \frac{S_3}{(1+r)^3} + \frac{S_4}{(1+r)^4} \tag{10.16}$$

or more concisely, using Σ notation:

$$PV = \sum \frac{S_t}{(1+r)^t} \tag{10.17}$$

[4]This present value example has only four terms, but in principle there can be any number of terms stretching into the future.

Table 10.20 The cash flows from an investment project

Year	Outlay or income		Discount factor	Discounted income
2005	Outlay	−1000		
2006	Income	300	0.893	267.86
2007		400	0.797	318.88
2008		450	0.712	320.30
2009		200	0.636	127.10
Total				1034.14

Note: The discount factors are calculated as $0.893 = 1/(1.12)$, $0.797 = 1/(1.12)^2$, etc.

Columns 3 and 4 of the table show the calculation of the present value. The discount factors, $1/(1 + r)^t$, are given in column 3. Multiplying column 2 by column 3 gives the individual elements of the *PV* calculation (as in equation (10.16)) and their sum is 1034.14, which is the present value of the returns. Since the *PV* is greater than the initial outlay of 1000, the investment generates a return of at least 12% and so is worthwhile.

An alternative investment criterion: the internal rate of return

The investment rule can be expressed in a different manner, using the **internal rate of return** (*IRR*). This is the rate of discount which makes the *NPV* equal to zero, i.e. the present value of the income stream is equal to the initial outlay. An *IRR* of 10% equates £1100 received next year to an outlay of £1000 today. Since the *IRR* is less than the market interest rate (12%), this indicates that the investment is not worthwhile: it only yields a rate of return of 10%. The rule 'invest if the *IRR* is greater than the market rate of interest' is similar to the rule 'invest if the net present value is positive, using the interest rate to discount future revenues'.

In general, it is mathematically difficult to find the *IRR* of even a simple project by hand – a computer needs to be used. The *IRR* is the value of *r* which sets the *NPV* equal to zero, i.e. it is the solution to

$$NPV = -S_0 + \sum \frac{S_t}{(1 + r)^t} = 0 \qquad (10.18)$$

where S_0 is the initial outlay. Fortunately, most spreadsheet programs have an internal routine for its calculation. This is illustrated in Figure 10.2 which shows the calculation of the *IRR* for the data in Table 10.20.

Cell C13 contains the formula '= *IRR*(C6:C10, 0.1)' – this can be seen just above the column headings – which is the function used in Excel to calculate the internal rate of return. The financial flows of the project are in cells C6:C10; the value 0.1 (10%) in the formula is an initial guess at the answer – Excel starts from this value and then tries to improve upon it. The *IRR* for this project is found to be 13.7% which is indeed above the market interest rate of 12%. The final two columns confirm that the *PV* of the income stream, when discounted using this internal rate of return, is equal to the initial outlay (as it should be). The discount factors in the penultimate column are calculated using $r = 13.7\%$.

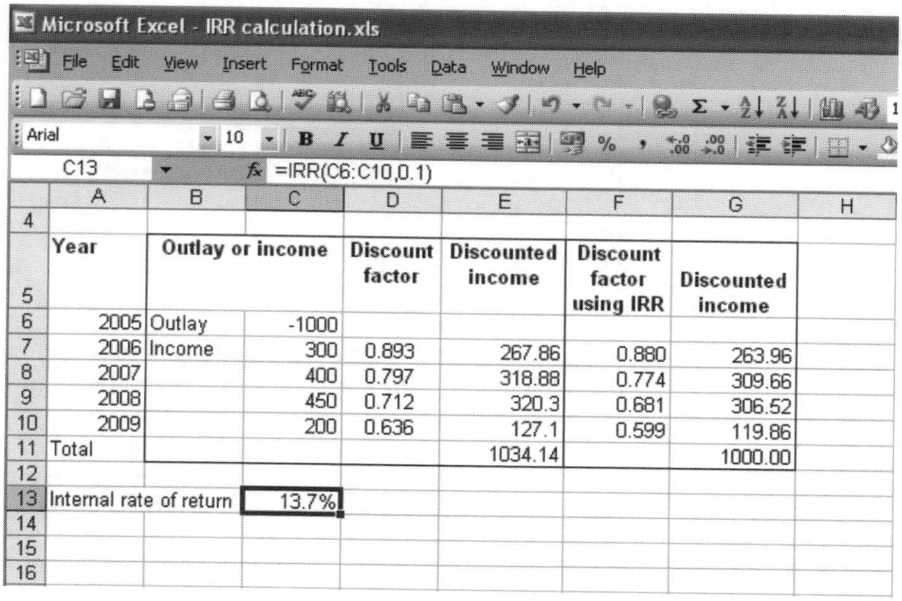

	A	B		C	D	E	F	G
5	Year	Outlay or income			Discount factor	Discounted income	Discount factor using IRR	Discounted income
6	2005	Outlay		-1000				
7	2006	Income		300	0.893	267.86	0.880	263.96
8	2007			400	0.797	318.88	0.774	309.66
9	2008			450	0.712	320.3	0.681	306.52
10	2009			200	0.636	127.1	0.599	119.86
11	Total					1034.14		1000.00
13	Internal rate of return			13.7%				

Figure 10.2 Calculation of the *IRR* using Excel

The *IRR* is particularly easy to calculate if the income stream is a constant monetary sum. If the initial outlay is S_0 and a sum S is received each year in perpetuity (like a bond), then the *IRR* is simply

$$IRR = \frac{S}{S_0}$$

For example, if an outlay of £1000 yields a permanent income stream of £120 p.a. then the *IRR* is 12%. This should be intuitively obvious, since investing £1000 at an interest rate of 12% would give you an annual income of £120.

Although the *NPV* and *IRR* methods are identical in the above example, this is not always the case in more complex examples. When comparing two investment projects of different sizes, it is possible for the two methods to come up with different rankings. Delving into this issue is beyond the scope of this text but, in general, the *NPV* method is the more reliable of the two.

Nominal and real interest rates

The above example took no account of possible inflation. If there were a high rate of inflation, part of the future returns to the project would be purely inflationary gains and would not reflect real resources. Is it possible our calculation is misleading under such circumstances?

There are two ways of dealing with this problem:

(a) use the actual cash flows and the nominal (market) interest rate to discount, or

(b) use real (inflation-adjusted) flows and the real interest rate.

These two methods should give the same answer.

If an income stream has already been deflated to real terms, then the present value should be obtained by discounting by the **real interest rate**, not the nominal

Table 10.21 Discounting a real income stream

Year		Cash flows (1)	Price index (2)	Real income (3)	Real discount factor (4)	Discounted sums (5)
2005	Outlay	−1000	100			
2006	Income	300	107.0	280.37	0.955	267.86
2007		400	114.5	349.38	0.913	318.88
2008		450	122.5	367.33	0.872	320.30
2009		200	131.1	152.58	0.833	127.10
Total						1034.14

(market) rate. Table 10.21 illustrates the principle. Column (1) repeats the income flows in cash terms from Table 10.20. Assuming an inflation rate of $i = 7\%$ p.a. gives the price index shown in column (2), based on 2005 = 100. This is used to deflate the cash series to real terms, shown in column (3). This is in constant (2005) prices. If we were presented only with the real income series and could not obtain the original cash flows we would have to discount the real series by the real interest rate r_r, defined by

$$1 + r_r = \frac{1 + r}{1 + i} \tag{10.19}$$

With a (nominal) interest rate of 12% and an inflation rate of 7% this gives

$$1 + r_r = \frac{1 + 0.12}{1 + 0.07} = 1.0467 \tag{10.20}$$

so that the real interest rate is 4.67% and in this example is the same every year. The one-year discount factor is therefore $1/1.0467 = 0.955$, for two years it is $1/1.0467^2 = 0.913$, etc. The discount factors used to discount the real income flows are shown in column (4) of the table, based on the real interest rate; the discounted sums are in column (5) and the present value of the real income series is £1034.14. This is the same as was found earlier, by discounting the cash figures by the nominal interest rate. Thus one can discount *either* the nominal (cash) values using the nominal discount rate, *or* the real flows by the real interest rate. Make sure you do not confuse the nominal and real interest rates.

The real interest rate can be approximated by subtracting the inflation rate from the nominal interest rate, i.e. 12% − 7% = 5%. This gives a reasonably accurate approximation for low values of the interest and inflation rates (below about 10% p.a.). Because of the simplicity of the calculation, this method is often preferred.

Exercise 10.7

?

(a) An investment of £100 000 yields returns of £25 000, £35 000, £30 000 and £15 000 in each of the subsequent four years. Calculate the present value of the income stream and compare to the initial outlay, using an interest rate of 10% p.a.

(b) Calculate the internal rate of return on this investment.

Exercise 10.8

?

(a) An investment of £50 000 yields cash returns of £20 000, £25 000, £30 000 and £10 000 in each subsequent year. The rate of inflation is a constant 5% and the rate of interest is constant at 9%. Use the rate of inflation to construct a price index and discount the cash flows to real terms.

(b) Calculate the real discount rate.

(c) Use the real discount rate to calculate the present value of the real income flows.

(d) Compare the result in part (c) to the answer obtained using the nominal cash flows and nominal interest rate.

Inequality indices

A separate set of index numbers is used specifically in the measurement of inequality, such as inequality in the distribution of income. We have already seen how we can measure the dispersion of a distribution (such as that of wealth) via the variance and standard deviation. This is based upon the deviations of the observations about the mean. An alternative idea is to measure the difference between *every pair* of observations, and this forms the basis of a statistic known as the **Gini coefficient**. This would probably have remained an obscure measure, due to the complexity of calculation, were it not for Konrad Lorenz, who showed that there is an attractive visual interpretation of it, now known as the **Lorenz curve**, and a relatively simple calculation of the Gini coefficient, based on this curve.

We start off by constructing the Lorenz curve, based on data for the UK income distribution in 2006 to 2007, and proceed then to calculate the Gini coefficient. We then use these measures to look at inequality both over time (in the United Kingdom) and across different countries.

We then examine another manifestation of inequality, in terms of market shares of firms. For this analysis we look at the calculation of **concentration ratios** and at their interpretation.

The Lorenz curve

Table 10.22 shows the data for the distribution of income in the United Kingdom based on data from the *Family Resources Survey* 2006 to 2007, published by the ONS. The data report the total weekly income of each household, which means that income

Table 10.22 The distribution of gross income in the United Kingdom, 2006–7

Range of weekly household income	Mid-point of interval	Numbers of households
0–	50	516
100–	150	3095
200–	250	3869
300–	350	3095
400–	450	2579
500–	550	2063
600–	650	2063
700–	750	1548
800–	850	1290
900–	950	1032
1000–	1250	4385
Total		25535

Figure 10.3
A typical Lorenz curve

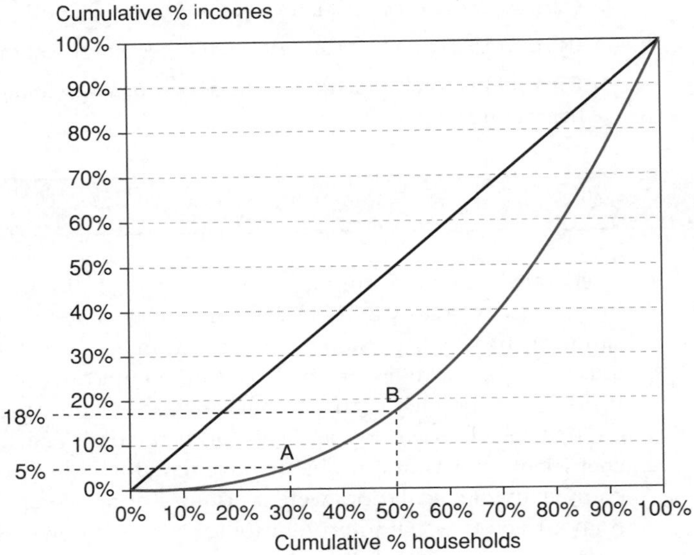

is recorded after any cash benefits from the state (e.g. a pension) have been received but before any taxes have been paid.

The table indicates a substantial degree of inequality. For example, the poorest 14% ((516 + 3095)/25,535) of households earn £200 per week or less, while the richest 17% earn more than £1000, five times as much. Although these figures give some idea of the extent of inequality, they relate only to relatively few households at the extremes of the distribution. A **Lorenz curve** is a way of graphically presenting the whole distribution. A typical Lorenz curve is shown in Figure 10.3.

Households are ranked along the horizontal axis, from poorest to richest, so that the median household, for example, is halfway along the axis. On the vertical axis is measured the cumulative share of income, which goes from 0% to 100%. A point such as A on the diagram indicates that the poorest 30% of households earn 5% of total income. Point B shows that the poorest half of the population earn only 18% of income (and hence the other half earn 82%). Joining up all such points maps out the Lorenz curve.

A few things are immediately obvious about the Lorenz curve:

- Since 0% of households earn 0% of income, and 100% of households earn 100% of income, the curve must run from the origin up to the opposite corner.
- Since households are ranked from poorest to richest, the Lorenz curve must lie below the 45° line, which is the line representing complete equality. The further below the 45° line is the Lorenz curve, the greater is the degree of inequality.
- The Lorenz curve must be concave from above: as we move to the right we encounter successively richer individuals, so the cumulative income grows faster.

Table 10.23 shows how to generate the data points from which the Lorenz curve may be drawn, for the data given in Table 10.22. The task is to calculate

Table 10.23 Calculation of the Lorenz curve coordinates

Range of income (1)	Mid-point (2)	Numbers of households (3)	Total income (4)	% Households (5)	% Cumulative households (x) (6)	% Income (7)	% Cumulative income (y) (8)
0–	50	516	25 792	2.0%	2.0%	0.2%	0.2%
100–	150	3 095	464 256	12.1%	14.1%	3.1%	3.3%
200–	250	3 869	967 200	15.2%	29.3%	6.5%	9.8%
300–	350	3 095	1 083 264	12.1%	41.4%	7.3%	17.1%
400–	450	2 579	1 160 640	10.1%	51.5%	7.8%	24.8%
500–	550	2 063	1 134 848	8.1%	59.6%	7.6%	32.5%
600–	650	2 063	1 341 184	8.1%	67.7%	9.0%	41.5%
700–	750	1 548	1 160 640	6.1%	73.7%	7.8%	49.3%
800–	850	1 290	1 096 160	5.1%	78.8%	7.4%	56.6%
900–	950	1 032	980 096	4.0%	82.8%	6.6%	63.2%
1000–	1 250	4 385	5 480 800	17.2%	100.0%	36.8%	100.0%
		25 535	14 894 880	100.0%		100.0%	

Notes:
Column (4) = column (2) × column (3)
Column (5) = column (3) ÷ 25 535
Column (6) = column (5) cumulated
Column (7) = column (4) ÷ 14 894 880
Column (8) = column (7) cumulated

the $\{x, y\}$ coordinates for the Lorenz curve, where y is the percentage share of total income earned by the poorest x% of the population. These values are given in columns (6) and (8), respectively, of the table. Their calculation is as follows:

- Column (3) gives the number of households in each income class and column (5) gives this as a percentage of all households.
- Column (6) cumulates the percentage shares in column (5), giving the percentage of households *below* a given income level. Thus, for example, 29.3% of households have income below £300 per week. These figures are the ones used for the x coordinates.
- The total income accruing to each income class is shown in column (4), obtained by multiplying the number of households by the mid-point of the class interval. The share of total income accruing to each class is then calculated in column (7), with these shares cumulated in column (8). These are the y coordinates for the chart.

Using columns (6) and (8) of the table we can see, for instance, that the poorest 2% of the population have about 0.2% of total income (one-tenth of their 'fair share'); the poorer half have about 25% of income (half of their fair share); and the top 20% have about 46% of total income. Figure 10.4 shows the Lorenz curve plotted, using the data in columns (6) and (8) of Table 10.23.

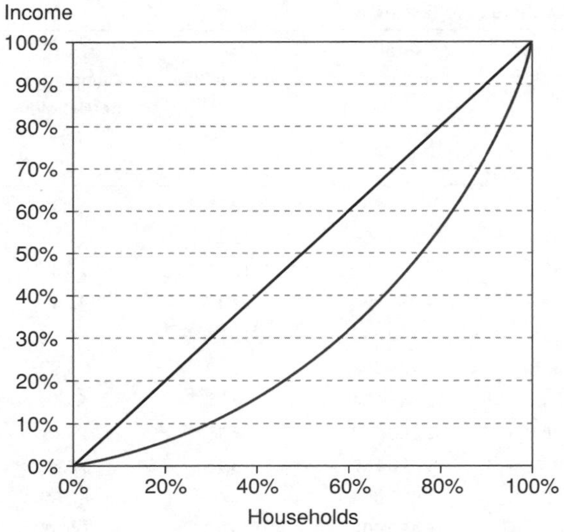

Figure 10.4
Lorenz curve for income data

The Gini coefficient

The Gini coefficient is a numerical representation of the degree of inequality in a distribution and can be derived directly from the Lorenz curve. The Lorenz curve is illustrated once again in Figure 10.5 and the Gini coefficient is simply the ratio of area A to the sum of areas A and B.

Denoting the Gini coefficient by G, we have

$$G = \frac{\text{A}}{\text{A} + \text{B}} \tag{10.21}$$

and it should be obvious that G must lie between 0 and 1. When there is total equality the Lorenz curve coincides with the 45° line, since 10% of the population

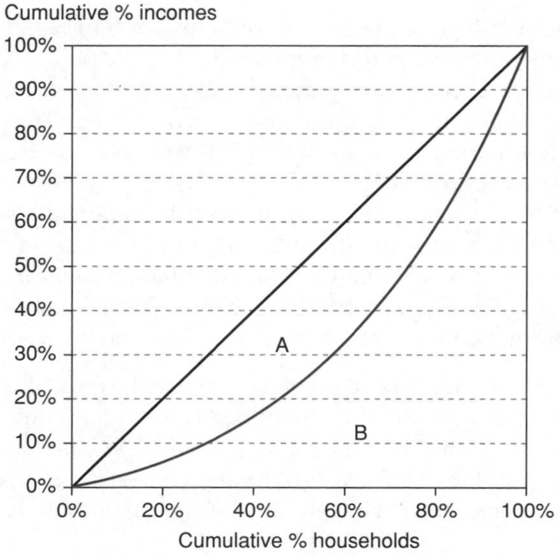

Figure 10.5
Calculation of the Gini coefficient from the Lorenz curve

earns 10% of the income, 20% earns 20%, etc. In this case, area A disappears, and $G = 0$ indicating no inequality. With total inequality (one household having all the income), area B disappears, and $G = 1$. Neither of these extremes is likely to occur in real life; instead one will get intermediate values, but the lower the value of G, the less inequality there is (though see the *caveats* listed below). One could compare two countries, for example, simply by examining the values of their Gini coefficients.

One interesting interpretation of the Gini is as follows. If you were to choose two people at random from the population, by how much would you expect their income to differ? Now the greater is inequality (greater G) the larger you would expect the difference to be. It can be shown that the expected difference is actually $2G$ as a proportion of the mean.

To give an example, if the Gini coefficient is 0.25 or 25%, then one would expect the difference between the two people to be 50% of mean income. In a country whose mean annual income is £30 000 (consistent with Table 10.22) one would expect a difference of £15 000, therefore.

The Gini coefficient may be calculated from the following rather cumbersome formulae for areas A and B, using the x and y coordinates from Table 10.23:

$$\begin{aligned} \mathrm{B} = \tfrac{1}{2}\{ &(x_1 - x_0) \times (y_1 + y_0) \\ &+ (x_2 - x_1) \times (y_2 + y_1) \\ &\quad\vdots \\ &+ (x_k - x_{k-1}) \times (y_k + y_{k-1})\} \end{aligned} \tag{10.22}$$

$x_0 = y_0 = 0$ and $x_k = y_k = 100$ represent the two end-points of the Lorenz curve and the other x and y values are the coordinates of the intermediate points. k is the number of classes for income in the frequency table. Area A is then given by:[5]

$$\mathrm{A} = 5000 - \mathrm{B} \tag{10.23}$$

and the Gini coefficient is then calculated as:

$$G = \frac{\mathrm{A}}{\mathrm{A} + \mathrm{B}} \text{ or } \frac{\mathrm{A}}{5000} \tag{10.24}$$

Thus for the data in Table 10.23 we have:

$$\begin{aligned} \mathrm{B} = \tfrac{1}{2} \times &(2.0 - 0) \times (0.2 + 0) \\ &+ (14.1 - 2.0) \times (3.3 + 0.2) \\ &+ (29.3 - 14.1) \times (9.8 + 3.3) \\ &+ (41.4 - 29.3) \times (17.1 + 9.8) \\ &+ (51.5 - 41.4) \times (24.8 + 17.1) \\ &+ (59.6 - 51.5) \times (32.5 + 24.8) \\ &+ (67.7 - 59.6) \times (41.5 + 32.5) \\ &+ (73.7 - 67.7) \times (49.3 + 41.5) \\ &+ (78.8 - 73.7) \times (56.6 + 49.3) \\ &+ (82.8 - 78.8) \times (63.2 + 56.6) \\ &+ (100 - 82.8) \times (100 + 63.2) \\ = {} &3210.5 \end{aligned} \tag{10.25}$$

[5]The value 5000 is correct if one uses percentages, as here (it is $100 \times 100 \times$ the area of the triangle). If one uses percentages expressed as decimals, then $\mathrm{A} = 0.5 - \mathrm{B}$.

Therefore, area A = 5000 − 3210.5 = 1789.5 and we obtain

$$G = \frac{1789.5}{5000} = 0.3579 \tag{10.26}$$

or approximately 36%. Based on the earlier discussion, this means that the expected income difference between two randomly selected people is about 70% of the mean, which equates to £21 000 approximately.

This method implicitly assumes that the Lorenz curve is made up of straight line segments connecting the observed points, which is in fact not quite true – it should be a smooth curve. Since the straight lines will lie inside the true Lorenz curve, area B is *over*-estimated and so the calculated Gini coefficient is biased downwards. The true value of the Gini coefficient is slightly greater than 36%, therefore. The bias will be greater (a) the fewer the number of observations and (b) the more concave is the Lorenz curve (i.e. the greater is inequality). The bias is unlikely to be substantial, however, so is best left untreated.

Is inequality increasing?

The Gini coefficient is only useful as a comparative measure, for looking at trends in inequality over time or for comparing different countries or regions. Table 10.24 shows the value of the Gini coefficient for the United Kingdom over the past 20 years or so and shows how it was affected by the tax system. The results are based on *equivalised* income, i.e. after making a correction for differences in family size.[6] For this reason there is a slight difference from the Gini coefficient calculated above, which uses unadjusted data.

Using equivalised income appears to make little difference in this case (compare the 'gross income' column with the earlier calculation).

The table shows essentially two things:

(1) The Gini coefficient changes little over time, suggesting that inequality is persistent. This is in spite of other recent evidence which tends to show growing inequality at the top of the income distribution.

(2) The biggest reduction in inequality comes through cash benefits paid out by the state, rather than through taxes. In fact, the tax system appears to *increase* inequality rather than to reduce it, primarily because of the effects of taxes other than the income tax.

Table 10.24 Gini coefficients for the United Kingdom, 1989–2009/10

	Original income	Gross income	Disposable income	Post-tax income
1989	49.7	36.3	34.4	37.8
1999/00	52.5	38.7	35.8	40.0
2009/10	52.0	37.2	33.5	37.1

Note: Gross income is original income plus certain state benefits, such as pensions. Taking off direct taxes gives disposable income and subtracting other taxes gives post-tax income.

Source: The effects of taxes and benefits on household income, 2009/10, Office of National Statistics, 2011.

[6] This is because a larger family needs more income to have the same living standard as a smaller one.

Table 10.25 Gini coefficients in past times

Year	Gini
1688	0.55
1801–3	0.56
1867	0.52
1913	0.43–0.63

Recent increases in inequality are a reversal of the historical trend. The figures presented in Table 10.25, from L. Soltow,[7] provide estimates of the Gini coefficient in earlier times. These figures suggest that a substantial decline in the Gini coefficient has occurred in the last century or so, perhaps related to the process of economic development. It is difficult to compare Soltow's figures directly with the modern ones because of such factors as the quality of data and different definitions of income.

Using the Gini coefficient to compare countries shows that the United Kingdom has one of the higher figures among advanced nations. OECD figures[8] show, for 2012, a Gini of 0.35 for the United Kingdom, 0.29 for Germany, 0.27 for Sweden and 0.33 for Italy, for example. The United States is one country that has more inequality than the UK, its Gini being 0.39. The average for all OECD countries is 0.32.

A simpler formula for the Gini coefficient

Kravis, Heston and Summers[9] provide estimates of 'world' GDP by decile and these figures, presented in Table 10.26, will be used to illustrate another method of calculating the Gini coefficient.

These figures show that the poorer half of the world population earns only about 10% of world income and that a third of world income goes to the richest 10% of the population. This suggests a higher degree of inequality than for a single country such as the United Kingdom, as one might expect.

When the class intervals contain equal numbers of households (for example, when the data are given for deciles of the income distribution, as here) formula (10.22) for area B simplifies to:

$$B = \frac{100}{2k}(y_0 + 2y_1 + 2y_2 + \cdots + 2y_{k-1} + y_k) = \frac{100}{k}\left(\sum_{i=0}^{i=k} y_i - 50\right) \qquad (10.27)$$

where k is the number of intervals (e.g. 10 in the case of deciles, 5 for quintiles). Thus you simply sum the y values, subtract 50,[10] and divide by the number

Table 10.26 The world distribution of income by decile

Decile	1	2	3	4	5	6	7	8	9	10
% GDP	1.5	2.1	2.4	2.4	3.3	5.2	8.4	17.1	24.1	33.5
Cumulative %	1.5	3.6	6.0	8.4	11.7	16.9	25.3	42.4	66.5	100.0

[7] Long run changes in British income inequality, *Economic History Review*, 21, 17–29, 1968.

[8] http://stats.oecd.org/Index.aspx?DataSetCode=IDD

[9] Real GDP per capita for more than one hundred countries, *Economic Journal*, 88 (349), 215–42, 1978.

[10] If using decimal percentages, subtract 0.5.

of classes k. The y values for the Kravis *et al.* data appear in the final row of Table 10.26, and their sum is 282.3. We therefore obtain

$$\mathrm{B} = \frac{100}{10}(282.3 - 50) = 2323 \tag{10.28}$$

Hence

$$\mathrm{A} = 5000 - 2323 = 2677 \tag{10.29}$$

and

$$G = \frac{2677}{5000} = 0.5354 \tag{10.30}$$

or about 53%. This is surprisingly similar to the figure for original income in the United Kingdom, but, of course, differences in definition, measurement, etc., may make direct comparison invalid. While the Gini coefficient may provide some guidance when comparing inequality over time or across countries, one needs to take care in its interpretation.

Exercise 10.9

(a) The same data as used in the text are presented below, but with fewer class intervals:

Range of income	Mid-point of interval	Numbers of households
0–	100	3 611
200–	300	6 964
400–	500	4 643
600–	700	3 611
800–	900	2 321
1 000–	1250	4 385
Total		25 535

Draw the Lorenz curve for these data.

(b) Calculate the Gini coefficient for these data and compare to that calculated earlier.

Exercise 10.10

Given shares of total income of 8%, 15%, 22%, 25% and 30% by each quintile of a country's population, calculate the Gini coefficient.

STATISTICS IN PRACTICE

Inequality and development

Table 10.27 presents figures for the income distribution in selected countries around the world. They are in approximately ascending order of national income.

The table shows that countries have very different experiences of inequality, even for similar levels of income (compare Bangladesh and Kenya, for example). Hungary, the only (former) communist country, shows the greatest equality, although whether income accurately measures people's access to resources in such a regime is perhaps debatable. Note that countries with fast growth (such as Korea and Hong Kong) do not have to have a high degree of inequality. Developed countries seem to have uniformly low Gini coefficients.

Table 10.27 Income distribution figures in selected countries

	Year	Quintiles					Top 10%	Gini
		1	2	3	4	5		
Bangladesh	1981–82	6.6	10.7	15.3	22.1	45.3	29.5	0.36
Kenya	1976	2.6	6.3	11.5	19.2	60.4	45.8	0.51
Côte d'Ivoire	1985–86	2.4	6.2	10.9	19.1	61.4	43.7	0.52
El Salvador	1976–77	5.5	10.0	14.8	22.4	47.3	29.5	0.38
Brazil	1972	2.0	5.0	9.4	17.0	66.6	50.6	0.56
Hungary	1982	6.9	13.6	19.2	24.5	35.8	20.5	0.27
Korea, Rep.	1976	5.7	11.2	15.4	22.4	45.3	27.5	0.36
Hong Kong	1980	5.4	10.8	15.2	21.6	47.0	31.3	0.38
New Zealand	1981–82	5.1	10.8	16.2	23.2	44.7	28.7	0.37
United Kingdom	1979	7.0	11.5	17.0	24.8	39.7	23.4	0.31
Netherlands	1981	8.3	14.1	18.2	23.2	36.2	21.5	0.26
Japan	1979	8.7	13.2	17.5	23.1	37.5	22.4	0.27

Source: World Development Report 2006.

Concentration ratios

Another type of inequality is the distribution of market shares of the firms in an industry. We all know that Microsoft currently dominates the software market with a large market share. In contrast, an industry such as bakery products has many different suppliers and there is little tendency towards dominance. The **concentration ratio** is a commonly used measure to examine the distribution of market shares among firms competing in a market. Of course, it would be possible to measure this using the Lorenz curve and Gini coefficient, but the concentration ratio has the advantage that it can be calculated on the basis of less information and also tends to focus attention on the largest firms in the industry. The concentration ratio is often used as a measure of the competitiveness of a particular market but, as with all statistics, it requires careful interpretation.

A market is said to be concentrated if most of the demand is met by a small number of suppliers. The limiting case is monopoly where the whole of the market is supplied by a single firm. We shall measure the degree of concentration by the **five-firm concentration ratio**, which is the proportion of the market held by the largest five firms, and it is denoted C_5. The larger is this proportion, the greater the degree of concentration and potentially the less competitive is that market. Table 10.28 gives the (imaginary) sales figures of all the firms in a particular industry.

For convenience the firms have already been ranked by size from A (the largest) to J (smallest). The output of the five largest firms is 482, out of a total of 569, so the five-firm concentration ratio is $C_5 = 84.7\%$, i.e. 84.7% of the market is supplied by the five largest firms.

Table 10.28 Sales figures for an industry (millions of units)

Firm	A	B	C	D	E	F	G	H	I	J
Sales	180	115	90	62	35	25	19	18	15	10

Without supporting evidence, it is hard to interpret this figure. Does it mean that the market is not competitive and the consumer being exploited? Some industries, such as the computer industry, have a very high concentration ratio, yet it is hard to deny that they are fiercely competitive. On the other hand, some industries with no large firms have restrictive practices, entry barriers, etc., which means that they are not very competitive (lawyers might be one example). A further point is that there may be a *threat* of competition from outside the industry which keeps the few firms acting competitively.

Concentration ratios can be calculated for different numbers of largest firms, e.g. the three-firm or four-firm concentration ratio. Straightforward calculation reveals them to be 67.7% and 78.6%, respectively for the data given in Table 10.28. There is little reason in general to prefer one measure to the others, and they may give different pictures of the degree of concentration in an industry.

The concentration ratio calculated above relates to the quantity of output produced by each firm, but it is possible to do the same with sales revenue, employment, investment or any other variable for which data are available. The interpretation of the results will be different in each case. For example, the largest firms in an industry, while producing the majority of output, might not provide the greater part of employment if they use more capital-intensive methods of production. Concentration ratios obviously have to be treated with caution, therefore, and are probably best combined with case studies of the particular industry before conclusions are reached about the degree of competition.

STATISTICS IN PRACTICE

Concentration in the banking industry

A study of the banking industry by Beck *et al.* looked at the relationship between concentration in the industry, measured by the three firm concentration ratio, and the existence of systemic banking crises, using data for 1980–97. A selection of their data is shown below:

Country	Concentration ratio	Crisis years
Germany	0.48	None
Italy	0.35	1990–95
Japan	0.24	1992–97
Sweden	0.89	1990–93
UK	0.57	None
USA	0.19	1980–92

The authors found that there was a negative relationship between the concentration ratio (based on a measure of banks' assets) and crises; in other words, countries with more concentrated industries were generally safer (despite the evidence from Sweden, above). However, they found that a more competitive banking industry need not be risky, if there were a good system of regulation in place. Note that this study was completed before the major financial crisis of 2007, so the conclusions might no longer be valid.

Source: T. Beck et al., Bank concentration and fragility: impact and dynamics, NBER Working Paper 11500, http://www.nber.org/papers/w11500.

Exercise 10.11

Total sales in an industry are \$400m. The largest five firms have sales of \$180m, \$70m, \$40m, \$25m and \$15m. Calculate the three- and five-firm concentration ratios.

Summary

- An index number summarises the variation of a variable over time or across space in a convenient way.
- Several variables can be combined into one index, providing an average measure of their individual movements. The consumer price index is an example.
- The Laspeyres price index combines the prices of many individual goods using base-year quantities as weights. The Paasche index is similar but uses current-year weights to construct the index.
- Laspeyres and Paasche quantity indices can also be constructed, combining a number of individual quantity series using prices as weights. Base-year prices are used in the Laspeyres index, current-year prices in the Paasche.
- A price index series multiplied by a quantity index series results in an index of expenditures. Rearranging this demonstrates that deflating (dividing) an expenditure series by a price series results in a volume (quantity) index. This is the basis of deflating a series in cash (or nominal) terms to one measured in real terms (i.e. adjusted for price changes).
- Two series covering different time periods can be spliced together (as long as there is an overlapping year) to give one continuous chain index.
- Discounting the future is similar to deflating but corrects for the rate of time preference rather than inflation. A stream of future income can thus be discounted and summarised in terms of its present value.
- An investment can be evaluated by comparing the discounted present value of the future income stream to the initial outlay. The internal rate of return of an investment is a similar but alternative way of evaluating an investment project.
- The Gini coefficient is a form of index number that is used to measure inequality (e.g. of incomes). It can be given a visual representation using a Lorenz curve diagram.
- For measuring the inequality of market shares in an industry, the concentration ratio is commonly used.

Key terms and concepts

base year
base-year weights
cash terms
chain index
concentration ratio
constant prices
Consumer Price Index (CPI)
current prices
current-year weights
deflating a data series
discount factor
discounting
expenditure or value index
five-firm concentration ratio
Gini coefficient
index number
internal rate of return
Laspeyres price index
Lorenz curve
net present value
Paasche index
present value

→

quantity indices	time preference
real interest rate	value (or expenditure) indices
real terms	volume series
reference year	volume terms
retail price index	weighted average

Reference

Soltow, L., Long Run Changes in British Income Inequality, *Economic History Review*, 21(1), 17–29, 1968.

Formulae used in this chapter

Formula	Description	Notes
$P_L^n = \frac{\sum p_n q_0}{\sum p_0 q_0} \times 100$	Laspeyres price index for year *n* with base year 0	
$P_L^n = \frac{\sum p_n}{\sum p_0} \times s_0 \times 100$	Laspeyres price index using expenditure weights *s* in base year	
$P_P^n = \frac{\sum p_n q_0}{\sum p_0 q_n} \times 100$	Paasche price index for year *n*	
$P_P^n = \frac{1}{\sum \frac{p_0}{p_n} \times s_n} \times 100$	Paasche price index using expenditure weights *s* in current year	
$Q_L^n = \frac{\sum q_n p_0}{\sum q_0 p_0} \times 100$	Laspeyres quantity index	
$Q_P^n = \frac{\sum q_n p_n}{\sum q_0 p_n} \times 100$	Paasche quantity index	
$E^n = \frac{\sum p_n q_n}{\sum p_0 q_0} \times 100$	Expenditure index	
$PV = \frac{S}{(1 + r)^t}$	Present value	The value now of a sum *S* to be received in *t* years' time, using discount rate *r*
$NPV = -S_0 + \sum \frac{S_t}{(1 + r)^t}$	Net present value	The value of an investment S_0 now, yielding S_t p.a., discounted at a constant rate *r*

Problems

Some of the more challenging problems are indicated by highlighting the problem number in colour.

10.1 The data below show exports and imports for the United Kingdom, 2005–10, in £bn at current prices.

	2005	2006	2007	2008	2009	2010
Exports	331.1	379.1	374.0	422.9	395.6	440.9
Imports	373.8	419.8	416.7	462.0	421.2	477.6

(a) Construct index number series for exports and imports, setting the index equal to 100 in 2005 in each case.

(b) Is it possible, using only the two indices constructed in part (a), to construct an index number series for the balance of trade (exports minus imports)? If so, do so; if not, why not?

10.2 The following data show the gross operating surplus of companies, 2005–10, in the United Kingdom, in £m.

2005	2006	2007	2008	2009	2010
224 811	244 309	257 995	255 260	233 436	243 036

(a) Turn the data into an index number series with 2005 as the reference year.

(b) Transform the series so that 2008 is the reference year.

(c) What increase has there been in profits between 2005 and 2010? Between 2008 and 2010?

10.3 The following tables show energy prices and consumption in 1999–2003 (analogous to the data in the chapter for the years 2006–10).

	Coal (£/tonne)	Petroleum (£/tonne)	Electricity (£/MWh)	Gas (£/MWh)
1999	34.77	104.93	36.23	5.46
2000	35.12	137.9	34.69	6.06
2001	38.07	148.1	31.35	8.16
2002	34.56	150.16	29.83	7.80
2003	34.5	140	28.44	8.07

Year	Coal (m. tonnes)	Petroleum (m. tonnes)	Electricity (m. MWh)	Gas (m. MWh)
1999	2.04	5.33	110.98	176.82
2000	0.72	5.52	114.11	183.44
2001	1.69	6.6	111.34	179.84
2002	1.1	5.81	112.37	165.42
2003	0.69	6.69	113.93	172.16

(a) Construct a Laspeyres price index using 1999 as the base year.

(b) Construct a Paasche price index. Compare this result with the Laspeyres index. Do they differ significantly?

(c) Construct Laspeyres and Paasche quantity indices. Check that they satisfy the conditions that $E^n = P_L \times Q_P$ etc.

10.4 The prices of different house types in south-east England are given in the table below:

Year	Terraced houses	Semi-detached	Detached	Bungalows	Flats
1991	59 844	77 791	142 630	89 100	47 676
1992	55 769	73 839	137 053	82 109	43 695
1993	55 571	71 208	129 414	82 734	42 746
1994	57 296	71 850	130 159	83 471	44 092

(a) If the numbers of each type of house in 1991 were 1898, 1600, 1601, 499 and 1702, respectively, calculate the Laspeyres price index for 1991–94, based on 1991 = 100.

(b) Calculate the Paasche price index, based on the following numbers of dwellings:

Year	Terraced houses	Semi-detached	Detached	Bungalows	Flats
1992	1903	1615	1615	505	1710
1993	1906	1638	1633	511	1714
1994	1911	1655	1640	525	1717

(c) Compare Paasche and Laspeyres price series.

10.5 (a) Using the data in Problem 10.3, calculate the expenditure shares on each fuel in 1999 and the individual price index number series for each fuel, with 1999 = 100.

(b) Use these data to construct the Laspeyres price index using the expenditures shares approach. Check that it gives the same answer as in Problem 10.3(a).

10.6 The following table shows the weights in the retail price index and the values of the index itself, for 1990 and 1994.

	Food	Alcohol and tobacco	Housing	Fuel and light	Household items	Clothing	Personal goods	Travel	Leisure
Weights									
1990	205	111	185	50	111	69	39	152	78
1994	187	111	158	45	123	58	37	162	119
Prices									
1990	121.0	120.7	163.7	115.9	116.9	115.0	122.7	121.2	117.1
1994	139.5	162.1	156.8	133.9	132.4	116.0	152.4	150.7	145.7

(a) Calculate the Laspeyres price index for 1994, based on 1990 = 100.

(b) Draw a bar chart of the expenditure weights in 1990 and 1994 to show how spending patterns have changed. What major changes have occurred? Do individuals seem to be responding to changes in relative prices?

(c) The pensioner price index is similar to the general index calculated above, except that it excludes housing. What effect does this have on the index? What do you think is the justification for this omission?

(d) If consumers spent, on average, £188 per week in 1990 and £240 per week in 1994, calculate the real change in expenditure on food.

(e) Do consumers appear rational, i.e. do they respond as one would expect to relative price changes? If not, why not?

10.7 Construct a chain index from the following data series:

	1998	2006	2000	2001	2006	2010	2008
Series 1	100	110	115	122	125		
Series 2			100	107	111	119	121

What problems arise in devising such an index and how do you deal with them?

10.8 Construct a chain index for 2001–10 using the following data, setting 2004 = 100.

2001	2002	2003	2004	2005	2006	2007	2008	2009	2010
87	95	100	105						
			98	93	100	104	110		
							100	106	112

10.9 Industry is always concerned about the rising price of energy. It demands to be compensated for any rise over 5% in energy prices between 2007 and 2008. How much would this compensation cost? Which price index should be used to calculate the compensation and what difference would it make? (Use the energy price data in the chapter.)

10.10 Using the data in Problem 10.6, calculate how much the average consumer would need to be compensated for the rise in prices between 1990 and 1994.

10.11 The following data show expenditure on the National Health Service (in cash terms), the GDP deflator, the NHS pay and prices index, population and population of working age:

Year	NHS expenditure (£m) (1)	GDP Deflator 1973 = 100 (2)	NHS pay and price index 1973 = 100 (3)	Population (000) (4)	Population of working age (000) (5)
1987	21 495	442	573	56 930	34 987
1988	23 601	473	633	57 065	35 116
1989	25 906	504	678	57 236	35 222
1990	28 534	546	728	57 411	35 300
1991	32 321	585	792	57 801	35 467

(In all the following answers, set your index to 1987 = 100.)

(a) Turn the expenditure cash figures into an index number series.

(b) Calculate an index of 'real' NHS expenditure using the GDP deflator. How does this alter the expenditure series?

(c) Calculate an index of the volume of NHS expenditure using the NHS pay and prices index. How and why does this differ from the answer arrived at in (b)?

(d) Calculate indices of real and volume expenditure *per capita*. What difference does this make?

(e) Suppose that those not of working age cost twice as much to treat, on average, as those of working age. Construct an index of the need for health care and examine how health care expenditures have changed relative to need.

(f) How do you think the needs index calculated in (e) could be improved?

10.12 (a) If w represents the wage rate and p the price level, what is w/p?

(b) If Δw represents the annual growth in wages and i is the inflation rate, what is $\Delta w - i$?

(c) What does $\ln(w) - \ln(p)$ represent? (ln= natural logarithm.)

10.13 A firm is investing in a project and wishes to receive a rate of return of at least 15% on it. The stream of net income is:

Year	1	2	3	4
Income	600	650	700	400

(a) What is the present value of this income stream?

(b) If the investment costs £1600, should the firm invest? What is the net present value of the project?

10.14 A firm uses a discount rate of 12% for all its investment projects. Faced with the following choice of projects, which yields the higher *NPV*?

Project	Outlay	Income stream					
		1	2	3	4	5	6
A	5600	1000	1400	1500	2100	1450	700
B	6000	800	1400	1750	2500	1925	1200

10.15 Calculate the internal rate of return for the project in Problem 10.13. Use either trial and error methods or a computer to solve.

10.16 Calculate the internal rates of return for the projects in Problem 10.14.

10.17 (a) Draw a Lorenz curve and calculate the Gini coefficient for the wealth data in Table 1.3 (Chapter 1).

(b) Why is the Gini coefficient typically larger for wealth distributions than for income distributions?

10.18 (a) Draw a Lorenz curve and calculate the Gini coefficient for the 1979 wealth data contained in Problem 1.5 (Chapter 1). Draw the Lorenz curve on the same diagram as you used in Problem 10.17.

(b) How does the answer compare to 2005 wealth data?

10.19 The following table shows the income distribution by quintile for the United Kingdom in 2006–7, for various definitions of income:

Quintile	Income measure			
	Original	Gross	Disposable	Post-tax
1 (bottom)	3%	7%	7%	6%
2	7%	10%	12%	11%
3	15%	16%	16%	16%
4	24%	23%	22%	22%
5 (top)	51%	44%	42%	44%

(a) Use equation (10.27) to calculate the Gini coefficient for each of the four categories of income.

(b) For the 'original income' category, draw a smooth Lorenz curve on a piece of gridded paper and calculate the Gini coefficient using the method of counting squares. How does your answer compare to that for part (a)?

10.20 For the Kravis, Heston and Summers data (Table 10.26), combine the deciles into quintiles and calculate the Gini coefficient from the quintile data. How does your answer compare with the answer given in the text, based on deciles? What do you conclude about the degree of bias?

10.21 Calculate the three-firm concentration ratio for employment in the following industry:

Firm	A	B	C	D	E	F	G	H
Employees	3350	290	440	1345	821	112	244	352

10.22 Compare the degrees of concentration in the following two industries. Can you say which is likely to be more competitive?

Firm	A	B	C	D	E	F	G	H	I	J
Sales	337	384	696	321	769	265	358	521	880	334
Sales	556	899	104	565	782	463	477	846	911	227

10.23 **(Project)** The World Development Report contains data on the income distributions of many countries around the world (by quintile). Use these data to compare income distributions across countries, focusing particularly on the differences between poor countries, middle-income and rich countries. Can you see any pattern emerging? Are there countries which do not fit into this pattern? Write a brief report summarising your findings.

Answers to exercises

Exercise 10.1

(a) 100, 111.9, 140.7, 163.4, 188.1.

(b) 61.2, 68.5, 86.1, 100, 115.1.

(c) $115.1/61.2 = 1.881$.

Exercise 10.2

	2002	2003	2004	2005	2006
(a) 2002 = 100	100	100.52	110.68	151.92	196.49
(b) 2004 = 100	90.35	90.83	100	137.26	177.53
(c) Using 2003 basket	100	100.71	110.75	151.94	196.48

Exercise 10.3

The Paasche index is:

2002	2003	2004	2005	2006
100	100.71	110.48	151.18	195.31

Exercise 10.4

(a) Expenditure shares in 2002 are:

	Expenditure	Share
Coal	66.92	1.2%
Petroleum	753.77	13.8%
Electricity	3360.35	61.4%
Gas	1288.33	23.6%

giving the Laspeyres index for 2003 as

$$P_1^n = \frac{34.03}{36.97} \times 0.012 + \frac{152.53}{132.24} \times 0.138 + \frac{28.68}{29.83} \times 0.614 + \frac{8.09}{7.80} \times 0.236$$
$$= 1.0052 \text{ or } 100.52.$$

The expenditure shares in 2003 are 0.3%, 8.9%, 46.3% and 44.4% which allows the 2003 Paasche index to be calculated as

$$P_1^n = \frac{1}{\frac{36.97}{34.03} \times 0.012 + \frac{132.24}{152.53} \times 0.149 + \frac{29.83}{28.68} \times 0.606 + \frac{7.80}{8.09} \times 0.233} \times 100$$
$$= 1.0071 \text{ or } 100.71.$$

Later years can be calculated in similar fashion.

Exercise 10.5

(a/b) The Laspeyres and Paasche quantity indexes are:

	Laspeyres index	Paasche index
2002	100	100
2003	101.95	102.14
2004	101.98	101.80
2005	103.33	102.83
2006	101.04	100.43

(c) The expenditure index is 100, 102.67, 112.67, 156.22 and 197.34.

(d) The Paasche quantity index times Laspeyres price index (or vice versa) gives the expenditure index.

Exercise 10.6

The full index is (using Laspeyres indexes):

				Chain index
1999	100			100
2000	101.01			101.01
2001	109.49			109.49
2002	99.40	100		99.40
2003	99.22	100.52		99.22
2004		110.68		109.25
2005		151.92		149.95
2006		196.49	100	193.95
2007			95.32	184.88
2008			126.17	244.71
2009			127.53	247.35
2010			121.72	236.08

Exercise 10.7

(a) The discounted figures are:

Year	Investment/yield	Discount factor	Discounted yield
0	−100 000		
1	25 000	0.9091	22 727.3
2	35 000	0.8264	28 925.6
3	30 000	0.7513	22 539.4
4	15 000	0.6830	10 245.2
Total			84 437.5

The present value is less than the initial outlay.

(b) The internal rate of return is 2.12%.

Exercise 10.8

(a) Deflating to real income gives:

Year	Investment/yield	Price index	Real income
0	−50 000	100	−50 000.0
1	20 000	105	19 047.6
2	25 000	110.250	22 675.7
3	30 000	115.763	25 915.1
4	10 000	121.551	8 227.0

(b) The real discount rate is 1.09/1.05 = 1.038 or 3.8% p.a.

(c/d)

Nominal values	Discount factor	Discounted value	Real values	Discount factor	Discounted value
−50 000			−50 000.0		
20 000	0.917	18 348.6	19 047.6	0.963	18 348.6
25 000	0.842	21 042.0	22 675.7	0.928	21 042.0
30 000	0.772	23 165.5	25 915.1	0.894	23 165.5
10 000	0.708	7 084.3	8 227.0	0.861	7 084.3
Totals		69 640.4			69 640.38

The present value is the same in both cases and exceeds the initial outlay.

Exercise 10.9

(b) Range of income	Mid-point	Number of households	Total income	% Households	% Cumulative households x	% Income	% Cumulative Income y
(1)	(2)	(3)	(4)	(5)	(6)	(7)	(8)
0–	100	3 611	361 100	14.1%	14.1%	2.4%	2.4%
200–	300	6 964	2 089 200	27.3%	41.4%	14.1%	16.5%
400–	500	4 643	2 321 500	18.2%	59.6%	15.6%	32.1%
600–	700	3 611	2 527 700	14.1%	73.7%	17.0%	49.1%
800–	900	2 321	2 088 900	9.1%	82.8%	14.0%	63.1%
1000–	1 250	4 385	5 481 250	17.2%	100.0%	36.9%	100.0%
Totals		25 535	14 869 650	100.0%		100.0%	

The Gini coefficient is then calculated as follows: B = 0.5 × {14.1 × (2.4 + 0) + 27.3 × (16.5 + 2.4) + 18.2 × (32.1 + 16.5) + 14.1 × (49.1 + 32.1) + 9.1 × (63.1 + 49.1) + 17.2 × (100 + 63.1) = 3201.} Area A = 5000 − 3301 = 1799. Hence Gini = 1799/5000 = 0.360, very similar to the value in the text using more categories of income.

Exercise 10.10

Using formula (10.28) we obtain B = 100/5 × (246 − 50) = 3920. Hence A = 1080 and Gini = 0.216. The sum of the cumulative y values is 246.

Exercise 10.11

$C_3 = 290/400 = 72.5\%$ and $C_5 = 82.5\%$.

Appendix Deriving the expenditure share form of the Laspeyres price index

We can obtain the expenditure share version of the formula from the standard formula given in equation (10.1):

$$P_L^n = \frac{\sum p_n q_0}{\sum p_0 q_0} = \frac{\sum \frac{p_n}{p_0} p_0 q_0}{\sum \frac{p_0}{p_0} p_0 q_0}$$

$$= \frac{\sum \frac{p_n}{p_0} \frac{p_0 q_0}{\sum p_0 q_0}}{\sum \frac{p_0}{p_0} \frac{p_0 q_0}{\sum p_0 q_0}} = \sum \frac{p_n}{p_0} \frac{p_0 q_0}{\sum p_0 q_0}$$

$$= \sum \frac{p_n}{p_0} \times s_0$$

which is equation (10.3) in the text (apart from the '× 100', omitted for simplicity in the derivation).

11 Seasonal adjustment of time-series data

Contents

Learning outcomes

By the end of this chapter you should be able to:

- recognise the different elements that make up a time series
- isolate the trend from a series, by either the additive or multiplicative method, or by using linear regression
- find the seasonal factors in a series
- use the seasonal factors to seasonally adjust the data
- forecast the series, taking account of seasonal factors
- appreciate the issues involved in the process of seasonal adjustment.

Introduction

'Economists noticed some signs in the data that suggest a turning point may be in the offing. The claimant count, although down, also showed February's figure had been revised to show a rise of 600 between January and February, the first occasion in 17 months that there had been an increase.'

The Guardian, 17 April 2008.

The quote above describes economists trying to spot a 'turning point' in the unemployment data, early in 2008. This is an extremely difficult task for several reasons:

- By definition, a turning point is a point at which a previous trend changes.
- Data are 'noisy', containing a lot of random movements.
- There may be seasonal factors involved (e.g. perhaps February's figures are usually considerably higher than January's, so what are we to make of a small increase?).

This chapter is concerned with the interpretation of time-series data, such as unemployment, retail sales, stock prices, etc. Agencies such as government, businesses and trade unions are interested in knowing how the economy is changing over time. Government may want to lower interest rates in response to an economic slowdown, businesses may want to know how much extra stock they need for Christmas, and trade unions will find pay bargaining more difficult if economic conditions worsen. For all of them, an accurate picture of the economy is important.

In this chapter we will show how to decompose a time series such as unemployment into its component parts: trend, cycle, seasonal and random. We then use this breakdown to **seasonally adjust** the original data, i.e. to remove any variation due solely to time of year effects (a vivid example would be the Christmas season in the case of retail sales). This allows us to more easily see any changes to the underlying data. Knowing the seasonal pattern to data also helps with forecasting: knowing that unemployment tends to be above trend in September can aid us in forecasting future levels in that month.

The methods used in this chapter are relatively straightforward compared to other, more sophisticated, methods that are available. However, they do illustrate the essential principles and give similar answers to the more advanced methods. Later in the chapter we discuss some of the more complex issues that can arise.

The components of a time series

Unemployment data will be used to illustrate the methods involved in decomposing a time series into its component parts. A similar analysis could be carried out for other time-series data, common examples being monthly sales data for a firm or quarterly data on the money supply. As always, one should begin by looking at the raw data, and the best way of doing this is via a time-series chart.

Table 11.1 presents the monthly unemployment figures for the period January 2012 to December 2014, and Figure 11.1 shows a plot of the data. The chart shows

Table 11.1 UK unemployment 2012–14

	2012	2013	2014
January	2612	2536	2220
February	2611	2509	2191
March	2565	2460	2105
April	2458	2407	2001
May	2472	2449	1992
June	2561	2495	2003
July	2584	2563	2013
August	2586	2558	2015
September	2573	2470	1998
October	2512	2325	1888
November	2442	2271	1771
December	2432	2245	1750

Note: The data are in 000s, so there were 2 612 000 people unemployed in January 2012.

Source: Adapted from data from the Office for National Statistics, UK unemployed aged over 16, not seasonally adjusted, licensed under the Open Government Licence v.1.0

Figure 11.1
Chart of unemployment data 2012–14

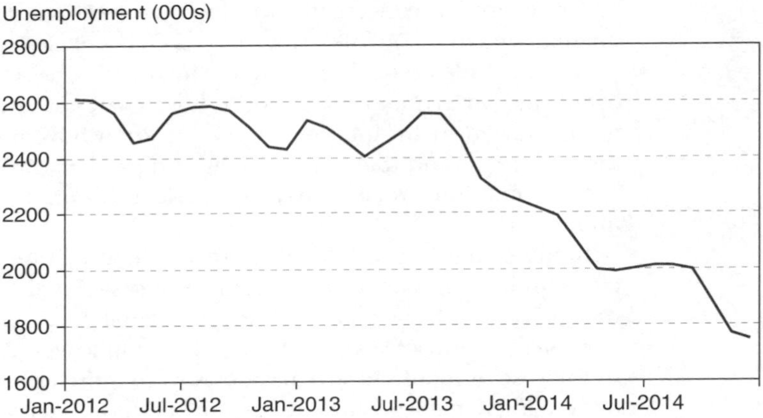

a downward trend to unemployment, particularly after mid-2013, around which there also appears to be a cycle of some kind.

Any time series such as this is made up of two types of elements:

(1) **systematic components**, such as a trend, cycle and seasonals, and
(2) **random elements**, which are by definition unpredictable.

It would be difficult to analyse a series which is completely random (such as the result of tossing a coin, see Figure 2.1 in Chapter 2). A look at the unemployment data, however, suggests that the series is definitely non-random – there is evidence of a downward trend and there does appear to be a seasonal component. The latter can be seen better if we superimpose each year on the same graph, as shown in Figure 11.2.

Note: 2011 and 2015 are also shown on the graph to provide more comparisons of the seasonal pattern.

The series generally show peaks around February and August/September, with dips around May and December time. The autumn peak occurs a little earlier in 2013. The other feature to note is that unemployment is generally falling over time, the lines for later years being below those for earlier years. If one wished to

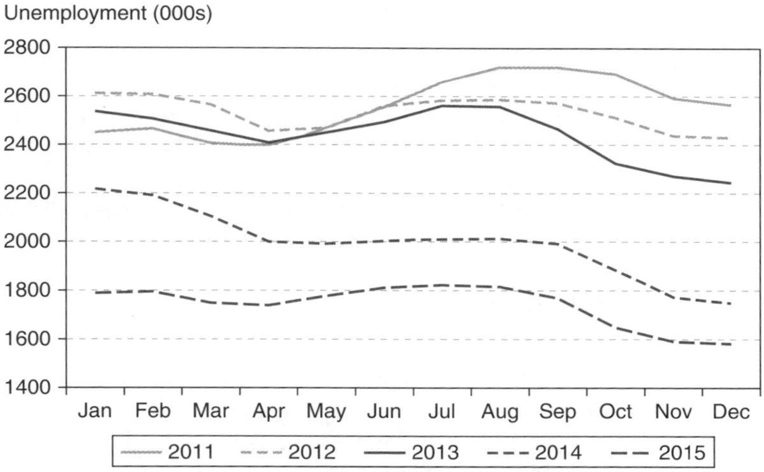

Figure 11.2 Superimposed time-series graphs of unemployment
Note: 2003 and 2007 have also been graphed to emphasise the similarity across several years.

predict unemployment for February 2016, the trend would be projected forward and account also taken of the fact that unemployment tends to be slightly above the trend in February. This also sheds some light upon the *Guardian* quote at the top of the chapter. A slight rise in unemployment in February is not surprising and may not indicate a longer-term increase in unemployment (though note the quote refers to the claimant count measure of unemployment, slightly different from the measure used in Table 11.1).

A time series can be decomposed into four components, three of them systematic and one random. These are:

(1) A **trend**: many economic variables are trended over time, as noted in Chapter 1 (the investment series). This measures the longer-term direction of the series, whether increasing, decreasing or unchanging.
(2) A **cycle**: most economies tend to progress unevenly, mixing periods of rapid growth with periods of relative stagnation. This business cycle can vary in length, which makes it difficult to analyse. Consequently, it is often ignored or combined together with the trend.
(3) A **seasonal component**: this is a regular, short-term (one year) cycle. Sales of ice cream vary seasonally, for obvious reasons. Since it is a regular cycle, it is relatively easy to isolate.
(4) A **random component**: this is what is left over after the above factors have been taken into account. By definition it cannot be predicted.

These four elements can be combined in either an **additive** or **multiplicative model**. The additive model of unemployment is

$$X_t = T + C + S + R \quad (11.1)$$

where X represents unemployment, T the trend component, C the cycle, S the seasonal component and R the random element.

The multiplicative model is

$$X_t = T \times C \times S \times R \quad (11.2)$$

There is little to choose between the two alternatives; the multiplicative formulation will be used in the rest of this chapter. This is the method generally used by the Office of National Statistics in officially published series.

The analysis of unemployment proceeds as follows:

(1) First the trend is isolated from the original data by the method of **moving averages**.
(2) Second, the actual employment figures are then compared to the trend to see which months tend to have unemployment above trend. This allows **seasonal factors** to be extracted from the data.
(3) Finally, the seasonal factors are used to **seasonally adjust** the data, so that the underlying movement in the figures can be observed.

Isolating the trend

There is a variety of methods for isolating the trend from time-series data. The method used here is that of moving averages, one of several methods of **smoothing** the data. These smoothing methods iron out the short-term fluctuations in the data by averaging successive observations. For example, to calculate the **three-month moving average** figure for the month of July, one would take the average of the unemployment figures for June, July and August. The three-month moving average for August would be the average of the July, August and September figures. The figures are therefore as follows (for 2012):

$$\text{July:} \quad \frac{2561 + 2584 + 2586}{3} = 2577$$

$$\text{August:} \quad \frac{2584 + 2586 + 2573}{3} = 2581$$

Note that two values (2584 and 2586) are common to the two calculations, so that the two averages tend to be similar and the data series is smoothed out. Thus the moving average is calculated by moving month by month through the data series, taking successive three-month averages.

The choice of the three-month moving average was arbitrary, it could just as easily have been a 4-, 5- or 12-month moving average process. How should the appropriate length of the moving average process be chosen? This depends upon the degree of smoothing of the data which is desired, and upon the nature of the fluctuations. The longer the period of the moving average process, the greater the smoothing of the data, since the greater is the number of terms in the averaging process. In the case of unemployment data, the fluctuations are probably fairly consistent from year to year since, for example, school leavers arrive on the unemployment register at the same time every year, causing a jump in the figures. A 12-month moving average process would therefore be appropriate to smooth this data series.

Table 11.2 shows how the 12-month moving average series is calculated. The calculation is the same in principle as the three-month moving average, but there is one slight complication, that of *centring* the data. The unemployment column (1) of the table repeats the raw data from Table 11.1. In column (2) is calculated the successive 12-month totals. Thus the total of the first 12 observations (Jan–Dec 2012) is 30 408 and this is placed in the middle of 2012, *between* the months of June and July. The sum of observations 2 to 13 is 30 332 and falls between July and August, and so on. Notice that it is impossible to calculate any total before June/July by the moving average process, using the data from the table. A similar effect occurs at the end of the series, in the second half of 2014. Values at the beginning and end of the period in question are always lost by this

Table 11.2 Calculation of the moving average series

Month	Unemployment (1)	12-month total (2)	Centred 12-month total (3)	Moving average (4)
2012 Jan	2612			
2012 Feb	2611			
2012 Mar	2565			
2012 Apr	2458			
2012 May	2472			
2012 Jun	2561	30 408		
2012 Jul	2584	30 332	30 370.0	2530.8
2012 Aug	2586	30 230	30 281.0	2523.4
2012 Sep	2573	30 125	30 177.5	2514.8
2012 Oct	2512	30 074	30 099.5	2508.3
2012 Nov	2442	30 051	30 062.5	2505.2
2012 Dec	2432	29 985	30 018.0	2501.5
2013 Jan	2536	29 964	29 974.5	2497.9
2013 Feb	2509	29 936	29 950.0	2495.8
2013 Mar	2460	29 833	29 884.5	2490.4
2013 Apr	2407	29 646	29 739.5	2478.3
2013 May	2449	29 475	29 560.5	2463.4
2013 Jun	2495	29 288	29 381.5	2448.5
2013 Jul	2563	28 972	29 130.0	2427.5
2013 Aug	2558	28 654	28 813.0	2401.1
2013 Sep	2470	28 299	28 476.5	2373.0
2013 Oct	2325	27 893	28 096.0	2341.3
2013 Nov	2271	27 436	27 664.5	2305.4
2013 Dec	2245	26 944	27 190.0	2265.8
2014 Jan	2220	26 394	26 669.0	2222.4
2014 Feb	2191	25 851	26 122.5	2176.9
2014 Mar	2105	25 379	25 615.0	2134.6
2014 Apr	2001	24 942	25 160.5	2096.7
2014 May	1992	24 442	24 692.0	2057.7
2014 Jun	2003	23 947	24 194.5	2016.2
2014 Jul	2013	23 517		
2014 Aug	2015			
2014 Sep	1998			
2014 Oct	1888			
2014 Nov	1771			
2014 Dec	1750			

Note: In column (2) are the 12-month totals, e.g. 30 408 is the sum of the values from 2612 to 2432. In column (3) these totals are centred on the appropriate month, e.g. 30 177.5 = (30 230 + 30 125)/2 The final column is column (3) divided by 12.

method of smoothing. The greater the length of the moving average process the greater the number of observations lost.

It is inconvenient to have this series falling between the months, so it is centred in column (3). This is done by averaging every two consecutive months' figures, so the June/July and July/August figures are averaged to give the July figure, as follows:

$$\frac{30\,408 + 30\,332}{2} = 30\,370$$

This centring problem always arises when the length of the moving average process is an even number. An alternative to having to centre the data is to use a

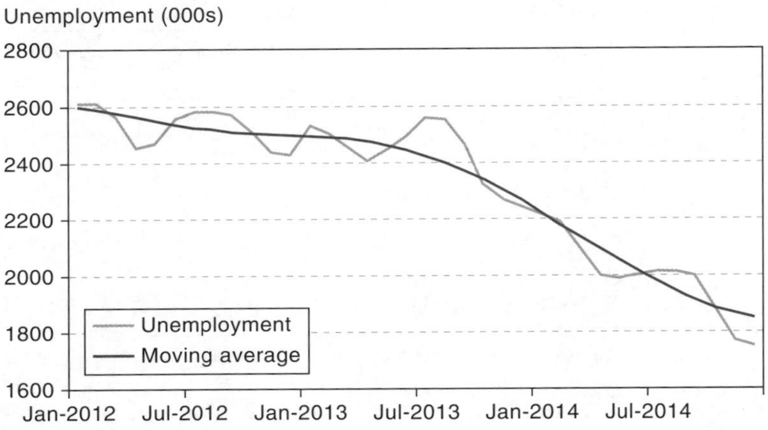

Figure 11.3 Unemployment and its moving average

13-month moving average. This gives virtually identical results, but it seems more natural to use a 12-month average for monthly data.

Column (4) of Table 11.2 is equal to column (3) divided by 12, and so gives the average of 12 consecutive observations, and this is the moving average series.

Comparison of the original data with the smoothed series shows the latter to be free of the short-term fluctuations present in the former. The two series are graphed together in Figure 11.3. Note that for this chart, we have used data from late 2011 and early 2015 (not shown in Table 11.2) to derive the moving average values at the beginning and end of the period, i.e. we have filled in the missing values in Table 11.2.

The chart shows the downward trend clearly, starting around January 2013, and also reveals how this trend appears to level off towards the end of 2014. Note that actual unemployment is still *decreasing* quite rapidly at that point. Actual unemployment stops falling in 2015 (not shown) and the moving average *anticipates* the movements in unemployment. This is not really so surprising since future values of unemployment are used in the calculation of the moving average figure for each month.

The moving average captures the trend and the cycle. How much of the cycle is included is debatable – the longer the period of the moving average, the less the cycle is captured (i.e. the smoother the series), in general. The difficulty of disentangling the cyclical element is that it is unclear how long the cycle is, or even whether it exists in the data. For the sake of argument, we will assume that our moving average fully captures both trend and cycle.

Exercise 11.1

Use the quarterly data below to calculate the (four quarter) moving average series and draw a graph of the two series for 2009–12:

	Q1	Q2	Q3	Q4
2008	–	–	152	149
2009	155	158	155	153
2010	159	166	160	155
2011	162	167	164	160
2012	170	172	172	165
2013	175	179	–	–

Isolating seasonal factors

Having obtained the trend-cycle (henceforth we refer to this as the trend, for brevity), the original data may be divided by the trend values (from Table 11.2, column (4)) to leave only the seasonal and random components. This can be understood by manipulating equation (11.2). Ignoring the cyclical component, we have

$$X = T \times S \times R \tag{11.3}$$

Dividing the original data series X by the trend values T therefore gives the seasonal and random components:

$$\frac{X}{T} = S \times R \tag{11.4}$$

Table 11.3 gives the results of this calculation where, once again, we have filled in the trend values for early 2012 and late 2014 using data from 2011 and 2015 for the calculation (calculation not shown but analogous to that used in Table 11.2).

The final column of the table shows the ratio of the actual unemployment level to the trend value. The value for January 2012, 1.005, shows the unemployment level in that month to be 0.5% above the trend. The July 2012 figure is 1.021, 2.1% above trend, etc. Other months' figures can be interpreted in the same way. Closer examination of the table shows that unemployment tends to be above its trend in the summer months (July to September), below trend in winter (November to January) and around the trend line during April to May. These figures reflect the seasonal pattern observed in Figure 11.2 earlier.

The next task is to disentangle the seasonal and random components which together make up the 'Ratio' value in the final column of the table. We make the assumption that the random component has a mean value of zero. Then, if we average the 'Ratio' values for a particular month, the random components should approximately cancel out, leaving just the seasonal component.

Hence the seasonal factor S can be obtained by averaging the three $S \times R$ components (for 2012, 2013, 2014) for each month. For example, for January, the seasonal component is obtained as follows:

$$S = \frac{1.005 + 1.015 + 0.999}{3} = 1.006 \tag{11.5}$$

The more years we have available entering this averaging process, the more accurate is the estimate of the seasonal component. The seasonal component in this case for January is therefore $1.006 - 1 = 0.006 = 0.6\%$, so January is typically 0.6% above trend. The random components are therefore (virtually) zero, positive and negative in 2012, 2013 and 2014, respectively.

Table 11.4 shows the calculation of the seasonal components for each month using the method described above.

Previous editions of this text calculated seasonal factors for earlier time periods, and Table 11.5 provides a comparison of four time periods. First, it should be stated that the definition and measurement of unemployment changed between pre- and post-2000 figures so one has to be wary of the comparison. It is noticeable,

Table 11.3 Isolating seasonal and random components

	Unemployment	Trend	Ratio
Jan 2012	2612	2600.3	1.005
Feb 2012	2611	2591.6	1.007
Mar 2012	2565	2579.6	0.994
Apr 2012	2458	2565.6	0.958
May 2012	2472	2551.7	0.969
Jun 2012	2561	2539.6	1.008
Jul 2012	2584	2530.8	1.021
Aug 2012	2586	2523.4	1.025
Sep 2012	2573	2514.8	1.023
Oct 2012	2512	2508.3	1.001
Nov 2012	2442	2505.2	0.975
Dec 2012	2 432	2501.5	0.972
Jan 2013	2536	2497.9	1.015
Feb 2013	2509	2495.8	1.005
Mar 2013	2460	2490.4	0.988
Apr 2013	2407	2478.3	0.971
May 2013	2449	2463.4	0.994
Jun 2013	2495	2448.5	1.019
Jul 2013	2563	2427.5	1.056
Aug 2013	2558	2401.1	1.065
Sep 2013	2470	2373.0	1.041
Oct 2013	2325	2341.3	0.993
Nov 2013	2271	2305.4	0.985
Dec 2013	2245	2265.8	0.991
Jan 2014	2220	2222.4	0.999
Feb 2014	2191	2176.9	1.006
Mar 2014	2105	2134.6	0.986
Apr 2014	2001	2096.7	0.954
May 2014	1992	2057.7	0.968
Jun 2014	2003	2016.2	0.993
Jul 2014	2013	1977.7	1.018
Aug 2014	2015	1943.3	1.037
Sep 2014	1998	1911.8	1.045
Oct 2014	1888	1886.0	1.001
Nov 2014	1771	1866.1	0.949
Dec 2014	1750	1849.2	0.946

Note: The 'Ratio' column is simply unemployment divided by its trend value, e.g. 1.005 = 2612/2600.3.

Table 11.4 Calculating the seasonal factors

	2012	2013	2014	Average
January	1.005	1.015	0.999	1.006
February	1.007	1.005	1.006	1.006
March	0.994	0.988	0.986	0.989
April	0.958	0.971	0.954	0.961
May	0.969	0.994	0.968	0.977
June	1.008	1.019	0.993	1.007
July	1.021	1.056	1.018	1.032
August	1.025	1.065	1.037	1.042
September	1.023	1.041	1.045	1.036
October	1.001	0.993	1.001	0.999
November	0.975	0.985	0.949	0.970
December	0.972	0.991	0.946	0.970

Table 11.5 Comparison of seasonal factors in different decades

	1982–84	1991–93	2004–6	2012–14
January	1.042	1.028	0.979	1.006
February	1.033	1.028	0.992	1.006
March	1.019	1.022	0.988	0.989
April	1.009	1.021	0.972	0.961
May	0.983	0.997	0.980	0.977
June	0.963	0.980	1.008	1.007
July	0.982	1.006	1.032	1.032
August	0.983	1.018	1.036	1.042
September	1.001	1.006	1.033	1.036
October	0.992	0.979	1.013	0.999
November	0.997	0.982	0.985	0.970
December	1.002	1.004	0.969	0.970

however, that the apparent seasonal pattern has become more pronounced post-2000. This demonstrates that a seasonal pattern is not necessarily fixed for all time but can be altered by factors such as changes in the law (unemployment benefit entitlements, school leaving age, etc.) and the changing pattern of the labour market in general.

Exercise 11.2

?

Using the data from Exercise 11.1, calculate the seasonal factors for each quarter.

Seasonal adjustment

Having found the seasonal factors, the original data can now be seasonally adjusted. This procedure eliminates the seasonal component from the original series leaving only the trend, cyclical and random components. It therefore removes the regular, month by month, differences and makes it easier to directly compare one month with another. Seasonal adjustment is now simple – the original data are divided by the seasonal factors shown in Table 11.4. Equation (11.6) demonstrates the principle:

$$\frac{X}{S} = T \times C \times R \tag{11.6}$$

Table 11.6 shows the calculation of the seasonally adjusted figures.

The final column of the table adds the official seasonally adjusted figures available from Eurostat. Although that uses slightly more sophisticated methods of adjustment, the results are similar to those we have calculated. Figure 11.4 graphs unemployment and the seasonally adjusted series.

Note that in some months the two series move in opposite directions. For example, in November 2013 the unadjusted series showed a fall in unemployment (of about 2.3%) yet the adjusted series rose slightly (by about 0.6%). In other words, the fall in unemployment was discounted as unemployment usually falls in November (compare October and November seasonal factors) and this observed fall was relatively small and expected.

Table 11.6 Seasonally adjusted unemployment

	Unemployment	Seasonal factor	Seasonally adjusted series	S.A. series from Eurostat
Jan 2012	2612	1.006	2596	2619
Feb 2012	2611	1.006	2594	2596
Mar 2012	2565	0.989	2592	2590
Apr 2012	2458	0.961	2557	2535
May 2012	2472	0.977	2530	2518
Jun 2012	2561	1.007	2543	2548
Jul 2012	2584	1.032	2505	2518
Aug 2012	2586	1.042	2481	2499
Sep 2012	2573	1.036	2483	2498
Oct 2012	2512	0.999	2516	2499
Nov 2012	2442	0.970	2518	2491
Dec 2012	2432	0.970	2508	2488
Jan 2013	2536	1.006	2520	2542
Feb 2013	2509	1.006	2493	2494
Mar 2013	2460	0.989	2486	2484
Apr 2013	2407	0.961	2504	2481
May 2013	2449	0.977	2507	2493
Jun 2013	2495	1.007	2478	2482
Jul 2013	2563	1.032	2485	2498
Aug 2013	2558	1.042	2454	2468
Sep 2013	2470	1.036	2383	2393
Oct 2013	2325	0.999	2328	2315
Nov 2013	2271	0.970	2342	2322
Dec 2013	2245	0.970	2315	2299
Jan 2014	2220	1.006	2206	2229
Feb 2014	2191	1.006	2177	2182
Mar 2014	2105	0.989	2127	2132
Apr 2014	2001	0.961	2082	2073
May 2014	1992	0.977	2039	2033
Jun 2014	2003	1.007	1989	1989
Jul 2014	2013	1.032	1951	1949
Aug 2014	2015	1.042	1933	1926
Sep 2014	1998	1.036	1928	1921
Oct 2014	1888	0.999	1891	1877
Nov 2014	1771	0.970	1826	1821
Dec 2014	1750	0.970	1805	1805

Note: The adjusted series is obtained by dividing the 'Unemployment' column by the 'Seasonal factor' column.

Figure 11.4 Unemployment and seasonally adjusted unemployment

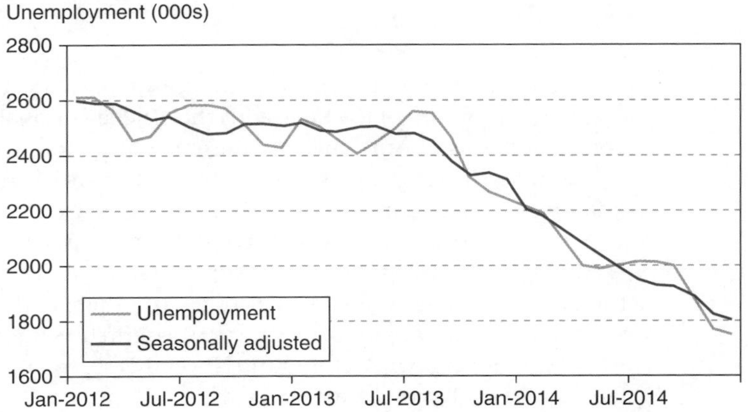

Fitting a moving average to a series using Excel

Many software programs can automatically produce a moving average of a data series. Microsoft Excel does this using a 12-period moving average which is not centred but located at the end of the averaged values. For example, the average of the Jan–Dec 2012 figures is placed against December 2012, not between June and July as was done above. This cuts off 11 observations at the beginning of the period but none at the end. Figure 11.5 compares the moving averages calculated by Excel and by the centred moving average method described earlier.

Figure 11.5
Excel version of moving average

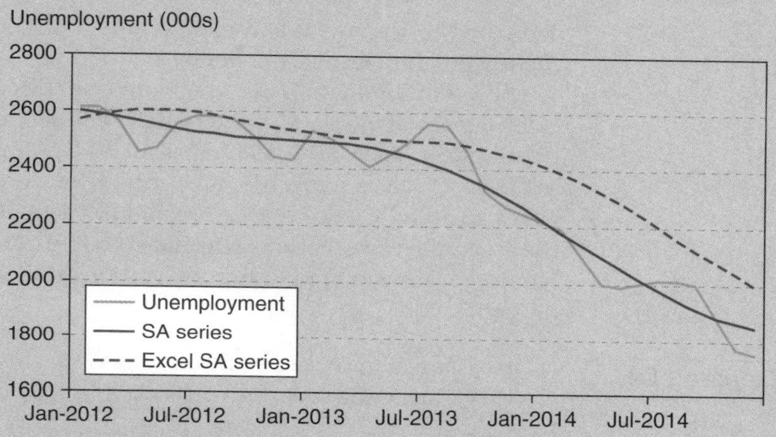

The Excel method appears much less satisfactory: it is always lagging behind the actual series, in contrast to the centred method. However, it has the advantage that the trend value for the latest month can always be calculated.

Exercise 11.3

Again using the data from Exercise 11.1, construct the seasonally adjusted series for 2009–12 and graph the unadjusted and adjusted series.

An alternative method for finding the trend

Chapter 9 on regression showed how a straight line could be fitted to a set of data as a means of summarising it. This offers an alternative means of smoothing data and finding a trend line. The dependent variable in the regression is unemployment, which is regressed on a time trend variable. This is simply measured by the values 1, 2, 3, . . . 36 and is denoted by the letter t. January 2012 is therefore represented by 1, February 2012 by 2, etc. Since the trend appears to be non-linear however, a fitted linear trend is unlikely to be accurate for forecasting. The regression equation can be made non-linear by including a t^2 term, for example. For January 2012 this would be 1, for February 2012 it would be 4, etc. The equation thus becomes

$$X_t = \mathrm{a} + \mathrm{b}t + \mathrm{c}t^2 + e_t \qquad (11.7)$$

where e_t is the error term which, in this case, is composed of the cyclical, seasonal and random elements of the cycle. The trend component is given by $\mathrm{a} + \mathrm{b}t + \mathrm{c}t^2$. The calculated regression equation is (calculation not shown)

$$X_t = 2527.2 + 10.84t - 0.90t^2 + e \qquad (11.8)$$

The trend values for each month can easily be calculated from this equation, by inserting the values $t = 1, 2, 3$, etc., as appropriate. January 2012, for example, is found by substituting $t = 1$ and $t^2 = 1^2$ into equation (11.8), giving

$$X_t = 2527.2 + 10.84 \times 1 - 0.90 \times 1^2 = 2537.14 \quad (11.9)$$

which compares to 2595.82 using the moving average method. For July 2012 ($t = 7$) we obtain

$$X_t = 2527.2 + 10.84 \times 7 - 0.90 \times 7^2 = 2559.06 \quad (11.10)$$

compared to the moving average estimate of 2504.93. The two methods give slightly different results, but not by a great deal.

The rest of the analysis can then proceed as before. The seasonal factors are calculated for each month and year by first dividing the actual value by the estimated trend value (hence $2612/2537.14 = 1.030$ for January 2012) and then averaging the January values across the three years gives the January seasonal factor. This is left as an exercise (see Exercise 11.4 and Problem 11.5) and gives similar results to the moving average method. One final point to note is that the regression method has the advantage of not losing observations at the beginning and end of the sample period.

Exercise 11.4

?

(a) Using the data from Exercise 11.1, calculate a regression of X on t and t^2 (and a constant) to find the trend cycle series. Use observations for 2009–12 only for the regression equation.

(b) Graph the original series and the calculated trend line.

(c) Use the new trend line to calculate the seasonal factors.

Forecasting

It is possible to forecast future levels of unemployment based on the methods outlined above. Each component of the series is forecast separately and the results multiplied together. As an example the level of unemployment for January 2015 will be forecast.

The trend can only be forecast using the regression method, since the moving average method requires future values of unemployment, which is what is being forecast. January 2015 corresponds to time period $t = 37$ so the forecast of the trend by the regression method is

$$X_t = 2527.2 + 10.84 \times 37 - 0.90 \times 37^2 = 1698.73 \quad (11.11)$$

The seasonal factor for January is 1.010 so the trend figure is multiplied by this, giving

$$1796.9 \times 0.988 = 1776.2 \qquad 1698.73 \times 1.010 = 1715.71 \quad (11.12)$$

The cyclical component is ignored and the random component set to a value of 1 (in the multiplicative model, zero in the additive model). This leaves 1715.71 as the forecast for January 2015. In the event the actual figure was 1790, so the forecast is not very good, with an error of approximately 4.2%. A chart of unemployment, the trend (using the regression method) and the forecast for the first six months of 2015 reveals the problem (see Figure 11.6).

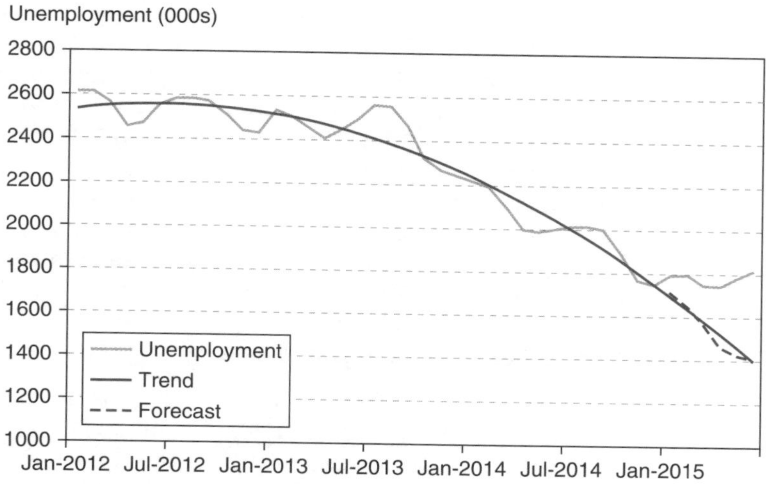

Figure 11.6 Forecasting unemployment

The forecast relentlessly follows the trend line downwards. Because it is only a quadratic trend (i.e. involving terms up to t^2) it cannot predict another turning point, which seems to have occurred around the end of 2014. Nevertheless, the error in the forecast for January 2015 would alert observers to the likelihood that some kind of change has occurred and that unemployment is no longer following its downward trend.

Exercise 11.5

Use the results of Exercise 11.4 to forecast the values of X for 2013Q1 and 2013Q2. How do they compare to the actual values?

Further issues

As stated earlier, official methods of seasonal adjustment are more sophisticated than those shown here, although with similar results. The main additional features that we have omitted are as follows:

- **Ad hoc adjustments** – the original data may be 'incorrect' for an obvious reason. A strike, for example, might lower output in a particular month. This not only gives an atypical figure for the month but will also affect the calculation of the seasonal factors. Hence, such an observation might be corrected in some way before the seasonal factors are calculated.
- **Calendar effects** – months are not all of the same length so retail sales, for example, might vary simply because there are more shopping days (especially if a month happens to have five weekends). Overseas trade statistics are routinely adjusted for the numbers of days in each month. Easter is another problem, because it is not on a regular date and so can affect monthly figures in different ways, depending where it falls.
- **Forecasting methods** – the trend is calculated by a mixture of regression and moving average methods, avoiding some of the problems exhibited above when forecasting.

The above analysis has taken a fairly mechanical approach to the analysis of time series, and has not sought the reasons *why* the data might vary seasonally.

The seasonal adjustment factors are simply used to adjust the original data for regular monthly effects, whatever the cause. Further investigation of the causes might be worthwhile as they might improve forecasting. For example, unemployment varies seasonally because of (among other things) greater employment opportunities in summer (e.g. deck chair attendants) and school leavers entering the unemployment register in September. The availability of summer jobs might be predictable (based on forecasts of the number of tourists, weather, etc.), and the number of school leavers next year can presumably be predicted by the number of pupils at present in their final year. These sorts of considerations should provide better forecasts rather than slavishly following the rules set out above.

Using adjusted or unadjusted data

Seasonal adjustment can also introduce problems into data analysis as well as resolve them. Although seasonal adjustment can help in interpreting figures, if the adjusted data are then used in further statistical analysis, they can mislead. It is well known, for example, that seasonal adjustment can introduce a cyclical component into a data series which originally had no cyclical element to it. This occurs because a large (random) deviation from the trend will enter the moving average process for 12 different months (or whatever is the length of the moving average process), and this tends to turn occasional, random disturbances into a cycle. Note also that the adjusted series will start to rise before the random shock in these circumstances.

The question then arises as to whether adjusted or unadjusted data are best used in, say, regression analysis. Use of unadjusted data means that the coefficient estimates may be contaminated by the seasonal effects; using adjusted data runs into the kind of problems outlined above. A suitable compromise is to follow the method outlined in Chapter 10: use unadjusted data with seasonal dummy variables. In this case the estimation of parameters and seasonal effects is dealt with simultaneously and generally gives the best results.

A further advantage of this regression method is that it allows the significance of the seasonal variations to be established. An F test for the joint significance of the seasonal coefficients will tell you whether any of the seasonal effects are statistically significant. If not, seasonal dummies need not be included in the regression equation.

Finally, it should be remembered that decomposing a time series is not a clear-cut procedure. It is often difficult to disentangle the separate effects, and different methods will give different results. The seasonally adjusted unemployment figures given by Eurostat are slightly different from the series calculated here, due to slightly different techniques being applied. The differences are not great and the direction of the seasonal effects are the same even if the sizes are slightly different.

Summary

- Seasonal adjustment of data allows us to see some of the underlying features, shorn of the distraction of seasonal effects (such as the Christmas effect on retail sales).
- The four components of a time series are the trend, the cycle, the seasonal component and the **random residual**.

- These components may be thought of either as being multiplied together or added together to make up the series. The former method is more common.
- The trend (possibly mixed with the cycle) can be identified by the method of moving averages or by the use of a regression equation.
- Removing the trend cycle values from a series leaves only the seasonal and random components.
- The residual component can then be eliminated by averaging the data over successive years (e.g. take the average of the January seasonal and random component over several years).
- Having isolated the seasonal effect in such a manner, it can be eliminated from the original series, leaving the seasonally adjusted series.
- Knowledge of the seasonal effects can be useful in forecasting future values of the series.

Key terms and concepts

additive model
calendar effects
cycle
forecasting
moving average
multiplicative model
random component
random elements
random residual
seasonal adjustment
seasonal component
seasonal factor
smoothing
systematic components
three-month moving average
trend

Problems

Some of the more challenging problems are indicated by highlighting the problem number in colour.

11.1 The following table contains data for consumers' non-durables expenditure in the United Kingdom, in constant 2003 prices.

(a) Graph the series and comment upon any apparent seasonal pattern. Why might it occur?

(b) Use the method of centred moving averages to find the trend values for 2000–14.

(c) Use the moving average figures to find the seasonal factors for each quarter (use the multiplicative model).

(d) By approximately how much does expenditure normally increase in the fourth quarter?

(e) Use the seasonal factors to obtain the seasonally adjusted series for non-durable expenditure.

(f) Were retailers happy or unhappy at Christmas in 2000? How about 2014?

	Q1	Q2	Q3	Q4
1999	–	–	153 888	160 187
2000	152 684	155 977	160 564	164 437
2001	156 325	160 069	165 651	171 281
2002	161 733	167 128	171 224	176 748
2003	165 903	172 040	176 448	182 769
2004	171 913	178 308	182 480	188 733
2013	175 174	180 723	184 345	191 763
2014	177 421	183 785	187 770	196 761
2007	183 376	188 955	–	–

Source: Data adapted from the Office for National Statistics licensed under the Open Government Licence v.1.0.

11.2 Repeat the exercise using the additive model. (In Problem 11.1(c), *subtract* the moving average figures from the original series. In (e), subtract the seasonal factors from the original data to get the adjusted series.) Is there a big difference between this and the multiplicative model?

11.3 The following data relate to car production in the United Kingdom (not seasonally adjusted).

	2003	2004	2005	2006	2007
January	–	141.3	136	119.1	124.2
February	–	141.1	143.5	131.2	115.6
March	–	163	153.3	159	138
April	–	129.6	139.8	118.6	120.4
May	–	143.1	132	132.3	127.4
June	–	155.5	144.3	139.3	137.5
July	146.3	140.5	130.2	117.8	129.7
August	91.4	83.2	97.1	73	–
September	153.5	155.3	149.9	122.3	–
October	153.4	135.1	124.8	116.1	–
November	142.9	149.3	149.7	128.6	–
December	112.4	109.7	95.3	84.8	–

Source: Data adapted from the Office for National Statistics licensed under the Open Government Licence v.1.0.

(a) Graph the data for 2004–14 by overlapping the three years (as was done in Figure 11.2) and comment upon any seasonal pattern.

(b) Use a 12-month moving average to find the trend values for 2004–14.

(c) Find the monthly seasonal factors (multiplicative method). Describe the seasonal pattern that emerges.

(d) By how much is the August production figure below the July figure in general?

(e) Obtain the seasonally adjusted series. Compare it with the original series and comment.

(f) Compare the seasonal pattern found with that for consumers' expenditure in Problem 11.1.

11.4 Repeat Problem 11.3 using the additive model and compare results.

11.5 (a) Using the data of Problem 11.1, fit a regression line through the data, using t and t^2 as explanatory variables (t is a time trend 1–36). Use only the observations from 2000 to 2014. Calculate the trend values using the regression.

(b) Calculate the seasonal factors (multiplicative model) based upon this trend. How do they compare to the values found in Problem 11.1?

(c) Predict the value of consumers' expenditure for 2007 Q4.

(d) Calculate the seasonal factors using the additive model and predict consumers' expenditure for 2007 Q4.

11.6 (a) Using the data from Problem 11.3 (2004–14 only), fit a linear regression line to obtain the trend values. By how much, on average, does car production increase per year?

(b) Calculate the seasonal factors (multiplicative model). How do they compare to the values in Problem 11.3?

(c) Predict car production for April 2007.

11.7 A computer will be needed to solve this and the next problem.

(a) Repeat the regression equation from Problem 11.5 but add three seasonal dummy variables (for quarters 2, 3 and 4) to the regressors. (The dummy for quarter 2 takes the value 1 in Q2, 0 in the other quarters. The Q3 dummy takes the value 1 in Q3, 0 otherwise, etc.) How does this affect the coefficients on the time trend variables? (Use data for 2000–14 only.)

(b) How do the t ratios on the time coefficients compare with the values found in Problem 11.5? Account for the difference.

(c) Compare the coefficients on the seasonal dummy variables with the seasonal factors found in Problem 11.5(d). Comment on your results.

11.8 (a) How many seasonal dummy variables would be needed for the regression approach to the data in Problem 11.3?

(b) Do you think the approach would bring as reliable results as it did for consumers' expenditure?

11.9 **(Project)** Obtain quarterly (unadjusted) data for a suitable variable (some suggestions are given below) and examine its seasonal pattern. Write a brief report on your findings. You should:

(a) Say what you *expect* to find, and why.

(b) Compare different methods of adjustment.

(c) Use your results to try to forecast the value of the variable at some future date.

(d) Compare your results, if possible, with the 'official' seasonally adjusted series. Some suitable variables are: the money stock, retail sales, rainfall, interest rates, house prices, corporate profits.

Answers to exercises

Exercise 11.1

The calculations are as follows:

Quarter	*X*	4th quarter total	Centred	Moving average
2008 Q3	152	–	–	–
2008 Q4	149	–	–	–
2009 Q1	155	614	615.5	153.875
2009 Q2	158	617	619.0	154.750
2009 Q3	155	621	623.0	155.750
2009 Q4	153	625	629.0	157.250
2010 Q1	159	633	635.5	158.875
2010 Q2	166	638	639.0	159.750
2010 Q3	160	640	641.5	160.375
2010 Q4	155	643	643.5	160.875
2011 Q1	162	644	646.0	161.500
2011 Q2	167	648	650.5	162.625
2011 Q3	164	653	657.0	164.250
2011 Q4	160	661	663.5	165.875
2012 Q1	170	666	670.0	167.500
2012 Q2	172	674	676.5	169.125
2012 Q3	172	679	681.5	170.375
2012 Q4	165	684	687.5	171.875
2013 Q1	175	691	693.0	173.250
2013 Q2	179	–	–	–

The chart of these data is:

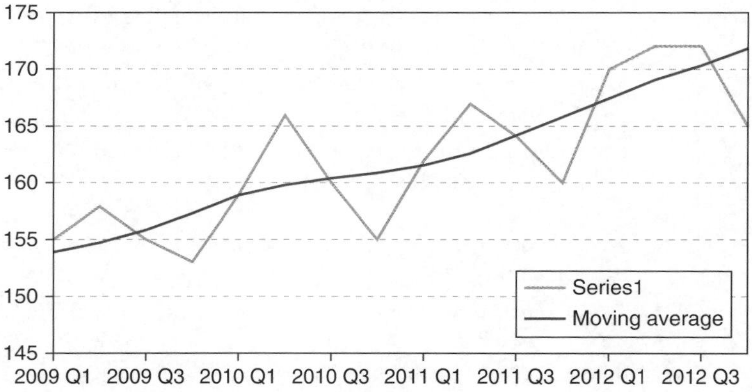

Exercise 11.2

The seasonal factors are calculated as follows:

	X	Moving average	Ratio	Seasonal factor
2009 Q1	155	153.875	1.007	1.007
2009 Q2	158	154.750	1.021	1.026
2009 Q3	155	155.750	0.995	1.000
2009 Q4	153	157.250	0.973	0.965
2010 Q1	159	158.875	1.001	1.007
2010 Q2	166	159.750	1.039	1.026
2010 Q3	160	160.375	0.998	1.000
2010 Q4	155	160.875	0.963	0.965
2011 Q1	162	161.500	1.003	1.007
2011 Q2	167	162.625	1.027	1.026
2011 Q3	164	164.250	0.998	1.000
2011 Q4	160	165.875	0.965	0.965
2012 Q1	170	167.500	1.015	1.007
2012 Q2	172	169.125	1.017	1.026
2012 Q3	172	170.375	1.010	1.000
2012 Q4	165	171.875	0.960	0.965

Note: The first seasonal factor, 1.007, is calculated as the average of 1.007, 1.001, 1.003 and 1.015.

Exercise 11.3

The adjusted series is calculated as follows:

Quarter	*X*	Seasonal factor	Seasonally adjusted figure
2009 Q1	155	1.007	153.994
2009 Q2	158	1.026	153.995
2009 Q3	155	1.000	154.967
2009 Q4	153	0.965	158.507
2010 Q1	159	1.007	157.968
2010 Q2	166	1.026	161.792
2010 Q3	160	1.000	159.966
2010 Q4	155	0.965	160.579
2011 Q1	162	1.007	160.949
2011 Q2	167	1.026	162.767
2011 Q3	164	1.000	163.965
2011 Q4	160	0.965	165.759
2012 Q1	170	1.007	168.897
2012 Q2	172	1.026	167.640
2012 Q3	172	1.000	171.963
2012 Q4	165	0.965	170.939

And the series are graphed as follows:

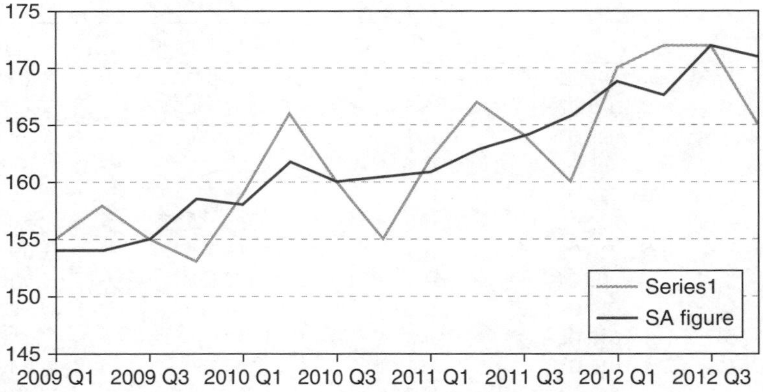

Exercise 11.4

(a) The regression equation is $X = 153.9 + 0.85t + 0.01t^2$ (note the coefficient on t^2 is very small, so this is virtually a straight line).

(b)

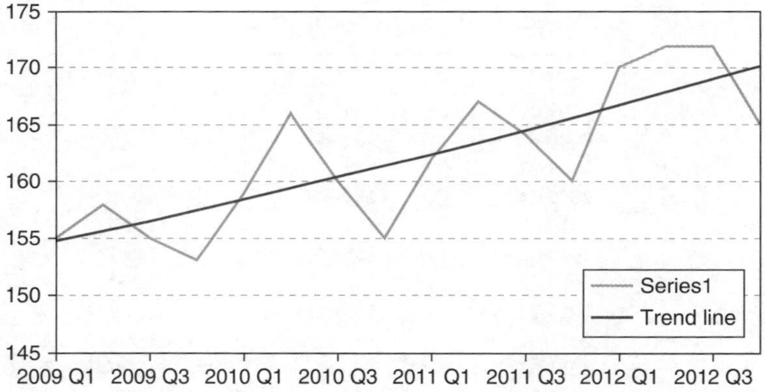

(c) The seasonal factors are calculated as follows:

Quarter	X	Predicted X	Ratio	Seasonal factor
2009 Q1	155	154.737	1.002	1.006
2009 Q2	158	155.617	1.015	1.026
2009 Q3	155	156.518	0.990	1.001
2009 Q4	153	157.440	0.972	0.967
2010 Q1	159	158.382	1.004	1.006
2010 Q2	166	159.345	1.042	1.026
2010 Q3	160	160.329	0.998	1.001
2010 Q4	155	161.333	0.961	0.967
2011 Q1	162	162.358	0.998	1.006
2011 Q2	167	163.404	1.022	1.026
2011 Q3	164	164.470	0.997	1.001
2011 Q4	160	165.557	0.966	0.967
2012 Q1	170	166.665	1.020	1.006
2012 Q2	172	167.793	1.025	1.026
2012 Q3	172	168.942	1.018	1.001
2012 Q4	165	170.112	0.970	0.967

Exercise 11.5

Substituting $t = 17$ and $t = 18$ into the regression equation gives predicted values of 171.302 and 172.513, for Q1 and Q2 respectively. Multiplying by the relevant seasonal factors (1.006 and 1.026) gives 172.304 and 177.005. These are close to, but slightly below, the actual values. The errors are 1.6% and 1.1%, respectively.

List of important formulae

Formula	Description	Notes
$\mu = \dfrac{\sum x}{N}$	Mean of a population	Use when all individual observations are available. N is the population size.
$\mu = \dfrac{\sum fx}{\sum f}$	Mean of a population	Use with grouped data. f represents the class or group frequencies, x represents the midpoint of the class interval.
$\overline{x} = \dfrac{\sum x}{n}$	Mean of a sample	n is the number of observations in the sample.
$\overline{x} = \dfrac{\sum fx}{\sum f}$	Mean of a sample	Use with grouped data.
$m = x_L + (x_U - x_L)\left\{\dfrac{\frac{N+1}{2} - F}{f}\right\}$	Median (where data are grouped)	x_L and x_U represent the lower and upper limits of the interval containing the median. F represents the cumulative frequency up to (but excluding) the interval.
$\sigma^2 = \dfrac{\sum (x - \mu)^2}{N}$	Variance of a population	N is the population size.
$\sigma^2 = \dfrac{\sum f(x - \mu)^2}{\sum f}$	Population variance (grouped data)	
$s^2 = \dfrac{\sum (x - \overline{x})^2}{n - 1}$	Sample variance	
$s^2 = \dfrac{\sum f(x - \overline{x})^2}{n - 1}$	Sample variance (grouped data)	
$c.v. = \dfrac{\sigma}{\mu}$	Coefficient of variation	The ratio of the standard deviation to the mean. A measure of dispersion.

Formula	Description	Notes
$z = \frac{x - \mu}{\sigma}$	z score	Measures the distance from observation x to the mean μ measured in standard deviations.
$\frac{\sum f(x - \mu)^3}{N\sigma^3}$	Coefficient of skewness	A positive value means the distribution is skewed towards the right (long tail to the right).
$g = \sqrt[T-1]{\frac{x_T}{x_1}} - 1$	Rate of growth	Measures the average annual rate of growth between years 1 and T.
$\sqrt[n]{\Pi x}$	Geometric mean (of n observations on x)	
$1 - \frac{1}{k^2}$	Chebyshev's inequality	Minimum proportion of observations lying within k standard deviations of the mean of any distribution.
$nCr = \frac{n!}{r!(n - r)!}$	Combinatorial formula	$n! = n \times (n - 1) \times \cdots \times 1$
$\Pr(r) = nCr \times P^r \times (1 - P)^{n-r}$	Binomial distribution	In shorthand notation, $r \sim B(n, P)$.
$\Pr(x) = \frac{1}{\sigma\sqrt{2\pi}} e^{-\frac{1}{2}\{\frac{x-\mu}{\sigma}\}^2}$	Normal distribution	In shorthand notation, $x \sim N(\mu, \sigma^2)$.
$\bar{z} = \frac{\bar{x} - \mu}{\sqrt{\sigma^2/n}}$	z score for the sample mean	Used to test hypotheses about the sample mean.
$\Pr(x) = \frac{\mu^x e^{-\mu}}{x!}$	Poisson distribution	Used when the probability of success is very small. The 'rare event' distribution.
$\bar{x} \pm 1.96\sqrt{s^2/n}$	95% confidence interval for the mean	Large samples, using Normal distribution.
$\bar{x} \pm t_\nu\sqrt{s^2/n}$	95% confidence interval for the mean	Small samples, using t distribution. t_ν is the critical value of the t distribution for $\nu = n - 1$ degrees of freedom.
$p \pm 1.96\sqrt{\frac{p(1 - p)}{n}}$	95% confidence interval for a proportion	Large samples only.

Formula	Description	Notes
$(\bar{x}_1 - \bar{x}_2) \pm 1.96\sqrt{\frac{s_1^2}{n_1} + \frac{s_2^2}{n_2}}$	95% confidence interval for the difference of two means	Large samples.
$(\bar{x}_1 - \bar{x}_2) \pm t_\nu\sqrt{\frac{S^2}{n_1} + \frac{S^2}{n_2}}$	95% confidence interval for the difference of two means	Small samples. The pooled variance is given by $S^2 = \frac{(n_1 - 1)s_1^2 + (n_2 - 1)s_2^2}{n_1 + n_2 - 2}$, $\nu = n_1 + n_2 - 2$.
$z = \frac{\bar{x} - \mu}{\sqrt{s^2/n}}$	Test statistic for H_0: mean $= \mu$	Large samples. For small samples, distributed as t with $\nu = n - 1$ degrees of freedom.
$z = \frac{p - \pi}{\sqrt{\frac{\pi(1 - \pi)}{n}}}$	Test statistic for H_0: true proportion $= \pi$	Large samples.
$z = \frac{(\bar{x}_1 - \bar{x}_2) - (\mu_1 - \mu_2)}{\sqrt{\frac{s_1^2}{n_1} + \frac{s_2^2}{n_2}}}$	Test statistic for $H_0{:}\mu_1 - \mu_2 = 0$	Large samples.
$t = \frac{(\bar{x}_1 - \bar{x}_2) - (\mu_1 - \mu_2)}{\sqrt{\frac{S^2}{n_1} + \frac{S^2}{n_2}}}$	Test statistic for $H_0{:}\mu_1 - \mu_2 = 0$	Small samples. S^2 as defined above. Degrees of freedom $\nu = n_1 + n_2 - 2$.
$z = \frac{(p_1 - p_2) - (\pi_1 - \pi_2)}{\sqrt{\frac{\hat{\pi}(1 - \hat{\pi})}{n_1} + \frac{\hat{\pi}(1 - \hat{\pi})}{n_2}}}$	Test statistic for $H_0{:}\pi_1 - \pi_2 = 0$	Large samples. $\hat{\pi} = \frac{n_1 p_1 + n_2 p_2}{n_1 + n_2}$
$\chi^2 = \sum \frac{(O - E)^2}{E}$	Test statistic for independence in a contingency table	$\nu = (r - 1) \times (c - 1)$, where r is the number of rows, c the number of columns.
$F = \frac{s_1^2}{s_2^2}$	Test statistic for H_0: $\sigma_1^2 = \sigma_2^2$	$\nu = n_1 - 1, n_2 - 1$. Place larger sample variance in the numerator to ensure rejection region is in right-hand tail of the F distribution.
$\sum_{j=1}^{n_i}\sum_{i=1}^{k} x_{ij}^2 - n\bar{x}^2$	Total sum of squares (ANOVA)	n is the total number of observations, k is the number of groups.

Formula	Description	Notes
$\sum_i n_i\bar{x}_i^2 - n\bar{x}^2$	Between sum of squares (ANOVA)	A n_i represents the number of observations in group i and $\bar{x}_i$ is the mean of the group.
$\sum_j \sum_i (x_{ij} - \bar{x}_i)^2$	Within sum of squares (ANOVA)	
$r = \frac{n\sum XY - \sum X \sum Y}{\sqrt{(n\sum X^2 - (\sum X)^2)(n\sum Y^2 - (\sum Y)^2)}}$	Correlation coefficient	$-1 \leq r \leq 1$.
$t = \frac{r\sqrt{n-2}}{\sqrt{1-r^2}}$	Test statistic for $H_0{:}\rho = 0$	$\nu = n - 2$.
$r_s = 1 - \frac{6\sum d^2}{n(n^2-1)}$	Spearman's rank correlation coefficient	$-1 \leq r_s \leq 1$. d is the difference in ranks between the two variables. Only works if there are no tied ranks. Otherwise use standard correlation formula.
$b = \frac{n\sum XY - \sum X \sum Y}{n\sum X^2 - (\sum X)^2}$	Slope of the regression line (simple regression)	
$a = \bar{Y} - b\bar{X}$	Intercept (simple regression)	
$\text{TSS} = \sum Y^2 - n\bar{Y}^2$	Total sum of squares	
$\text{ESS} = \sum Y^2 - a\sum Y - b\sum XY$	Error sum of squares	
$\text{RSS} = \text{TSS} - \text{ESS}$	Regression sum of squares	
$R^2 = \frac{\text{RSS}}{\text{TSS}}$	Coefficient of determination	
$s_e^2 = \frac{\text{ESS}}{n-2}$	Variance of the error term in regression	Replace $n - 2$ by $n - k - 1$ in multiple regression.
$s_b^2 = \frac{s_e^2}{\sum(X - \bar{X})^2}$	Variance of the slope coefficient in simple regression	
$s_a^2 = s_e^2 \times \sqrt{\frac{1}{n} + \frac{\bar{X}^2}{\sum(X - \bar{X})^2}}$	Variance of the intercept in simple regression	

Formula	Description	Notes
$b \pm t_v \times s_b$	Confidence interval estimate for b in simple regression	t_v is the critical value of the t distribution with $v = n - 2$ degrees of freedom.
$fpc = 1 - n/N$	Finite population correction for the variance of $\bar{x}$	
$n = \dfrac{Z_\alpha^2 \times S^2}{p^2}$	Required sample size to obtain desired confidence interval	p is the desired accuracy (half the width of the CI), Z_α is the critical value from the Normal distribution (depends on confidence level specified).
$P_L^n = \dfrac{\sum p_n q_0}{\sum p_0 q_0} \times 100$	Laspeyres price index for year n with base year 0	
$P_L^n = \dfrac{\sum p_n}{\sum p_0} \times s_0 \times 100$	Laspeyres price index using expenditure weights s in base year	
$P_P^n = \dfrac{\sum p_n q_n}{\sum p_0 q_n} \times 100$	Paasche price index for year n	
$P_P^n = \dfrac{1}{\sum \dfrac{p_0}{p_n} \times s_n} \times 100$	Paasche price index using expenditure weights s in current year	
$Q_L^n = \dfrac{\sum q_n p_0}{\sum q_0 p_0} \times 100$	Laspeyres quantity index	
$Q_P^n = \dfrac{\sum q_n p_n}{\sum q_0 p_n} \times 100$	Paasche quantity index	
$E^n = \dfrac{\sum p_n q_n}{\sum p_0 q_0} \times 100$	Expenditure index	
$PV = \dfrac{S}{(1 + r)^t}$	Present value	The value now of a sum S to be received in t years' time, using discount rate r

Formula	Description	Notes
$NPV = -S_0 + \sum \frac{S_t}{(1+r)^t}$	Net present value	The value of an investment S_0 now, yielding S_t per annum, discounted at a constant rate r.
$t = \frac{b - \beta}{s_b}$	Test statistic for $H_0: \beta = 0$	$\nu = n - 2$ in simple regression, $n - k - 1$ in multiple regression.
$F = \frac{\text{RSS}/1}{\text{ESS}/(n-2)}$	Test statistic for $H_0: R^2 = 0$	$\nu = k, n - k - 1$ in multiple regression.
$\hat{Y} \pm t_\nu \times s_e \sqrt{\frac{1}{n} + \frac{(X_P - \overline{X})^2}{\sum (X - \overline{X})^2}}$	Confidence interval for a prediction (simple regression) at $X = X_P$	$\nu = n - 2.$
$\hat{Y} \pm t_\nu \times s_e \sqrt{1 + \frac{1}{n} + \frac{(X_P - \overline{X})^2}{\sum (X - \overline{X})^2}}$	Confidence interval for an observation on Y at $X + X_P$	$\nu = n - 2.$
$F = \frac{(\text{ESS}_P - \text{ESS}_1)/n_2}{\text{ESS}_1/(n_1 - k - 1)}$	Chow test for a prediction	First n_1 observations used for estimation, last n_2 for prediction.
$DW = \frac{\sum (e_t - e_{t-1})^2}{\sum e_t^2}$	Durbin–Watson statistic for testing autocorrelation	
$F = \frac{(\text{ESS}_R - \text{ESS}_U)/q}{\text{ESS}_U/(n - k - 1)}$	Test statistic for testing q restrictions in the regression model	$\nu = q, n - k - 1.$

Appendix: Tables

Table A1 Random number table

This table contains 1000 random numbers within the range 0 to 99. Each number within the range has an equal probability of occurrence. The range may be extended by combining successive entries in the table. Thus 7399 becomes the first of 500 random numbers in the range 0 to 9999. To obtain a sample of random numbers, choose an arbitrary starting point in the table and go down the columns collecting successive values until the required sample is obtained. If the population has been numbered, this method can be used to select a random sample from the population. Alternatively, the method can simulate sampling experiments such as the tossing of a coin (an even number representing a head and an odd number a tail).

73	23	41	53	38	87	71	79	3	55	24	7	7	17	19	70
99	13	91	13	90	72	84	15	64	90	56	68	38	40	73	78
97	16	58	2	67	3	92	83	50	53	59	60	33	75	44	95
73	10	29	14	9	92	35	47	21	47	82	25	71	68	87	53
99	79	29	68	44	90	65	33	55	85	7	57	77	84	83	5
71	97	98	60	62	18	49	80	4	51	8	74	81	64	29	45
41	26	41	30	82	38	52	81	89	64	17	10	49	28	72	99
60	87	77	81	91	57	6	1	30	47	93	82	81	67	4	3
95	84	74	92	15	10	37	52	8	10	96	38	69	9	65	41
59	19	2	61	40	67	80	25	31	18	1	36	54	31	100	27
35	3	54	83	62	28	21	23	91	46	73	85	11	63	63	49
66	18	31	17	72	15	8	46	10	3	64	22	100	62	85	16
3	4	42	8	4	6	40	73	97	0	37	34	91	56	48	98
28	20	23	98	86	41	41	13	53	61	16	92	95	31	79	36
74	49	86	5	74	82	12	58	80	14	94	4	88	95	9	32
80	80	2	47	91	100	76	84	0	57	17	69	87	29	52	39
65	67	0	39	11	10	54	80	74	56	55	91	94	52	32	18
67	44	89	50	7	73	70	52	18	28	89	43	54	60	20	10
48	33	61	66	2	71	74	91	31	45	63	2	97	62	30	90
3	18	54	19	17	87	3	91	41	64	78	10	99	24	1	20
69	35	12	53	97	30	96	69	59	55	65	64	30	3	100	17
15	0	33	86	93	73	52	57	77	77	83	10	64	54	85	18
87	79	51	68	5	23	50	15	68	67	14	59	42	61	83	2
69	52	34	86	34	34	78	51	48	65	57	91	8	74	72	36
11	1	11	43	51	85	6	47	72	43	34	54	20	56	31	81
59	14	78	32	94	24	19	44	16	49	65	16	30	86	0	65
18	86	62	47	96	46	73	67	79	40	45	82	96	61	34	60
99	63	2	81	58	93	81	37	53	20	64	87	3	27	19	55
34	55	14	29	10	59	7	69	13	8	54	97	56	7	57	16
88	90	6	98	32	55	37	17	35	93	31	66	67	84	15	14
78	30	30	78	41	59	79	77	21	89	76	59	30	9	64	9
67	10	37	14	62	3	85	2	16	74	40	85	30	83	29	5
93	50	83	76	42	86	92	41	27	73	31	70	25	40	11	88

Table A1 Continued

35	68	98	18	67	22	95	34	19	27	21	90	66	20	32	48
32	52	29	78	68	96	94	44	38	95	27	85	53	76	63	78
92	100	75	77	26	39	61	33	88	66	77	76	25	67	90	1
40	73	28	5	50	73	92	32	82	23	78	30	26	52	28	94
57	41	64	50	78	35	12	60	25	4	5	82	82	57	68	43
82	41	67	79	30	43	15	72	98	48	6	22	46	92	43	41
100	11	21	44	43	51	76	89	4	90	48	31	19	89	97	45
94	8	20	67	32	42	39	6	38	25	97	10	18	85	9	60
21	59	27	39	13	81	2	47	83	12	17	54	84	68	56	29
63	62	36	6	57	96	6	36	24	13	70	32	90	92	81	86
91	42	57	99	55	31	58	21	21	65	70	4	37	28	59	9
91	27	61	86	36	57	11	35	92	15	79	30	19	85	39	49
97	39	12	28	35	37	90	93	88	20	99	76	81	61	95	70
64	89	32	80	9	66	73	71	84	69	70	12	10	56	59	56
45	34	1	32	80	99	39	52	25	87	76	91	22	26	46	67
21	65	14	1	78	35	35	63	21	66	34	3	47	51	24	37
85	64	69	93	47	82	55	87	22	56	53	85	43	66	23	66
21	37	62	29	44	39	4	4	99	3	6	82	67	53	14	0
23	8	62	9	19	31	81	92	63	10	65	78	79	96	65	33
84	14	92	85	9	16	51	70	26	60	7	7	55	66	5	51
70	37	11	7	93	63	48	12	35	95	32	5	64	5	63	28
80	27	32	92	81	27	55	98	71	22	66	64	78	79	34	73
66	13	16	48	74	51	78	83	42	31	97	72	25	75	34	40
1	51	47	84	82	27	77	40	99	13	66	52	56	27	2	19
84	26	0	38	55	30	45	80	50	20	17	78	87	4	88	86
95	28	57	33	51	39	18	12	37	100	89	63	22	50	10	22
45	76	48	43	18	24	19	1	65	93	16	48	8	60	32	76

Table A2 **The standard Normal distribution**

The table shows the area in the upper tail of the standard Normal distribution, for the *z* score shown in the margins.

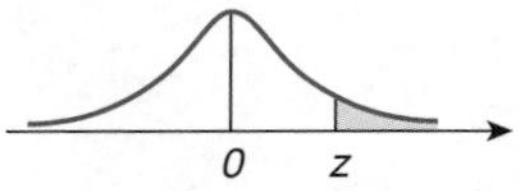

z	0.00	0.01	0.02	0.03	0.04	0.05	0.06	0.07	0.08	0.09
0.0	0.5000	0.4960	0.4920	0.4880	0.4840	0.4801	0.4761	0.4721	0.4681	0.4641
0.1	0.4602	0.4562	0.4522	0.4483	0.4443	0.4404	0.4364	0.4325	0.4286	0.4247
0.2	0.4207	0.4168	0.4129	0.4090	0.4052	0.4013	0.3974	0.3936	0.3897	0.3859
0.3	0.3821	0.3783	0.3745	0.3707	0.3669	0.3632	0.3594	0.3557	0.3520	0.3483
0.4	0.3446	0.3409	0.3372	0.3336	0.3300	0.3264	0.3228	0.3192	0.3156	0.3121
0.5	0.3085	0.3050	0.3015	0.2981	0.2946	0.2912	0.2877	0.2843	0.2810	0.2776
0.6	0.2743	0.2709	0.2676	0.2643	0.2611	0.2578	0.2546	0.2514	0.2483	0.2451
0.7	0.2420	0.2389	0.2358	0.2327	0.2296	0.2266	0.2236	0.2206	0.2177	0.2148
0.8	0.2119	0.2090	0.2061	0.2033	0.2005	0.1977	0.1949	0.1922	0.1894	0.1867
0.9	0.1841	0.1814	0.1788	0.1762	0.1736	0.1711	0.1685	0.1660	0.1635	0.1611
1.0	0.1587	0.1562	0.1539	0.1515	0.1492	0.1469	0.1446	0.1423	0.1401	0.1379
1.1	0.1357	0.1335	0.1314	0.1292	0.1271	0.1251	0.1230	0.1210	0.1190	0.1170
1.2	0.1151	0.1131	0.1112	0.1093	0.1075	0.1056	0.1038	0.1020	0.1003	0.0985
1.3	0.0968	0.0951	0.0934	0.0918	0.0901	0.0885	0.0869	0.0853	0.0838	0.0823
1.4	0.0808	0.0793	0.0778	0.0764	0.0749	0.0735	0.0721	0.0708	0.0694	0.0681
1.5	0.0668	0.0655	0.0643	0.0630	0.0618	0.0606	0.0594	0.0582	0.0571	0.0559
1.6	0.0548	0.0537	0.0526	0.0516	0.0505	0.0495	0.0485	0.0475	0.0465	0.0455
1.7	0.0446	0.0436	0.0427	0.0418	0.0409	0.0401	0.0392	0.0384	0.0375	0.0367
1.8	0.0359	0.0351	0.0344	0.0336	0.0329	0.0322	0.0314	0.0307	0.0301	0.0294
1.9	0.0287	0.0281	0.0274	0.0268	0.0262	0.0256	0.0250	0.0244	0.0239	0.0233
2.0	0.0228	0.0222	0.0217	0.0212	0.0207	0.0202	0.0197	0.0192	0.0188	0.0183
2.1	0.0179	0.0174	0.0170	0.0166	0.0162	0.0158	0.0154	0.0150	0.0146	0.0143
2.2	0.0139	0.0136	0.0132	0.0129	0.0125	0.0122	0.0119	0.0116	0.0113	0.0110
2.3	0.0107	0.0104	0.0102	0.0099	0.0096	0.0094	0.0091	0.0089	0.0087	0.0084
2.4	0.0082	0.0080	0.0078	0.0075	0.0073	0.0071	0.0069	0.0068	0.0066	0.0064
2.5	0.0062	0.0060	0.0059	0.0057	0.0055	0.0054	0.0052	0.0051	0.0049	0.0048
2.6	0.0047	0.0045	0.0044	0.0043	0.0041	0.0040	0.0039	0.0038	0.0037	0.0036
2.7	0.0035	0.0034	0.0033	0.0032	0.0031	0.0030	0.0029	0.0028	0.0027	0.0026
2.8	0.0026	0.0025	0.0024	0.0023	0.0023	0.0022	0.0021	0.0021	0.0020	0.0019
2.9	0.0019	0.0018	0.0018	0.0017	0.0016	0.0016	0.0015	0.0015	0.0014	0.0014
3.0	0.0013	0.0013	0.0013	0.0012	0.0012	0.0011	0.0011	0.0011	0.0010	0.0010

Table A3 Percentage points of the t distribution

The table gives critical values of the *t* distribution cutting off an area α in each tail, shown by the top row of the table.

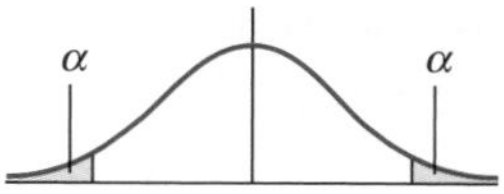

Area (α) in each tail

v	0.4	0.25	0.1	0.05	0.025	0.01	0.005	0.0025	0.001	0.0005
1	0.325	1.000	3.078	6.314	12.706	31.821	63.657	127.320	318.310	636.620
2	0.289	0.816	1.886	2.920	4.303	6.965	9.925	14.089	22.327	31.598
3	0.277	0.765	1.638	2.353	3.182	4.541	5.841	7.453	10.214	12.924
4	0.271	0.741	1.533	2.132	2.776	3.747	4.604	5.598	7.173	8.610
5	0.267	0.727	1.476	2.015	2.571	3.365	4.032	4.773	5.893	6.869
6	0.265	0.718	1.440	1.943	2.447	3.143	3.707	4.317	5.208	5.959
7	0.263	0.711	1.415	1.895	2.365	2.998	3.499	4.029	4.785	5.408
8	0.262	0.706	1.397	1.860	2.306	2.896	3.355	3.833	4.501	5.041
9	0.261	0.703	1.383	1.833	2.262	2.821	3.250	3.690	4.297	4.781
10	0.260	0.700	1.372	1.812	2.228	2.764	3.169	3.581	4.144	4.587
11	0.260	0.697	1.363	1.796	2.201	2.718	3.106	3.497	4.025	4.437
12	0.259	0.695	1.356	1.782	2.179	2.681	3.055	3.428	3.930	4.318
13	0.259	0.694	1.350	1.771	2.160	2.650	3.012	3.372	3.852	4.221
14	0.258	0.692	1.345	1.761	2.145	2.624	2.977	3.326	3.787	4.140
15	0.258	0.691	1.341	1.753	2.131	2.602	2.947	3.286	3.733	4.073
16	0.258	0.690	1.337	1.746	2.120	2.583	2.921	3.252	3.686	4.015
17	0.257	0.689	1.333	1.740	2.110	2.567	2.898	3.222	3.646	3.965
18	0.257	0.688	1.330	1.734	2.101	2.552	2.878	3.197	3.610	3.922
19	0.257	0.688	1.328	1.729	2.093	2.539	2.861	3.174	3.579	3.883
20	0.257	0.687	1.325	1.725	2.086	2.528	2.845	3.153	3.552	3.850
21	0.257	0.686	1.323	1.721	2.080	2.518	2.831	3.135	3.527	3.819
22	0.256	0.686	1.321	1.717	2.074	2.508	2.819	3.119	3.505	3.792
23	0.256	0.685	1.319	1.714	2.069	2.500	2.807	3.104	3.485	3.767
24	0.256	0.685	1.318	1.711	2.064	2.492	2.797	3.091	3.467	3.745
25	0.256	0.684	1.316	1.708	2.060	2.485	2.787	3.078	3.450	3.725
26	0.256	0.684	1.315	1.706	2.056	2.479	2.779	3.067	3.435	3.707
27	0.256	0.684	1.314	1.703	2.052	2.473	2.771	3.057	3.421	3.690
28	0.256	0.683	1.313	1.701	2.048	2.467	2.763	3.047	3.408	3.674
29	0.256	0.683	1.311	1.699	2.045	2.462	2.756	3.038	3.396	3.659
30	0.256	0.683	1.310	1.697	2.042	2.457	2.750	3.030	3.385	3.646
40	0.255	0.681	1.303	1.684	2.021	2.423	2.704	2.971	3.307	3.551
60	0.254	0.679	1.296	1.671	2.000	2.390	2.660	2.915	3.232	3.460
120	0.254	0.677	1.289	1.658	1.980	2.358	2.617	2.860	3.160	3.373
∞	0.253	0.674	1.282	1.645	1.960	2.326	2.576	2.807	3.090	3.291

Table A4 Critical values of the χ^2 distribution

The values in the table give the critical values of χ^2 which cut off the area in the right-hand tail given at the top of the column.

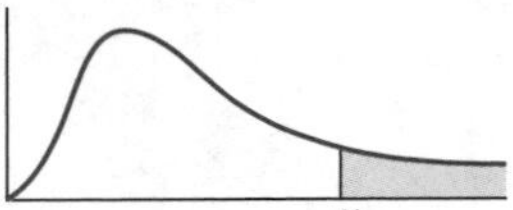

	Area in right-hand tail						
v	0.995	0.990	0.975	0.950	0.900	0.750	0.500
1	392704.10^{-10}	157088.10^{-9}	982069.10^{-9}	393214.10^{-8}	0.0157908	0.1015308	0.454936
2	0.0100251	0.0201007	0.0506356	0.102587	0.210721	0.575364	1.38629
3	0.0717218	0.114832	0.215795	0.351846	0.584374	1.212534	2.36597
4	0.206989	0.297109	0.484419	0.710723	1.063623	1.92256	3.35669
5	0.411742	0.554298	0.831212	1.145476	1.61031	2.67460	4.35146
6	0.675727	0.872090	1.23734	1.63538	2.20413	3.45460	5.34812
7	0.989256	1.239043	1.68987	2.16735	2.83311	4.25485	6.34581
8	1.34441	1.64650	2.17973	2.73264	3.48954	5.07064	7.34412
9	1.73493	2.08790	2.70039	3.32511	4.16816	5.89883	8.34283
10	2.15586	2.55821	3.24697	3.94030	4.86518	6.73720	9.34182
11	2.60322	3.05348	3.81575	4.57481	5.57778	7.58414	10.3410
12	3.07382	3.57057	4.40379	5.22603	6.30380	8.43842	11.3403
13	3.56503	4.10692	5.00875	5.89186	7.04150	9.29907	12.3398
14	4.07467	4.66043	5.62873	6.57063	7.78953	10.1653	13.3393
15	4.60092	5.22935	6.26214	7.26094	8.54676	11.0365	14.3389
16	5.14221	5.81221	6.90766	7.96165	9.31224	11.9122	15.3385
17	5.69722	6.40776	7.56419	8.67176	10.0852	12.7919	16.3382
18	6.26480	7.01491	8.23075	9.39046	10.8649	13.6753	17.3379
19	6.84397	7.63273	8.90652	10.1170	11.6509	14.5620	18.3377
20	7.43384	8.26040	9.59078	10.8508	12.4426	15.4518	19.3374
21	8.03365	8.89720	10.28293	11.5913	13.2396	16.3444	20.3372
22	8.64272	9.54249	10.9823	12.3380	14.0415	17.2396	21.3370
23	9.26043	10.19567	11.6886	13.0905	14.8480	18.1373	22.3369
24	9.88623	10.8564	12.4012	13.8484	15.6587	19.0373	23.3367
25	10.5197	11.5240	13.1197	14.6114	16.4734	19.9393	24.3266
26	11.1602	12.1981	13.8439	15.3792	17.2919	20.8434	25.3365
27	11.8076	12.8785	14.5734	16.1514	18.1139	21.7494	26.3363
28	12.4613	13.5647	15.3079	16.9279	18.9392	22.6572	27.3362
29	13.1211	14.2565	16.0471	17.7084	19.7677	23.5666	28.3361
30	13.7867	14.9535	16.7908	18.4927	20.5992	24.4776	29.3360
40	20.7065	22.1643	24.4330	26.5093	29.0505	33.6603	39.3353
50	27.9907	29.7067	32.3574	34.7643	37.6886	42.9421	49.3349
60	35.5345	37.4849	40.4817	43.1880	46.4589	52.2938	59.3347
70	43.2752	45.4417	48.7576	51.7393	55.3289	61.6983	69.3345
80	51.1719	53.5401	57.1532	60.3915	64.2778	71.1445	79.3343
90	59.1963	61.7541	65.6466	69.1260	73.2911	80.6247	89.3342
100	67.3276	70.0649	74.2219	77.9295	82.3581	90.1332	99.3341

v	0.250	0.100	0.050	0.025	0.010	0.005	0.001
1	1.32330	2.70554	003.84146	5.02389	6.63490	7.87944	10.828
2	2.77259	4.60517	5.99146	7.37776	9.21034	10.5966	13.816
3	4.10834	6.25139	7.81473	9.34840	11.3449	12.8382	16.266
4	5.38527	7.77944	9.48773	11.1433	13.2767	14.8603	18.467
5	6.62568	9.23636	11.0705	12.8325	15.0863	16.7496	20.515
6	7.84080	10.6446	12.5916	14.4494	16.8119	18.5476	22.458
7	9.03715	12.0170	14.0671	16.0128	18.4753	20.2777	24.322
8	10.2189	13.3616	15.5073	17.5345	20.0902	21.9550	26.125
9	11.3888	14.6837	16.9190	19.0228	21.6660	23.5894	27.877
10	12.5489	15.9872	18.3070	20.4832	23.2093	25.1882	29.588
11	13.7007	17.2750	19.6751	21.9200	24.7250	26.7568	31.264
12	14.8454	18.5493	21.0261	23.3367	26.2170	28.2995	32.909
13	15.9839	19.8119	22.3620	24.7356	27.6882	29.8195	34.528
14	17.1169	21.0641	23.6848	26.1189	29.1412	31.3194	36.123
15	18.2451	22.3071	24.9958	27.4884	30.5779	32.8013	37.697
16	19.3689	23.5418	26.2962	28.8454	31.9999	34.2672	29.252
17	20.4887	24.7690	27.5871	30.1910	33.4087	35.7185	40.790
18	21.6049	25.9894	28.8693	31.5264	34.8053	37.1565	42.312
19	22.7178	27.2036	30.1435	32.8523	36.1909	38.5823	43.820
20	23.8277	28.4120	31.4104	34.1696	37.5662	39.9968	45.315
21	24.9348	29.6151	32.6706	35.4789	38.9322	41.4011	46.797
22	26.40393	30.8133	33.9244	36.7807	40.2894	42.7957	48.268
23	27.1413	32.0069	35.1725	38.0756	41.6384	44.1813	49.728
24	28.2412	33.1962	36.4150	39.3641	42.9798	45.5585	51.179
25	29.3389	34.3816	37.6525	40.6465	44.3141	46.9279	52.618
26	30.4346	35.5632	38.8851	41.9232	45.6417	48.2899	54.052
27	31.5284	36.7412	40.1133	43.1945	46.9629	49.6449	55.476
28	32.6205	37.9150	41.3371	44.4608	48.2782	50.9934	56.892
29	33.7109	39.0875	42.5570	45.7223	49.5879	52.3356	58.301
30	34.7997	40.2560	43.7730	46.9792	50.8922	53.6720	59.703
40	45.6160	51.8051	55.7585	59.3417	63.6907	66.7660	73.402
50	56.3336	63.1671	67.5048	71.4202	76.1539	79.4900	86.661
60	66.9815	74.3970	79.0819	83.2977	88.3794	91.9517	99.607
70	77.5767	85.5270	90.5312	95.0232	100.425	104.215	112.317
80	88.1303	96.5782	101.879	106.629	112.329	116.321	124.839
90	98.6499	107.565	113.145	118.136	124.116	128.299	137.208
100	109.141	118.498	124.342	129.561	135.807	140.169	149.449

Table A5(a) Critical values of the *F* distribution (upper 5% points)

The entries in the table give the critical values of *F* cutting off 5% in the right-hand tail of the distribution. v_1 gives the degrees of freedom in the numerator, v_2 those in the denominator.

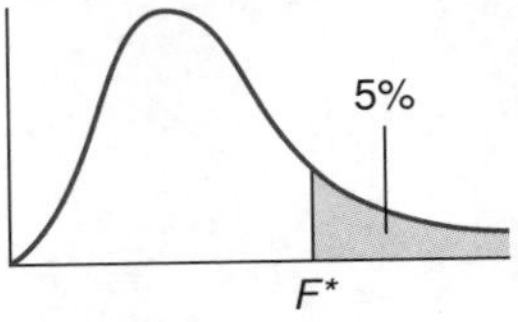

v_2 \ v_1	1	2	3	4	5	6	7	8	9
1	161.45	199.50	215.71	224.58	230.16	230.99	236.77	238.88	240.54
2	18.513	19.000	19.164	19.247	19.296	19.330	19.353	19.371	19.385
3	10.128	9.5521	9.2766	9.1172	9.0135	8.9406	8.8867	8.8452	8.8123
4	7.7086	6.9443	6.5914	6.3882	6.2561	6.1631	6.0942	6.0410	5.9988
5	6.6079	5.7861	5.4095	5.1922	5.0503	4.9503	4.8759	4.8183	4.7725
6	5.9874	5.1433	4.7571	4.5337	4.3874	4.2839	4.2067	4.1468	4.0990
7	5.5914	4.7374	4.3468	4.1203	3.9715	3.8660	3.7870	3.7257	3.6767
8	5.3177	4.4590	4.0662	3.8379	3.6875	3.5806	3.5005	3.4381	3.3881
9	5.1174	4.2565	3.8625	3.6331	3.4817	3.3738	3.2927	3.2296	3.1789
10	4.9646	4.1028	3.7083	3.4780	3.3258	3.2172	3.1355	3.0717	3.0204
11	4.8443	3.9823	3.5874	3.3567	3.2039	3.0946	3.0123	2.9480	2.8962
12	4.7472	3.8853	3.4903	3.2592	3.1059	2.9961	2.9134	2.8486	2.7964
13	4.6672	3.8056	3.4105	3.1791	3.0254	2.9153	2.8321	2.7669	2.7144
14	4.6001	3.7389	3.3439	3.1122	2.9582	2.8477	2.7642	2.6987	2.6458
15	4.5431	3.6823	3.2874	3.0556	2.9013	2.7905	2.7066	2.6408	2.5876
16	4.4940	3.6337	3.2389	3.0069	2.8524	2.7413	2.6572	2.5911	2.5377
17	4.4513	3.5915	3.1968	2.9647	2.8100	2.6987	2.6143	2.5480	2.4943
18	4.4139	3.5546	3.1599	2.9277	2.7729	2.6613	2.5767	2.5102	2.4563
19	4.3807	3.5219	3.1274	2.8951	2.7401	2.6283	2.5435	2.4768	2.4227
20	4.3512	3.4928	2.0984	2.8661	2.7109	2.5990	2.5140	2.4471	2.3928
21	4.3248	3.4668	3.0725	2.8401	2.6848	2.5727	2.4876	2.4205	2.3660
22	4.3009	3.4434	3.0491	2.8167	2.6613	2.5491	2.4638	2.3965	2.3419
23	4.2793	3.4221	3.0280	2.7955	2.6400	2.5277	2.4422	2.3748	2.3201
24	4.2597	3.4028	3.0088	2.7763	2.6307	2.5082	2.4226	2.3551	2.3002
25	4.2417	3.3852	2.9912	2.7587	2.6030	2.4904	2.4047	2.3371	2.2821
26	4.2252	3.3690	2.9752	2.7426	2.5868	2.4741	2.3883	2.3205	2.2655
27	4.2100	3.3541	2.9604	2.7278	2.5719	2.4591	2.3732	2.3053	2.2501
28	4.1960	3.3404	2.9467	2.7141	2.5581	2.4453	2.3593	2.2913	2.2360
29	4.1830	3.3277	2.9340	2.7014	2.5454	2.4324	2.3463	2.2783	2.2229
30	4.1709	3.3158	2.9223	2.6896	2.5336	2.4205	2.3343	2.2662	2.2107
40	4.0847	3.2317	2.8387	2.6060	2.4495	2.3359	2.2490	2.1802	2.1240
60	4.0012	3.1504	2.7581	2.5252	2.3683	2.2541	2.1665	2.0970	2.0401
120	3.9201	3.0718	2.6802	2.4472	2.2899	2.1750	2.0868	2.0164	1.9588
∞	3.8415	2.9957	2.6049	2.3719	2.2141	2.0986	2.0096	1.9384	1.8799

v_2 \ v_1	10	12	15	20	24	30	40	60	120	∞
1	241.88	243.91	245.95	248.01	249.05	250.10	251.14	252.20	253.25	254.31
2	19.396	19.413	19.429	19.446	19.454	19.462	19.471	19.479	19.487	19.496
3	8.7855	8.7446	8.7029	8.6602	8.6385	8.6166	8.5944	8.5720	8.5494	8.5264
4	5.9644	5.9117	5.8578	5.8025	5.7744	5.7459	5.7170	5.6877	5.6581	5.6281
5	4.7351	4.6777	4.6188	4.5581	4.5272	4.4957	4.4638	4.4314	4.3985	4.3650
6	4.0600	3.9999	3.9381	3.8742	3.8415	3.8082	3.7743	3.7398	3.7047	3.6689
7	3.6365	3.5747	3.5107	3.4445	3.4105	3.3758	3.3404	3.3043	3.2674	3.2298
8	3.3472	3.2839	3.2184	3.1503	3.1152	3.0794	3.0428	3.0053	2.9669	2.9276
9	3.1373	3.0729	3.0061	2.9365	2.9005	2.8637	2.8259	2.7872	2.7475	2.7067
10	2.9782	2.9130	2.8450	2.7740	2.7372	2.6996	2.6609	2.6211	2.5801	2.5379
11	2.8536	2.7876	2.7186	2.6464	2.6090	2.5705	2.5309	2.4901	2.4480	2.4045
12	2.7534	2.6866	2.6169	2.5436	2.5055	2.4663	2.4259	2.3842	2.3410	2.2962
13	2.6710	2.6037	2.5331	2.4589	2.4202	2.3803	2.3392	2.2966	2.2524	2.2064
14	2.6022	2.5342	2.4630	2.3879	2.3487	2.3082	2.2664	2.2229	2.1778	2.1307
15	2.5437	2.4753	2.4034	2.3275	2.2878	2.2468	2.2043	2.1601	2.1141	2.0658
16	2.4935	2.4247	2.3522	2.2756	2.2354	2.1938	2.1507	2.1058	2.0589	2.0096
17	2.4499	2.3807	2.3077	2.2304	2.1898	2.1477	2.1040	2.0584	2.0107	1.9604
18	2.4117	2.3421	2.2686	2.1906	2.1497	2.1071	2.0629	2.0166	1.9681	1.9168
19	2.3779	2.3080	2.2341	2.1555	2.1141	2.0712	2.0264	1.9795	1.9302	1.8780
20	2.3479	2.2776	2.2033	2.1242	2.0825	2.0391	1.9938	1.9464	1.8963	1.8432
21	2.3210	2.2504	2.1757	2.0960	2.0540	2.0102	1.9645	1.9165	1.8657	1.8117
22	2.2967	2.2258	2.1508	2.0707	2.0283	1.9842	1.9380	1.8894	1.8380	1.7831
23	2.2747	2.2036	2.1282	2.0476	2.0050	1.9605	1.9139	1.8648	1.8128	1.7570
24	2.2547	2.1834	2.1077	2.0267	1.9838	1.9390	1.8920	1.8424	1.7896	1.7330
25	2.2365	2.1649	2.0889	2.0075	1.9643	1.9192	1.8718	1.8217	1.7684	1.7110
26	2.2197	2.1479	2.0716	1.9898	1.9464	1.9010	1.8533	1.8027	1.7488	1.6906
27	2.2043	2.1323	2.0558	1.9736	1.9299	1.8842	1.8361	1.7851	1.7306	1.6717
28	2.1900	2.1179	2.0411	1.9586	1.9147	1.8687	1.8203	1.7689	1.7138	1.6541
29	2.1768	2.1045	2.0275	1.9446	1.9005	1.8543	1.8055	1.7537	1.6981	1.6376
30	2.1646	2.0921	2.0148	1.9317	1.8874	1.8409	1.7918	1.7396	1.6835	1.6223
40	2.0772	2.0035	1.9245	1.8389	1.7929	1.7444	1.6928	1.6373	1.5766	1.5089
60	1.9926	1.9174	1.8364	1.7480	1.7001	1.6491	1.5943	1.5343	1.4673	1.3893
120	1.9105	1.8337	1.7505	1.6587	1.6084	1.5543	1.4952	1.4290	1.3519	1.2539
∞	1.8307	1.7522	1.6664	1.5705	1.5173	1.4591	1.3940	1.3180	1.2214	1.0000

Table A5(b) Critical values of the *F* distribution (upper 2.5% points)

The entries in the table give the critical values of *F* cutting off 2.5% in the right-hand tail of the distribution. v_1 gives the degrees of freedom in the numerator, v_2 in the denominator.

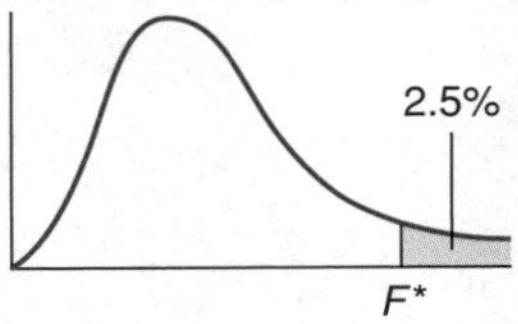

v_2 \ v_1	1	2	3	4	5	6	7	8	9
1	647.79	799.50	864.16	899.58	921.85	937.11	948.22	956.66	963.28
2	38.506	39.000	39.165	39.248	39.298	39.331	39.355	39.373	39.387
3	17.443	16.044	15.439	15.101	14.885	14.735	14.624	14.540	14.473
4	12.218	10.649	9.9792	9.6045	9.3645	9.1973	9.0741	8.9796	8.9047
5	10.007	8.4336	7.7636	7.3879	7.1464	6.9777	6.8531	6.7572	6.6811
6	8.8131	7.2599	6.5988	6.2272	5.9876	5.8198	5.6955	5.5996	5.5234
7	8.0727	6.5415	5.8898	5.5226	5.2852	5.1186	4.9949	4.8993	4.8232
8	7.5709	6.0595	5.4160	5.0526	4.8173	4.6517	4.5286	4.4333	4.3572
9	7.2093	5.7147	5.0781	4.7181	4.4844	4.3197	4.1970	4.1020	4.0260
10	6.9367	5.4564	4.8256	4.4683	4.2361	4.0721	3.9498	3.8549	3.7790
11	6.7241	5.2559	4.6300	4.2751	4.0440	3.8807	3.7586	3.6638	3.5879
12	6.5538	5.0959	4.4742	4.1212	3.8911	3.7283	3.6065	3.5118	3.4358
13	6.4143	4.9653	4.3472	3.9959	3.7667	3.6043	3.4827	3.3880	3.3120
14	6.2979	4.8567	4.2417	3.8919	3.6634	3.5014	3.3799	3.2853	3.2093
15	6.1995	4.7650	4.1528	3.8043	3.5764	3.4147	3.2934	3.1987	3.1227
16	6.1151	4.6867	4.0768	3.7294	3.5021	3.3406	3.2194	3.1248	3.0488
17	6.0420	4.6189	4.0112	3.6648	3.4379	3.2767	3.1556	3.0610	2.9849
18	5.9781	4.5597	3.9539	3.6083	3.3820	3.2209	3.0999	3.0053	2.9219
19	5.9216	4.5075	3.9034	3.5587	3.3327	3.1718	3.0509	2.9563	2.8801
20	5.8715	4.4613	3.8587	3.5147	3.2891	3.1283	3.0074	2.9128	2.8365
21	5.8266	4.4199	3.8188	3.4754	3.2501	3.0895	2.9686	2.8740	2.7977
22	5.7863	4.3828	3.7829	3.4401	3.2151	3.0546	2.9338	2.8392	2.7628
23	5.7498	4.3492	3.7505	3.4083	3.1835	3.0232	2.9023	2.8077	2.7313
24	5.7166	4.3187	3.7211	3.3794	3.1548	2.9946	2.8738	2.7791	2.7027
25	5.6864	4.2909	3.6943	3.3530	3.1287	2.9685	2.8478	2.7531	2.6766
26	5.6586	4.2655	3.6697	3.3289	3.1048	2.9447	2.8240	2.7293	2.6528
27	5.6331	4.2421	3.6472	3.3067	3.0828	2.9228	2.8021	2.7074	2.6309
28	5.6096	4.2205	3.6264	3.2863	3.0626	2.9027	2.7820	2.6872	2.6106
29	5.5878	4.2006	3.6072	3.2674	3.0438	2.8840	2.7633	2.6686	2.5919
30	5.5675	4.1821	3.5894	3.2499	3.0265	2.8667	2.7460	2.6513	2.5746
40	5.4239	4.0510	3.4633	3.1261	2.9037	2.7444	2.6238	2.5289	2.4519
60	5.2856	3.9253	3.3425	3.0077	2.7863	2.6274	2.5068	2.4117	2.3344
120	5.1523	3.8046	3.2269	2.8943	2.6740	2.5154	2.3948	2.2994	2.2217
∞	5.0239	3.6889	3.1161	2.7858	2.5665	2.4082	2.2875	2.1918	2.1136

v_2 \ v_1	10	12	15	20	24	30	40	60	120	∞
1	968.63	976.71	984.87	993.10	997.25	1001.4	1005.6	1009.8	1014.0	1018.3
2	39.398	39.415	39.431	39.448	39.456	39.465	39.473	39.481	39.400	39.498
3	14.419	14.337	14.253	14.167	14.124	14.081	14.037	13.992	13.947	13.902
4	8.8439	8.7512	8.6565	8.5599	8.5109	8.4613	8.4111	8.3604	8.3092	8.2573
5	6.6192	6.5245	6.4277	6.3286	6.2780	6.2269	6.1750	6.1225	6.069?	6.0153
6	5.4613	5.3662	5.2687	5.1684	5.1172	5.0652	5.0125	4.9589	4.9044	4.8491
7	4.7611	4.6658	4.5678	4.4667	4.4150	4.3624	4.3089	4.2544	4.1989	4.1423
8	4.2951	4.1997	4.1012	3.9995	3.9472	3.8940	3.8398	3.7844	3.7279	3.6702
9	3.9639	3.8682	3.7694	3.6669	3.6142	3.5604	3.5055	3.4493	3.3918	3.3329
10	3.7168	3.6209	3.5217	3.4185	3.3654	3.3110	3.2554	3.1984	3.1399	3.0798
11	3.5257	3.4296	3.3299	3.2261	3.1725	3.1176	3.0613	3.0035	2.9441	2.8828
12	3.3736	3.2773	3.1772	3.0728	3.0187	2.9633	2.9063	2.8478	2.7874	2.7249
13	3.2497	3.1532	3.0527	2.9477	2.8932	2.8372	2.7797	2.7204	2.6590	2.5955
14	3.1469	3.0502	2.9493	2.8437	2.7888	2.7324	2.6742	2.6142	2.5519	2.4872
15	3.0602	2.9633	2.8621	2.7559	2.7006	2.6437	2.5850	2.5242	2.4611	2.3953
16	2.9862	2.8890	2.7875	2.6808	2.6252	2.5678	2.5085	2.4471	2.3831	2.3163
17	2.9222	2.8249	2.7230	2.6158	2.5598	2.5020	2.4422	2.3801	2.3153	2.2474
18	2.8664	2.7689	2.6667	2.5590	2.5027	2.4445	2.3842	2.3214	2.2558	2.1869
19	2.8172	2.7196	2.6171	2.5089	2.4523	2.3937	2.3329	2.2696	2.2032	2.1333
20	2.7737	2.6758	2.5731	2.4645	2.4076	2.3486	2.2873	2.2234	2.1562	2.0853
21	2.7348	2.6368	2.5338	2.4247	2.3675	2.3082	2.2465	2.1819	2.1141	2.0422
22	2.6998	2.6017	2.4984	2.3890	2.3315	2.2718	2.2097	2.1446	2.0760	2.0032
23	2.6682	2.5699	2.4665	2.3567	2.2989	2.2389	2.1763	2.1107	2.0415	1.9677
24	2.6396	2.5411	2.4374	2.3273	2.2693	2.2090	2.1460	2.0799	2.0099	1.9353
25	2.6135	2.5149	2.4110	2.3005	2.2422	2.1816	2.1183	2.0516	1.9811	1.9055
26	2.5896	2.4908	2.3867	2.2759	2.2174	2.1565	2.0928	2.0257	1.9545	1.8781
27	2.5676	2.4688	2.3644	2.2533	2.1946	2.1334	2.0693	2.0018	1.9299	1.8527
28	2.5473	2.4484	2.3438	2.2324	2.1735	2.1121	2.0477	1.9797	1.9072	1.8291
29	2.5286	2.4295	2.3248	2.2131	2.1540	2.0923	2.0276	1.9591	1.8861	1.8072
30	2.5112	2.4120	2.3072	2.1952	2.1359	2.0739	2.0089	1.9400	1.8664	1.7867
40	2.3882	2.2882	2.1819	2.0677	2.0069	1.9429	1.8752	1.8028	1.7242	1.6371
60	2.2702	2.1692	2.0613	1.9445	1.8817	1.8152	1.7440	1.6668	1.5810	1.4821
120	2.1570	2.0548	1.9450	1.8249	1.7597	1.6899	1.6141	1.5299	1.4327	1.3104
∞	2.0483	1.9447	1.8326	1.7085	1.6402	1.5660	1.4835	1.3883	1.2684	1.0000

Table A5(C) Critical values of the *F* distribution (upper 1% points)

The entries in the table give the critical values of *F* cutting off 1% in the right-hand tail of the distribution. v_1 gives the degrees of freedom in the numerator, v_2 in the denominator.

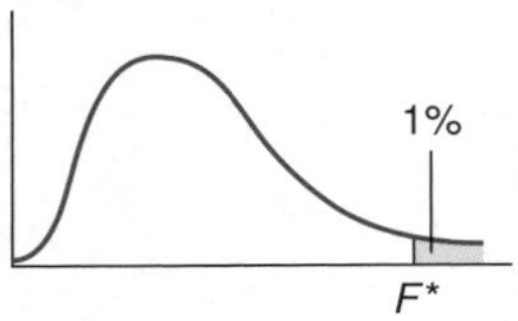

v_1 / v_1	1	2	3	4	5	6	7	8	9
1	4052.2	4999.5	5403.4	5624.6	5763.6	5859.0	5928.4	5981.1	6022.5
2	98.503	99.000	99.166	99.249	99.299	99.333	99.356	99.374	99.388
3	34.116	30.817	29.457	28.710	28.237	27.911	27.672	27.489	27.345
4	21.198	18.000	16.694	15.977	15.522	15.207	14.976	14.799	14.659
5	16.258	13.274	12.060	11.392	10.967	10.672	10.456	10.289	10.158
6	13.745	10.925	9.7795	9.1483	8.7459	8.4661	8.2600	8.1017	7.9761
7	12.246	9.5466	8.4513	7.8466	7.4604	7.1914	6.9928	6.8400	6.7188
8	11.259	8.6491	7.5910	7.0061	6.6318	6.3707	6.1776	6.0289	5.9106
9	10.561	8.0215	6.9919	6.4221	6.0569	5.8018	5.6129	5.4671	5.3511
10	10.044	7.5594	6.5523	5.9943	5.6363	5.3858	5.2001	5.0567	4.9424
11	9.6460	7.2057	6.2167	5.6683	5.3160	5.0692	4.8861	4.7445	4.6315
12	9.3302	6.9266	5.9525	5.4120	5.0643	4.8206	4.6395	4.4994	4.3875
13	9.0738	6.7010	5.7394	5.2053	4.8616	4.6204	4.4410	4.3021	4.1911
14	8.8618	6.5149	5.5639	5.0354	4.6950	4.4558	4.2779	4.1399	4.0297
15	8.6831	6.3589	5.4170	4.8932	4.5556	4.3183	4.1415	4.0045	3.8948
16	8.5310	6.2262	5.2922	4.7726	4.4374	4.2016	4.0259	3.8896	3.7804
17	8.3997	6.1121	5.1850	4.6690	4.3359	4.1015	3.9267	3.7910	3.6822
18	8.2854	6.0129	5.0919	4.5790	4.2479	4.0146	3.8406	3.7054	3.5971
19	8.1849	5.9259	5.0103	4.5003	4.1708	3.9386	3.7653	3.6305	3.5225
20	8.0960	5.8489	4.9382	4.4307	4.1027	3.8714	3.6987	3.5644	3.4567
21	8.0166	5.7804	4.8740	4.3688	4.0421	3.8117	3.6396	3.5056	3.3981
22	7.9454	5.7190	4.8166	4.3134	3.9880	3.7583	3.5867	3.4530	3.3458
23	7.8811	5.6637	4.7649	4.2636	3.9392	3.7102	3.5390	3.4057	3.2986
24	7.8229	5.6136	4.7181	4.2184	3.8951	3.6667	3.4959	3.3629	3.2560
25	7.7698	5.5680	4.6755	4.1774	3.8550	3.6272	3.4568	3.3439	3.2172
26	7.7213	5.5263	4.6366	4.1400	3.8183	3.5911	3.4210	3.2884	3.1818
27	7.6767	5.4881	4.6009	4.1056	3.7848	3.5580	3.3882	3.2558	3.1494
28	7.6356	5.4529	4.5681	4.0740	3.7539	3.5276	3.3581	3.2259	3.1195
29	7.5977	5.4204	4.5378	4.0449	3.7254	3.4995	3.3303	3.1982	3.0920
30	7.5625	5.3903	4.5097	4.0179	3.6990	3.4735	3.3045	3.1726	3.0665
40	7.3141	5.1785	4.3126	3.8283	3.5138	3.2910	3.1238	2.9930	2.8876
60	7.0771	4.9774	4.1259	3.6490	3.3389	3.1187	2.9530	2.8233	2.7185
120	6.8509	4.7865	3.9491	3.4795	3.1735	2.9559	2.7918	2.6629	2.5586
∞	6.6349	4.6052	3.7816	3.3192	3.0173	2.8020	2.6393	2.5113	2.4073

v_2 \ v_1	10	12	15	20	24	30	40	60	120	∞
1	6055.8	6106.3	6157.3	6208.7	6234.6	6260.6	6286.8	6313.0	6339.4	6365.9
2	99.399	99.416	99.433	99.449	99.458	99.466	99.474	99.482	99.491	99.499
3	27.229	27.052	26.872	26.690	26.598	26.505	26.411	26.316	26.221	26.125
4	14.546	14.374	14.198	14.020	13.929	13.838	13.745	13.652	13.558	13.463
5	10.051	9.8883	9.7222	9.5526	9.4665	9.3793	9.2912	9.2020	9.1118	9.0204
6	7.8741	7.7183	7.5590	7.3958	7.3127	7.2285	7.1432	7.0567	6.9690	6.8800
7	6.6201	6.4691	6.3143	6.1554	6.0743	5.9920	5.9084	5.8236	5.7373	5.6495
8	5.8143	5.6667	5.5151	5.3591	5.2793	5.1981	5.1156	5.0316	4.9461	4.8588
9	5.2565	5.1114	4.9621	4.8080	4.7290	4.6486	4.5666	4.4831	4.3978	4.3105
10	4.8491	4.7059	4.5581	4.4054	4.3269	4.2469	4.1653	4.0819	3.9965	3.9090
11	4.5393	4.3974	4.2509	4.0990	4.0209	3.9411	3.8596	3.7761	3.6904	3.6024
12	4.2961	4.1553	4.0096	3.8584	3.7805	3.7008	3.6192	3.5355	3.4494	3.3608
13	4.1003	3.9603	3.8154	3.6646	3.5868	3.5070	3.4253	3.3413	3.2548	3.1654
14	3.9394	3.8001	3.6557	3.5052	3.4274	3.3476	3.2656	3.1813	3.0942	3.0040
15	3.8049	3.6662	3.5222	3.3719	3.2940	3.2141	3.1319	3.0471	2.9595	2.8684
16	3.6909	3.5527	3.4089	3.2587	3.1808	3.1007	3.0182	2.9330	2.8447	2.7528
17	3.5931	3.4552	3.3117	3.1615	3.0835	2.0032	2.9205	2.8348	2.7459	2.6530
18	3.5082	3.3706	3.2273	3.0771	2.9990	2.9185	2.8354	2.7493	2.6597	2.5660
19	3.4338	3.2965	3.1533	3.0031	2.9249	2.8442	2.7608	2.6742	2.5839	2.4893
20	3.3682	3.2311	3.0880	2.9377	2.8594	2.7785	2.6947	2.6077	2.5168	2.4212
21	3.3098	3.1730	3.0300	2.8796	2.8010	2.7200	2.6359	2.5484	2.4568	2.3603
22	3.2576	3.1209	2.9779	2.8274	2.7488	2.6675	2.5831	2.4951	2.4029	2.3055
23	3.2106	3.0740	2.9311	2.7805	2.7017	2.6202	2.5355	2.4471	2.3542	2.2558
24	3.1681	3.0316	2.8887	2.7380	2.6591	2.5773	2.4923	2.4035	2.3100	2.2107
25	3.1294	2.9931	2.8502	2.6993	2.6203	2.5383	2.4530	2.3637	2.2696	2.1694
26	3.0941	2.9578	2.8150	2.6640	2.5848	2.5026	2.4170	2.3273	2.2325	2.1315
27	3.0618	2.9256	2.7827	2.6316	2.5522	2.4699	2.3840	2.2938	2.1985	2.0965
28	3.0320	2.8959	2.7530	2.6017	2.5223	2.4397	2.3535	2.2629	2.1670	2.0642
29	3.0045	2.8685	2.7256	2.5742	2.4946	2.4118	2.3253	2.2344	2.1379	2.0342
30	2.9791	2.8431	2.7002	2.5487	2.4689	2.3860	2.2992	2.2079	2.1108	2.0062
40	2.8005	2.6648	2.5216	2.3689	2.2880	2.2034	2.1142	2.0194	1.9172	1.8047
60	2.6318	2.4961	2.3523	2.1978	2.1154	2.0285	1.9360	1.8363	1.7263	1.6006
120	2.4721	2.3363	2.1915	2.0346	1.9500	1.8600	1.7628	1.6557	1.5330	1.3805
∞	2.3209	2.1847	2.0385	1.8783	1.7908	1.6964	1.5923	1.4730	1.3246	1.0000

Table A5(d) Critical values of the *F* distribution (upper 0.5% points)

The entries in the table give the critical values of *F* cutting off 0.5% in the right-hand tail of the distribution. v_1 gives the degrees of freedom in the numerator, v_2 in the denominator.

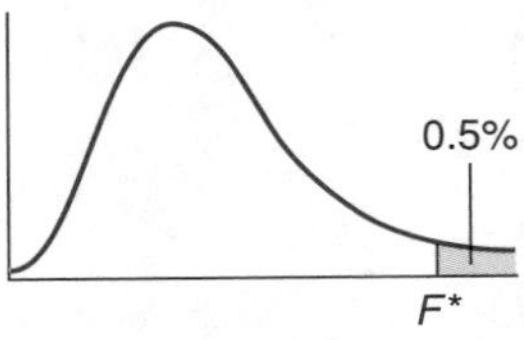

v_1 / v_1	1	2	3	4	5	6	7	8	9
1	16211	20000	21615	22500	23056	23437	23715	23925	24091
2	198.50	199.00	199.17	199.25	199.30	199.33	199.36	199.37	199.39
3	55.552	49.799	47.467	46.195	45.392	44.838	44.434	44.126	43.882
4	31.333	26.284	24.259	23.155	22.456	21.975	21.622	21.352	21.139
5	22.785	18.314	16.530	15.556	14.940	14.513	14.200	13.961	13.772
6	18.635	14.544	12.917	12.028	11.464	11.073	10.786	10.566	10.391
7	16.236	12.404	10.882	10.050	9.5221	9.1553	8.8854	8.6781	8.5138
8	14.688	11.042	9.5965	8.8051	9.3018	7.9520	7.6941	7.4959	7.3386
9	13.614	10.107	8.7171	7.9559	7.4712	7.1339	6.8849	6.6933	6.5411
10	12.826	9.4270	8.0807	7.3428	6.8724	6.5446	6.3025	6.1159	5.9676
11	12.226	8.9122	7.6004	6.8809	6.4217	6.1016	5.8648	5.6821	5.5368
12	11.754	8.5096	7.2258	6.5211	6.0711	5.7570	5.5245	5.3451	5.2021
13	11.374	8.1865	6.9258	6.2335	5.7910	5.4819	5.2529	5.0761	4.9351
14	11.060	7.9216	6.6804	5.9984	5.5623	5.2574	5.0313	4.8566	4.7173
15	10.798	7.7008	6.4760	5.8029	5.3721	5.0708	4.8473	4.6744	3.5364
16	10.575	7.5138	6.3034	5.6378	5.2117	4.9134	4.6920	4.5207	4.3838
17	10.384	7.3536	6.1556	5.4967	5.0746	4.7789	4.5594	4.3894	4.2535
18	10.218	7.2148	6.0278	5.3746	3.9560	4.6627	4.4448	3.2759	4.1410
19	10.073	7.0935	5.9161	5.2681	4.8526	4.5614	4.3448	4.1770	4.0428
20	9.9439	6.9865	5.8177	5.1743	4.7616	4.4721	4.2569	4.0900	3.9564
21	9.8295	6.8914	5.7304	5.0911	4.6809	4.3931	4.1789	4.0128	3.8799
22	9.7271	6.8064	5.6524	5.0168	4.6088	4.3225	4.1094	3.9440	3.8116
23	9.6348	6.7300	5.5823	4.9500	3.5441	4.2591	4.0469	3.8822	3.7502
24	9.5513	6.6609	5.5190	4.8898	4.4857	4.2019	3.9905	3.8264	3.6949
25	9.4753	6.5982	5.4615	4.8351	4.4327	4.1500	3.9394	3.7758	3.6447
26	9.4059	6.5409	5.4091	4.7852	4.3844	4.1027	3.8928	3.7297	3.5989
27	9.3423	6.4885	5.3611	4.7396	4.3402	4.0594	3.8501	3.6875	3.5571
28	9.2838	6.4403	5.3170	4.6977	4.2996	4.0197	3.8110	3.6487	3.5186
29	9.2297	6.3958	5.2764	4.6591	4.2622	3.9831	3.7749	3.6131	3.4832
30	9.1797	6.3547	5.2388	4.6234	4.2276	3.9492	3.7416	3.5801	3.4504
40	8.8279	6.0664	4.9758	4.3738	3.9860	3.7129	3.5088	3.3498	3.2220
60	8.4946	5.7950	4.7290	4.1399	3.7599	3.4918	3.2911	3.1344	3.0083
120	8.1788	5.5393	4.4972	3.9207	3.5482	3.2849	3.0874	2.9330	2.8083
∞	7.894	5.2983	4.2794	3.7151	3.3499	3.0913	2.8968	2.7444	2.6210

v_2 \ v_1	10	12	15	20	24	30	40	60	120	∞
1	24224	24426	24630	24836	24940	25044	25148	25253	25359	25464
2	199.40	199.42	199.43	199.45	199.46	199.47	199.47	199.48	199.49	199.50
3	43.686	43.387	43.085	42.778	42.622	42.466	42.308	42.149	41.989	41.828
4	20.967	20.705	20.438	20.167	20.030	19.892	19.752	19.611	19.468	19.325
5	13.618	13.384	13.146	12.903	12.780	12.656	12.530	12.402	12.274	12.144
6	10.250	10.034	9.8140	9.5888	9.4742	9.3582	9.2408	9.1219	9.0015	8.8793
7	8.3803	8.1764	7.9678	7.7540	7.6450	7.5345	7.4224	7.3088	7.1933	7.0760
8	7.2106	7.0149	6.8143	6.6082	6.5029	6.3961	6.2875	6.1772	6.0649	5.9506
9	6.4172	6.2274	6.0325	5.8318	5.7292	5.6248	5.5186	5.4104	5.3001	5.1875
10	5.8467	5.6613	5.4707	5.2740	5.1732	5.0706	4.9659	4.8592	4.7501	4.6385
11	5.4183	5.2363	5.0489	4.8552	4.7557	4.6543	4.5508	4.4450	4.3367	4.2255
12	5.0855	4.9062	4.7213	4.5299	4.4314	4.3309	4.2282	5.1229	4.0149	3.9039
13	4.8199	4.6429	4.4600	4.2703	4.1726	4.0727	3.9704	3.8655	3.7577	3.6465
14	4.6034	4.4281	4.2468	4.0585	3.9614	3.8619	3.7600	3.6552	3.5473	3.4359
15	4.4235	4.2497	4.0698	3.8826	3.7859	3.6867	3.5850	3.4803	3.3722	3.2602
16	4.2719	4.0994	3.9205	3.7342	3.6378	3.5389	3.4372	3.3324	3.2240	3.1115
17	4.1424	3.9709	3.7929	3.6073	3.5112	3.4124	3.3108	3.2058	3.0971	2.9839
18	4.0305	3.8599	3.6827	3.4977	3.4017	3.3030	3.2014	3.0962	2.9871	2.8732
19	3.9329	3.7631	4.5866	3.4020	3.3062	3.2075	3.1058	3.0004	2.8908	2.7762
20	3.8470	3.6779	3.5020	3.3178	3.2220	3.1234	3.0215	2.9159	2.8058	2.6904
21	3.7709	3.6024	3.4270	3.2431	3.1474	3.0488	2.9467	2.7408	2.7302	2.6140
22	3.7030	3.5350	3.3600	3.1764	3.0807	2.9821	2.8799	2.7736	2.6625	2.5455
23	3.6420	3.4745	3.2999	3.1165	3.0208	2.9221	2.8197	2.7132	2.6015	2.4837
24	3.5870	3.4199	3.2456	3.0624	2.9667	2.8679	2.7654	2.6585	2.5463	2.4276
25	3.5370	3.3704	3.1963	3.0133	2.9176	2.8187	2.7160	2.6088	2.4961	2.3765
26	3.4916	3.3252	3.1515	2.9685	2.8728	2.7738	2.6709	2.5633	2.4501	2.3297
27	3.4499	3.2839	3.1104	2.9275	2.8318	2.7327	2.6296	2.5217	2.4079	2.2867
28	3.4117	3.2460	3.0727	2.8899	2.7941	2.6949	2.5916	2.4834	2.3690	2.2470
29	3.3765	3.2110	3.0379	2.8551	2.7594	2.6600	2.5565	2.4479	2.3331	2.2102
30	3.3440	3.1787	3.0057	2.8230	2.7272	2.6278	2.5241	2.4151	2.2998	2.1760
40	3.1167	2.9531	2.7811	2.5984	2.5020	2.4015	2.2958	2.1838	2.0636	1.9318
60	2.9042	2.7419	2.5705	2.3872	2.2898	2.1874	2.0789	1.9622	1.8341	1.6885
120	2.7052	2.5439	2.3727	2.1881	2.0890	1.9840	1.8709	1.7469	1.6055	1.4311
∞	2.5188	2.3583	2.1868	1.9998	1.8983	1.7891	1.6691	1.5325	1.3637	1.0000

Table A6 Critical values of Spearman's rank correlation coefficient

Entries in the table show critical values of Spearman's rank correlation coefficient. The value at the top of each column shows the significance level for a two-tailed test. For a one-tailed test, the significance level is half that shown.

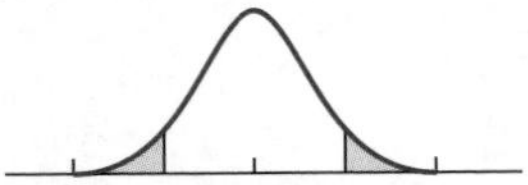

N	10%	5%	2%	1%
5	0.900	–	–	–
6	0.829	0.886	0.943	–
7	0.714	0.786	0.893	–
8	0.643	0.738	0.833	0.881
9	0.600	0.683	0.783	0.833
10	0.564	0.648	0.745	0.818
11	0.523	0.623	0.763	0.794
12	0.497	0.591	0.703	0.780
13	0.475	0.566	0.673	0.746
14	0.457	0.545	0.646	0.716
15	0.441	0.525	0.623	0.689
16	0.425	0.507	0.601	0.666
17	0.412	0.490	0.582	0.645
18	0.399	0.476	0.564	0.625
19	0.388	0.462	0.549	0.608
20	0.377	0.450	0.534	0.591
21	0.368	0.438	0.521	0.576
22	0.359	0.428	0.508	0.562
23	0.351	0.418	0.496	0.549
24	0.343	0.409	0.485	0.537
25	0.336	0.400	0.475	0.526
26	0.329	0.392	0.465	0.515
27	0.323	0.385	0.456	0.505
28	0.317	0.377	0.448	0.496
29	0.311	0.370	0.440	0.487
30	0.305	0.364	0.432	0.478

Source: ***Annals of Statistics***, 1936 and 1949.

Table A7 Critical values for the Durbin–Watson test at 5% significance level

Sample size	Number of explanatory variables									
	1		2		3		4		5	
n	d_L	d_U	d_L	d_U	d_L	d_U	d_L	d_U	d_L	d_U
10	0.879	1.320	0.697	1.641	0.525	2.016	0.376	2.414	0.243	2.822
11	0.927	1.324	0.758	1.604	0.595	1.928	0.444	2.283	0.316	2.645
12	0.971	1.331	0.812	1.579	0.658	1.864	0.512	2.177	0.379	2.506
13	1.010	1.340	0.861	1.562	0.715	1.816	0.574	2.094	0.445	2.390
14	1.045	1.350	0.905	1.551	0.767	1.779	0.632	2.030	0.505	2.296
15	1.077	1.361	0.946	1.543	0.814	1.750	0.685	1.977	0.562	2.220
20	1.201	1.411	1.100	1.537	0.998	1.676	0.894	1.828	0.792	1.991
25	1.288	1.454	1.206	1.550	1.123	1.654	1.038	1.767	0.953	1.886
30	1.352	1.489	1.284	1.567	1.214	1.650	1.143	1.739	1.071	1.833
35	1.402	1.519	1.343	1.584	1.283	1.653	1.222	1.726	1.160	1.803
40	1.442	1.544	1.391	1.600	1.338	1.659	1.285	1.721	1.230	1.786
50	1.503	1.585	1.462	1.628	1.421	1.674	1.378	1.721	1.335	1.771
100	1.654	1.694	1.634	1.715	1.613	1.736	1.592	1.758	1.571	1.780
200	1.758	1.778	1.748	1.789	1.738	1.799	1.728	1.810	1.718	1.820

Answers and Commentary on Problems

Answers to Chapter 1

Problem 1.1

(a)

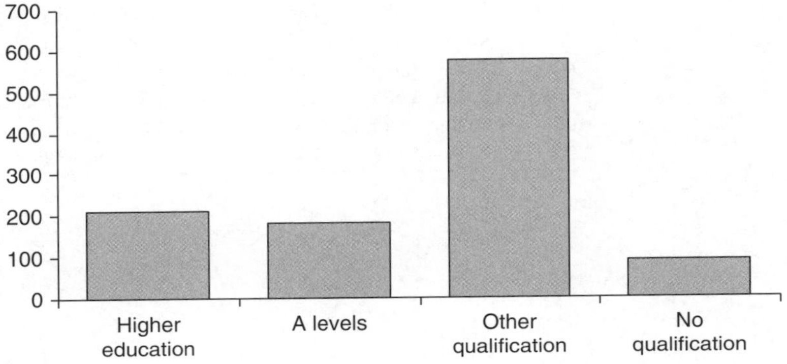

Comparison is complicated because the data for women only are from a survey, hence the absolute numbers are much smaller. Also, we compare women here with the total for men and women in the text. We have to infer what the difference between men and women is. The major difference apparent is that there are relatively more women in the 'Other qualification' category, and relatively fewer in the higher education category.

(b)

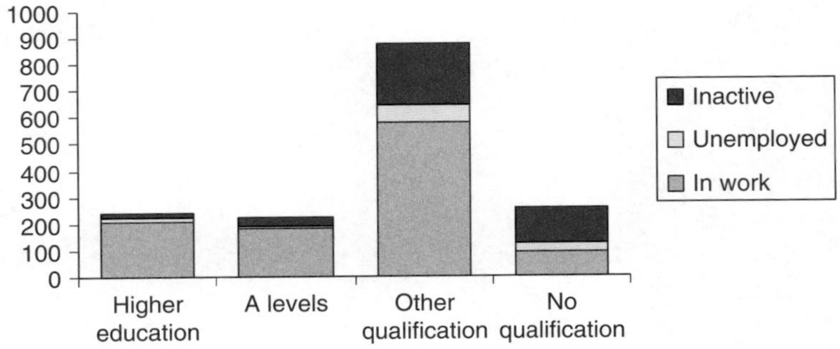

There is not a great difference apparent between this graph and the one in the text. Closer inspection of the figures suggests a higher proportion of women are 'inactive' and a lower proportion in work, but this detail is difficult to discern from the graph alone.

(c)

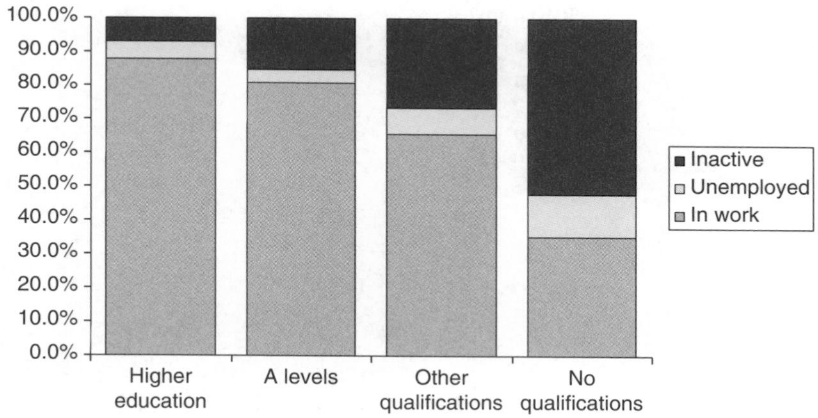

This chart brings out a little more clearly that there is a slightly higher degree of inactivity amongst women, particularly in the lower education categories.

(d)

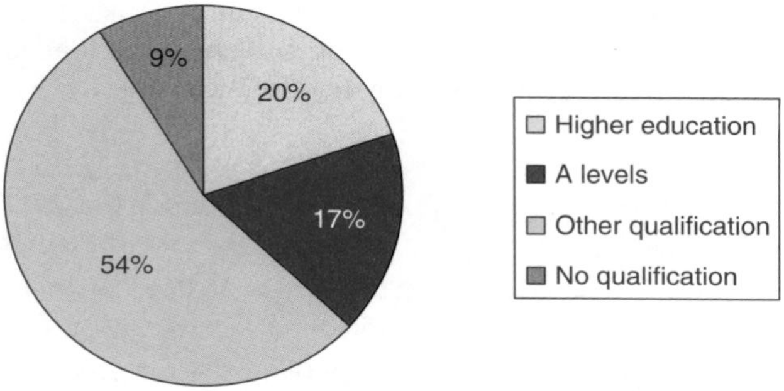

This again brings out the higher proportion with 'other' qualifications and the fewer with higher education.

Problem 1.3

(a) Higher education, 88%.

(b) Those in work, 20%.

Problem 1.5

Bar chart

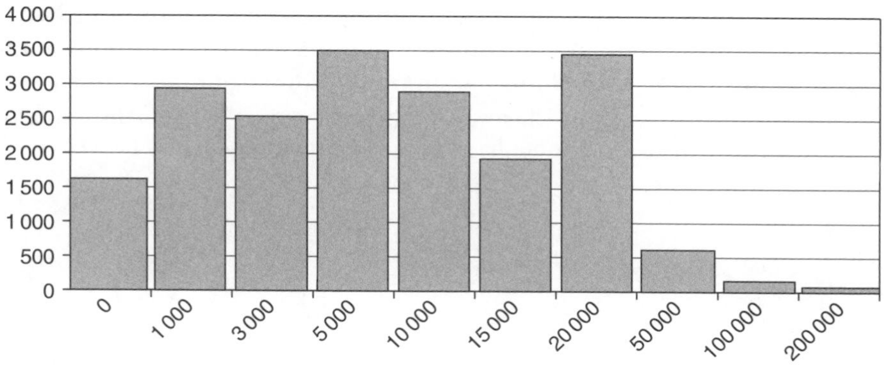

Histogram

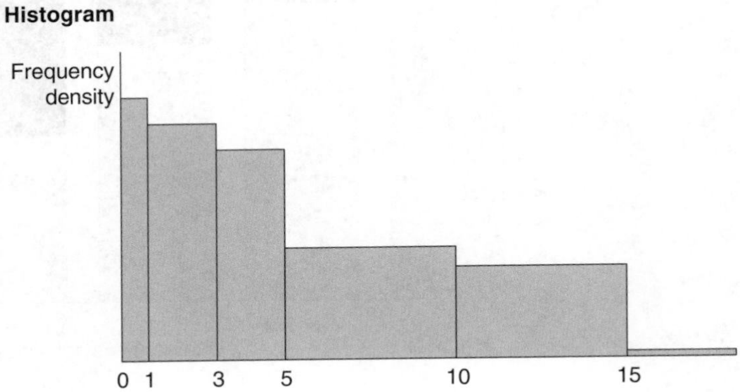

The difference between bar chart and histogram is similar to that for the 2005 distribution. The overall shape of the histogram is similar (heavily skewed to right). Comparison is difficult because of different wealth levels (due to inflation), and grouping into classes can affect the precise shape of the graph.

Problem 1.7

(a) Mean 16.399 (£000); median 8.92; mode 0–1 (£000) group has the greatest frequency density. They differ because of skewness in the distribution.

(b) Q1 = 3.295, Q3 = 18.399, IQR = 15.044; variance = 652.88; s.d. = 25.552; coefficient of variation = 1.56.

(c) $95{,}469.32/25.55^3 = 5.72 > 0$ as expected. Ask your class whether this number is revealing to them. Is there any intuition?

(d) Comparison in text.

(e) This would increase the mean substantially (to 31.12), but the median and mode would be unaffected.

Problem 1.9

$$(33 \times 134 + 40 \times 139 + 25 \times 137)/(33 + 40 + 25) = 136.8 \text{ pence/litre}$$

Problem 1.11

(a) $z = 1.5\,(= (83 - 65)/\sqrt{144})$ and $-1.5(= (47 - 65/\sqrt{144})$ respectively.

(b) Using Chebyshev's theorem with $k = 1.5$, we have that at least $(1 - 1/1.5^2) = 0.56$ (56%) lies within 1.5 standard deviations of the mean, so at most 0.44 (44 students) lies outside the range.

(c) As Chebyshev's theorem applies to *both* tails, we cannot answer this part. You cannot halve 0.44 as the distribution may be skewed.

Problem 1.13

(a)

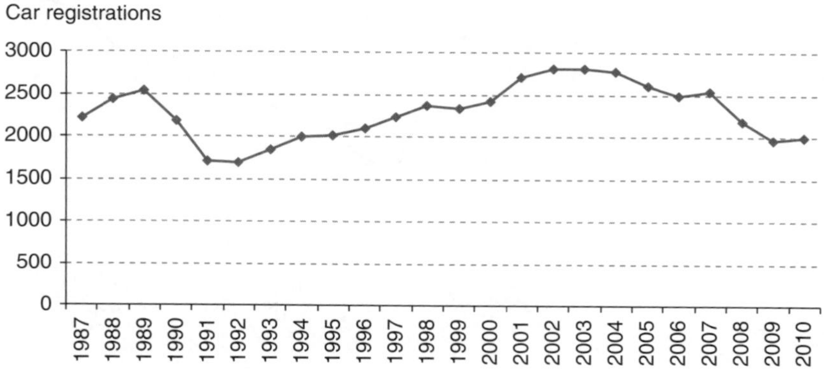

The series shows an initial peak in 1989, then a slow recovery from 1991 to 2003. After this the market turns down again. The series is quite volatile with long upward and downward swings.

(b)

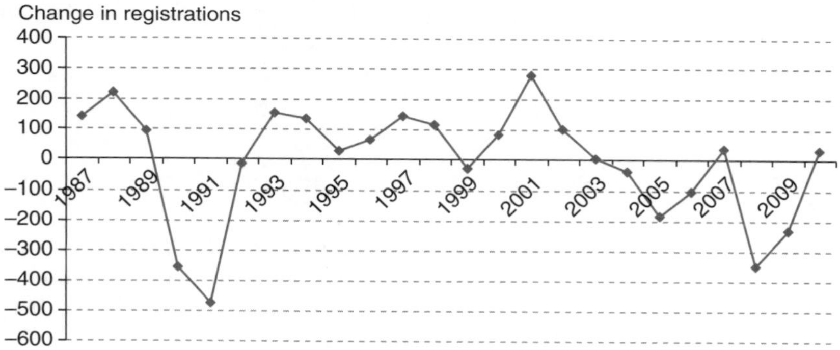

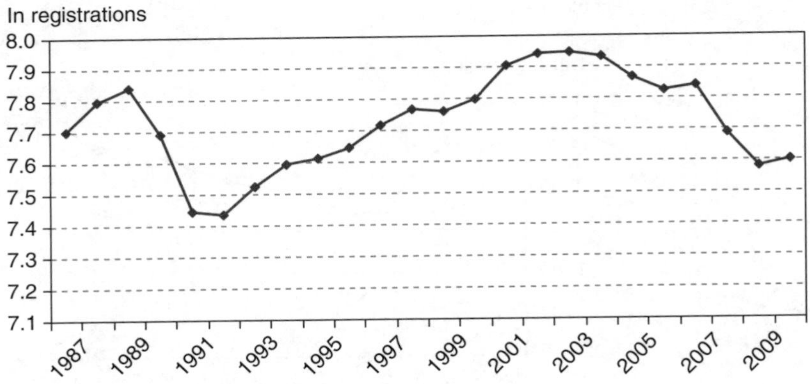

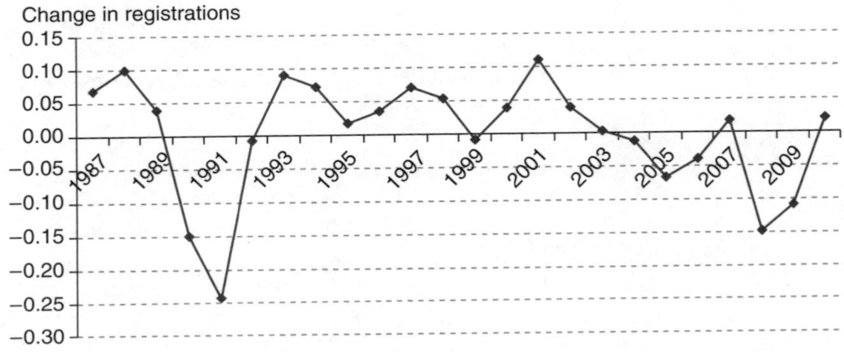

This form of the graphs shows that recessions can be quite severe for the vehicle market, while upswings show smaller per annum increases. The log graphs are very similar to the levels graphs. There is not always an advantage in drawing these.

Problem 1.15

(a) $(1994.6/2212.6)^{1/23} = -0.0045$ or -0.45% p.a. (note that this is not particularly representative – most years had positive growth but there were sharp falls at the beginning and end of the period).

(b) 0.084 (around the geometric mean).

(c) The standard deviations of the two growth rates are similar (it was 0.0766 for investment), so this suggests a similar level of volatility. Note that, since the means are very different (6.37% p.a. for investment) and for registrations is actually negative, there is a big difference in volatility if one relies upon the coefficient of variation. This latter is misleading in this case, however – it reflects differences in the means rather than in volatility.

Problem 1.17

(a) Non-linear, upward trend. It is likely to be positively autocorrelated. Variation around the trend is likely to grow over time (heteroscedasticity).

(b) Similar to (a), except that the trend would be shallower after deflation. Probably there will be less heteroscedasticity because price variability has been removed, which may also increase the autocorrelation of the series.

(c) Unlikely to show a trend in the very long run, but there might be one over, say, five years, if inflation is increasing. Likely to be homoscedastic, with some degree of autocorrelation.

Problem 1.19

(a) Rearranging $S_t = S_0(1 + r)^t$, we can obtain $S_0 = S_t/(1 + r)^t$. Setting $S_t = 1000$, $r = 0.07$, and $t = 5$ gives $S_0 = 712.99$. Price after two years: 816.30 (use $t = 3$ to obtain this). If r rose after two years to 10%, the bond would fall to $1000/(1 + 0.10)^3 = 751.31$

(b) The income stream should be discounted to the present using

$$\frac{200}{1+r} + \frac{200}{(1+r)^2} + \frac{200}{(1+r)^3} + \frac{200}{(1+r)^4} + \frac{200}{(1+r)^5} = 820.04,$$

so the bond should sell for £820.04. It is worth more than the previous bond because the return is obtained earlier.

Problem 1.21

(a) 17.9% p.a. for BMW, 14% p.a. for Mercedes

(b) Depreciated values are as shown in this table,

BMW 525i	22 275	18 284	15 008	12 319	10 112	8 300
Merc 200E	21 900	18 833	16 196	13 928	11 977	10 300

which are close to actual values. Depreciation is initially slower than the average, then speeds up, for both cars.

Problem 1.23

$$E(x + k) = \frac{\sum(x + k)}{n} = \frac{\sum x + nk}{n} = \frac{\sum x}{n} + k = E(x) + k$$

Problem 1.25

The mistake is comparing non-comparable averages. A first-time buyer would have an above-average mortgage and purchase a below-average priced house, hence the amount of buyer's equity would be small. The original argument came from the *Morning Star* newspaper many years ago. With reasoning like this, no wonder communism failed.

Answers to exercises on Σ notation

Problem 1A.1

20, 90, 400, 5, 17, 11.

Problem 1A.2

40, 360, 1600, 25, 37, 22.

Problem 1A.3

88, 372, 16, 85.

Problem 1A.4

352, 2976, 208, 349.

Problem 1A.5

113, 14, 110.

Problem 1A.6

56, 8, 48.

Problem 1A.7

$$\frac{\sum f(x-k)}{\sum f} = \frac{\sum fx - k\sum f}{\sum f} = \frac{\sum fx}{\sum f} - k$$

Problem 1A.8

$$\frac{\sum f(x-\mu)^2}{\sum f} = \frac{\sum f(x^2 - 2\mu x + \mu^2)}{\sum f} = \frac{\sum fx^2 - 2\mu\sum fx + \mu^2\sum f}{\sum f}$$

$$= \frac{\sum fx^2}{\sum f} - 2\mu^2 + \mu^2 = \frac{\sum fx^2}{\sum f} - \mu^2$$

Answers to exercises on logarithms

Problem 1C.1

−0.8239, 0.17609, 1.17609, 2.17609, 3.17609, 1.92284, 0.96142, impossible.

Problem 1C.2

−0.09691, 0.90309, 1.90309, 0.60206, 1.20412, impossible.

Problem 1C.3

−1.89712, 0.40547, 2.70705, 5.41610, impossible.

Problem 1C.4

01.20397, 1, 1.09861, 3.49651, impossible.

Problem 1C.5

0.15, 12.58925, 125.8925, 1258.925, 10^{12}

Problem 1C.6

0.8, 199.5262, 1995.2623, 1995262.3

Problem 1C.7

15, 40.77422, 2.71828, 22026.4658

Problem 1C.8

33, 1,202,604.284, 3,269,017.372, 0.36788

Problem 1C.9

3.16228, 1.38692, 1.41421, 0.0005787, 0.008

Answers to Chapter 2

Problem 2.1

(a) 4/52 or 1/13. (There are four aces in a pack of 52 cards.)

(b) 12/52 or 3/13. Again, this is calculated simply by counting the possibilities, 12 court cards (three in each suit)

(c) 1/2.

(d) $4/52 \times 3/52 \times 2/52 = 3/17576(0.017)$.

(e) $(4/52)^3 = 0.000455$.

Problem 2.3

(a) 0.25 (three to one against means one win for every three losses, so one win in four races), 0.4, 5/9.

(b) 'Probabilities' are 0.33, 0.4, 0.5, which sum to 1.23. These cannot be real probabilities, therefore. The difference leads to a (expected) gain to the bookmaker.

(c) Suppose the true probabilities of winning are proportional to the odds, i.e. 0.33/1.23, 0.4/1.23,0.5/1.23, or 0.268, 0.325, 0.407. If £1 was bet on each horse, then the bookie would expect to pay out $0.268 \times 3 + 0.325 \times 1.5 + 0.407 \times 0.8 = 1.6171$, plus one of the £1 stakes i.e., £2.62 in total. He would thus gain 38 pence on every £3 bet, or about 12.7%.

Problem 2.5

A number of factors might help: statistical ones such as the ratio of exports to debt interest, the ratio of GDP to external debt, the public sector deficit, etc., and political factors such as the policy stance of the government. More insight could be gained by looking at the current interest rate on the debt. A high interest rate suggests investors believe there is a greater chance of default, other things equal.

Problem 2.7

(a) Is the more probable, since it encompasses her being active or not active in the feminist movement. Many people get this wrong, which shows how one's preconceptions can mislead. People tend to read part (a) as 'Judy is a bank clerk, not active in the feminist movement'. A simple way of stating this mathematically is $\Pr(B) \geq \Pr(B \text{ and } A)$ where B indicates a bank clerk and A indicates and activist. It has to be true since $\Pr(A) \leq 1$.

Problem 2.9

The advertiser is a trickster and guesses at random. Every correct guess ($P = 0.5$) nets a fee, every wrong one costs nothing except reimbursing the fee. The trickster would thus keep half the money sent in. You should be wary of such advertisements!

Problem 2.11

(a) E(winning) $= 0.520 \times$ £1billion $+ (1 - 0.520) \times -$ £100 $=$ £853.67

(b) Despite the positive expected value, most would not play because of their aversion to risk. Would you? The size of the prize is also not credible – would they actually pay up?

Problem 2.13

(a), (b) and (d) are independent, though legend says that rain on St Swithin's Day means rain for the next 40 days, so (d) is arguable according to legend.

Problem 2.15

(a) There are 15 ways where a 4–2 score could be arrived at, of which this is one. Hence the probability is 1/15.

(b) Six of the routes through the tree diagram involve a 2–2 score at some stage, so the probability is 6/15.

Problem 2.17

Pr(guessing all six) $= 6/50 \times 5/49 \times \cdots 1/45 = 1/15890700$.

Pr(six from 10 guesses) $= 10/50 \times 9/49 \times \cdots \times 5/45 = 151200/11441304000$. This is exactly 210 times the first answer, so there is no discount for bulk gambling.

Problem 2.19

	Prior	Likelihood	Prior × likelihood	Posterior
Fair coin	0.5	0.25	0.125	0.2
Two heads	0.5	1.00	0.500	0.8
			0.625	

Problem 2.21

(a) Write the initial probability of guilt as $\Pr(G) = 1/2$. The probability the witness says the defendant is guilty, given they are guilty, is $\Pr(W|G) = p$. Using Bayes' theorem, the probability of guilt, given the witness's statement, Pr(G|W) is

$$\frac{\Pr(W|G) \times \Pr(G)}{\Pr(W|G) \times \Pr(G) + \Pr(W|\textit{not-G}) \times \Pr(\textit{not-G})} = \frac{p \times 0.5}{p \times 0.5 + (1-p) \times 0.5} = p$$

(b) Again using Bayes' theorem, and writing $\Pr(2W|G)$ for the probability that both witnesses claim the defendant is guilty, etc., we obtain $\Pr(G|2W)$ as

$$\frac{\Pr(2W|G) \times \Pr(G)}{\Pr(2W|G) \times \Pr(G) + \Pr(2W|\textit{not-G}) \times \Pr(\textit{not-G})}$$

$$= \frac{p^2 \times 0.5}{p^2 \times 0.5 + (1-p)^2 \times 0.5} = \frac{p^2}{p^2 + (1-p)^2}$$

(c) If $p < 0.5$, then the value in part (b) is less than the value in (a). The agreement of the second witness *reduces* the probability that the defendant is guilty. Intuitively, this seems unlikely. The fallacy is that they can lie in many different ways, so Bayes' theorem is not applicable here.

Problem 2.23

(a) Expected Values are 142, 148.75 and 146, respectively. Hence B is chosen.

(b) The minima are 100, 130, 110, so B has the greatest minimum. The maxima are 180, 170, 200, so C is chosen.

(c) The regret table is

	low	middle	high	Max
A	30	5	20	30
B	0	0	30	30
C	20	15	0	20

so C has the minimax regret figure.

(d) The EV assuming perfect information is 157.75, against an EV of 148.75 for project B, so the value of information is 9.

Problem 2.25

The probability of no common birthday is $365/365 \times 364/365 \times 363/365 \times \cdots \times 341/365 = 0.43$. (This is obtained by noting that the probability of the second person having a birthday on a different day from the first is $364/365$, the probability that the third has a birthday different from the first two is $363/365$, etc.) Hence the probability of at least one birthday in common is 0.57, or greater than one-half. Most people underestimate this probability by a large amount. (This result could form the basis of a useful source of income at parties.)

Problem 2.27

(a) Choose low confidence. The expected score is $0.6 \times 1 + 0.4 \times 0 = 0.6$. For medium confidence the score would be $0.6 \times 2 + 0.4 \times -2 = 0.4$, and for high confidence $0.6 \times 3 + 0.4 \times -6 = -0.6$.

(b) Let p be the probability desired. We require $\text{E(score}|\text{medium)} = p \times 2 + (1 - p) \times \; -2 > p = \text{E(score}|\text{low)}$—the payoff to medium confidence should be greater than low confidence. Hence $4p - 2 > p \Rightarrow p > 2/3$. Similarly, we also require $p \times 2 + (1 - p) \times -2 > p \times 3 + (1 - p) \times -6$ (the payoff to medium has to exceed the payoff to high). This implies $p < 4/5$. Hence $2/3 < p < 4/5$.

(c) Given $p = 0.85$, the expected scores are $p = 0.85$ (low), $2p - 2(1 - p) = 1.4$ (medium) and $3p - 6(1 - p) = 1.65$. Hence the expected loss would by $1.65 - 1.4 = 0.25$ if opting for medium confidence and $1.65 - 0.85 = 0.8$ if opting for low confidence.

Answers to Chapter 3

Problem 3.1

The graph looks like a pyramid, centred on the value of 7, which is the mean of the distribution. The probabilities of scores of 2,3, . . . ,12 are (out of 36): 1, 2, 3, 4, 5, 6, 5, 4, 3, 2, 1, respectively. The probability that the sum is nine or greater is therefore $10/36(4 + 3 + 2 + 1$, divided by 36). The variance is 5.83.

Problem 3.3

The distribution should be sharply peaked (at or just after the departure time as shown in the time table) and should be skewed to the right.

Problem 3.5

Similar to the train departure time, except it is a discrete distribution. The mode would be zero accidents, and the probability above one accident per day very low indeed.

Problem 3.7

The probabilities are 0.33, 0.40, 0.20, 0.05, 0.008, 0.000, 0.000 of 0–6 sixes, respectively, using the Binomial formula. For example, 0.33 is calculated as (5/6)6.

Problem 3.9

(a) $\Pr(0) = 0.915 = 0.21$, $\Pr(1) = 15 \times 0.914 \times 0.1 = 0.34$, hence $\Pr(0 \text{ or } 1) = 0.55$.

(b) By taking a larger sample or tightening the acceptance criteria, e.g. only accepting if the sample is defect free.

(c) $\Pr(0 \text{ or } 1) = 0.9715 + 15 \times 0.9714 \times 0.03 = 0.927$ so the probability of sending it back is 7.3%.

(d) The assumption of a large batch means that the probability of a defective component being selected does not alter significantly as the sample is drawn.

Problem 3.11

The y-axis coordinates are:

0.05 0.13 0.24 0.35 0.40 0.35 0.24 0.13 0.05

This gives the outline of the central part of the Normal distribution.

Problem 3.13

(a) 5%

(b) 30.85%

(c) 93.32%

(d) 91.04%

(e) Zero. (You must have an *area* for a probability.)

Problem 3.15

$z = \frac{12 - 10}{\sqrt{9}} = 0.67$, area = 25%; $z = -1$, area = 15.87%, $Z_L = -0.67$, $Z_U = 1.67$, area = 70.11%, zero again.

Problem 3.17

(a) $\overline{IQ} \sim N(100, 16^2/10)$;

(b) $z = 1.98$, Pr = 2.39%.

(c) 2.39% (same as (b)).

(d) 95.22%. It is much greater than the previous answer. This question refers to the distribution of sample means, which is less dispersed than the population.

(e) Since the marginal student has an IQ of 108 (see previous question), nearly all university students will have an IQ above 110 and so, to an even greater extent, the sample mean will be above 110. Note that the distribution of students' IQ is not Normal but skewed to the right, since it is taken from the upper tail of a Normal distribution. The small sample size means we cannot safely use the Central Limit Theorem here.

(f) 105, *not* 100. The expected value of the last nine is 100, so the average is 105.

Problem 3.19

(a) $r \sim B(10, ^1\!/_2)$.

(b) $r \sim N(5, 2.5)$.

(c) Binomial: Pr = 0.828; Normal: 73.57% (82.9% using the continuity correction).

Problem 3.21

(a) By the Binomial, Pr(no errors) = 0.99^{100} = 36.6%. By the Poisson, $nP = 1$, so

$$\Pr(x = 0) = \frac{1^0 e^{-1}}{0!} = 36.8\%$$

(b) $\Pr(r = 1) = 100 \times 0.99^{99} \times 0.01 = 0.370$; $\Pr(r = 2) = 100C2 \times 0.99^{98} \times 0.01^2 = 0.185$. Hence $\Pr(r \leq 2) = 0.921$. Poisson method: $\Pr(x = 1) = 1^1 \times e^{-1}/1. = 0.368$; $\Pr(x = 2) = 1^2 \times e^{-1}/2. = 0.184$. Hence $\Pr(x \leq 2) = 0.920$. Hence the probability of more than two errors is about 8% using either method. Using the Normal method, we would have $x \sim N(1,0.99)$. So, the probability of $x > 2.5$ (taking account of the continuity correction) is given by $z = (2.5 - 1)/\sqrt{0.99} = 1.51$, giving an answer of 6.55%, a significant underestimate of the true value.

Problem 3.23

(a) Normal.

(b) Uniform distribution between zero and one (look up the = *RAND*() function in your software documentation).

(c) mean = 0.5, variance = 5/12 = 0.42 for parent, mean = 0.5, variance = 0.42/5 for sample means (Normal distribution)

Problem 3.25

Looking at goals scored in the English Premier League 2010–11 (http://en.wikipedia.org/wiki/2010%E2%80%9311_Premier_League#Scoring) reveals the following frequency table:

Goals	Actual	Expected
0	25	23.17
1	54	64.81
2	95	90.65
3	93	84.53
4	61	59.11
5	26	33.07
6	19	15.42
7	4	6.16
8	3	2.15
9	0	0.67
Totals	380	379.75

Thus 25 games saw no goals, 54 had one goal, etc. The 'Expected' column gives the expected frequency based on the Poisson formula (3.26). The expected values reveal a close match to the actual numbers.

Answers to Chapter 4

Problem 4.1

(a) It gives the reader some idea of the reliability of an estimate, around the point value of the estimate.

(b) The population variance (or its sample estimate) and the sample size.

Problem 4.3

An estimator is the rule used to find the estimate or a parameter. A good estimator does not guarantee a good estimate, only that it is correct on average (if the estimator is unbiased) and close to the true value (if precise).

Problem 4.5

$E(w_1x_1 + w_2x_2) = w_1E(x_1) + w_2E(x_2) = w_1\mu + w_2\mu = \mu$ if $w_1 + w_2 = 1$.

Problem 4.7

$40 \pm 2.57 \times \sqrt{10^2/36} = [35.71, 44.28]$. If $n = 20$, the t distribution should be used, giving $40 \pm 2.861 \times \sqrt{10^2/20} = [33.60, 46.40]$

Problem 4.9

$0.4 \pm 2.57 \times \sqrt{0.4 \times 0.6/50} = [0.22, 0.58]$

Problem 4.11

$(25 - 22) \pm 1.96 \times \sqrt{12^2/80 + 18^2/100} = [-1.40, 7.40]$

Problem 4.13

$$(0.67 - 0.62) \pm 2.57 \times \sqrt{\frac{0.67 \times 0.33}{150} + \frac{0.62 \times 0.38}{120}} = [-0.10, 0.20]$$

Problem 4.15

$30 \pm 2.131 \times \sqrt{5^2/16} = [27.34, 32.66]$.

Problem 4.17

$(45 - 52) \pm 2.048 \times \sqrt{40.32/12 + 40.32/18} = [-2.15, -11.85]$.
40.32 is the pooled variance.

Answers to Chapter 5

Problem 5.1

(a) False. You can alter the sample size. Increasing n will reduce the probability of both types of error by shrinking the width of the distributions under H_0 and H_1.

(b) True.

(c) False. You need to consider the Type II error probability also. Reducing the Type I error probability will increase the Type II probability (for a given sample size).

(d) True. The probability of a Type II error falls, therefore increasing power.

(e) False. The significance level is the probability of a Type I error.

(f) False. The confidence level is one – Pr(Type I error) or the probability of accepting H_0 when true.

Problem 5.3

H_0: fair coin, ($Pr(H) = 1/2$), H_1: two heads ($Pr(H) = 1$). A Type I error is two heads from a fair coin, so Pr(Type I error) $= (1/2)^2 = 1/4$; A Type II error is less than two heads from a two-headed coin, which is impossible, so Pr(Type II error) $= 0$.

Problem 5.5

(a) Rejecting a good batch or accepting a bad batch.

(b) H_0: $\mu = 0.01$ and H_1: $\mu = 0.10$ are the hypotheses. Under H_0, Pr(zero or one defective in sample) $= 0.99^{50} \times 0.99^{49} \times 0.01 \times 50 = 0.911 = 91.1\%$, hence an 8.9% chance of rejecting a good batch. Under H_1, Pr(0 or 1) $= 0.90^{50} \times 0.90^{49} \times 0.10 \times 50 = 0.034$, hence a 3.4% chance of accepting a bad batch. One could also use the Normal approximation to the Binomial, giving probabilities of 7.78% and 4.95%.

(c) Pr(Type I error) = 26%; Pr(Type II error) = 4.2%.

(d) (i) Try to avoid faulty batches, hence increase the risk of rejecting good batches, the significance level of the test. (ii) Since there are alternative suppliers, it can again increase the risk of rejecting good batches (which upsets its supplier). (iii) Avoid accepting bad batches.

Problem 5.7

$z = (12 - 10)/(6/\sqrt{30}) = 1.83$, hence Prob-value is 3.36%

Problem 5.9

Power is 1 – Pr(Type II error). The Type II error probability is zero (see the answer to 5.3 above), so the power of this test is 100%. You will always reject H_0 when false.

Problem 5.11

$z = (15 - 12)/\sqrt{(270/30)} = 1 < 1.96$, the critical value at the 95% confidence level. The hypothesis is not rejected. The rejection region is the area in the right hand tail, beyond $z = 1$.

Problem 5.13

$z = (0.45 - 0.5)/\sqrt{(0.5 \times (1 - 0.5)/35)} = 0.59 < 1.96$, not significant, so do not reject H_0.

Problem 5.15

$z = (115 - 105)/\sqrt{21^2/49 + 23^2/63} = 2.4 > 1.64$, the critical value, hence reject with 95% confidence.

Problem 5.17

(a) $$z = \frac{0.57 - 0.47}{\sqrt{\frac{0.513 \times (1 - 0.513)}{180} + \frac{0.513 \times (1 - 0.513)}{225}}} = 2.12.$$

This is significant using either a one- or a two-tail test. The pooled variance in this case is calculated using $\hat{\pi} = 0.513 = 208/405$, the overall proportion of people who pass. Whether you used a one- or a two-tail test reveals something about your prejudices. In the United Kingdom, the proportions passing are the actual ones for 1992, on the basis of 1.85 million tests altogether. You might have an interesting class discussion about what these statistics prove. Insurance statistics, on the other hand, suggest women are safer drivers.

(b) You should get the same *t* statistic. The proportion of men in the 'Pass' group is 49.5% and in the 'Fail' group is 39.1%. The overall proportion of men is 44.4%. Putting these values into the same formula as in part (a) gives the same *t* value. The same result is obtained if you use the proportion of women in each group.

(c) This result is inevitable. This can be demonstrated algebraically, though it is lengthy and not included here.

Problem 5.19

(a) $t = -5\sqrt{10^2/20} = -2.24 < -2.093$, the critical value, hence reject H_0.

(b) The parent distribution is Normal.

Problem 5.21

$$s^2 = \frac{(12-1)\times 50 + (15-1)\times 30}{12+15-2} = 38.8, \text{ and } t = \frac{150-130}{\sqrt{\dfrac{38.8}{12}+\dfrac{38.8}{15}}} = 8.29,$$

so H_0 is rejected. The critical value is $t^* = 2.060$ (95% confidence, 25 degrees of freedom).

Problem 5.23

(a) For the 'Before' sample, the mean is 55.5 and the variance 12.3. For the 'After' sample, the mean is 57.5 with variance 14.1. Hence the pooled variance is 13.16 and the test statistic is $t_{20} = 1.18$.

(b) Measuring the differences between the before and after figures, the mean is 1.8 and the variance 13.8. This gives a test statistic of $t_{10} = 1.63$. Neither is significant at the 5% level, though the latter is closer. Note that only one worker performs worse, but this one does substantially worse, perhaps because of other factors.

Problem 5.25

(a) It would be important to check *all* the predictions of the astrologer. Too often, correct predictions are highlighted ex-post and incorrect ones ignored.

(b) Like astrology, a fair test is important, in which it is possible to pass or fail, with known probabilities. Then performance can be judged.

(c) Samples of both taken-over companies and independent companies should be compared, with as little difference in other respects between samples as possible.

Problem 5.27

(a) The sample means should be Normally distributed so approximately 5% of the z scores should lie beyond ± 1.96.

(b) Your outcome should be similar to that described in part (a). You might get a slightly different result due the vagaries of random sampling.

(c) Your graph should look approximately like a Normal distribution.

Answers to Chapter 6

Problem 6.1

The 95% c.i. for the variance is $\frac{39 \times 20^2}{59.34} \leq \sigma^2 \leq \frac{39 \times 20^2}{24.43}$ in which 24.43 and 59.34 are the limits cutting off 2.5% in each tail of the χ^2 distribution with 40 degrees of freedom (close to the correct 39), so the c.i. for σ is [16.21, 25.29].

Problem 6.3

Observed values are 833 and 942 with expected values of 887.5 in each case. Hence we obtain $\chi^2 = 6.69 > 3.84$, the 95% critical value, and there is an apparent difference between quarters I/III and II/IV. (Note that the previous edition of this book used data from 2006, which found a different result.)

Problem 6.5

Using the data as presented, with expected values of 565 (the average of the four numbers), yields $\chi^2(3) = 1.33$, not significant. However, with the additional information, adding the dissatisfied customers (24, 42, 20, 54) and constructing a contingency table yields $\chi^2(3) = 22.94$, highly significant. The differences between the small numbers of dissatisfied customers adds most to the test statistic. The former result should be treated with suspicion since it is fairly obvious that there would be small numbers of dissatisfied customers.

Problem 6.7

$\chi^2(4) = 8.12$ which is not significant at the 5% significance level. There appears no relationship between size and profitability.

Problem 6.9

(a) The correct observed and expected values are

47.0 (54.9)	72.0 (64.1)
86.0 (76.6)	80.0 (89.4)
4.0 (5.5)	8.0 (6.5)

and this yields a χ^2 value of 5.05 against a critical value of 5.99 (5% significance level, 2 degrees of freedom).

(b) Omitting the non-responses leads to a 2×2 contingency table with a test statistic of 4.22, against a critical value of 3.84, yielding a significant result. It is debatable whether it is better to omit the non-responses. The real problem here is that the test statistic is close to the critical value, so it is easy for our decision to change with small changes to the data.

Problem 6.11

$F = 55/48 = 1.15 < 2.76$, the 1% critical value for 24 and 29 degrees of freedom. There is no significant difference, therefore.

Problem 6.13

(a) Between Sum of Squares = 335.8, Within Sum of Squares = 2088, Total Sum of Squares = 2424. $F = (335.8/3)/(2088/21) = 1.126 < 3.07$, the critical value for 3, 21 degrees of freedom. There appears to be no significant difference between classes.

(b) A significant result would indicate some difference between the classes, but this could be because of any number of factors which have not been controlled for, e.g. different teachers, different innate ability, different gender ratios and so on.

Answers to Chapter 7

Problem 7.1

(a)

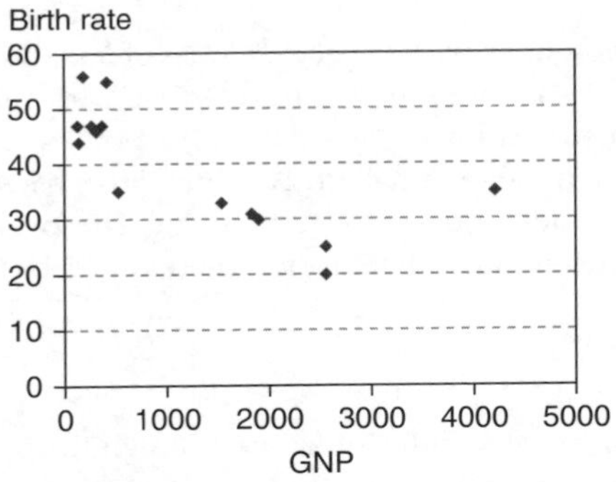

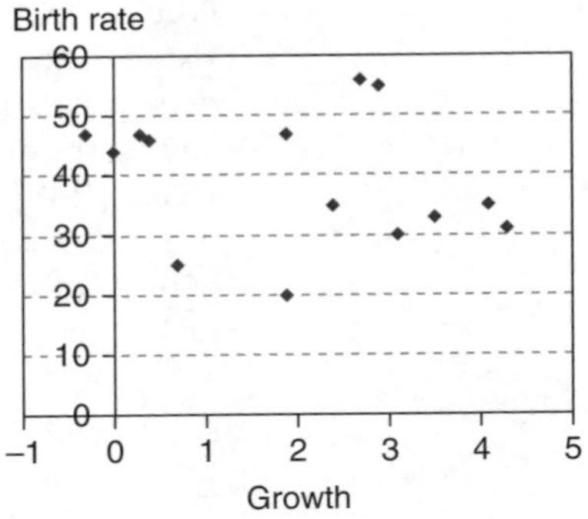

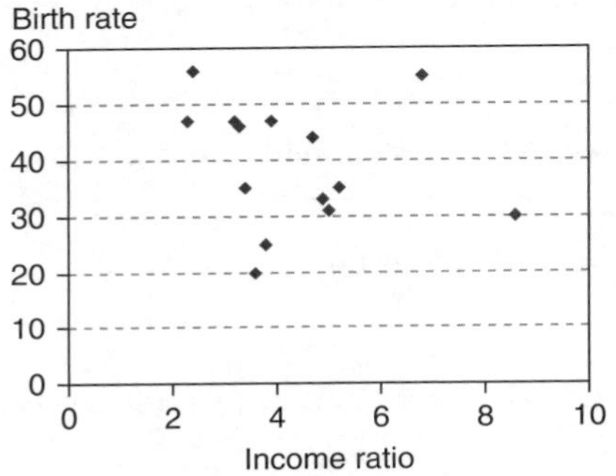

(b) We would expect similar slopes to those using Todaro's data, but the graphs for growth and the income ratio do not look promising. It is difficult to discern any relationship between them. The GNP graph suggests a negative relationship.

(c) There seems to exist a psychological propensity to overestimate the degree of correlation when viewing a graph. See (d) below to see if you did so.

(d) $r = -0.73, -0.25, -0.22$ for GNP, growth and the income ratio, respectively. Note that $r < 0$ for the income ratio, in contrast with the result in the text.

(e) $t = -3.7, -0.89, -0.78$, so only the first is significant. The critical value is 1.78 (for a one-tail test) or 2.18 if using a two-tail test.

Problem 7.3

(a) Very high and positive. This is because (i) income is a major determinant of people's consumption, (ii) measuring the variables at national level means a great deal of smoothing of individual variation and (iii) the effects of inflation have not been corrected for.

(b) A medium degree of negative correlation (bigger countries can provide a greater variety of goods and services for themselves so need to import less).

(c) Theoretically negative, but empirically the association tends to be weak, especially using the real interest rate.

There might be lags in (a) and (c). (b) would best be estimated in cross section.

Problem 7.5

Rank correlations are:

	birth rate
GNP	−0.77
Growth	−0.27
Inc ratio	−0.37

These are similar to the ordinary *r* values. Only, the first is significant at 5%.

Problem 7.7

(a) The regression results for the three equations are:

	GNP	growth	income ratio
a	47.18	42.88	45.46
b	−0.006	−1.77	−1.40
R^2	0.53	0.061	0.047

These results are quite different from what was found before. GNP appears the best, not worst, explanatory variable, using the R^2 value as a guide.

(b)–(c)

	GNP	growth	income ratio
S_e	7.83	11.11	11.20
S_b	0.0017	1.99	1.82
t	−3.71	−0.89	−0.77
F	13.74	0.78	0.59

Only in the case of GNP is the *t* ratio significant. Same is true for the *F* statistic.

(d) You should be starting to have serious doubts. Two samples produce quite different results. We should think more carefully about how to model the birth rate and about what data to apply it to.

(e) Using all 26 observations gives:

	GNP	growth	IR
a	42.40	43.06	36.76
b	−0.01	−2.77	−0.20
R	0.31	0.28	0.002
F	11.02	9.52	0.04
s_b	0.00	0.90	1.01
t	−3.32	−3.08	−0.20

The results seem quite sensitive to the data employed; they are not very robust. This suggests something is missing from our model(s). Perhaps we need to take account of each country's cultural context or geographical location, or something.

Problem 7.9

(a) 27.91 (34.71)

(b) 37.58 (32.61)

(c) 35.67 (33.76)

(predictions in brackets obtained from Todaro's data). Again, different samples, different results, which does not inspire confidence. The predictions from this set of data are more diverse than from those using Todaro's data.

Problem 7.11

A good source of data for this project is the World Bank (http://data.worldbank.org/topic). From this, the following relationship between the birth rate and GDP per capita was found (data for 182 countries, for 2009):

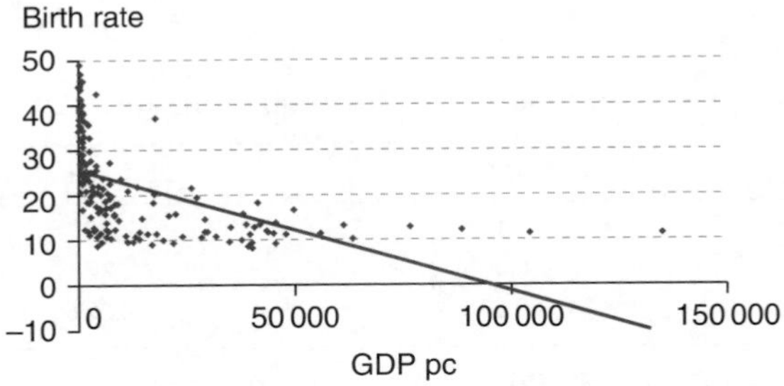

with regression

$$\text{BR} = 25.6 - 0.00027\ \text{GDPpc} + e$$
$$t(31.5)(-7.78)$$
$$R^2 = 0.25.$$

Although demonstrating a negative relationship and highly significant, it is evident from the graph that the wrong functional form has been used. Taking logs of both variables, we obtain:

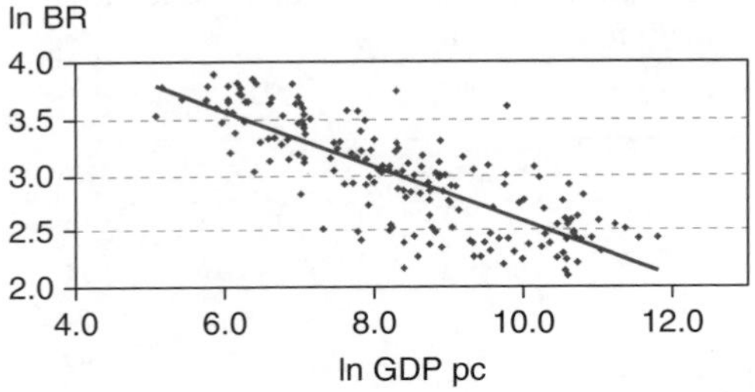

with regression

$$\ln \text{BR} = 5.05 - 0.25 \ln \text{GDPpc} + e$$
$$t(43.2)\ (-18.0)$$
$$R^2 = 0.64.$$

This seems a better characterisation of the data. Analyses with the other explanatory variables should be done in similar manner and might yield better results, perhaps not.

Answers to Chapter 8

Problem 8.1

(a) $B = 46.77 - 0.0064\,\text{GNP} - 0.55\,\text{GROWTH} + 0.34\,\text{IR}$

s.e. (0.002) (1.66) (1.54)

$R^2 = 0.54, F = 3.91$

(i) Calculating elasticities: GNP: $-\ 0.0064 \times 17050/551 = -0.19$; *growth*: $-0.55 \times 27.9/551 = -0.03$; IR $-0.34 \times 61.1/551 = 0.04$. All seem quite small, although the first suggests a 10% rise in GNP should lower the birth rate by 2%.

(ii) Only GNP appears significant. Note that the R^2 value is only just greater than in the simple regression on GNP.

(iii) The F statistic is just significant at the 5% level (3.71 is the critical value).

(b) It looks like the growth and income ratio variables should be dropped.

(c) We need to compare the regression in part (a) above with the restricted regression $B = b_0 + b_1\text{GNP}$ (i.e. the coefficients on GROWTH and IR are both equal to zero). From the former, we obtain error sum of squares of $\text{ESS}_\text{U} = 726.96$; from the latter (equation estimated as part of problem 7.7), $\text{ESS}_\text{R} = 736.17$. The restricted model seems to fit almost as well. To check this, we conduct an F test: $F = \dfrac{(736.17 - 726.96)/2}{726.96/10} = 0.063$, less than the critical value ($F^*(2, 10) = 4.10$) so the variables can be omitted.

(d) $B = 43.61 - 0.005\,\text{GNP} - 2.026\,\text{GROWTH} + 0.69\,\text{IR}$

(0.0017) (0.845) (0.81)

$R^2 = 0.48, F = 6.64, n = 26.$

The GNP and IR coefficients are of similar orders of magnitude and significance levels. The GROWTH coefficient changes markedly and is now significant. The F-statistic for exclusion is 3.365, against a critical value of 3.44 at 5%, suggesting both could be excluded, in spite of the significant t-ratio on the growth variable.

(e) Not much progress has been made. More planning of the research is needed.

(f) Women's education, religion and health expenditures are possibilities.

Problem 8.3

28.29 from 14 countries; 27.93 from all 26 countries.

Problem 8.5

(a) A set of dummy variables, one for each class.

(b) Difficult, because crime is so heterogeneous. One could use the number of recorded offences, but this would equate a murder with bicycle theft. It would be better to model the different types of crime separately.

(c) A proxy variable could be constructed, using such factors as the length of time for which the bank governor is appointed, whether appointed by the government, and so on. This would be somewhat arbitrary, but possibly better than nothing.

Problem 8.7

(a) Time-series data, since the main interest is in *movements* of the exchange rate in response to changes in the money supply. The relative money stock movements in the two countries might be needed.

(b) Cross-section (cross-country) data would be affected by enormous cultural and social differences, which would be hard to measure. Regional (within country) data might not yield many observations and might simply vary randomly. Time series data might be better, but it would still be difficult to measure the gradual change in cultural and social influences. Best would be cross-section data on couples (both divorced and still married).

(c) Cross-section data would be of more interest. There would be many observations, with substantial variation across hospitals. This rich detail would not be so easily observable in time-series data.

Problem 8.9

Suitable models would be:

(a) $C = b_0 + b_1P + b_2F + b_3L + b_4W$ where C: total costs, P: passenger miles flown; F: freight miles flown; L: % of long haul flights; W: wage rates faced by the firm. This would be estimated using cross-section data, each airline constituting an observation. P^2 and F^2 terms could be added, to allow the cost function to be non-linear. Alternatively, it could be estimated in logs to get elasticity estimates. One would expect $b_1, b_2, b_4 > 0, b_3 < 0$.

(b) $\text{IM} = b_0 + b1\,\text{GNP} + b_2\text{FEMED} + b_3\text{HLTHEXP}$ where IM: infant mortality (deaths per thousand births); *FEMED*: a measure of female education (e.g. the literacy rate); HLTHEXP: health expenditure (ideally on women, as % of GNP). One would expect $b_1, b_2, b_3 < 0$. This would be a cross-country study. There is likely to be a 'threshold' effect of GNP, so a non-linear (e.g. log) form should be estimated.

(c) $\text{BP} = b_0 + b_1\,\Delta\text{GNP} + b_2\,R$ where BP: profits; ΔGNP: growth; R: the interest rate. This would be estimated on time-series data. BP and growth should be measured in real terms, but the nominal interest rate might be appropriate. Bank profits depend upon the spread of interest rates, which tends to be greater when the rate is higher.

Problem 8.11

(a) Higher U reduces the demand for imports; higher OECD income raises the demand for UK exports; higher materials prices (the UK imports materials) lowers demand, but the effect on expenditure (and hence the BOP) depends upon the elasticity. Here, higher P leads to a greater BoP deficit, implying inelastic demand; higher C (lower competitiveness) worsens the BoP.

(b) (iii) higher material prices.

(c) U: linear, Y: non-linear

(d) Since B is sometimes negative, a log transformation cannot be performed. This means elasticity estimates cannot be obtained directly. Since B is sometimes positive, sometimes negative, an elasticity estimate would be hard to interpret.

(e) 1.80, a surplus.

Problem 8.13

Suitable explanatory variables might be: personal disposable income, company profits (because of company cars), car price index (ideally adjusted for quality), interest rates, petrol prices and lagged sales. Sales might follow a cycle if, for example, firms tend to replace their cars after three years.

Answers to Chapter 9

Problem 9.1

Should you use: GNP or GDP; gross or net national product; measured at factor cost or market prices; coverage (United Kingdom, Great Britain, England and Wales); and current or constant prices are some of the issues.

Problem 9.3

The following are measures of UK and US GDP, both at year 2000 prices. The UK figures are in £bn, the US figures are in $bn. Your own figures may be slightly different but should be highly correlated with these numbers.

	1995	1996	1997	1998	1999	2000	2001	2002	2003
UK	821.4	843.6	875.0	897.7	929.7	961.9	979.2	997.5	1023.2
US	8031.7	8328.9	8703.5	9066.9	9470.3	9817.0	9890.7	10074.8	10381.3

Problem 9.5

$n = 1.96^2 \times 400/2^2 = 385$

Answers to Chapter 10

Problem 10.1

(a)

Year	2005	2006	2007	2008	2009	2010
Exports	100.0	114.5	113.0	127.7	119.5	133.2
Imports	100.0	112.3	111.5	123.6	112.7	127.8

(b) No. Using the indices, information about the *levels* of imports and exports is lost.

Problem 10.3

(a)–(c)

Year	E	P_L	P_P	Q_L	Q_P
1999	100	100	100	100	100
2000	104.28	101.99	102.03	102.20	102.24
2001	106.83	103.07	104.04	102.68	103.64
2002	98.87	99.01	98.76	100.10	99.86
2003	99.53	96.14	96.48	103.16	103.52

Problem 10.5

(a)

Year	Coal	Petroleum	Electricity	Gas
1999	100	100	100	100
2000	101.01	131.42	95.75	110.99
2001	109.49	141.14	86.53	149.45
2002	99.40	143.10	82.34	142.86
2003	99.22	133.42	78.50	147.80
Shares	1.3%	10.0%	71.6%	17.2%

(b) Answer as in Problem 10.3(a).

Problem 10.7

The chain index is 100, 110, 115, 123.1, 127.7, 136.9, 139.2, using 2000 as the common year. Using one of the other years to chain yields a slightly different index. There is no definitive right answer because series 1 and 2 are likely to based on slightly different baskets of goods, rising in price at different rates. Hence the date at which they are spliced will make a slight difference to the chained index.

Problem 10.9

Expenditure on energy in 2007 was £9741.04m. The Laspeyres price index increased from 95.32 to 126.17 between 2007 and 2008, an increase of 32.4%. Hence compensation of 32.4% of £9741.04 is needed, amounting to £3156.10m. The Paasche index rose by 31.8% over that period, implying compensation of 31.8% of £9741.04m, i.e. £3097.65m. This is about £58m cheaper than using the Laspeyres index.

Problem 10.11

The index number series are as follows:

	Cash expenditure (a)	Real expenditure (b)	Volume of expenditure (c)	Real expenditure per capita (d)	Volume of expenditure per capita (e)	Needs index (f)	Spending deflated by need. (g)
1987	100.0	100.0	100.0	100.0	100.0	100.0	100.0
1988	109.8	102.6	99.4	102.4	99.2	100.2	99.2
1989	120.5	105.7	101.9	105.1	101.3	100.5	101.4
1990	132.7	107.5	104.5	106.6	103.6	100.8	103.6
1991	150.4	113.6	108.8	111.9	107.1	101.6	107.1

(a) $109.8 = 23601/21495 \times 100; 120.5 = 25906/21495 \times 100$; and so on.

(b) This series is obtained by dividing column 1 by column 2 (and setting 1987 as the reference year). Clearly, much of the increase in column 1 is because of inflation.

(c) This series is column 1 divided by column 3. Since the NHS price index rose faster than the GDP deflator, the volume of expenditure rises more slowly than the real figure.

(d) Per capita figures are obtained by dividing by the population in column 4.

(e) A Needs index can be calculated as working population + 2 × non-working population, and then set to 100 in 1987. Once the needs index is calculated, it can be used to deflate the volume expenditure, giving the final column above. This shows little difference to the per capita series.

(f) Needs index could be improved by finding the true cost of treating people of different ages.

Problem 10.13

(a) 1702.20. (b) Yes, 102.20.

Problem 10.15

18.3%. There can be multiple solutions to the Internal Rate of Return calculation, which can be problematic and students should be aware of this. It should not be a problem if there is a single initial outlay followed by a stream of returns – in this case a unique solution occurs.

Problem 10.17

(a) The x and y co-ordinate values for the Lorenz curve are:

Households	8.9	16.0	22.3	25.8	29.2	35.0	41.4	58.9	71.7	87.2	95.1	98.5	99.5	100.0
Income	0.2	0.9	2.0	2.8	3.8	6.0	9.1	20.8	32.8	53.5	70.4	83.9	92.4	100.0

and from this we can calculate:

Area A = 0.227, B = 0.273, Gini = 0.546.

(b) The elderly have had a lifetime to accumulate wealth, whereas the young have not. This does not apply to income. Also, via inheritance, some families can build up wealth over time, increasing the disparity.

Problem 10.19

(a) The Gini coefficients are 45.2%, 33.2%, 33.2% and 36.0%, respectively.

(b) These differ from the values given in Table 10.24 of the text, substantially so in the case of original and gross. The figures based on quintiles are all lower than the figures in Table 10.24, as expected, although the bias is large in some cases.

Problem 10.21

79.3%.

Answers to Chapter 11

Problem 11.1

(a)

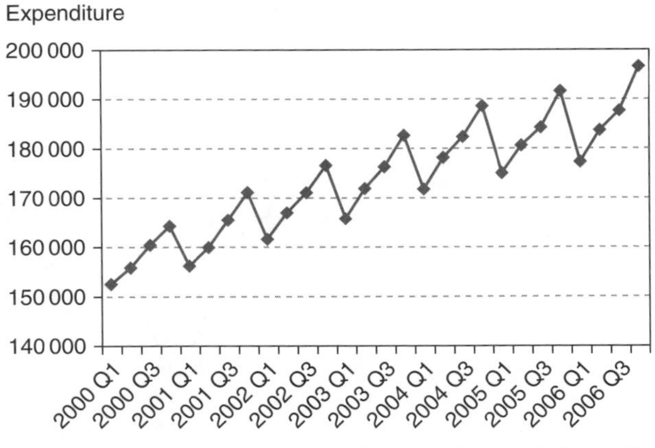

There is an obvious increase throughout each year, followed by a fall in the subsequent Q1. High spending in Q4 is likely a Christmas effect, followed by a lack of spending afterwards, in Q1.

(b) and (c) The calculations are:

	Non-durable expenditure	4-qtr total	Centred 4-qtr total	Moving average	Ratio	Seasonal factor	Adjusted series
2000 Q1	152684	629412	626074	156519	0.976	0.967	157967
2000 Q2	155977	633662	631537	157884	0.988	0.990	157607
2000 Q3	160564	637303	635483	158871	1.011	1.008	159291
2000 Q4	164437	641395	639349	159837	1.029	1.037	158576
2001 Q1	156325	646482	643939	160985	0.971	0.967	161733
2001 Q2	160069	653326	649904	162476	0.985	0.990	161741
2001 Q3	165651	658734	656030	164008	1.010	1.008	164337
2001 Q4	171281	665793	662264	165566	1.035	1.037	165176
2002 Q1	161733	671366	668580	167145	0.968	0.967	167329
2002 Q2	167128	676833	674100	168525	0.992	0.990	168874
2002 Q3	171224	681003	678918	169730	1.009	1.008	169866
2002 Q4	176748	685915	683459	170865	1.034	1.037	170448
2003 Q1	165903	691139	688527	172132	0.964	0.967	171643
2003 Q2	172040	697160	694150	173537	0.991	0.990	173837
2003 Q3	176448	703170	700165	175041	1.008	1.008	175049
2003 Q4	182769	709438	706304	176576	1.035	1.037	176254
2004 Q1	171913	715470	712454	178114	0.965	0.967	177861
2004 Q2	178308	721434	718452	179613	0.993	0.990	180171
2004 Q3	182480	724695	723065	180766	1.009	1.008	181033
2004 Q4	188733	727110	725903	181476	1.040	1.037	182006
2005 Q1	175174	728975	728043	182011	0.962	0.967	181235
2005 Q2	180723	732005	730490	182623	0.990	0.990	182611
2005 Q3	184345	734252	733129	183282	1.006	1.008	182883
2005 Q4	191763	737314	735783	183946	1.042	1.037	184928
2006 Q1	177421	740739	739027	184757	0.960	0.967	183559
2006 Q2	183785	745737	743238	185810	0.989	0.990	185705
2006 Q3	187770	751692	748715	187179	1.003	1.008	186281
2006 Q4	196761	756862	754277	188569	1.043	1.037	189748

(d) From the table above, the seasonal factor for Q4 is 1.037, so Q4 is approximately 3.7% above the rest of the year.

(e) See table above, final column.

(f) Christmas 2000 was a poor one (SA figure declines from the previous quarter) whereas 2006 proved a good year.

Problem 11.3

Car production

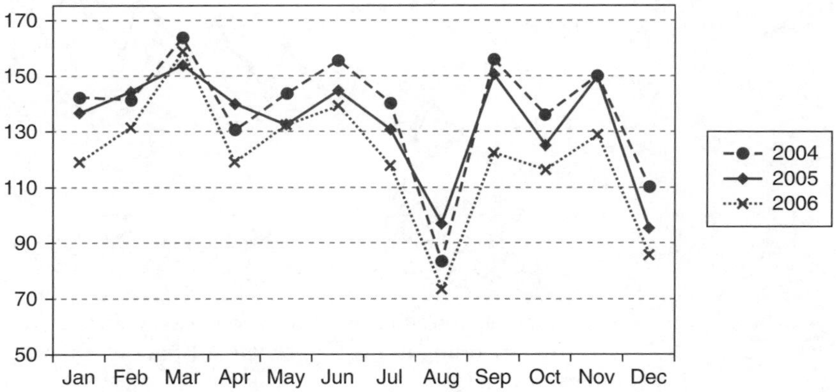

(a) There are obvious troughs in production in August (summer holidays) and December (Christmas break).

(b) and (c) The calculations are:

	Production	12-month total	Centred 12-month total	Moving average	Ratio	Seasonal factor	Adjusted series
Jan-04	141.3	1667.7	1670.6	139.2	1.015	0.984	143.6
Feb-04	141.1	1659.5	1663.6	138.6	1.018	1.039	135.8
Mar-04	163.0	1661.3	1660.4	138.4	1.178	1.196	136.3
Apr-04	129.6	1643.0	1652.2	137.7	0.941	0.981	132.1
May-04	143.1	1649.4	1646.2	137.2	1.043	1.037	138.0
Jun-04	155.5	1646.7	1648.1	137.3	1.132	1.122	138.5
Jul-04	140.5	1641.4	1644.1	137.0	1.026	0.996	141.1
Aug-04	83.2	1643.8	1642.6	136.9	0.608	0.652	127.5
Sep-04	155.3	1634.1	1639.0	136.6	1.137	1.105	140.5
Oct-04	135.1	1644.3	1639.2	136.6	0.989	0.978	138.1
Nov-04	149.3	1633.2	1638.8	136.6	1.093	1.115	133.9
Dec-04	109.7	1622.0	1627.6	135.6	0.809	0.757	144.9
Jan-05	136.0	1611.7	1616.9	134.7	1.009	0.984	138.2
Feb-05	143.5	1625.6	1618.7	134.9	1.064	1.039	138.2
Mar-05	153.3	1620.2	1622.9	135.2	1.134	1.196	128.2
Apr-05	139.8	1609.9	1615.1	134.6	1.039	0.981	142.5
May-05	132.0	1610.3	1610.1	134.2	0.984	1.037	127.3
Jun-05	144.3	1595.9	1603.1	133.6	1.080	1.122	128.6
Jul-05	130.2	1579.0	1587.5	132.3	0.984	0.996	130.7
Aug-05	97.1	1566.7	1572.9	131.1	0.741	0.652	148.8
Sep-05	149.9	1572.4	1569.6	130.8	1.146	1.105	135.6
Oct-05	124.8	1551.2	1561.8	130.2	0.959	0.978	127.6
Nov-05	149.7	1551.5	1551.4	129.3	1.158	1.115	134.2
Dec-05	95.3	1546.5	1549.0	129.1	0.738	0.757	125.9
Jan-06	119.1	1534.1	1540.3	128.4	0.928	0.984	121.0
Feb-06	131.2	1510.0	1522.1	126.8	1.034	1.039	126.3
Mar-06	159.0	1482.4	1496.2	124.7	1.275	1.196	133.0
Apr-06	118.6	1473.7	1478.1	123.2	0.963	0.981	120.9
May-06	132.3	1452.6	1463.2	121.9	1.085	1.037	127.5
Jun-06	139.3	1442.1	1447.4	120.6	1.155	1.122	124.1
Jul-06	117.8	1447.2	1444.7	120.4	0.979	0.996	118.3
Aug-06	73.0	1431.6	1439.4	120.0	0.609	0.652	111.9
Sep-06	122.3	1410.6	1421.1	118.4	1.033	1.105	110.7
Oct-06	116.1	1412.4	1411.5	117.6	0.987	0.978	118.7
Nov-06	128.6	1407.5	1410.0	117.5	1.095	1.115	115.3
Dec-06	84.8	1405.7	1406.6	117.2	0.723	0.757	112.0

(d) $0.652/0.996 = 0.655$, about a 35% decline.

(e) See table above, final column, for the adjusted series. This is much smoother than the original series.

(f) It is obviously a monthly seasonal pattern, rather than the quarterly one for consumer expenditure. Consumer expenditure shows no decline in summer, unlike car production, and the Christmas effect is negative for cars, but positive for consumer expenditures.

Problem 11.5

(a) The regression equation is $C = 153420.6 + 1603.011t - 11.206t^2$. (The t^2 term is not significant in the regression, so a simpler regression using t only might suffice instead.). Fitted values (abridged) are:

2000q1	155 012.5
2000q2	156 581.8
2000q3	158 128.8
2000q4	159 653.4
2001q1	161 155.5
⋮	⋮
2005q4	185 438.2
2006q1	186 492.1
2006q2	187 523.6
2006q3	188 532.7
2006q4	189 519.4

(b) Seasonal factors are:

Q1	Q2	Q3	Q4
0.967	0.990	1.008	1.036

(c) For 2007 Q4 ($t = 32$) we have: $C = (153420.6 + 1603.011 \times 32 - 11.206 \times 32^2) \times 1.036 = 200154.65$. (The actual value in that quarter was 202 017.)

(d) For the additive model we use the same regression equation, but subtract the predicted values from the actual ones. Averaging by quarter then gives the following seasonal factors:

Q1	Q2	Q3	Q4
−5769.78	−1802.08	1270.17	6301.69

The predicted value from the regression line for 2007 Q4 is 193 242.0 and adding 6301.69 to this gives 199 543.69. The actual value is 202 017, so there is an error of −1.2%.

Problem 11.7

(a) The regression equation is:

	Coef.	Std. Err.	T-ratio
trend	1 528.55	129.89	11.77
trendsq	−11.28	4.34	−2.60
Q2	4 044.27	716.40	5.65
Q3	7 193.25	718.40	10.01
Q4	12 301.66	721.91	17.04
cons	148 637.1	903.60	164.49

The coefficients on t and t^2 are similar to the previous values. This is because the additional dummy variables are not correlated with the trend or its square (the correlation coefficient is less than 0.05).

(b) The *t* ratios are bigger in this equation. The seasonal dummies take care of seasonal effects and allows more precise estimates of the coefficients on trend and trend squared.

The *t* ratios in the equation without quarterly dummies are 3.39 and −0.71, compared with the values 11.77 and −2.60 when the quarterly dummies are included.

(c) The seasonal factors from 11.5(d) above are:

Q1	Q2	Q3	Q4
−5769.78	−1802.08	1270.17	6301.69

If we adjust these so that Q1 is set as zero (as is effectively done in the regression, by adding 5769.78) we get:

Q1	Q2	Q3	Q4
−5 769.78	−1 802.08	1 270.17	6 301.69
0.00	3 967.70	7 039.95	12 071.47

These values are very close to the coefficients in the regression equation in part (a) above.

Index